T0260224

CRC HANDBOOK OF
LIE GROUP ANALYSIS OF DIFFERENTIAL EQUATIONS

VOLUME 3

NEW TRENDS IN THEORETICAL DEVELOPMENTS AND COMPUTATIONAL METHODS

EDITED BY
N. H. IBRAGIMOV

R. L. ANDERSON, V. A. BAIKOV, R. K. GAZIZOV
W. HEREMAN, N. H. IBRAGIMOV,
F. M. MAHOMED, S. V. MELESHKO, M. C. NUCCI
P. J. OLVER, M. B. SHEFTEL', A. V. TURBINER,
E. M. VOROB'EV

CRC Press
Taylor & Francis Group
Boca Raton London New York

CRC Press is an imprint of the
Taylor & Francis Group, an **informa** business

LIMITED WARRANTY

CRC Press warrants the physical diskette(s) enclosed herein to be free of defects in materials and workmanship for a period of thirty days from the date of purchase. If within the warranty period CRC Press receives written notification of defects in materials or workmanship, and such notification is determined by CRC Press to be correct, CRC Press will replace the defective diskette(s).

The entire and exclusive liability and remedy for breach of this Limited Warranty shall be limited to replacement of defective diskette(s) and shall not include or extend to any claim for or right to cover any other damages, including but not limited to, loss of profit, data, or use of the software, or special, incidental, or consequential damages or other similar claims, even if CRC Press has been specifically advised of the possibility of such damages. In no event will the liability of CRC Press for any damages to you or any other person ever exceed the lower suggested list price or actual price paid for the software, regardless of any form of the claim.

CRC Press SPECIFICALLY DISCLAIMS ALL OTHER WARRANTIES, EXPRESS OR IMPLIED, INCLUDING BUT NOT LIMITED TO, ANY IMPLIED WARRANTY OF MERCHANTABILITY OR FITNESS FOR A PARTICULAR PURPOSE. Specifically, CRC Press makes no representation or warranty that the software is fit for any particular purpose and any implied warranty of merchantability is limited to the thirty-day duration of the Limited Warranty covering the physical diskette(s) only (and not the software) and is otherwise expressly and specifically disclaimed.

Since some states do not allow the exclusion of incidental or consequential damages, or the limitation on how long an implied warranty lasts, some of the above may not apply to you.

Library of Congress Cataloging-in-Publication Data
(Revised for v. 3)

CRC handbook of Lie group analysis of differential equations.

Includes bibliographical references and indexes.
Contents: v. 1. Symmetries, exact solutions, and conservation laws / W.F. Ames ... [et al.] — v. 2. Applications in engineering and physical sciences / A.V. Aksenov ... [et al.] — v. 3. New trends in theoretical developments and computational methods / R.L. Anderson ... [et al.].
1. Differential equations—Numerical solutions. 2. Lie groups. I. Ibragimov, N. Kh. (Nail Khairullovich) II. Title: Lie group analysis of differential equations.
QA372.C73 1994   515'.35   92-46289
ISBN 0-8493-4488-3 (v. 1) ISBN 0-8493-2864-0 (v. 2 : acid-free) ISBN 0-8493-9419-8 (v. 3 : acid-free)

Library of Congress Card Number 92-46289

CRC Press
Taylor & Francis Group
6000 Broken Sound Parkway NW, Suite 300
Boca Raton, FL 33487-2742

© 1996 by Taylor & Francis Group, LLC
CRC Press is an imprint of Taylor & Francis Group, an Informa business

No claim to original U.S. Government works

ISBN 13: 978-0-8493-9419-5 (hbk)

Visit the Taylor & Francis Web site at
http://www.taylorandfrancis.com

and the CRC Press Web site at
http://www.crcpress.com

CRC HANDBOOK OF

LIE GROUP ANALYSIS OF DIFFERENTIAL EQUATIONS

EDITED BY N. H. IBRAGIMOV

VOLUME 1
SYMMETRIES, EXACT SOLUTIONS,
AND CONSERVATION LAWS

VOLUME 2
APPLICATIONS IN ENGINEERING
AND PHYSICAL SCIENCES

VOLUME 3
NEW TRENDS IN
THEORETICAL DEVELOPMENTS
AND COMPUTATIONAL METHODS

Preface to Volume 3

Nowadays, Lie group theoretic approach to differential equations has been extended to new situations and has become applicable to the majority of equations that frequently occur in applied sciences. The intention of this volume is to present newly developed theoretical and computational methods.

The book is addressed to both beginners and experienced researchers in the Lie group analysis. It is written by the leading world experts in the field and contains a compilation of classical results as well as new results by the contributors of this volume based on their current research. Consequently, this volume covers a variety of topics and is up-to-date with all the latest information on recent theoretical developments and symbolic software packages. The book is provided with a floppy disk containing the programs to assist the reader in his efforts to extend the knowledge of computer algebra and use it in his own problems.

Chapter 1 was written by R. L. Anderson and N. H. Ibragimov.

Chapters 2 and 9 were written by V. A. Baikov, R. K. Gazizov, and N. H. Ibragimov.

Chapters 3 and 6 were written by N. H. Ibragimov.

Chapter 4 and 7 were prepared by M. B. Sheftel', then revised and recast by N. H. Ibragimov.

S. V. Meleshko collected material on symmetries of integro-differential equations and on differential constraints. This material was revised, essentially condensed and rewritten as Chapters 5 and 10 by N. H. Ibragimov.

Chapter 8 was written by N. H. Ibragimov and F. M. Mahomed.

Chapter 11 was written by P. J. Olver and E. M. Vorob'ev.

Chapter 12 was written by A. V. Turbiner.

Chapter 13 was written by W. Hereman.

Chapter 14 was written by M. C. Nucci.

In this book, the first and second volumes of the Handbook are referred to as [H1] and [H2], respectively.

I am greatly indebted to the contributors of the Handbook for their excellent work.

This volume was prepared during my stay as a Visiting Professor in the Department of Computational and Applied Mathematics, The University of Witwatersrand, Johannesburg, South Africa. I acknowledge financial support from the department and university, and I would like to register my appreciation to the members of the department for their warm hospitality.

I am grateful to my wife for her lasting encouragement and invaluable assistance throughout various stages of the work of the Handbook.

Finally, thanks are due to the staff of CRC Press for their skill and cooperation. In particular, I express my gratitude to Ms. Svetlana Landau, Acquiring Editor, who persuaded me to undertake this venture and was unfailing in her assistance and courtesy.

Nail Ibragimov
Johannesburg

Contents

Part I

Apparatus of Group Analysis

1

Lie-Bäcklund Symmetries: Representation by Formal Power Series

This chapter[1] describes the general structure of Lie-Bäcklund transformation groups. The emphasis in Sections 1.1 and 1.2 is on the role of the space $\mathcal{A}$ of differential functions in the theory. Section 1.3 contains the main theorems and algorithms used for computation of Lie-Bäcklund symmetries of differential equations, while in Section 1.4 we present the recent application of formal power series to the construction of invariants.

We follow the internal logic of the subject. For a historical survey of the development of the theory of Lie-Bäcklund transformation groups starting with the works of Lie and Bäcklund and ending with the modern contributions, the reader is referred to [H1], Chapter 5 (see also Chapter 6, Section 6.2.1 in this volume, and Anderson and Ibragimov [1979]). A detailed discussion of the theory and many applications are to be found in Ibragimov [1983].

Lie-Bäcklund transformation groups and their invariants are represented by formal series. The question of the convergence of these series is not treated here. The reader who is interested in this aspect of the problem is referred to the classical book by Hardy [1949] and for a modern approach see Flato, Pinczon, and Simon [1977].

1.1. The universal space of modern group analysis

The prolongation theory of Lie point and Lie contact transformation groups requires the introduction of functions depending not only on the independent variables x and dependent variables u, but also on derivatives of finite orders.

[1]Research supported in part by NATO Collaborative Research Grant 930117 and travel funds from the University of Georgia.

This prolongation is sufficient in the context of classical Lie theory. However, it is insufficient for the natural generalization of the classical theory given by Lie-Bäcklund transformation groups. In this generalization, one deals with transformations acting on intrinsically infinite-dimensional spaces. This new approach mandates the space $\mathcal{A}$ of differential functions as the universal space of modern group analysis.

1.1.1.   The space $\mathcal{A}$

The space $\mathcal{A}$ of differential functions was previously discussed in this Handbook (see [H1], Section 5.2, and [H2], Section 1.2.1). But, for the convenience of the reader, we reproduce here the necessary notation.

Let[2]

$$x = \{x^i\}, \quad u = \{u^\alpha\}, \quad u_{(1)} = \{u_i^\alpha\}, \quad u_{(2)} = \{u_{ij}^\alpha\}, ..., \qquad (1.1)$$

where $\alpha = 1, ..., m; \ i, j = 1, ..., n$. These variables are connected by the total differentiations

$$D_i = \frac{\partial}{\partial x^i} + u_i^\alpha \frac{\partial}{\partial u^\alpha} + u_{ij}^\alpha \frac{\partial}{\partial u_j^\alpha} + ..., \ i = 1, ..., n, \qquad (1.2)$$

as follows:

$$u_i^\alpha = D_i(u^\alpha), \ u_{ij}^\alpha = D_j(u_i^\alpha) = D_j D_i(u^\alpha), ... \qquad (1.3)$$

The variables x are called *the independent variables*, and the variables u are known as *the differential variables* with the successive *derivatives* $u_{(1)}, u_{(2)},...$. In the particular case, when $n = m = 1$, these derivatives will be denoted by $u_1, u_2, ...,$ so that $u_1 = u_x$, $u_2 = u_{xx}$, etc.

DEFINITION 1.1.   *A locally analytic function (i.e., locally expandable in a Taylor series with respect to all arguments) of a finite number of the variables 1.1 is called a differential function. The highest order of derivatives appearing in the differential function is called the order of this function. The vector space of all differential functions of finite order is denoted by $\mathcal{A}$.*

The space $\mathcal{A}$ has the intrinsic property of being closed under the derivation given by the total derivatives D_i.

---

[2]In [H1] and [H2], a slightly different notation for derivatives was used, namely, $\underset{1}{u}, \underset{2}{u}$ instead of $u_{(1)}, u_{(2)}$, etc.

1.1.2. Lie-Bäcklund operators

DEFINITION 1.2. *A differential operator of the form*

$$X = \xi^i \frac{\partial}{\partial x^i} + \eta^\alpha \frac{\partial}{\partial u^\alpha} + \zeta_i^\alpha \frac{\partial}{\partial u_i^\alpha} + \zeta_{i_1 i_2}^\alpha \frac{\partial}{\partial u_{i_1 i_2}^\alpha} + \cdots, \qquad (1.4)$$

where $\xi^i, \eta^\alpha \in \mathcal{A}$, and

$$\zeta_i^\alpha = D_i(\eta^\alpha - \xi^j u_j^\alpha) + \xi^j u_{ij}^\alpha, \quad \zeta_{i_1 i_2}^\alpha = D_{i_1} D_{i_2}(\eta^\alpha - \xi^j u_j^\alpha) + \xi^j u_{j i_1 i_2}^\alpha, \ldots \quad (1.5)$$

is called a Lie-Bäcklund operator. The abbreviated operator

$$X = \xi^i \frac{\partial}{\partial x^i} + \eta^\alpha \frac{\partial}{\partial u^\alpha} \qquad (1.6)$$

is also referred to as a Lie-Bäcklund operator provided that its prolonged action given by Formulas 1.4 and 1.5 is implied.[3]

Operator 1.4 is a formal sum. However, it truncates when acting on any differential function. Hence, *the action of Lie-Bäcklund operators is defined on the space $\mathcal{A}$.*

1.1.3. Lie-Bäcklund algebra $L_\mathcal{B}$

Let

$$X_\nu = \xi_\nu^i \frac{\partial}{\partial x^i} + \eta_\nu^\alpha \frac{\partial}{\partial u^\alpha}, \quad \nu = 1, 2,$$

be two Lie-Bäcklund operators 1.6. Their *commutator* (Lie bracket) is defined by the usual formula:

$$[X_1, X_2] = X_1 X_2 - X_2 X_1.$$

THEOREM 1.1. *The commutator $[X_1, X_2]$ is the Lie-Bäcklund operator given by*

$$[X_1, X_2] = \left(X_1(\xi_2^i) - X_2(\xi_1^i)\right) \frac{\partial}{\partial x^i} + \left(X_1(\eta_2^\alpha) - X_2(\eta_1^\alpha)\right) \frac{\partial}{\partial u^\alpha} + \cdots, \quad (1.7)$$

---

[3]Since the prolongations of Lie group generators and Lie-Bäcklund operators are uniquely determined by the prolongation formulas 1.5, there is no necessity to use a special notation for Operator 1.4 to indicate the prolongation.

where the terms denoted by dots are obtained by prolonging the coefficients of $\partial/\partial x^i$ and $\partial/\partial u^\alpha$ in accordance with Equations 1.4 and 1.5.

As a consequence, the following statement is valid.

THEOREM 1.2.   *The set of all Lie-Bäcklund operators is an infinite dimensional Lie algebra with respect to the Lie bracket 1.7. We call it the Lie-Bäcklund algebra and denote it by $L_\mathcal{B}$.*

1.1.4.   Properties of $L_\mathcal{B}$. Canonical operators

The Lie-Bäcklund algebra is endowed with the following properties (see, e.g., Ibragimov [1983]):

I. The total derivation 1.2 is a Lie-Bäcklund operator, i.e., $D_i \in L_\mathcal{B}$. Furthermore,

$$X_* = \xi_*^i D_i \in L_\mathcal{B} \tag{1.8}$$

for any $\xi_*^i \in \mathcal{A}$.

II. Let L_* be the set of all Lie-Bäcklund operators of the form 1.8. Then L_* is an ideal of $L_\mathcal{B}$, i.e., $[X, X_*] \in L_*$ for any $X \in L_\mathcal{B}$. Indeed,

$$[X, X_*] = \left(X(\xi_*^i) - X_*(\xi^i)\right)D_i \in L_*.$$

III. In accordance with Property II, two operators $X_1, X_2 \in L_\mathcal{B}$ are said to be *equivalent* (i.e., $X_1 \sim X_2$) if $X_1 - X_2 \in L_*$. In particular, every operator $X \in L_\mathcal{B}$ is equivalent to an operator 1.4 with $\xi^i = 0, i = 1, ..., n$. Namely,

$$X \sim X - \xi^i D_i \equiv (\eta^\alpha - \xi^i u_i^\alpha)\frac{\partial}{\partial u^\alpha} + \cdots. \tag{1.9}$$

DEFINITION 1.3.   *Operators of the form*

$$X = \eta^\alpha \frac{\partial}{\partial u^\alpha} + \cdots, \quad \eta^\alpha \in \mathcal{A}, \tag{1.10}$$

are called canonical Lie-Bäcklund operators.

Thus, Property III can be formulated as follows.

THEOREM 1.3.   *Any operator $X \in L_\mathcal{B}$ is equivalent to a canonical Lie-Bäcklund operator.*

Canonical operators leave invariant the independent variables x^i. Therefore, the use of the canonical form is convenient, e.g., for investigating symmetries of integro-differential equations and for finding recursion operators (cf. Chapters 4 and 5; see also Section 1.2.5). It is important to note that the canonical form for the generator of a Lie point (or contact) transformation group is, in general, intrinsically a Lie-Bäcklund operator. Therefore, the canonical form was not used in classical Lie theory.

IV. The following statements describe all Lie-Bäcklund operators equivalent to generators of Lie point and Lie contact transformation groups.

THEOREM 1.4. *Operator 1.4 is equivalent to the infinitesimal operator of a one-parameter point transformation group if and only if its coordinates assume the form*

$$\xi^i = \xi_1^i(x, u) + \xi_*^i, \quad \eta^\alpha = \eta_1^\alpha(x, u) + \left(\xi_2^i(x, u) + \xi_*^i\right)u_i^\alpha,$$

where $\xi_*^i \in \mathcal{A}$ is an arbitrary differential function and $\xi_1^i, \xi_2^i, \eta_1^\alpha$ are arbitrary functions of x and u.

THEOREM 1.5. *Let $m = 1$. Then Operator 1.4 is equivalent to the infinitesimal operator of a one-parameter contact transformation group if and only if its coordinates assume the form*

$$\xi^i = \xi_1^i(x, u, u_{(1)}) + \xi_*^i, \quad \eta = \eta_1(x, u, u_{(1)}) + \xi_*^i u_i,$$

where $\xi_*^i \in \mathcal{A}$ is an arbitrary differential function and ξ_1^i, η_1 are arbitrary functions of x, u and $u_{(1)}$.

EXAMPLES

To illustrate the notion of the canonical Lie-Bäcklund operators, we give here the generators of familiar point transformation groups written in both the classical and the canonical Lie-Bäcklund forms.

Translations. Let $n = m = 1$, $u_1 = u_x$. The generator of the group of translations along the x-axis,

$$X = \frac{\partial}{\partial x},$$

is shifted, by Equation 1.9, into the following canonical Lie-Bäcklund operator:

$$u_x \frac{\partial}{\partial u}.$$

Dilations. Let x, y be the independent variables, and $k, c = $ const. The generator of non-homogeneous dilations and its canonical Lie-Bäcklund form are as follows:

$$X = x\frac{\partial}{\partial x} + ky\frac{\partial}{\partial y} + cu\frac{\partial}{\partial u} \sim (cu - xu_x - kyu_y)\frac{\partial}{\partial u} + \cdots.$$

Galilean Boosts. Here,

$$X = t\frac{\partial}{\partial x} + \frac{\partial}{\partial u} \sim (1 - tu_x)\frac{\partial}{\partial u} + \cdots,$$

where t, x are the independent variables.

1.2.   Formal transformation groups

For the sake of brevity, we introduce the sequence

$$z = (x, u, u_{(1)}, u_{(2)}, \ldots) \tag{1.11}$$

with elements z^v, $v \geq 1$. Here, all that is essential is that some ordering be established within the sets $x, u, u_{(1)}, \ldots$. For x and u we will use the natural ordering so that

$$z^i = x^i, \; 1 \leq i \leq n, \quad z^{n+\alpha} = u^\alpha, \; 1 \leq \alpha \leq m.$$

Denote by $[z]$ any finite subsequence of z. Then elements of $\mathcal{A}$ are written as $f([z])$.

1.2.1.   The representation space $[[\mathcal{A}]]$

Consider formal power series in one symbol a:

$$f(z, a) = \sum_{k=0}^{\infty} f_k([z])a^k, \quad f_k([z]) \in \mathcal{A}. \tag{1.12}$$

Let f and g be formal power series, where f is defined by Formula 1.12 and g is given by a similar formula, viz.

$$g(z, a) = \sum_{k=0}^{\infty} g_k([z])a^k, \quad g_k([z]) \in \mathcal{A}.$$

Their linear combination $\lambda f([z]) + \mu g([z])$ with constant coefficients λ, μ and product $f([z]) \cdot g([z])$, respectively, are defined as follows:

$$\lambda \sum_{k=0}^{\infty} f_k([z])a^k + \mu \sum_{k=0}^{\infty} g_k([z])a^k = \sum_{k=0}^{\infty} \big(\lambda f_k([z]) + \mu g_k([z])\big)a^k, \quad (1.13)$$

and

$$\left(\sum_{p=0}^{\infty} f_p([z])a^p\right) \cdot \left(\sum_{q=0}^{\infty} g_q([z])a^q\right) = \sum_{k=0}^{\infty} \left(\sum_{p+q=k} f_p([z])g_q([z])\right)a^k. \quad (1.14)$$

The space of all formal power series 1.12 endowed with the addition and multiplication operations defined by Equations 1.13 and 1.14, respectively, is denoted by $[[\mathcal{A}]]$.

Lie point and Lie contact transformations, together with their prolongations of all orders, are represented by elements of the space $[[\mathcal{A}]]$. The utilization of this space is *a fortiori* necessary in the theory of Lie-Bäcklund transformation groups. Therefore, we call $[[\mathcal{A}]]$ *the representation space of modern group analysis.*

1.2.2. Lie-Bäcklund equations

DEFINITION 1.4. *Given an Operator 1.4, the Lie-Bäcklund equations are*

$$\frac{d}{da}\bar{x}^i = \xi^i([\bar{z}]), \quad \frac{d}{da}\bar{u}^\alpha = \eta^\alpha([\bar{z}]),$$

$$\frac{d}{da}\bar{u}_i^\alpha = \zeta_i^\alpha([\bar{z}]), \quad \frac{d}{da}\bar{u}_{ij}^\alpha = \zeta_{ij}^\alpha([\bar{z}]), \dots, \quad (1.15)$$

where $\alpha = 1, \dots, m$ and $i, j, \dots = 1, \dots, n$.

In the case of canonical operators 1.10, the infinite system of equations 1.15 simplify to the finite system:

$$\frac{d}{da}\bar{u}^\alpha = \eta^\alpha([\bar{z}]), \quad \alpha = 1, \dots, m. \quad (1.16)$$

Then the transformations of the successive derivatives are obtained by the total differentiation:

$$\overline{u}_i^\alpha = D_i(\overline{u}^\alpha), \quad \overline{u}_{ij}^\alpha = D_i D_j(\overline{u}^\alpha), \dots . \tag{1.17}$$

1.2.3. Definition of a formal group

Consider sequences of elements

$$f^\nu(z, a) \in [[\mathcal{A}]], \quad \nu \geq 1,$$

such that

$$f^\nu(z, a) = \sum_{k=0}^{\infty} f_k^\nu([z]) a^k, \quad f_0^\nu([z]) = z^\nu, \quad \nu = 1, 2, \dots, \tag{1.18}$$

where $f_k^\nu([z])$, $k = 1, 2, \dots$, are arbitrary elements of A.

DEFINITION 1.5.   *Define a transformation of Sequences 1.11 by*

$$\overline{z}^\nu = f^\nu(z, a), \quad \nu \geq 1. \tag{1.19}$$

DEFINITION 1.6.   *Transformation 1.19 is called a formal one-parameter group if the coefficients in the formal series 1.18 obey the property:*

$$f_k^\nu([\overline{z}]) = \sum_{l=0}^{\infty} \frac{(k+l)!}{k!\,l!} f_{k+l}^\nu([z]) a^l, \quad \nu \geq 1, \ k = 0, 1, 2, \dots . \tag{1.20}$$

THEOREM 1.6.   *Equations 1.20 are equivalent to the usual group property:*

$$f^\nu(\overline{z}, b) = f^\nu(z, a + b), \quad \nu \geq 1, \tag{1.21}$$

written for formal power series.

For the proof, see Ibragimov [1983], Section 15.1.

This theorem requires more explanation. By definition, the left hand sides of Equation 1.21 are formal power series in one symbol b:

$$f^\nu(\overline{z}, b) = \sum_{l=0}^{\infty} f_l^\nu([\overline{z}]) b^l.$$

In these series, $\bar{z}$ is given by power series in one symbol a, hence after rearrangement we can write them as formal power series in two symbols a and b:

$$\sum_{l=0}^{\infty} f_l^{\nu}([\bar{z}])b^l = \sum_{k,l=0}^{\infty} g_{kl}^{\nu}([\bar{z}])a^k b^l. \tag{1.22}$$

Equation 1.21 asserts that the right hand side of Equation 1.22 can be rewritten as a formal power series in one symbol $(a + b)$, viz.

$$\sum_{k=0}^{\infty} f_k^{\nu}([\bar{z}])(a + b)^k = f^{\nu}(z, a + b), \quad \nu = 1, 2, \ldots.$$

REMARK. The group property, in its most general form

$$f^{\nu}(\bar{z}, b) = f^{\nu}(z, \phi(a, b)), \quad \nu \geq 1,$$

with the group composition law $\phi(a, b)$, can be transformed to Equation 1.21 by a suitable change of the group parameter a. The proof is similar to the classical case.

1.2.4. Integration of Lie-Bäcklund equations

Consider Equations 1.15 with the initial conditions given in Section 1.2.3:

$$\frac{d}{da}\bar{x}^i = \xi^i([\bar{z}]), \quad \bar{x}^i\big|_{a=0} = x^i,$$

$$\frac{d}{da}\bar{u}^{\alpha} = \eta^{\alpha}([\bar{z}]), \quad \bar{u}^{\alpha}\big|_{a=0} = u^{\alpha}, \tag{1.23}$$

$$\cdot \quad \cdot \quad \cdot \quad \cdot \quad \cdot \quad \cdot \quad \cdot \quad \cdot \quad \cdot \quad \cdot \quad \cdot$$

The formal integrability of the infinite system of Equations 1.23 is proved, e.g., in Ibragimov [1983], Section 15.1. It is also discussed in [H1], Section 5.2. However, for completeness we restate the relevant theorem here.

THEOREM 1.7. *Lie-Bäcklund equations 1.23 have a solution in the space* $[[\mathcal{A}]]$. *The solution is unique and is given by a sequence of formal power series of the form*

$$\bar{x}^i = x^i + \sum_{k=0}^{\infty} A_k^i([z])a^k, \quad A_k^i([z]) \in \mathcal{A},$$

$$\overline{u}^\alpha = u^\alpha + \sum_{k=0}^{\infty} B_k^\alpha([z])a^k, \quad B_k^\alpha([z]) \in \mathcal{A}, \tag{1.24}$$

. . . .

The coefficients $A_k^i([z])$, $B_k^\alpha([z])$, ... satisfy the group property 1.20.

1.2.5. Exponential map

The question immediately arises as to how one calculates the coefficients in Series 1.24. The fundamental idea is provided by the exponential map in Lie group theory. For the generator X of a Lie transformation group of points $x = (x^1, ..., x^n)$, the group transformation is given by

$$\overline{x}^i = \exp(aX)(x^i), \quad i = 1, ..., n, \tag{1.25}$$

where

$$\exp(aX) = 1 + aX + \frac{a^2}{2!}X^2 + \frac{a^3}{3!}X^3 + \cdots . \tag{1.26}$$

For example, let's take, in the case $n = 1$, the generator

$$X = x^2 \frac{\partial}{\partial x}.$$

Equation 1.25 yields:

$$\overline{x} = x + ax^2 + a^2 x^3 + \cdots . \tag{1.27}$$

Similarly and more generally, as in the case of Lie group theory, one can easily prove that the solution 1.24 to the Lie-Bäcklund equations 1.23 is given by the exponential map

$$\overline{x}^i = \exp(aX)(x^i), \quad \overline{u}^\alpha = \exp(aX)(u^\alpha), \quad \overline{u}_i^\alpha = exp(aX)(u_i^\alpha), \ldots, \tag{1.28}$$

where $\exp(aX)$ is given by Equation 1.26 written for the Lie-Bäcklund operator 1.4.

As we discussed before, we can restrict our consideration to canonical operators 1.10. Then Equations 1.23 reduce to the finite system of equations

$$\frac{d}{da}\overline{u}^\alpha = \eta^\alpha([\overline{z}]), \quad \overline{u}^\alpha\big|_{a=0} = u^\alpha \tag{1.29}$$

supplemented by the transformation formulas 1.17. It follows that formal groups of Lie-Bäcklund transformations can be constructed by virtue of the following theorem.

THEOREM 1.8. *Given a canonical Lie-Bäcklund operator,*

$$X = \eta^\alpha \frac{\partial}{\partial u^\alpha} + \cdots,$$

the corresponding formal one-parameter group is represented by the series

$$\bar{u}^\alpha = u^\alpha + a\eta^\alpha + \frac{a^2}{2!}X(\eta^\alpha) + \cdots + \frac{a^n}{n!}X^{n-1}(\eta^\alpha) + \cdots \qquad (1.30)$$

together with its differential consequences

$$\bar{u}_i^\alpha = u_i^\alpha + aD_i(\eta^\alpha) + \frac{a^2}{2!}X(D_i(\eta^\alpha)) + \cdots + \frac{a^n}{n!}X^{n-1}(D_i(\eta^\alpha)) + \cdots, \qquad (1.31)$$

and

$$\bar{u}_{i_1\cdots i_s}^\alpha = u_{i_1\cdots i_s}^\alpha + aD_{i_1}\cdots D_{i_s}(\eta^\alpha) + \cdots + \frac{a^n}{n!}X^{n-1}(D_{i_1}\cdots D_{i_s}(\eta^\alpha)) + \cdots \qquad (1.32)$$

for any s > 1.

PROOF. This result can be proved directly without using the general exponential map 1.28. Indeed, because of the uniqueness of the solution, one needs to only show that the formal power series 1.30 satisfy Equations 1.29. By definition, $d\bar{u}^\alpha/da$ is obtained by differentiating the series 1.30 term by term:

$$\frac{d\bar{u}^\alpha}{da} = \eta^\alpha + aX(\eta^\alpha) + \frac{a^2}{2!}X^2(\eta^\alpha) + \cdots + \frac{a^{n-1}}{(n-1)!}X^{n-1}(\eta^\alpha) + \cdots.$$

The right hand side of this equation is identical with the formal power expansion (with respect to a) of the function $\eta^\alpha([\bar{z}])$, where $[\bar{z}]$ is given by the series 1.30 to 1.32. At last, we note that the series 1.31 for the transformation of the first-order derivatives and the similar series 1.32 for the higher-order derivatives are obtained from the series 1.30 via Formulas 1.17. Indeed, since the total derivatives D_i commute with any canonical Lie-Bäcklund operator X, one has

$$D_i \exp(aX) = \exp(aX)D_i.$$

Hence, Equation 1.30 and Formulas 1.17 yield the equations 1.31 and 1.32.

EXAMPLES

We illustrate an application of Theorem 1.8 by the following examples with $n = m = 1$.

Example 1. Let

$$X = u_1 \frac{\partial}{\partial u} + u_2 \frac{\partial}{\partial u_1} + \cdots .$$

We have here $\eta = u_1$. Therefore,

$$X(\eta) = u_2, \ X^2(\eta) = u_3, \ldots, \ X^{n-1}(\eta) = u_n .$$

Hence, Transformation 1.30 of the corresponding formal group has the form

$$\bar{u} = u + \sum_{n=1}^{\infty} \frac{a^n}{n!} u_n .$$

Example 2. Let

$$X = u_p \frac{\partial}{\partial u} + u_{p+1} \frac{\partial}{\partial u_1} + \cdots . \tag{1.33}$$

Here, $\eta = u_p$ with an arbitrary positive integer p. We have:

$$X(\eta) = u_{2p}, \ X^2(\eta) = u_{3p}, \ldots, \ X^{n-1}(\eta) = u_{np} .$$

Hence, Transformation 1.30 is given by the following formal power series:

$$\bar{u} = u + \sum_{n=1}^{\infty} \frac{a^n}{n!} u_{np} . \tag{1.34}$$

Example 3. Let

$$X = u_1^2 \frac{\partial}{\partial u} + 2u_1 u_2 \frac{\partial}{\partial u_1} + \cdots .$$

Here, $\eta = u_1^2$. Successively, we find:

$$X(\eta) = 4u_1^2 u_2, \quad X^2(\eta) = 8(u_1^3 u_3 + 3u_1^2 u_2^2),$$

$$X^3(\eta) = 16(u_1^4 u_4 + 12u_1^3 u_2 u_3 + 12u_1^2 u_2^3), \ldots .$$

Hence, Transformation 1.30 has the form:

$$\bar{u} = u + au_1^2 + 2a^2 u_1^2 u_2 + \frac{4}{3}a^3(u_1^3 u_3 + 3u_1^2 u_2^2)$$

$$+\frac{2}{3}a^4(u_1^4 u_4 + 12u_1^3 u_2 u_3 + 12u_1^2 u_2^3) + \cdots .$$

1.2.6. Representations: Lie versus Lie-Bäcklund

Let us begin with an example, e.g., the series 1.27. It is convergent for $|ax| < 1$. In this case, the group transformation can be represented by the following analytic function:

$$\bar{x} = \frac{x}{1 - ax} . \tag{1.35}$$

In fact, Equation 1.35 is arrived at directly by integrating the Lie equation

$$\frac{d\bar{x}}{da} = \bar{x}^2, \quad \bar{x}\big|_{a=0} = x. \tag{1.36}$$

For any given x, the solution of the Cauchy problem 1.36 exists for sufficiently small a, namely for $|a| < 1/|x|$. This solution is unique and is given by Equation 1.35.

This approach, Lie's approach, of solving the Lie equations leads one to local Lie groups of transformations represented by functions similar to Equation 1.35.

However, in light of the problems posed by solving the Lie-Bäcklund equations 1.15, we emphasize an alternative direction for developing representations by power series similar to Equation 1.27, namely, representation given by the exponential map 1.28. In this latter approach, one immediately arrives at the theory of transformation groups represented by *formal* power series. For instance, in our example we can consider the transformation 1.27 without any restriction on the group parameter a. Then the representation is given by a divergent series for $|ax| \geq 1$. This is the direction taken in the book by Ibragimov [1983]. The essential guiding principle is that formal power series representations merit in their own right. In fact, they are necessary to develop any comprehensive, sensible theory of Lie-Bäcklund transformation groups. This situation pertains because the convergence problem for these formal power series cannot be universally solved and must be treated separately for each type of Lie-Bäcklund operators.

We note that series representations can, in principle, be used for constructing representations by functions. For instance, Equation 1.27 can be rewritten as

$$\bar{x} = x(1 + ax + a^2 x^2 + \cdots).$$

One recognizes the terms in the parentheses as the Taylor expansion of the function $1/(1 - ax)$.

In practice, in the case of Lie groups, it is easier to solve the Lie equations than it is to sum the exponential map. While, at the present stage of development, for Lie-Bäcklund transformations there is no difference in difficulty between solving the Lie-Bäcklund equations and employing the exponential map.

Finally, we remark that the origin of the fundamental difference between Lie and Lie-Bäcklund transformation groups is that the former are determined by ordinary differential equations (Lie equations) while the latter are determined by evolutionary partial differential equations (Lie-Bäcklund equations) for $\bar{u} = \bar{u}(x, a)$:

$$\frac{\partial}{\partial a}\bar{u}^\alpha = \eta^\alpha(x, \bar{u}, \bar{u}_{(1)}, ..., \bar{u}_{(k)}), \quad \alpha = 1, ..., m.$$

EXAMPLE

Consider the canonical Lie-Bäcklund operator 1.33 from Example 2 of Section 1.2.5. The corresponding group transformations are given by the formal power series 1.34. These series converge in the disk $|a| < r$, if the sequence $(u, u_1, u_2, ...)$ satisfies the inequalities

$$|u_{pk}| \leq Ck!r^{-k}, \quad C = \text{const.}, \quad k = 0, 1, 2, ... \tag{1.37}$$

Then the transformed sequence $(\bar{u}, \bar{u}_1, \bar{u}_2, ...)$ satisfies the same inequalities 1.37. It follows that the representation 1.34 gives a local group of transformations in the class of *entire functions $u(x)$ of order $p/(p - 1)$* determined by the conditions 1.37 (see Ibragimov [1983], Section 16.2).

Titov [1990] generalized this result to Operators 1.10 such that their coordinates η^α are linear functions of $u, u_{(1)}, ..., u_{(k)}, k > 1$, with coefficients depending on x.

Furthermore, it is proved in Ibragimov [1977] that the ideal $L_* \subset L_B$ (see Section 1.1.4, Property II) generates an infinite local group of transformations acting in the *scale of Banach spaces of locally analytic functions $u(x)$*. Hence, the convergence problem concerns the quotient algebra L_B/L_*. More precisely, it concerns only the canonical Lie-Bäcklund operators 1.10 with coordinates η^α depending (nonlinearly) on derivatives $u_{(k)}$ with $k > 1$.

1.3. Invariant differential equations

Let $F \in \mathcal{A}$ be an arbitrary differential function of order $k \geq 1$, i.e., $F = F(x, u, u_{(1)}, ..., u_{(k)})$. Consider the kth-order differential equation(s) (F may be vector valued):

$$F(x, u, u_{(1)}, ..., u_{(k)}) = 0. \tag{1.38}$$

Denote by $[F]$ the set of Sequences 1.11 determined by the infinite system of equations

$$[F]: \qquad F = 0, \ D_i F = 0, \ D_i D_j F = 0, ... \tag{1.39}$$

The set $[F]$ is called *the extended frame* of the differential equation 1.38 (cf. [H2], Chapter 1).

1.3.1. Determining equation for Lie-Bäcklund symmetries

DEFINITION 1.7. *Equation 1.38 admits a Lie-Bäcklund transformation group G if the extended frame 1.39 is invariant under G. Then G is also called a symmetry group of Equation 1.38.*

Here, we present the infinitesimal criteria proved in Ibragimov [1983].

THEOREM 1.9. *Let G be a Lie-Bäcklund transformation group with the generator X. Equation 1.38 admits G if and only if*

$$XF\big|_{[F]} = 0, \quad XD_i F\big|_{[F]} = 0, ...$$

This criterion requires an infinite number of infinitesimal tests. However, it can be simplified and reduced to a finite number of tests.

LEMMA. *The equation*

$$XF\big|_{[F]} = 0$$

yields the infinite series of equations

$$XD_i F\big|_{[F]} = 0, \quad XD_i D_j F\big|_{[F]} = 0, ...$$

Thus, we have the following finite criterium for calculating Lie-Bäcklund symmetries of differential equations:

THEOREM 1.10. *Equation 1.38 admits the formal group G generated by the Lie-Bäcklund operator 1.4,*

$$X = \xi^i \frac{\partial}{\partial x^i} + \eta^\alpha \frac{\partial}{\partial u^\alpha} + \zeta_i^\alpha \frac{\partial}{\partial u_i^\alpha} + \zeta_{i_1 i_2}^\alpha \frac{\partial}{\partial u_{i_1 i_2}^\alpha} + \cdots,$$

if and only if

$$XF|_{[F]} = 0, \tag{1.40}$$

where $|_{[F]}$ *means evaluated on the extended frame* $[F]$ *defined by the infinite system of Equations 1.39.*

DEFINITION 1.8.   *Equation 1.40 is called the determining equation for Lie-Bäcklund symmetries of the differential equation 1.38.*

REMARK.   The main difference between calculating Lie and Lie-Bäcklund symmetries lies in the fact that in the Lie case one deals with generators whose coefficients ξ^i and η^α are differential functions of order *zero* (point symmetries) and *one* (contact symmetries), while in the Lie-Bäcklund case one has no *a priori* restriction on the order. As a result of this situation, additional machinery is necessary (see Ibragimov [1983] and Chapter 4 of this volume; cf. also [H1], Section 5.1.3).

1.3.2.   Trivial symmetries

In investigating the Lie-Bäcklund symmetries, it is useful to take into account the following properties of the determining equation 1.40:

I. The ideal $L_* \subset L_B$ (see Section 1.1.4, Property II) is admitted by any system of differential equations. Therefore, in applications to differential equations, the quotient algebra L_B/L_* is the object of study.

Indeed, let $X_* \in L_*$. It has the form 1.8,

$$X_* = \xi_*^i D_i .$$

Therefore, $X_* F = \xi_*^i D_i F$. Hence, X_* satisfies the determining equation 1.40,

$$X_* F|_{[F]} = 0,$$

by virtue of the definition of the extended frame $[F]$ by Equations 1.39.

II. Equation 1.38 admits any canonical Lie-Bäcklund operator 1.10,

$$X = \eta^\alpha \frac{\partial}{\partial u^\alpha} + \cdots,$$

with arbitrary coordinates η^α satisfying the condition

$$\eta^\alpha|_{[F]} = 0, \quad \alpha = 1, ..., m. \tag{1.41}$$

Indeed, for a locally analytic η^α, Equation 1.41 yields

$$\eta^\alpha = \eta_0^\alpha F + \eta_1^{\alpha,i} D_i F + \eta_2^{\alpha,ij} D_i D_j F + \cdots + \eta_k^{\alpha,i_1\ldots i_k} D_{i_1} \cdots D_{i_k} F, \quad k < \infty.$$

Hence, the determining equation 1.40 is satisfied. Here, the coefficients η_0^α, $\eta_1^{\alpha,i}$, $\eta_2^{\alpha,ij}$, ..., $\eta_k^{\alpha,i_1\ldots i_k} \in \mathcal{A}$ are arbitrary differential functions regular on the extended frame of the differential equation 1.38. Note that these coefficients may depend on F and its derivatives. However, the regularity requires that this dependence must be such that the functions η^α are determined, finite and locally analytic on $[F]$.

The general case of Lie-Bäcklund operators 1.4 such that

$$\xi^i\big|_{[F]} = 0, \quad \eta^\alpha\big|_{[F]} = 0,$$

reduces, by virtue of the equivalence relation 1.9, to a canonical operator satisfying the condition 1.41.

Lie-Bäcklund symmetries given by the properties I and II are naturally considered as *trivial* symmetries of a given differential equation 1.38. In group analysis we are interested in nontrivial symmetries only.

EXAMPLE

Consider an arbitrary pth-order scalar evolution equation in one space variable x:

$$u_t = f(x, u, u_1, \ldots, u_p),$$

where $u_1 = u_x$, etc. For this equation, nontrivial Lie-Bäcklund symmetries can be taken in the form

$$X = \eta(t, x, u, u_1, \ldots, u_k)\frac{\partial}{\partial u} + \cdots.$$

For instance, consider the group of translations in t. Its generator $\partial/\partial t$ is equivalent to the following nontrivial Lie-Bäcklund symmetry:

$$X = u_t\frac{\partial}{\partial u} + \cdots$$

and can be rewritten, by using the differential equation under consideration, in the form

$$X = f(x, u, u_1, \ldots, u_p)\frac{\partial}{\partial u} + \cdots.$$

1.4. Invariants

In the group analysis, invariants of Lie point symmetry groups are used, e.g., for an explicit representation of invariant solutions (Ovsiannikov [1962]). The prolongation theory allows one to construct invariant differential equations via differential invariants of a given group prolonged to the space $\mathcal{A}$.

Lie-Bäcklund transformation groups do not have enough invariants in the space $\mathcal{A}$. However, they have as many differential invariants as any one-parameter Lie point transformation group, if one allows formal sums of elements from $\mathcal{A}$.

1.4.1. Differential invariants in $\mathcal{A}$

DEFINITION 1.9.   *Let G be a formal one-parameter group of transformations 1.24 acting in the space of Sequences 1.11. Let $F([z]) \in \mathcal{A}$ be a differential function of order k. Then F is called a differential invariant of order k for the Lie-Bäcklund group G if*

$$F([\bar{z}]) = F([z])$$

for any z given by Sequence 1.11 and for its image $\bar{z}$ under Transformation 1.24.

This definition subsumes Lie's differential invariants of all orders and generalizes the notion to Lie-Bäcklund groups.

THEOREM 1.11.   *The function $F([z]) \in \mathcal{A}$ is a differential invariant of the Lie-Bäcklund group with the generator X if and only if*

$$XF = 0.$$

The proof can be carried out using the exponential map 1.28 and is similar to the Lie case.

1.4.2. Differential invariants of the Maxwell group

Consider the evolutionary part of the system of Maxwell's equations, viz.

$$\frac{\partial \mathbf{E}}{\partial t} = \text{curl}\mathbf{B}, \qquad \frac{\partial \mathbf{B}}{\partial t} = -\text{curl}\mathbf{E}. \tag{1.42}$$

We treat this system as a Lie-Bäcklund equation 1.16. The corresponding Lie-Bäcklund transformation group with the group parameter $a = t$ is called *the*

Maxwell group. Its Lie-Bäcklund operator is:

$$X = \sum_{i=1}^{3}\left[(\text{curl}\mathbf{B})^i\,\frac{\partial}{\partial E^i} - (\text{curl}\mathbf{E})^i\,\frac{\partial}{\partial B^i}\right].$$

THEOREM 1.12. *The Maxwell group has the following basis of differential invariants of order 0 and 1 (Ibragimov [1983], Section 17.2):*

$$\mathbf{x} = (x^1, x^2, x^3), \quad \text{div}\mathbf{E}, \quad \text{div}\mathbf{B},$$

where $\mathbf{x}$ *denotes the three spatial coordinates. Furthermore,[4] all other (higher-order) differential invariants are functions of these basic invariants and of the successive derivatives of* $\text{div}\mathbf{E}$ *and* $\text{div}\mathbf{B}$ *with respect to* x.

It follows that, e.g., equations

$$\text{div}\mathbf{E} = f(\mathbf{x}), \quad \text{div}\mathbf{B} = g(\mathbf{x}), \tag{1.43}$$

with arbitrary functions f and g, are invariant under the action of the Maxwell group. Hence, if Equations 1.43 are satisfied at time $t = 0$, then they are satisfied at all subsequent times t. Here, the physics dictates the choice. Note that with the choice $f = g = 0$, we come to Maxwell's equations in a vacuum.

1.4.3. Lack of differential invariants in the one-dimensional case

Let us discuss the one-dimensional case. Namely, consider transformation groups involving one independent variable x and one dependent (i.e., differential) variable u together with successive derivatives $u_1, u_2, ...,$ such that $u_{i+1} = D(u_i)$, $u_0 = u$, where

$$D \equiv D_x = \frac{\partial}{\partial x} + u_1\frac{\partial}{\partial u} + u_2\frac{\partial}{\partial u_1} + \cdots.$$

It is well known that local Lie groups of transformations have an infinite number of differential invariants. For Lie-Bäcklund transformation groups the corresponding problem is open, if we consider invariants to be from the space $\mathcal{A}$. More precisely, the situation in the one-dimensional case is described by the following statement.

---

[4]N.H. Ibragimov, unpublished.

THEOREM 1.13. *Let*

$$X = \eta(x, u, u_1, ..., u_l)\frac{\partial}{\partial u} + \cdots$$

be a proper Lie-Bäcklund operator (not a generator of a Lie point or contact transformation group). Then it has no differential invariants in $\mathcal{A}$ (in the sense of Definition 1.9) except arbitrary functions of x.

PROOF.   The invariance of a differential function $F(x, u, ..., u_k) \in \mathcal{A}$ of order k, under the action of the operator X is determined by the equation

$$XF \equiv \eta\frac{\partial F}{\partial u} + D(\eta)\frac{\partial F}{\partial u_1} + D^2(\eta)\frac{\partial F}{\partial u_2} + \cdots + D^k(\eta)\frac{\partial F}{\partial u_k} = 0. \qquad (1.44)$$

The highest-order derivative involved in this equation is u_{k+l}. It appears in the expression $D^k(\eta)$. Hence, Equation 1.44 yields

$$\frac{\partial F}{\partial u_k} = 0.$$

Consequently,

$$\frac{\partial F}{\partial u_{k-1}} = 0, ..., \frac{\partial F}{\partial u} = 0.$$

This implies $F = F(x)$. Hence, X has only one functionally independent invariant in the space $\mathcal{A}$, namely x.

1.4.4.   Formal invariants in $[[\mathcal{A}]]$: Outline of the method

In this section, we consider again the one-dimensional case of the preceding section. We show that one has for Lie-Bäcklund transformation groups the same situation as for Lie point transformation groups if one considers an approximation of invariants by divergent power series with coefficients from the space $\mathcal{A}$. For a detailed discussion of the method sketched here, see Anderson and Ibragimov [1994].

DEFINITION 1.10. *Two Lie-Bäcklund operators,*

$$X_1 = \xi_1(x, u, u_1, ..., u_k)\frac{\partial}{\partial x} + \eta_1(x, u, u_1, ..., u_l)\frac{\partial}{\partial u} + \cdots,$$

and

$$X_2 = \xi_2(y, v, v_1, ..., v_r)\frac{\partial}{\partial y} + \eta_2(y, v, v_1, ..., v_s)\frac{\partial}{\partial v} + \cdots,$$

are said to be similar if there exists an invertible formal transformation in [[A]], *namely given by the formal power series in one symbol* ε:

$$y = x + \varepsilon Y_1(x, u, u_1, ..., u_{m_1}) + \varepsilon^2 Y_2(x, u, u_1, ..., u_{m_2}) + \cdots, \quad Y_i \in A, \quad (1.45)$$

$$v = u + \varepsilon V_1(x, u, u_1, ..., u_{n_1}) + \varepsilon^2 V_2(x, u, u_1, ..., u_{n_2}) + \cdots, \quad V_i \in A, \quad (1.46)$$

such that X_1, *written in the new variables* y, v, *coincide with* X_2, *i.e.,*

$$X_1 \equiv X_1(y)\frac{\partial}{\partial y} + X_1(v)\frac{\partial}{\partial v} = X_2. \quad (1.47)$$

This definition subsumes the usual definition of similarity in Lie theory.

We shall use the well-known fact that the generator X of any one-parameter Lie transformation group is similar to the generator of translations:

$$X_1 = \frac{\partial}{\partial x}. \quad (1.48)$$

Consider two Lie-Bäcklund operators: the operator X_1 given in the form 1.48 and a canonical Lie-Bäcklund operator

$$X_2 = \eta(y, v, v_1, ..., v_s)\frac{\partial}{\partial v} + \cdots, \quad (1.49)$$

where $v_{i+1} = D_y(v_i)$. The method for approximating invariants of the Lie-Bäcklund operator X_2 consists of three steps.

FIRST, we introduce the operator

$$X_\varepsilon = (1 - \varepsilon)\frac{\partial}{\partial y} + \varepsilon X_2, \quad (1.50)$$

where X_2 is the operator 1.49. The operator X_ε continuously connects the operators 1.48 and 1.49:

$$X_{\varepsilon=0} = X_1, \quad X_{\varepsilon=1} = X_2.$$

SECOND, we establish the following result on similarity:

THEOREM 1.14. *Operator 1.50 is similar to the operator 1.48.*

The proof is direct. We find the similarity transformations 1.45 and 1.46, by solving Equation 1.47, in particular

$$X_1(y) = 1 - \varepsilon, \quad X_1(v) = \varepsilon\eta(y, v, v_1, ..., v_s). \tag{1.51}$$

THIRD, take the known basis of all differential invariants of Operator 1.48, viz.

$$u, u_1, u_2, ..., \tag{1.52}$$

and subject them to the similarity transformation obtained by solving Equations 1.51.

As a result, one obtains invariants represented by elements of the space $[[\mathcal{A}]]$, in accordance with the following definition.

DEFINITION 1.11. *An element of $[[\mathcal{A}]]$, namely a formal power series in one symbol ε,*
$$F(y, v, v_1, ...; \varepsilon) = F_0 + \varepsilon F_1 + \varepsilon^2 F_2 + \cdots$$

is called an invariant of the formal group with the Lie-Bäcklund operator 1.50 if the formal series $X_\varepsilon F$ vanishes:

$$X_\varepsilon F = 0.$$

Here F_0, F_1, F_2, ... depend on finite numbers of the variables y, v, v_1, ... and are elements of $\mathcal{A}$.

1.4.5.   Example to Section 1.4.4

Consider the Lie-Bäcklund operator

$$X_2 = v_1 \frac{\partial}{\partial v} + \cdots .$$

In this case, Operator 1.50 has the form

$$X_\varepsilon = (1 - \varepsilon)\frac{\partial}{\partial y} + \varepsilon v_1 \frac{\partial}{\partial v} + \varepsilon v_2 \frac{\partial}{\partial v_1} + \varepsilon v_3 \frac{\partial}{\partial v_2} + \cdots . \tag{1.53}$$

We select "the simplest nontrivial" solution of Equations 1.51 at each step and obtain the following transformation:

$$y = (1 - \varepsilon)x + \varepsilon u,$$

$$v = u + \varepsilon x u_1 + \varepsilon^2 [(2 - u_1) x u_1 + \frac{1}{2} x^2 u_2]$$

$$+ \varepsilon^3 [4 x u_1 + 2 x^2 u_2 - \frac{3}{2} x^2 u_1 u_2 + x u_1^3 - 4 x u_1^2 + \frac{1}{3!} x^3 u_3] + \cdots.$$

One can easily find the transformations of derivatives, e.g.,

$$v_1 = u_1 + \varepsilon [2 u_1 + x u_2 - u_1^2] + \varepsilon^2 [4 u_1 - 4 u_1^2 + 4 x u_1 u_2 + \frac{1}{2} x^2 u_3 + u_1^3] + \cdots,$$

$$v_2 = u_2 + \varepsilon x u_1 + \varepsilon^2 [4 u_2 - 3 u_1 u_2 + x u_3] + \cdots, \qquad v_3 = u_3 + \cdots.$$

The reason for the diminishing orders of ε with the ascending orders of the derivatives is to illustrate which powers of ε are required to invert the above transformations up to order $o(\varepsilon^3)$.

In this process, one finds

$$u = v - \varepsilon x v_1 + \frac{1}{2} \varepsilon^2 x^2 v_2 + o(\varepsilon^2),$$

which, together with the expression for y, yields the following transcendental expression for x:

$$y - (1 - \varepsilon) x = \varepsilon v - \varepsilon^2 x v_1 + \frac{1}{2} \varepsilon^3 x^2 v_2 + o(\varepsilon^3).$$

These approximate equations can be calculated with any degree of precision. The general result is as follows (Anderson and Ibragimov [1994]):

The Lie-Bäcklund operator 1.53,

$$X_\varepsilon = (1 - \varepsilon) \frac{\partial}{\partial y} + \varepsilon v_1 \frac{\partial}{\partial v} + \varepsilon v_2 \frac{\partial}{\partial v_1} + \varepsilon v_3 \frac{\partial}{\partial v_2} + \cdots,$$

is similar to the generator of the group of translations 1.48,

$$X_1 = \frac{\partial}{\partial x},$$

in accordance with Theorem 1.14. The similarity map 1.45 – 1.46 is given by

$$y = (1 - \varepsilon) x + \varepsilon u, \qquad u = \sum_{i=0}^{\infty} \frac{(-\varepsilon x)^i}{i!} v_i, \qquad (1.54)$$

where $v_{i+1} = D_y(v_i)$, $v_0 = v$.

With this result in hand, we can map the infinite number of differential invariants 1.52 of Operator 1.48 into equivalent ones for the Lie-Bäcklund operator X_ε. Thus,

THEOREM 1.15.   *The Lie-Bäcklund operator 1.53 has a countable basis of invariants determined by functions*

$$u(y, v, v_1, v_2, ...), \quad u_1(y, v, v_1, v_2, ...), \quad u_2(y, v, v_1, v_2, ...), ... \qquad (1.55)$$

where $y, v, v_1, v_2, ...$ *are defined recursively via Equations 1.54 using the differentiation formula*

$$D_x \equiv D_x(y)D_y = (1 - \varepsilon(1 - u_1))\, D_y. \qquad (1.56)$$

Using this theorem, one can easily find, e.g., the invariant u with the second order of precision:

$$u = v - \varepsilon y v_1 + \varepsilon^2 (v v_1 - y v_1 + \frac{1}{2} y^2 v_2) + o(\varepsilon^2).$$

1.5.   Bäcklund transformations for evolution equations

1.5.1.   Definition

The majority of the Bäcklund transformations that appear in applications satisfy the following definition (Anderson and Ibragimov [1978], Fokas and Anderson [1979]).

DEFINITION 1.12.   *Consider the following two evolution equations:*

$$u_t = F(x, u, u_1, ..., u_n) \qquad (1.57)$$

and

$$v_t = H(x, v, v_1, ..., v_n). \qquad (1.58)$$

Here, $u_1, ..., u_n$ *and* $v_1, ..., v_n$ *are successive derivatives of u and v with respect to the spatial variable x. Equations 1.57 and 1.58 are said to be related by a Bäcklund transformation if there exists an ordinary differential equation*

$$\Phi(x, u, u_1, ..., u_r, v, v_1, ..., v_s) = 0 \qquad (1.59)$$

which is invariant under the Lie-Bäcklund transformations group given by the coupled evolution equations 1.57 and 1.58 (cf. Section 1.2.6), i.e., generated by the canonical Lie-Bäcklund operator

$$X = F(x, u, ..., u_n)\frac{\partial}{\partial u} + H(x, v, ..., v_n)\frac{\partial}{\partial v} + \cdots . \tag{1.60}$$

If $F \equiv H$, one says that Equation 1.59 defines an auto-Bäcklund transformation of Equation 1.57.

This definition naturally generalizes to systems of evolution equation. Namely, consider Equations 1.57 and 1.58 with vector-valued functions

$$u = (u^1, ..., u^m), \quad v = (v^1, ..., v^m)$$

and

$$F = (F^1, ..., F^m), \quad H = (H^1, ..., H^m).$$

Then a Bäcklund transformation is given by Definition 1.12 where Equation 1.59 is a system of m equations with a vector-valued function

$$\Phi = (\Phi^1, ..., \Phi^m)$$

and Operator 1.60 has the form:

$$X = \sum_{\alpha=1}^{m} \left(F^\alpha(x, u, ..., u_n)\frac{\partial}{\partial u^\alpha} + H(x, v, ..., v_n)\frac{\partial}{\partial v} \right) + \cdots .$$

1.5.2. Illustrative examples

Example 1. The Burgers equation

$$u_t = uu_x + u_{xx}$$

and the heat equation

$$v_t = v_{xx}$$

are related by the following Bäcklund transformation (known as the Hopf-Cole transformation; see, e.g., [H1], p. 182):

$$v_x - \frac{1}{2}uv = 0.$$

Example 2. The Korteweg-de Vries equation

$$u_t = uu_x + u_{xxx}$$

and the modified Korteweg-de Vries equation

$$v_t = \frac{1}{6}v^2 v_x - v_{xxx}$$

are related by the Bäcklund transformation (known as the Miura transformation; see, e.g., [H1], p. 189):

$$v_x - \gamma(u + v^2) = 0, \quad \gamma = \pm 1.$$

Example 3. The Korteweg-de Vries equation

$$u_t = u_{xxx} + 6uu_x$$

has the auto-Bäcklund transformation (see, e.g., [H1], p.189):

$$u_x + v_x + (u - v)(\lambda - 2(u + v))^{1/2} = 0, \ \lambda = \text{const.}$$

Example 4. The classical example is the Bonnet equation

$$2u_{\xi\eta} = \sin(2u).$$

It is invariant under the Bianchi-Lie transformation[5]

$$u(\xi, \eta) \mapsto v(\xi, \eta)$$

determined by the system of first-order equations:

$$u_\xi + v_\xi = \sin(u - v), \quad u_\eta - v_\eta = \sin(u + v).$$

Bianchi-Lie transformation satisfies Definition 1.12 if we rewrite the Bonnet equation, in new independent variables $t = \xi + \eta$, $x = \xi - \eta$ and dependent variables u^1, u^2, as an evolutionary system:

$$u_t^1 = u^2, \quad u_t^2 = u_{xx}^1 + \sin u^1 \cos u^1.$$

---

[5]Historical remarks are to be found in Anderson and Ibragimov [1978].

Then the Bianchi-Lie transformation is an auto–Bäcklund transformation of this system and is given by the following two equations 1.59:

$$v_x^1 + u^2 - \sin u^1 \cos v^1 = 0, \quad u_x^1 + v^2 + \sin v^1 \cos u^1 = 0.$$

Acknowledgments

We acknowledge financial support from the University of Georgia, and in particular we would like to thank Vice President for Research Joe L. Key for arranging for this support. In addition, we would like to thank the members of the Department of Physics and Astronomy for their hospitality during NHI's visit in the summer of 1994 when this chapter was written.

2

Approximate Transformation Groups and Deformations of Symmetry Lie Algebras

In applied sciences, differential equations depending on a small parameter are of frequent occurrence. Therefore, a theory based on the new concept of an approximate group was developed for tackling differential equations with a small parameter — particularly those relating to applications.

One-parameter approximate transformation groups were introduced in Baikov, Gazizov, and Ibragimov [1987a], [1988a], [1989a]. Recently, the theory was evolved in Baikov, Gazizov, and Ibragimov [1993], and Lie theorems were extended to multi-parameter approximate transformation groups.

Though the new concept maintains the essential features of the Lie group theory, it has certain peculiarities both in theory and applications.

This chapter provides a concise introduction to the theory of approximate transformation groups and *regular* approximate symmetries of differential equations with a small parameter.

2.1. Preliminaries

In what follows, all the functions under consideration are assumed to be locally analytic in their arguments.

2.1.1. Notation

Let $z = (z^1, ..., z^N)$ be an independent variable and ε a small parameter. We write

$$F(z, \varepsilon) = o(\varepsilon^p),$$

if

$$\lim_{\varepsilon \to 0} \frac{F(z, \varepsilon)}{\varepsilon^p} = 0,$$

or equivalently, if

$$F(z, \varepsilon) = \varepsilon^{p+1} \varphi(z, \varepsilon)$$

where $\varphi(z, \varepsilon)$ is an analytic function defined in a neighborhood of $\varepsilon = 0$ and p is a positive integer.

If

$$f(z, \varepsilon) - g(z, \varepsilon) = o(\varepsilon^p),$$

we write

$$f(z, \varepsilon) = g(z, \varepsilon) + o(\varepsilon^p),$$

or, briefly

$$f \approx g$$

when there is no ambiguity. In words, f *is approximately equal to* g.

In theoretical discussions, approximate equalities are considered with an error $o(\varepsilon^p)$ of an arbitrary order $p \geq 1$. However, in most of the applications and examples (Sections 2.2.4, 2.2.5, 2.4.8, 2.5.3 to 2.5.6, 2.6.2, 2.7), the theory is simplified to the case $p = 1$.

2.1.2. Approximate Cauchy problem and its solution

An approximate Cauchy problem,

$$\frac{dz}{dt} \approx f(z, \varepsilon), \tag{2.1}$$

$$z\big|_{t=0} \approx \alpha(\varepsilon), \tag{2.2}$$

is defined as follows.

The approximate differential equation 2.1 denotes the class of differential equations

$$\frac{dz}{dt} = g(z, \varepsilon) \tag{2.3}$$

with functions $g(z, \varepsilon) \approx f(z, \varepsilon)$.

Similarly, the approximate initial condition 2.2 is the class of conditions

$$z\big|_{t=0} = \beta(\varepsilon) \tag{2.4}$$

with $\beta(\varepsilon) \approx \alpha(\varepsilon)$.

The following known statement on a continuous dependence of a parameter hints the natural definition of the solution to Problem 2.1 – 2.2 and furnishes with the theorem on the existence and uniqueness of the solution of an arbitrary approximate Cauchy problem.

THEOREM 2.1. *Let the functions* $f(z, \varepsilon)$, $g(z, \varepsilon)$ *be analytic in a neighborhood of a given point* $(z_0, 0)$*. Let*

$$f(z, \varepsilon) = g(z, \varepsilon) + o(\varepsilon^p),$$

and

$$\alpha(\varepsilon) = \beta(\varepsilon) + o(\varepsilon^p), \quad \alpha(0) = \beta(0) = z_0.$$

Then there exist the solutions $z(t, \varepsilon)$ *and* $w(t, \varepsilon)$ *of the problems*

$$\frac{dz}{dt} = f(z, \varepsilon), \quad z\big|_{t=0} = \alpha(\varepsilon),$$

and

$$\frac{dw}{dt} = g(w, \varepsilon), \quad w\big|_{t=0} = \beta(\varepsilon),$$

respectively. These solutions are locally analytic, unique and approximately equal, viz.

$$z(t, \varepsilon) = w(t, \varepsilon) + o(\varepsilon^p).$$

Thus, the solutions to all the problems of the form 2.3 – 2.4 coincide in a given precision. Therefore, *the solution of the approximate Cauchy problem 2.1 – 2.2 is defined as the class of functions* $z(t, \varepsilon)$ *approximately equal to the solution of any specified problem 2.3 – 2.4.* According to the above theorem, this definition does not depend on the choice of Problem 2.3 – 2.4, and the solution of the approximate Cauchy problem 2.1 – 2.2 is determined uniquely. In applications, it is convenient to identify the solution of the approximate Cauchy problem with the solution of a specified problem 2.3 – 2.4.

2.1.3. Completely integrable systems

A system of approximate equations

$$\frac{\partial z^i}{\partial a^\alpha} \approx \psi_\alpha^i(z, a, \varepsilon), \quad i = 1, \ldots, N, \ \alpha = 1, \ldots, r, \tag{2.5}$$

are said to be completely integrable, if

$$\frac{\partial}{\partial a^\beta}\left(\frac{\partial z^i}{\partial a^\alpha}\right) \approx \frac{\partial}{\partial a^\alpha}\left(\frac{\partial z^i}{\partial a^\beta}\right)$$

whenever z satisfies Equation 2.5.

The completely integrable system 2.5 with arbitrary initial conditions

$$z^i\big|_{a=0} \approx z_0^i$$

has the unique approximate solution of the form

$$z^i \approx f_0^i(a) + \varepsilon f_1^i(a) + \cdots + \varepsilon^p f_p^i(a), \quad i = 1, \ldots, N.$$

2.2.   One-parameter approximate groups

A detailed discussion of the material presented here is to be found in Baikov, Gazizov, and Ibragimov [1989a].

2.2.1.   Definition

Consider a set of smooth vector-functions $f_0(z, a)$, $f_1(z, a), \ldots, f_p(z, a)$ with coordinates

$$f_0^i(z, a), \ f_1^i(z, a), \ldots, f_p^i(z, a), \quad i = 1, \ldots, N.$$

Let us define the one-parameter family G of *approximate transformations*

$$\bar{z}^i \approx f_0^i(z, a) + \varepsilon f_1^i(z, a) + \cdots + \varepsilon^p f_p^i(z, a), \quad i = 1, \ldots, N, \qquad (2.6)$$

of points $z = (z^1, \ldots, z^N) \in R^N$ into points $\bar{z} = (\bar{z}^1, \ldots, \bar{z}^N) \in R^N$ as the class of invertible transformations

$$\bar{z} = f(z, a, \varepsilon) \qquad (2.7)$$

with vector-functions $f = (f^1, \ldots, f^N)$ such that

$$f^i(z, a, \varepsilon) \approx f_0^i(z, a) + \varepsilon f_1^i(z, a) + \cdots + \varepsilon^p f_p^i(z, a).$$

Here, a is a real parameter, and the following condition is imposed:

$$f(z, 0, \varepsilon) \approx z.$$

Furthermore, it is assumed that Transformation 2.7 is defined for any value of a from a small neighborhood of $a = 0$, and that, in this neighborhood, the equation $f(z, a, \varepsilon) \approx z$ yields $a = 0$.

The set G of Transformations 2.6 is called a (local) one-parameter approximate transformation group if

$$f(f(z, a, \varepsilon), b, \varepsilon) \approx f(z, a + b, \varepsilon)$$

for all Transformations 2.7.

Here, f does not necessarily denote the same function at each occurrence.

Example. Let us take $N = 1$ and set $z = x$. Consider the following two functions:

$$f(x, a, \varepsilon) = x + a(1 + \varepsilon x + \frac{1}{2}\varepsilon a)$$

and

$$g(x, a, \varepsilon) = x + a(1 + \varepsilon x)(1 + \frac{1}{2}\varepsilon a).$$

They are equal in the first order of precision, viz.

$$g(x, a, \varepsilon) = f(x, a, \varepsilon) + \varepsilon^2 \varphi(x, a), \quad \varphi(x, a) = \frac{1}{2}a^2 x,$$

and satisfy the approximate group property. Indeed,

$$f(g(x, a, \varepsilon), b, \varepsilon) = f(x, a + b, \varepsilon) + \varepsilon^2 \phi(x, a, b, \varepsilon),$$

where

$$\phi(x, a, b, \varepsilon) = \frac{1}{2}a(ax + ab + 2bx + \varepsilon abx).$$

2.2.2. Approximate group generator

The generator of the approximate group G of Transformations 2.6 is the class of first-order linear differential operators

$$X = \xi^i(z, \varepsilon)\frac{\partial}{\partial z^i} \tag{2.8}$$

such that

$$\xi^i(z, \varepsilon) \approx \xi_0^i(z) + \varepsilon \xi_1^i(z) + \cdots + \varepsilon^p \xi_p^i(z),$$

where the vector fields $\xi_0, \xi_1, \ldots, \xi_p$ are given by

$$\xi_\nu^i(z) = \frac{\partial f_\nu^i(z, a)}{\partial a}\Big|_{a=0}, \quad \nu = 0, \ldots, p; \; i = 1, \ldots, N.$$

In what follows, an approximate group generator is written as

$$X \approx \left(\xi_0^i(z) + \varepsilon\xi_1^i(z) + \; \ldots \; + \varepsilon^p\xi_p^i(z)\right) \frac{\partial}{\partial z^i} \; .$$

It is written also in a specified form, viz.

$$X = \xi^i(z, \varepsilon)\frac{\partial}{\partial z^i} \equiv \left(\xi_0^i(z) + \varepsilon\xi_1^i(z) + \cdots + \varepsilon^p\xi_p^i(z)\right) \frac{\partial}{\partial z^i} \; . \qquad (2.9)$$

2.2.3.   The approximate Lie equation

For approximate groups, the Lie theorem can be proved in the following modification.

THEOREM 2.2.   *Let G be an approximate group given by Transformations 2.6, and let X be its generator 2.9. Then the function $f(z, a, \varepsilon)$ satisfies the equation*

$$\frac{\partial f(z, a, \varepsilon)}{\partial a} \approx \xi(f(z, a, \varepsilon), \varepsilon).$$

Conversely, for any Operator 2.9, the solution of the approximate Cauchy problem

$$\frac{d\bar{z}}{da} \approx \xi(\bar{z}, \varepsilon), \qquad (2.10)$$

$$\bar{z}\big|_{a=0} \approx z, \qquad (2.11)$$

determines an approximate one-parameter group of Transformations 2.6 with the group parameter a.

Equation 2.10 is called *the approximate Lie equation.*

2.2.4.   The first-order approximation to the Lie equation

Consider the approximate generator 2.9 with $p = 1$:

$$X = (\xi_0(z) + \varepsilon\xi_1(z)) \frac{\partial}{\partial z} \; .$$

The corresponding approximate group transformations have the form:

$$\overline{z} \approx f_0(z, a) + \varepsilon f_1(z, a).$$

In this case, the approximate Lie equation 2.10 is written:

$$\frac{d(f_0 + \varepsilon f_1)}{da} \approx \xi_0(f_0 + \varepsilon f_1) + \varepsilon \xi_1(f_0 + \varepsilon f_1).$$

It follows that the approximate Cauchy problem 2.10 – 2.11 reduces to the usual Cauchy problem of the form:

$$\frac{df_0}{da} = \xi_0(f_0), \qquad \frac{df_1}{da} = \xi_0'(f_0)f_1 + \xi_1(f_0), \tag{2.12}$$

$$f_0\big|_{a=0} = z, \qquad f_1\big|_{a=0} = 0. \tag{2.13}$$

Here,

$$\xi_0'(z) = \left\| \frac{\partial \xi_0^i(z)}{\partial z^j} \right\|.$$

In coordinates, the differential equations 2.12 are written:

$$\frac{df_0^i}{da} = \xi_0^i(f_0), \qquad \frac{df_1^i}{da} = \sum_{k=1}^{N} \frac{\partial \xi_0^i(z)}{\partial z^k}\bigg|_{z=f_0} f_1^k + \xi_1^i(f_0).$$

REMARK.   The case of an arbitrary p is treated in Baikov, Gazizov, and Ibragimov [1987a], [1988a], [1989a].

2.2.5.   Solution of the approximate Lie equation in the first-order precision (examples)

Example 1. Let $N = 1$, $z = x$, and let

$$X = (1 + \varepsilon x)\frac{\partial}{\partial x}.$$

Here, $\xi_0(x) = 1$, $\xi_1(x) = x$, and the corresponding Cauchy problem 2.12 – 2.13 is written:

$$\frac{df_0}{da} = 1, \qquad \frac{df_1}{da} = f_0,$$

$$f_0|_{a=0} = x, \quad f_1|_{a=0} = 0.$$

Its solution has the form

$$f_0 = x + a, \quad f_1 = xa + \frac{a^2}{2}.$$

Hence, the approximate group is given by (cf. the example in Section 2.2.1)

$$\bar{x} \approx x + a + \varepsilon \left(xa + \frac{a^2}{2} \right).$$

This approximate transformation is certainly contained in the Taylor expansion (in ε) of the exact group transformations generated by X:

$$\bar{x} = x \exp(a\varepsilon) + \frac{\exp(a\varepsilon) - 1}{\varepsilon} = x + a + \varepsilon \left(xa + \frac{a^2}{2} \right) + \varepsilon^2 \left(\frac{a^2}{2}x + \frac{a^3}{6} \right) + \cdots .$$

Example 2. Let $N = 2$, $z = (x, y)$, and let

$$X = (1 + \varepsilon x^2) \frac{\partial}{\partial x} + \varepsilon xy \frac{\partial}{\partial y} .$$

Here, $\xi_0(x, y) = (1, 0)$, $\xi_1(x, y) = (x^2, xy)$, and the Cauchy problem 2.12 – 2.13 is written:

$$\frac{df_0^1}{da} = 1, \quad \frac{df_0^2}{da} = 0, \quad \frac{df_1^1}{da} = (f_0^1)^2, \quad \frac{df_1^2}{da} = f_0^1 f_0^2,$$

$$f_0^1|_{a=0} = x, \quad f_0^2|_{a=0} = y, \quad f_1^1|_{a=0} = 0, \quad f_1^2|_{a=0} = 0.$$

The solution of this problem yields the following approximate group transformations:

$$\bar{x} \approx x + a + \varepsilon \left(x^2 a + xa^2 + \frac{a^3}{3} \right), \quad \bar{y} \approx y + \varepsilon \left(xya + \frac{ya^2}{2} \right).$$

The exact group transformations are

$$\bar{x} = \frac{\delta x \cos \delta a + \sin \delta a}{\delta (\cos \delta a - \delta x \sin \delta a)}, \quad \bar{y} = \frac{y}{\cos \delta a - \delta x \sin \delta a},$$

where $\delta = \sqrt{\varepsilon}$.

2.3. Approximate Lie algebras

Current developments in the theory of multi-parameter approximate groups require a modification of basic notions of Lie algebras. This section, based on Baikov, Gazizov, and Ibragimov [1993], is aimed to introduce the minimum of the necessary modifications.

2.3.1. Approximate operators

Let

$$\xi_0^i(z), \; \xi_1^i(z), \ldots, \; \xi_p^i(z), \quad i = 1, \ldots, N,$$

be a set of smooth vector fields. *An approximate operator* is the class of first-order differential operators (cf. Section 2.2.2),

$$X = \xi^i(z, \varepsilon)\frac{\partial}{\partial z^i}$$

such that

$$\xi^i(z, \varepsilon) \approx \xi_0^i(z) + \varepsilon\xi_1^i(z) + \cdots + \varepsilon^p\xi_p^i(z), \quad i = 1, \ldots, N.$$

According to this definition, X does not necessarily denote here the same operator at each occurrence. Thus, in approximate group theory, we deal with "unspecified" operators.

2.3.2. Approximate commutator

An approximate commutator of the operators X_1 and X_2 is an approximate operator denoted by $[X_1, X_2]$ and given by

$$[X_1, X_2] \approx X_1X_2 - X_2X_1.$$

Example. For the operators

$$X_1 = \frac{\partial}{\partial x} + \varepsilon x\frac{\partial}{\partial y}, \quad X_2 = \frac{\partial}{\partial y} + \varepsilon y\frac{\partial}{\partial x},$$

the approximate commutator is equal to zero up to the error ε^2, while, in the second order of precision, it is equal to

$$X_3 = \varepsilon^2 \left(x \frac{\partial}{\partial x} - y \frac{\partial}{\partial y} \right).$$

2.3.3. Properties of the commutator

The approximate commutator adopts the usual properties, viz.
a) the linearity:

$$[aX_1 + bX_2, X_3] \approx a[X_1, X_3] + b[X_2, X_3], \quad a, b = \text{const.},$$

b) the skew-symmetry:

$$[X_1, X_2] \approx -[X_2, X_1],$$

c) the Jacobi identity:

$$[[X_1, X_2], X_3] + [[X_2, X_3], X_1] + [[X_3, X_1], X_2] \approx 0.$$

2.3.4. Approximate Lie algebra of approximate operators

A vector space L of approximate operators is called *an approximate Lie algebra of operators*, if it is closed (in the approximation of a given order p) under the approximate commutator, i.e., if

$$[X_1, X_2] \in L$$

for any $X_1, X_2 \in L$. Here the approximate commutator $[X_1, X_2]$ is calculated to the precision indicated.

2.3.5. Linear independence

Approximate operators X_α, $\alpha = 1, \ldots, r$, are said to be linearly independent if the approximate equation

$$C^\alpha X_\alpha \approx 0$$

with constants coefficients C^α (they are assumed to be independent of ε) yields

$$C^\alpha = 0, \quad \alpha = 1, \ldots, r.$$

Example. The approximate operators

$$X_1 = \frac{\partial}{\partial t}, \quad X_2 = \varepsilon \frac{\partial}{\partial t}$$

are linearly dependent in the zero order of precision, while they are linearly independent in the first order of precision.

2.3.6. Basis of an approximate Lie algebra

A basis of an approximate Lie algebra L is a set of linearly independent approximate operators that span the vector space L. If L has a finite basis consisting of $r < \infty$ operators, then it is said to be finite-dimensional (r-dimensional) and is often denoted by L_r.

2.3.7. Essential operators

Let L be an approximate Lie algebra, and let $\{X_\alpha\}$ be a set of linearly independent approximate operators from L, considered in a given approximation of order p. Denote by $\{X'_\beta\}$ the set of approximate operators obtained by multiplying elements of $\{X_\alpha\}$ by $\varepsilon, \varepsilon^2, \ldots, \varepsilon^p$ and neglecting the terms of order $o(\varepsilon^p)$. If the set of approximate operators

$$\{X_\alpha, X'_\beta\}$$

provides a basis of L, then the operators X_α are said to be *essential operators* of the approximate Lie algebra L.

An approximate Lie algebra is completely determined, in a given precision, by its essential operators.

2.3.8. Structure constants of an approximate Lie algebra

Let L_r be an r-dimensional approximate Lie algebra, and let $\{X_\alpha\}$, $\alpha = 1, \ldots, r$, be its basis. Then

$$[X_\alpha, X_\beta] \approx c_{\alpha\beta}^\gamma X_\gamma, \quad \alpha, \beta = 1, \ldots, r,$$

with constant (and independent of ε) coefficients $c_{\alpha\beta}^\gamma$, $\quad \alpha, \beta, \gamma = 1, \ldots, r$.

The coefficients $c_{\alpha\beta}^\gamma$ are termed *the structure constants* of L_r.

2.3.9. Example of an approximate Lie algebra and its essential operators

Consider the following operators:

$$X_1 = \frac{\partial}{\partial x} + \varepsilon x \frac{\partial}{\partial y}, \qquad X_2 = \frac{\partial}{\partial y} + \varepsilon y \frac{\partial}{\partial x},$$

$$X_3 = \varepsilon \left(x \frac{\partial}{\partial x} + y \frac{\partial}{\partial y} \right), \quad X_4 = \varepsilon \frac{\partial}{\partial x}, \quad X_5 = \varepsilon \frac{\partial}{\partial y}.$$

Their linear span is not a Lie algebra in the usual (exact) sense. For instance, the (exact) commutator

$$[X_1, X_2] = \varepsilon^2 \left(x \frac{\partial}{\partial x} - y \frac{\partial}{\partial y} \right)$$

is not a linear combination of the above operators.

However, these operators span an approximate Lie algebra in the first order of precision. Indeed, in this approximation, we have the following non-vanishing commutators:

$$[X_1, X_3] \approx -[X_3, X_1] \approx X_4, \quad [X_2, X_3] \approx -[X_3, X_2] \approx X_5.$$

Thus our operators span the 5-dimensional approximate Lie algebra L_5 with the following non-zero structure constants:

$$c_{13}^4 = -c_{31}^4 = 1, \quad c_{23}^5 = -c_{32}^5 = 1.$$

The operators X_1, X_2, and X_3 are the essential operators of L_5.

2.4. Multi-parameter approximate groups

In this section we recount the results of current developments. For the proofs of the main theorems, the reader is referred to Baikov, Gazizov, and Ibragimov [1993].

2.4.1. Definition

Consider approximate transformations (cf. Section 2.2.1):

$$\bar{z} = f(z, a, \varepsilon) \approx f_0(z, a) + \varepsilon f_1(z, a) + \ldots + \varepsilon^p f_p(z, a), \tag{2.14}$$

where $z = (z^1, \ldots, z^N) \in R^N$ and $a = (a^1, \ldots, a^r)$ is a vector-parameter. It is assumed that

$$f(z, 0, \varepsilon) \approx z.$$

The set G_r of Transformations 2.14 is called a (local) r-parameter approximate group of transformations in R^N if

$$f(f(z, a, \varepsilon), b, \varepsilon) \approx f(z, c, \varepsilon)$$

with the vector-parameter $c = (c^1, \ldots, c^r)$ given by

$$c^\mu = \varphi^\mu(a, b), \quad \mu = 1, \ldots, r, \tag{2.15}$$

where $\varphi^\mu(a, b)$ are smooth functions defined for sufficiently small vector-parameters a and b.

Functions 2.15 define a group composition law $\varphi = (\varphi^1, \ldots, \varphi^r)$ and satisfy the following conditions:

$$\varphi(a, 0) = a, \quad \varphi(0, b) = b.$$

2.4.2. Infinitesimal generators

Let G_r be an approximate group. Its infinitesimal generators are the approximate operators (see Section 2.3.1)

$$X_\alpha = \xi_\alpha^i(z, \varepsilon) \frac{\partial}{\partial z^i} \approx (\xi_{\alpha,0}^i(z) + \varepsilon \xi_{\alpha,1}^i(z) + \ldots + \varepsilon^p \xi_{\alpha,p}^i(z)) \frac{\partial}{\partial z^i}, \tag{2.16}$$

where $\alpha = 1, \ldots, r$. The vector fields

$$\xi_{\alpha,\nu}(z, \varepsilon) = (\xi_{\alpha,\nu}^1(z, \varepsilon), \ldots, \xi_{\alpha,\nu}^N(z, \varepsilon)), \quad \alpha = 1, \ldots, r; \ \nu = 1, \ldots, p,$$

are determined by the group transformations 2.14 as follows:

$$\xi_{\alpha,\nu}^i(z, \varepsilon) = \left. \frac{\partial f_\nu^i(z, a, \varepsilon)}{\partial a^\alpha} \right|_{a=0}, \quad i = 1, \ldots, N. \tag{2.17}$$

2.4.3.   The direct first Lie theorem

Let G_r be an approximate group of Transformations 2.14 with the composition law 2.15. We set

$$A^\alpha_\beta(a) = \left.\frac{\partial \varphi^\alpha(a, b)}{\partial b^\beta}\right|_{b=0}, \quad \alpha, \beta = 1, \ldots, r.$$

It follows from the properties of the group composition law $\varphi(a, b)$ that (see Section 2.4.1):

$$A^\alpha_\beta(0) = \delta^\alpha_\beta$$

where δ^α_β is the Kronecker symbol. Hence, the matrix $A = \left(A^\alpha_\beta(a)\right)$ has the inverse matrix $V = \left(V^\beta_\gamma(a)\right)$ when a is sufficiently small. Lie's direct first theorem is valid for the approximate group of Transformations 2.14 in the following modification.

THEOREM 2.3.   *The functions $f^i(z, a, \varepsilon)$ satisfy the equations*

$$\frac{\partial f^i}{\partial a^\alpha} \approx \xi^i_\beta(f, \varepsilon) V^\beta_\alpha(a), \quad i = 1, \ldots, N; \; \alpha = 1, \ldots, r, \qquad (2.18)$$

called the approximate Lie equations. Furthermore, the infinitesimal generators 2.16 span an r-dimensional approximate Lie algebra.

2.4.4.   The inverse first Lie theorem

Consider an arbitrary set of r linearly independent (to a given precision, see Section 2.3.5) vector fields $\xi_\alpha(z, \varepsilon) = (\xi^1_\alpha(z, \varepsilon), \ldots, \xi^N_\alpha(z, \varepsilon))$, $\alpha = 1, \ldots, r$, and a matrix $V(a) = (V^\beta_\alpha(a))$ of the rank r. Consider, for the given ξ_α and $V(a)$, the system of the approximate equations 2.18. Assume that this system is completely integrable (Section 2.1.3). Then the solution of the equations 2.18, under the initial conditions

$$f\big|_{a=0} \approx z,$$

yields a local r-parameter approximate group G_r of Transformations 2.14.

2.4.5.   Structure constants of an approximate group

Under the assumptions of Section 2.4.4, the coefficients $V^\beta_\alpha(a)$ of Equations 2.18 satisfy the Maurer-Cartan equations:

$$\frac{\partial V^\sigma_\beta}{\partial a^\alpha} - \frac{\partial V^\sigma_\alpha}{\partial a^\beta} = c^\sigma_{\gamma\mu} V^\mu_\beta V^\gamma_\alpha$$

with constant coefficients $c_{\gamma\mu}^{\sigma}$, $\gamma, \mu, \sigma = 1, \ldots, r$.

The coefficients $c_{\gamma\mu}^{\sigma}$ are termed *the structure constants* of the approximate group G_r given in Section 2.4.4.

2.4.6. The second Lie theorem

THEOREM 2.4. *Consider a given local r-parameter approximate group G_r. Let $c_{\gamma\mu}^{\sigma}$, $\gamma, \mu, \sigma = 1, \ldots, r$, be its structure constants. Then the infinitesimal generators 2.16 of G_r are linearly independent (Section 2.3.5) and span an r-dimensional approximate Lie algebra L_r. The structure constants of the algebra L_r (Section 2.3.8) are identical with those of the approximate group G_r. The approximate Lie algebra L_r is called the Lie algebra of the approximate group G_r.*

Conversely, let r linearly independent approximate operators X_α, $\alpha = 1, \ldots, r$, span an approximate Lie algebra L_r. Then there exists a local r-parameter approximate group G_r such that its Lie algebra is identical to the approximate Lie algebra L_r.

Given an approximate Lie algebra L_r, the group G_r is determined uniquely. One can determine the approximate transformations of the group G_r by constructing the Lie equations 2.18 or by composing an r-parameter group from r one-parameter groups generated by the basic operators of the algebra L_r. Both ways were previously discussed in this Handbook in the case of classical Lie group theory (see [H2], Sections 1.5.3 and 1.5.5) and can be easily modified to the case of approximate group theory. The second way (the composition of G_r from one-parameter approximate group transformations) is simple. The first way is more complicated for the practical use. However, it is of fundamental theoretical value. Therefore, for the convenience of the reader, we sketch it in the next section.

2.4.7. Approximate Lie equations for a given approximate Lie algebra

Given an r-dimensional approximate Lie algebra L_r with the basic operators 2.16, the problem of determining the completely integrable system of the approximate Lie equations 2.18 reduces to the solution of the Maurer-Cartan equations (Section 2.4.5). The solution of the latter problem is given by the following construction.

Let $c_{\alpha\beta}^{\gamma}$, $\alpha, \beta, \gamma = 1, \ldots, r$ be the structure constants of the approximate Lie algebra L_r. Consider the Cauchy problem

$$\frac{d\theta_\alpha^\beta}{dt} = \delta_\alpha^\beta + c_{\nu\mu}^\beta \lambda^\mu \theta_\alpha^\nu, \qquad \theta_\alpha^\beta\Big|_{t=0} = 0, \qquad (2.19)$$

where δ_α^β is the Kronecker symbol and λ^μ is a system of r parameters. The solution of the problem 2.19 has the form

$$\theta_\alpha^\beta = t h_\alpha^\beta (\lambda^1 t, \ldots, \lambda^r t).$$

Define the functions $V_\alpha^\beta(b)$ as follows:

$$V_\alpha^\beta(b) = h_\alpha^\beta(b^1, \ldots, b^r), \qquad b^\nu = \lambda^\nu t. \tag{2.20}$$

Then the functions $V_\alpha^\beta(b)$ solve the Maurer-Cartan equations, and hence the system of the approximate Lie equations 2.18 is completely integrable.

2.4.8.   Example (the first order of precision)

Consider the approximate Lie algebra L_r discussed in Section 2.3.9. In this case, the Cauchy problem 2.19 has the following solution:

$$\theta_\alpha^\alpha = t, \quad \alpha = 1, \ldots, 5, \quad \theta_1^4 = \frac{1}{2}\lambda^3 t^2, \quad \theta_2^5 = \frac{1}{2}\lambda^3 t^2,$$

and $\theta_\alpha^\beta = 0$ for other values of α and β. Formula 2.20 yields:

$$V_\alpha^\alpha = 1, \quad \alpha = 1, \ldots, 5; \quad V_1^4 = \frac{1}{2}b^3; \quad V_2^5 = \frac{1}{2}b^3,$$

and $V_\alpha^\beta = 0$ for other values of α and β. Thus, by setting $b^3 = 2a^3$, one can write the approximate Lie equations 2.18 in the form:

$$\frac{\partial \bar{x}}{\partial a^1} = 1 + \varepsilon a^3, \quad \frac{\partial \bar{x}}{\partial a^2} = \varepsilon \bar{y}, \quad \frac{\partial \bar{x}}{\partial a^3} = \varepsilon \bar{x}, \quad \frac{\partial \bar{x}}{\partial a^4} = \varepsilon, \quad \frac{\partial \bar{x}}{\partial a^5} = 0;$$

$$\frac{\partial \bar{y}}{\partial a^1} = \varepsilon \bar{x}, \quad \frac{\partial \bar{y}}{\partial a^2} = 1 + \varepsilon a^3, \quad \frac{\partial \bar{y}}{\partial a^3} = \varepsilon \bar{y}, \quad \frac{\partial \bar{y}}{\partial a^4} = 0, \quad \frac{\partial \bar{y}}{\partial a^5} = \varepsilon.$$

This system, together with the initial conditions

$$\bar{x}\big|_{a=0} = x, \qquad \bar{y}\big|_{a=0} = y,$$

yield

$$\bar{x} \approx x + a^1 + \varepsilon \left(a^4 + a^3 x + a^2 y + \frac{1}{2}(a^2)^2 + a^1 a^3 \right),$$

$$\bar{y} \approx y + a^2 + \varepsilon \left(a^5 + a^3 y + a^1 x + \frac{1}{2}(a^1)^2 + a^2 a^3 \right).$$

2.5. Equations with a small parameter: Deformations of symmetry Lie algebras

The purpose of this section is to carry over Lie's infinitesimal method (cf. [H2], Sections 1.1.8, 1.2.4, and 1.3.3) to approximate group analysis of differential equations with a small parameter. We sketch the main algorithms for calculating approximate symmetries. A detailed presentation with the proofs of the basic statements is to be found in Baikov, Gazizov, and Ibragimov [1989a].

2.5.1. Approximate invariance and the determining equation

Consider a set of smooth s-dimensional vector-functions

$$F_0(z), F_1(z), \ldots, F_q(z)$$

with coordinates

$$F_0^\sigma(z), F_1^\sigma(z), \ldots, F_q^\sigma(z), \quad \sigma = 1, \ldots, s.$$

It is supposed that $s \leq N$.

Let G be an approximate group of Transformations 2.6 with the given order of approximation $p \geq q$. The approximate equation

$$F(z, \varepsilon) \equiv F_0(z) + \varepsilon F_1(z) + \cdots + \varepsilon^q F_q(z) = o(\varepsilon^q) \tag{2.21}$$

is said to be approximately invariant with respect to G if

$$F(f(z, a, \varepsilon), \varepsilon) = o(\varepsilon^q)$$

whenever $z = (z^1, \ldots, z^N)$ satisfies Equation 2.21.

THEOREM 2.5. *Let*

$$\text{rank} \left\| \frac{\partial F_0^\sigma(z)}{\partial z^i} \right\|_{F_0(z)=0} = s .$$

Let X be the Generator 2.8 of the group G. Then Equation 2.21 is approximately invariant under the approximate group G if and only if

$$XF(z, \varepsilon)\Big|_{(2.21)} = o(\varepsilon^p). \tag{2.22}$$

Equation 2.22 is called *the determining equation* for approximate symmetries. If the determining equation 2.22 is satisfied we also say that *Equation 2.21 admits the approximate operator X.*

2.5.2.   Deformations of a symmetry Lie algebra. Stable symmetries

Theorem 2.5 yields the following simple result which is, however, of fundamental importance.

THEOREM 2.6.   *Let the equation 2.21 be approximately invariant under the approximate group with the generator 2.9:*

$$X = \xi^i(z, \varepsilon) \frac{\partial}{\partial z^i} \equiv \left(\xi_0^i(z) + \varepsilon \xi_1^i(z) + \cdots + \varepsilon^p \xi_p^i(z) \right) \frac{\partial}{\partial z^i},$$

such that $\xi_0(z) = (\xi_0^1(z), \ldots, \xi_0^N(z)) \neq 0$. *Then the (exact) operator*

$$X^0 = \xi_0^i(z) \frac{\partial}{\partial z^i} \tag{2.23}$$

is the generator of an exact symmetry group for the equation

$$F_0(z) = 0. \tag{2.24}$$

In what follows, Equation 2.24 is treated as an *unperturbed equation*, and Equation 2.21 is termed a *perturbed equation*. Under the conditions of Theorem 2.6, the exact symmetry generator X^0 is called a *stable symmetry* of the unperturbed equation 2.24. The corresponding approximate symmetry generator X for the perturbed equation 2.21 is called a *deformation* (of order p) of the operator X^0 caused by the perturbation of order q, viz.

$$\varepsilon F_1(z) + \cdots + \varepsilon^q F_q(z).$$

The notions of a stable symmetry and of its deformations apply to any symmetry Lie algebra of Equation 2.24. In particular, it may happen that the most general symmetry Lie algebra of Equation 2.24 is stable. In this particular case we say that the perturbed equation 2.21 *inherits the symmetries of the unperturbed equation.*

2.5.3. First-order deformations

Here, we simplify the equations of Section 2.5.1 by letting $p = q = 1$. Then Equation 2.21 and an approximate group generator 2.9 become

$$F_0(z) + \varepsilon F_1(z) \approx 0$$

and

$$X = X^0 + \varepsilon X^1 \equiv \xi_0^i(z)\frac{\partial}{\partial z^i} + \varepsilon \xi_1^i(z)\frac{\partial}{\partial z^i},$$

respectively. The determining equation 2.22 for approximate symmetries reduces to the following:

$$(X^0 + \varepsilon X^1)(F_0(z) + \varepsilon F_1(z))\Big|_{F_0(z)+\varepsilon F_1(z)=0} = o(\varepsilon).$$

The determining equation can also be written, by using undeterminate coefficients, in the form similar to the case of exact Lie symmetries (cf. [H2], Section 1.3.3, Equation 1.50'). This form is given by the following theorem (see Baikov, Gazizov, and Ibragimov [1989a], Equations 4.23 to 4.25). For the sake of brevity, we limit here the discussion to scalar equations. That is, we let $s = 1$.

THEOREM 2.7.   *In the first order of precision, the determining equation for approximate symmetries can be written as follows:*

$$X^0 F_0(z) = \lambda(z) F_0(z), \tag{2.25}$$

$$X^1 F_0(z) + X^0 F_1(z) = \lambda(z) F_1(z). \tag{2.26}$$

Here the factor $\lambda(z)$ is determined by Equation 2.25 and afterwards substituted into Equation 2.26 where Equation 2.26 itself must hold for the solutions z of the unperturbed equation 2.24, viz. $F_0(z) = 0$.

2.5.4. Algorithm for calculating the first-order approximate symmetries

Theorems 21 and 22 provide a simple and convenient algorithm for calculating (both "by hand" and by using symbolic software) the first-order approximate symmetries of equations with a small parameter. The algorithm consists of the following three steps.

1st step. Find the exact symmetry generators X^0 of the unperturbed Equation 2.24, e.g., by solving *the determining equation for exact symmetries*:

$$X^0 F_0(z)\Big|_{F_0(z)=0} = 0.$$

2nd step. Given X^0 and a perturbation $\varepsilon F_1(z)$, calculate *the auxiliary function H* given by (cf. Equations 2.25 and 2.26)

$$H \approx \frac{1}{\varepsilon} X^0(F_0(z) + \varepsilon F_1(z)) \Big|_{F_0(z)+\varepsilon F_1(z)=0}.$$

3rd step. Find the first-order deformations (i.e., the operators X^1) from *the determining equation for deformations:*[1]

$$X^1 F_0(z) \Big|_{F_0(z)=0} + H = 0.$$

In what follows, this algorithm applies to differential equations with a small parameter. We note that, in the approximate group analysis of differential equations with a small parameter, the prolongation formulas are the same as in the classical Lie group theory (see, e.g., [H2], Chapter 1).

2.5.5. An example of the implementation of the algorithm

Here, the algorithm of the previous section is illustrated by its applying to the nonlinear wave equation

$$u_{tt} + \varepsilon u_t = \left(u^\sigma u_x\right)_x, \quad \sigma \neq 0, \tag{2.27}$$

with a small dissipation (cf. Section 9.2.1).

Approximate group generators are written:

$$X = X^0 + \varepsilon X^1 \equiv \left(\xi_0^0 + \varepsilon\xi_1^0\right)\frac{\partial}{\partial t} + \left(\xi_0^1 + \varepsilon\xi_1^1\right)\frac{\partial}{\partial x} + (\eta_0 + \varepsilon\eta_1)\frac{\partial}{\partial u},$$

where ξ_ν^0, ξ_ν^1, η_ν ($\nu = 0, 1$) depend on t, x, and u. After the prolongation to the

---

[1]This equation, unlike determining equations for exact symmetries, is *inhomogeneous.*

derivatives involved in Equation 2.27, the operator X has the form:[2]

$$X = \left(\xi_0^0 + \varepsilon\xi_1^0\right)\frac{\partial}{\partial t} + \left(\xi_0^1 + \varepsilon\xi_1^1\right)\frac{\partial}{\partial x} + (\eta_0 + \varepsilon\eta_1)\frac{\partial}{\partial u} + \left(\zeta_{0,0} + \varepsilon\zeta_{1,0}\right)\frac{\partial}{\partial u_t}$$

$$+ \left(\zeta_{0,1} + \varepsilon\zeta_{1,1}\right)\frac{\partial}{\partial u_x} + + \left(\zeta_{0,00} + \varepsilon\zeta_{1,00}\right)\frac{\partial}{\partial u_{tt}} + \left(\zeta_{0,11} + \varepsilon\zeta_{1,11}\right)\frac{\partial}{\partial u_{xx}}.$$

Here, the prolongation formulas are

$$\zeta_{\nu,0} = D_t(\eta_\nu) - u_t D_t(\xi_\nu^0) - u_x D_t(\xi_\nu^1),$$

$$\zeta_{\nu,1} = D_x(\eta_\nu) - u_t D_x(\xi_\nu^0) - u_x D_x(\xi_\nu^1),$$

$$\zeta_{\nu,00} = D_t(\zeta_{\nu,0}) - u_{tt} D_t(\xi_\nu^0) - u_{tx} D_t(\xi_\nu^1),$$

$$\zeta_{\nu,11} = D_x(\zeta_{\nu,1}) - u_{tx} D_x(\xi_\nu^0) - u_{xx} D_x(\xi_\nu^1),$$

where $\nu = 0, 1$. The operators D_t and D_x are the total differentiations 1.2 (see Chapter 1) with respect to the independent variables t and x, viz.

$$D_t = \frac{\partial}{\partial t} + u_t\frac{\partial}{\partial u} + \dots, \qquad D_x = \frac{\partial}{\partial x} + u_x\frac{\partial}{\partial u} + \dots.$$

The algorithm of Section 2.5.4 requires the calculation of exact symmetries for the unperturbed equation

$$u_{tt} = \left(u^\sigma u_x\right)_x, \quad \sigma \neq 0. \tag{2.28}$$

We will use the available result of the group classification of nonlinear wave equations due to Ames, Lohner, and Adams [1981] (see also [H1], Section 12.4.1). This classification singles out three types of Equations 2.28 corresponding to the following values of σ: (i) $\sigma \neq 0$ is an arbitrary constant, (ii) $\sigma = -4/3$, (iii) $\sigma = -4$. Let's implement the algorithm in these three distinct cases.

---

[2]Editor's note: In contemporary group analysis based on the Lie, Lie-Bäcklund, and approximate group theories, infinitesimal generators act in the *universal space* $\mathcal{A}$ of differential functions (see Chapter 1). Given an operator

$$X = \xi^i \frac{\partial}{\partial x^i} + \eta^\alpha \frac{\partial}{\partial u^\alpha}, \quad \xi^i, \eta^\alpha \in \mathcal{A},$$

its extension to derivatives (of any order) of u is uniquely determined by the prolongation formulas. Therefore, there is no ambiguity if one uses (as S. Lie did) the same symbol X for the prolonged action of this operator. It's no advantage introducing special symbols to designate prolonged groups and their infinitesimal generators.

I. $\sigma \neq 0$ **is an arbitrary constant.**

1st step. The most general Lie symmetry generator for Equation 2.28 is

$$X^0 = (C_1 + C_3 t)\frac{\partial}{\partial t} + (C_2 + C_3 x + C_4 x)\frac{\partial}{\partial x} + C_4\frac{2u}{\sigma}\frac{\partial}{\partial u}$$

where $C_1, \ldots, C_4$ are arbitrary constants. Hence, the unperturbed equation admits a 4-dimensional Lie algebra L_4.

2nd step. Using this generator X^0, one readily finds the auxiliary function

$$H = C_3 u_t.$$

3rd step. Thus, we arrive at the following determining equation for deformations:

$$X^1\left(u_{tt} - u^\sigma u_{xx} + \sigma u^{\sigma-1}u_x^2\right)\Big|_{(2.28)} + C_3 u_t = 0,$$

where X^1 is the operator

$$X^1 = \xi_1^0\frac{\partial}{\partial t} + \xi_1^1\frac{\partial}{\partial x} + \eta_1\frac{\partial}{\partial u},$$

extended to the derivatives of u involved in Equation 2.28. To solve this equation, we apply the same approach as in the case of determining equations for Lie symmetries (cf. [H2], Section 1.3.4). Namely, we isolate, in the left-hand side of our determining equation, the terms containing u_{tx}, u_{xx}, u_t, u_x. As a result, we arrive at a polynomial in the variables u_{tx}, u_{xx}, u_t, u_x. Then we set the coefficients of this polynomial equal to zero. It follows:

$$\xi_1^0 = \xi_1^0(t), \ \xi_1^1 = \xi_1^1(x), \ \eta_1 = \frac{2u}{\sigma}((\xi_1^1)_x - (\xi_1^0)_t),$$

$$\left(\sigma + \frac{4}{3}\right)(\xi_1^1)_{xx} = 0, \tag{2.29}$$

$$\left(\frac{4}{\sigma} + 1\right)(\xi_1^0)_{tt} = C_3, \ (\xi_1^0)_{ttt} = 0, \ (\xi_1^1)_{xxx} = 0.$$

The general solution of Equations 2.29 is

$$\xi_1^0 = \frac{\sigma}{2(\sigma+4)}C_3 t^2 + A_1 + A_3 t, \quad \xi_1^1 = A_2 + A_3 x + A_4 x,$$

$$\eta_1 = \frac{2u}{\sigma}\left(A_4 - \frac{\sigma}{\sigma+4}C_3 t\right).$$

Thus, if $\sigma \neq 0$ is an arbitrary constant, Equation 2.27 has the following first-order approximate symmetry generators:

$$X_1 = \frac{\partial}{\partial t}, \quad X_2 = \frac{\partial}{\partial x},$$

$$X_3 = \left(t + \frac{\varepsilon \sigma t^2}{2(\sigma + 4)}\right)\frac{\partial}{\partial t} + x\frac{\partial}{\partial x} - \frac{2\varepsilon t u}{\sigma + 4}\frac{\partial}{\partial u}, \quad X_4 = x\frac{\partial}{\partial x} + \frac{2u}{\sigma}\frac{\partial}{\partial u},$$

$$X_5 = \varepsilon\frac{\partial}{\partial t}, \quad X_6 = \varepsilon\frac{\partial}{\partial x},$$

$$X_7 = \varepsilon\left(t\frac{\partial}{\partial t} + x\frac{\partial}{\partial x}\right), \quad X_8 = \varepsilon\left(x\frac{\partial}{\partial x} + \frac{2u}{\sigma}\frac{\partial}{\partial u}\right).$$

These operators span an approximate 8-dimensional Lie algebra L_8. Its essential operators (see Section 2.3.7) are X_1, X_2, X_3, X_4. Further, the approximate Lie algebra L_8 is a first-order deformation of the most general symmetry Lie algebra L_4 for Equation 2.28. That is, the algebra L_4 is stable. Hence, according to Section 2.5.2, the perturbed equation 2.27 inherits the symmetries of the unperturbed equation 2.28 with arbitrary σ.

II. $\sigma = -4/3$.

1st step. In this case, the most general Lie symmetry generator for Equation 2.28 is

$$X^0 = (C_1 + C_3t)\frac{\partial}{\partial t} + (C_2 + C_3x + C_4x + C_5x^2)\frac{\partial}{\partial x} - (\frac{3}{2}C_4u + 3C_5xu)\frac{\partial}{\partial u}$$

where $C_1, \ldots, C_5$ are arbitrary constants. Hence, the unperturbed equation admits a 5-dimensional Lie algebra L_5.

2nd step. Here, the auxiliary function has the form $H = C_3u_t$.

3rd step. The determining equation for deformations is given by the System 2.29 with $\sigma = -4/3$. It follows:

$$\xi_1^0 = -\frac{C_3}{4}t^2 + A_1 + A_3t, \quad \xi_1^1 = A_2 + A_3x + A_4x + A_5x^2,$$

$$\eta_1 = -\frac{3}{4}C_3tu - 3A_5xu - \frac{3}{2}A_4u.$$

Thus, Equation 2.27 with $\sigma = -4/3$ admits a 10-dimensional approximate Lie algebra L_{10}. It is a first-order deformation of the symmetry Lie algebra L_5 for the corresponding unperturbed equation 2.28. The essential operators of L_{10} are

$$X_1 = \frac{\partial}{\partial t}, \quad X_2 = \frac{\partial}{\partial x}, \quad X_3 = (t - \frac{\varepsilon}{4}t^2)\frac{\partial}{\partial t} + x\frac{\partial}{\partial x} - \frac{3}{4}\varepsilon tu\frac{\partial}{\partial u},$$

$$X_4 = x\frac{\partial}{\partial x} - \frac{3}{2}u\frac{\partial}{\partial u}, \quad X_5 = x^2\frac{\partial}{\partial x} - 3xu\frac{\partial}{\partial u}.$$

In this case, the maximal symmetry algebra L_5 is also stable, i.e., the perturbed equation inherits the symmetries of the unperturbed equation.

III. $\sigma = -4.$

1st step. The most general Lie symmetry generator for Equation 2.28 is

$$X^0 = (C_1 + C_3t + C_5t^2)\frac{\partial}{\partial t} + (C_2 + C_3x + C_4x)\frac{\partial}{\partial x} + (-\frac{C_4}{2}u + C_5tu)\frac{\partial}{\partial u}$$

where $C_1, \ldots, C_5$ are arbitrary constants. Hence, the unperturbed equation admits a 5-dimensional Lie algebra L_5.

2nd step. The auxiliary function has the form $H = C_3u_t + 2C_5tu_t + C_5u.$

3rd step. The determining equation for deformations reduces to the following system:

$$\xi_1^0 = \xi_1^0(t), \; \xi_1^1 = \xi_1^1(x), \; \eta_1 = -\frac{u}{2}\left((\xi_1^1)_x - (\xi_1^0)_t\right),$$

$$\frac{C_3}{2} + C_5t = 0, \; (\xi_1^1)_{xx} = 0, \; (\xi_1^0)_{ttt} = -C_5.$$

The general solution of this system is

$$\xi_1^0 = A_1 + A_3t + A_5t^2, \; \xi_1^1 = A_2 + A_3x + A_4x,$$

$$\eta_1 = -\frac{u}{2}(A_4 - 2A_5t), \; C_3 = C_5 = 0.$$

It follows that the operators

$$X_3^0 = t\frac{\partial}{\partial t} + x\frac{\partial}{\partial x}, \; X_5^0 = t^2\frac{\partial}{\partial t} + tu\frac{\partial}{\partial u}$$

are not stable (see Section 2.5.2).

Thus, Equation 2.27 with $\sigma = -4$ admits an 8-dimensional approximate Lie algebra L_8. Its essential operators are

$$X_1 = \frac{\partial}{\partial t}, \; X_2 = \frac{\partial}{\partial x}, \; X_3 = \varepsilon\left(t\frac{\partial}{\partial t} + x\frac{\partial}{\partial x}\right), \; X_4 = x\frac{\partial}{\partial x} - \frac{u}{2}\frac{\partial}{\partial u},$$

$$X_5 = \varepsilon\left(t^2\frac{\partial}{\partial t} + tu\frac{\partial}{\partial u}\right).$$

In this case, the maximal symmetry algebra L_5 is not stable. Hence, the perturbed equation does not inherit the symmetries of the unperturbed equation.

2.5.6. Approximate equivalence transformations

In Lie group classification of differential equations, equivalence transformations are of great importance (see [H2], Chapter 2). In the case of differential equations with a small parameter, a natural modification of equivalence transformations is used that involves approximate transformations. The calculation of approximate equivalence transformations is similar to that used in the case of exact equivalence transformation groups (cf. [H2], Section 2.2.1; for detailed calculations see Akhatov, Gazizov, and Ibragimov [1989]). Here, the calculation of approximate equivalence transformations is outlined by means of an example.

The example is furnished by the second-order ordinary differential equation

$$y'' = \varepsilon f(y) \tag{2.30}$$

with the independent variable x and the dependent variable y. Here f is an arbitrary function of the one variable y. An approximate equivalence transformation is a nondegenerate (at $\varepsilon = 0$) change of variables of the form:

$$\tilde{x} = \varphi_0(x, y) + \varepsilon \varphi_1(x, y), \quad \tilde{y} = \psi_0(x, y) + \varepsilon \psi_1(x, y),$$

such that, in the precision indicated, Equation 2.30 is written

$$\frac{d^2 \tilde{y}}{d\tilde{x}^2} = \varepsilon \tilde{f}(\tilde{y}) + o(\varepsilon)$$

with a function $\tilde{f}$ depending only on $\tilde{y}$. That is, the form of Equation 2.30 is unalterable.

A modification of the usual infinitesimal technique applies to approximate equivalence transformation groups as follows. Equation 2.30 is rewritten as the system

$$y'' = \varepsilon f, \quad \varepsilon f_x = 0. \tag{2.31}$$

Here, y is a differential variable with the independent variable x, whereas f is treated as a differential variable with the two independent variables x and y. The generator of a one-parameter approximate group of equivalence transformations is sought in the form

$$Y = Y^0 + \varepsilon Y^1 \equiv (\xi_0 + \varepsilon \xi_1)\frac{\partial}{\partial x} + (\eta_0 + \varepsilon \eta_1)\frac{\partial}{\partial y} + \phi \frac{\partial}{\partial f},$$

where the unknown coefficients ξ_ν and η_ν ($\nu = 0, 1$) depend on x, y, and the coefficient ϕ depends on x, y, and f. After the prolongation to derivatives of the differential variables y and f, the operator Y has the form:

$$Y = (\xi_0 + \varepsilon\xi_1)\frac{\partial}{\partial x} + (\eta_0 + \varepsilon\eta_1)\frac{\partial}{\partial y} + \phi\frac{\partial}{\partial f} + \zeta_1\frac{\partial}{\partial y'} + \zeta_{11}\frac{\partial}{\partial y''} + \omega_1\frac{\partial}{\partial f_x}. \quad (2.32)$$

Here, the coefficients ζ_1 and ζ_{11} are given by the prolongation formulas of Section 2.5.5, and the coefficient ω_1 is obtained by the similar formula applied to the differential function f with the independent variables x, y, viz.

$$\omega_1 = \tilde{D}_x(\phi) - f_x\tilde{D}_x(\xi_0 + \varepsilon\xi_1) - f_y\tilde{D}_x(\eta_0 + \varepsilon\eta_1),$$

where (cf. D_x of Section 2.5.5)

$$\tilde{D}_x = \frac{\partial}{\partial x} + f_x\frac{\partial}{\partial f} + \dots .$$

The infinitesimal approximate invariance criterion of System 2.31 has the form:

$$Y(y'' - \varepsilon f)\big|_{(2.31)} = o(\varepsilon), \quad Y(\varepsilon f_x)\big|_{(2.31)} = o(\varepsilon), \quad\quad (2.33)$$

where Y is the operator 2.32.

In the zero order of precision, the first equation 2.33 yields:

$$Y(y'')\big|_{y''=0} = 0.$$

This is the determining equation for exact symmetries of the unperturbed equation $y'' = 0$. Hence (see, e.g., [H1], Section 8.4),

$$\xi_0 = (C_1x + C_2)y + C_3x^2 + C_4x + C_5,$$

$$\eta_0 = C_1y^2 + C_3xy + C_6y + C_7x + C_8,$$

where $C_1, \dots, C_8$ are arbitrary constants.

Further, in the first order of precision, the second equation 2.33 yields:

$$\phi_x = 0, \quad (\eta_0)_x = 0.$$

Then the first equation 2.33 reduces to the following system:

$$(\xi_1)_{yy} = 0, \quad (\eta_1)_{yy} - 2(\xi_1)_{xy} = 0, \quad 2(\eta_1)_{xy} - (\xi_1)_{xx} - 3(\xi_0)_y f = 0,$$

$$\phi = (\eta_1)_{xx} + ((\eta_0)_y - 2(\xi_0)_x)f.$$

Since the coefficients ξ and η do not depend on the differential variable f, the above equations yield:

$$C_1 = 0, \quad C_2 = 0, \quad C_3 = 0, \quad C_7 = 0,$$

$$\xi_1 = (A_1 x + A_2)y + 2B_1 x^3 + A_3 x^2 + A_4 x + A_5,$$

$$\eta_1 = A_1 y^2 + 3B_1 x^2 y + A_3 xy + A_6 y + B_2 x^2 + A_7 x + A_8,$$

$$\phi = 6B_1 y + 2B_2 + (C_6 - 2C_4)f.$$

Thus, Equation 2.30 admits a 14-dimensional approximate Lie algebra L_{14} of infinitesimal generators of approximate equivalence transformations. The algebra L_{14} is spanned by

$$Y_1 = x\frac{\partial}{\partial x} - 2f\frac{\partial}{\partial f}, \qquad Y_2 = \frac{\partial}{\partial x}, \qquad Y_3 = y\frac{\partial}{\partial y} + f\frac{\partial}{\partial f},$$

$$Y_4 = \frac{\partial}{\partial y}, \quad Y_5 = 2\varepsilon x^3\frac{\partial}{\partial x} + 3\varepsilon x^2 y\frac{\partial}{\partial y} + 6y\frac{\partial}{\partial f}, \quad Y_6 = \varepsilon x^2\frac{\partial}{\partial y} + 2\frac{\partial}{\partial f},$$

$$Y_7 = \varepsilon\left(xy\frac{\partial}{\partial x} + y^2\frac{\partial}{\partial y}\right), \quad Y_8 = \varepsilon y\frac{\partial}{\partial x}, \quad Y_9 = \varepsilon\left(x^2\frac{\partial}{\partial x} + xy\frac{\partial}{\partial y}\right),$$

$$Y_{10} = \varepsilon x\frac{\partial}{\partial x}, \quad Y_{11} = \varepsilon\frac{\partial}{\partial x}, \quad Y_{12} = \varepsilon y\frac{\partial}{\partial y}, \quad Y_{13} = \varepsilon x\frac{\partial}{\partial y}, \quad Y_{14} = \varepsilon\frac{\partial}{\partial y}.$$

The corresponding 14-parameter approximate equivalence transformation group is given by:

$$\tilde{x} = a_1 x + a_2 + \varepsilon(2a_1 a_5 x^3 + a_1 a_7 xy + a_8 y + a_1 a_9 x^2 + a_{10} x + a_{11}),$$

$$\tilde{y} = a_3 y + a_4 + \varepsilon(3a_3 a_5 x^2 y + a_6 x^2 + a_3 a_7 y^2 + a_3 a_9 xy + a_{12} y + a_{13} x + a_{14}),$$

$$\tilde{f} = \frac{a_3}{a_1^2}f + 6\frac{a_3 a_5}{a_1^2}y + 2\frac{a_6}{a_1^2}.$$

2.6. Approximate conservation laws

2.6.1. Adaptation of the Noether theorem

Here, the discussion is restricted to the case of Lagrangians $L(x, u, u_{(1)}, \varepsilon)$ depending on an independent variable $x = (x^1, \ldots, x^n)$, a dependent variable $u = (u^1, \ldots, u^m)$, and the first-order derivatives $u_{(1)} = \{u_i^\alpha\}$ of u with respect to x. Thus, $L \in \mathcal{A}$ is a differential function of the first order (see Section 1.1.1) depending also on a small parameter ε. Further restriction is that approximate groups are limited by point transformations. The general case involving higher-order Lagrangians and approximate Lie-Bäcklund transformation groups can be treated in a similar way.

THEOREM 2.8.   *(Cf. [H1], Section 6.2.) Consider an approximate Euler-Lagrange equation, viz.*

$$\frac{\delta L}{\delta u^\alpha} \equiv \frac{\partial L}{\partial u^\alpha} - D_i\left(\frac{\partial L}{\partial u_i^\alpha}\right) = o(\varepsilon^q). \tag{2.34}$$

Let Equation 2.34 be invariant under the approximate (of an arbitrary order p) group of point transformations with the approximate generator

$$X \approx \xi^i(x, u, \varepsilon)\frac{\partial}{\partial x^i} + \eta^\alpha(x, u, \varepsilon)\frac{\partial}{\partial u^\alpha},$$

where (cf. Operator 2.9)

$$\xi^i(x, u, \varepsilon) = \xi_0^i(x, u) + \varepsilon\xi_1^i(x, u) + \cdots + \varepsilon^p\xi_p^i(x, u),$$

$$\eta^\alpha(x, u, \varepsilon) = \eta_0^\alpha(x, u) + \varepsilon\eta_1^\alpha(x, u) + \cdots + \varepsilon^p\eta_p^\alpha(x, u).$$

Let X be an approximate Noether symmetry, i.e.,

$$XL + LD_i(\xi^i) = D_i(B^i) + o(\varepsilon^p) \tag{2.35}$$

with $B^i \in \mathcal{A}$, $i = 1, \ldots, n$. Then the differential functions

$$C^i = L\xi^i + (\eta^\alpha - \xi^j u_j^\alpha)\frac{\partial L}{\partial u_i^\alpha} - B^i + o(\varepsilon^p) \tag{2.36}$$

satisfy the approximate conservation law for Equation 2.34:

$$D_i(C^i)\big|_{(2.34)} = o(\varepsilon^p).\qquad(2.37)$$

2.6.2.   An application (the first order of precision)

The equation

$$u_{tt} + \varepsilon u_t - u_{xx} - u_{yy} = 0$$

has the Lagrangian

$$L = \frac{1}{2}e^{\varepsilon t}(u_t^2 - u_x^2 - u_y^2)$$

and admits the approximate operator (see Section 9.2.2.1)

$$X = (t^2 + x^2 + y^2)\frac{\partial}{\partial t} + 2tx\frac{\partial}{\partial x} + 2ty\frac{\partial}{\partial y} - \left(t + \frac{\varepsilon}{2}(t^2 + x^2 + y^2)\right)u\frac{\partial}{\partial u}.$$

The condition 2.35 holds with the functions

$$B^1 \approx -(1 + \varepsilon t)u^2/2,\quad B^2 \approx \varepsilon x u^2/2,\quad B^3 \approx \varepsilon y u^2/2.$$

Formula 2.36 yields the following approximate conservation law:

$$D_t\left(e^{\varepsilon t}[-\frac{s^2}{2}(u_t^2 + u_x^2 + u_y^2) - u_t(tu + \frac{\varepsilon}{2}s^2u + 2txu_x + 2tyu_y)] + \frac{1+2\varepsilon t}{2}u^2\right)$$

$$+D_x\left(e^{\varepsilon t}[tx(u_t^2 + u_x^2 - u_y^2) + u_x(tu + \frac{\varepsilon}{2}s^2u + s^2u_t - \frac{\varepsilon}{2}xu^2 + 2tyu_y)]\right)$$

$$+D_y\left(e^{\varepsilon t}[ty(u_t^2 - u_x^2 + u_y^2) + u_y(tu + \frac{\varepsilon}{2}s^2u + s^2u_t - \frac{\varepsilon}{2}yu^2 + 2txu_x)]\right) = o(\varepsilon),$$

where $s^2 = t^2 + x^2 + y^2$.

2.7.   Approximately invariant solutions (the first order of precision)

Cf. [H1], Section 4.1 or [H2], Section 1.5.11.

2.7.1. An illustrative example

The equation

$$u_{tt} + \varepsilon u_t = (u^\sigma u_x)_x, \quad \sigma \neq -4, \tag{2.38}$$

admits the approximate operator (see Section 2.5.5)

$$X = X^0 + \varepsilon X^1 \equiv t\frac{\partial}{\partial t} - \frac{2u}{\sigma}\frac{\partial}{\partial u} + \varepsilon\left(\frac{\sigma}{2\sigma + 8}t^2\frac{\partial}{\partial t} - \frac{2}{\sigma + 4}tu\frac{\partial}{\partial u}\right). \tag{2.39}$$

It is sufficient, for purposes of illustration, to consider *regular invariant solutions* (see [H2], Section 1.5.11) written via invariants. Approximate invariants

$$J(t, x, u, \varepsilon) \approx J_0(t, x, u) + \varepsilon J_1(t, x, u)$$

for the operator 2.39 are determined by the equation

$$XJ = o(\varepsilon),$$

or equivalently,

$$X^0 J_0 + \varepsilon(X^1 J_0 + X^0 J_1) = 0.$$

This equation splits into the system:

$$X^0 J_0 = 0, \quad X^0 J_1 = -X^1 J_0.$$

It follows that the operator X has two functionally independent approximate invariants given by

$$J^1 \approx x + \varepsilon\alpha(x, ut^{2/\sigma}), \quad J^2 \approx ut^{2/\sigma} + \varepsilon\left[\frac{ut^{2/\sigma+1}}{\sigma + 4} + \beta(x, ut^{2/\sigma})\right]$$

with arbitrary functions α and β.

In the simple case when $\alpha = \beta = 0$, an approximately invariant solution given by the equation $J^2 \approx \varphi(J^1)$ has the form

$$u \approx t^{-2/\sigma}\left(1 - \frac{\varepsilon t}{\sigma + 4}\right)\varphi(x). \tag{2.40}$$

The substitution of the function 2.40 into Equation 2.38 yields

$$(\varphi^\sigma \varphi')' = \frac{2\sigma + 4}{\sigma^2}\varphi. \tag{2.41}$$

Equation 2.41 can be integrated by quadrature,

$$x + x_0 = \int \frac{d\varphi}{\sqrt{C\varphi^{-2\sigma} + 4\varphi^{2-\sigma}/\sigma^2}},$$

where C and x_0 are arbitrary constants.

2.7.2. The use of equivalence transformations

Here, a possibility for constructing approximately invariant solutions is discussed, offered by equivalence transformations.

Consider the equation

$$u_{tt} + \varepsilon u_t = (u^\sigma u_x)_x \qquad (2.42)$$

with an arbitrary constant $\sigma \neq 0$. It admits, e.g., the one-parameter group of equivalence transformations with the generator

$$X = t\frac{\partial}{\partial t} - \frac{2u}{\sigma}\frac{\partial}{\partial u} - \varepsilon\frac{\partial}{\partial \varepsilon}. \qquad (2.43)$$

Operator 2.43 has the following three functionally independent invariants:

$$J^1 = ut^{2/\sigma}, \quad J^2 = x, \quad J^3 = \varepsilon t.$$

The corresponding approximately invariant solution can be sought in the form

$$u = t^{-2/\sigma}\psi(\varepsilon t, x).$$

In the first order of precision, it is written

$$u \approx t^{-2/\sigma}(\psi_1(x) + \varepsilon t\psi_2(x)). \qquad (2.44)$$

Substitution of the function 2.44 into Equation 2.42 yields the following system of second-order ordinary differential equations:

$$\frac{2}{\sigma}\left(\frac{2}{\sigma} + 1\right)\psi_1 = (\psi_1^\sigma \psi_1')', \qquad (2.45)$$

$$\frac{2(\sigma - 2)}{\sigma^2}\psi_2 + \frac{2}{\sigma}\psi_1 = -(\psi_1^\sigma \psi_2)''. \qquad (2.46)$$

Equation 2.45 coincides with Equation 2.41. Given any solution of Equation 2.45, the integration of the linear equation 2.46 with respect to ψ_2 provides an approximate solution 2.44, depending on two arbitrary constants. In the particular case when

$$\psi_2 = -\frac{\psi_1}{\sigma + 4},$$

the system 2.45 – 2.46 provides a solution of the type 2.44 approximately invariant with respect to the operator 2.39. However, in general, the solutions of the form 2.44 can not be obtained as approximately invariant solutions by using the approximate symmetries given in Section 2.5.5. This is due to the fact that, according to Section 2.5.5, the symmetry

$$X^0 = t\frac{\partial}{\partial t} - \frac{2u}{\sigma}\frac{\partial}{\partial u}$$

is not stable for Equation 2.42 with $\sigma = -4$.

Note that Operator 2.43 generates a group of *exact* equivalence transformations. However, the same approach applies to approximate equivalence transformations.

2.8. Formal symmetries

The generalization of the previous considerations to infinite-order approximate symmetries leads to what is called *a formal symmetry* and *a formal Bäcklund transformation* (Baikov, Gazizov, and Ibragimov [1987b], [1988d], [1989b]). Here, the approach is applied to evolution equations of the form $u_t = h(u)u_1 + \varepsilon H(t, x, u, u_1, \ldots, u_n)$. In particular, a new viewpoint to Lie-Bäcklund symmetries offered by formal Bäcklund transformations is sketched and illustrated by the Kortgeweg-de Vries equation.

The notation is taken from Chapter 1, Sections 1.1.1 and 1.4.3. Namely: t and x are independent variables, u is a differential variable with successive derivatives (with respect to x) $u_1 = u_x$, $u_2 = u_{xx}, \ldots,$ so that $u_{i+1} = D(u_i)$, $u_0 = u$, where

$$D = \frac{\partial}{\partial x} + u_1\frac{\partial}{\partial u} + \ldots$$

is the total differentiation with respect to x. $\mathcal{A}$ is the space of differential functions depending on the variables $t, x, u, u_1, \ldots$. In addition, the following abbreviations are used for derivatives of differential functions η:

$$\eta_t = \frac{\partial\eta}{\partial t}, \quad \eta_x = \frac{\partial\eta}{\partial x}, \quad \eta_i = \frac{\partial\eta}{\partial u_i}, \quad \eta_* = \sum_{i \geq 0} \eta_i D^i.$$

2.8.1. Formal symmetries for the equation $u_t = h(u)u_1 + \varepsilon H$

All Lie-Bäcklund symmetries of the first-order evolutionary equation

$$u_t = h(u)u_1 \tag{2.47}$$

with an arbitrary function $h(u)$ are stable under the perturbation εH with any differential function $H \in \mathcal{A}$. That is, a perturbed equation of the form

$$u_t = h(u)u_1 + \varepsilon H, \quad H \in \mathcal{A}, \tag{2.48}$$

inherits the symmetries of the unperturbed equation 2.47 as defined in Section 2.5.2. Moreover, the following statement about the infinite-order stability is valid (for the proof, see, e.g., Baikov, Gazizov, and Ibragimov [1989a], Section 6.1).

THEOREM 2.9. *Given a canonical Lie-Bäcklund operator*

$$X^0 = \eta^0 \frac{\partial}{\partial u} + \dots, \qquad \eta^0 \in \mathcal{A},$$

admitted by the unperturbed equation 2.47, the perturbed equation 2.48 admits an infinite-order deformation

$$X = \eta \frac{\partial}{\partial u} + \dots,$$

where η is an element of the space $[[\mathcal{A}]]$ of formal power series (see Chapter 1, Section 1.2.1), i.e., it has the form

$$\eta = \sum_{\nu=0}^{\infty} \varepsilon^\nu \eta^\nu, \qquad \eta^\nu \in \mathcal{A}. \tag{2.49}$$

In this theorem, the coefficients η^0 of Lie-Bäcklund symmetries X^0 of Equation 2.47 are determined by the equation

$$\eta_t^0 - h(u)\eta_x^0 + \sum_{i \geq 1}[D^i(hu_1) - hu_{i+1}]\eta_i^0 - h'(u)u_1\eta^0 = 0. \tag{2.50}$$

By an inspection the general solution of Equation 2.50 may be verified to be

$$\eta^0 = u_1\alpha\left(h(u), \ x + t\,h(u), \ t + \frac{1}{h'(u)u_1}, \dots\right) \tag{2.51}$$

with an arbitrary function $\alpha \in \mathcal{A}$ (see [H1], Section 11.3). Given η^0, the coefficients η^ν, $\nu \geq 1$ of the corresponding power series 2.49 are determined recursively by the following infinite system of linear first-order partial differential equations:

$$\eta_t^\nu - h(u)\eta_x^\nu + \sum_{i \geq 1}[D^i(hu_1) - hu_{i+1}]\eta_i^\nu - h'(u)u_1\eta^i$$

$$= \sum_{i \geq 1}[D^i(\eta^{\nu-1})H_i - \eta_i^{\nu-1}D^i(H)], \qquad \nu = 1, \ldots . \qquad (2.52)$$

Furthermore, if the solution $\eta^0 = \eta^0(t, x, u, ..., u_k)$ of Equation 2.50 and the perturbation $H = H(t, x, u, ..., u_n)$ are differential functions of the orders k and n, respectively, then the system 2.52 is solved by a sequence of differential functions η^ν of the orders $k_\nu = \nu(n - 1) + k$.

An *infinite-order deformation X determined by the formal power series 2.49 is called a formal symmetry for Equation 2.48.*

2.8.2. Lie-Bäcklund via formal symmetries

It is clear that if one truncates arbitrarily the series 2.49 and takes a finite sum

$$\eta = \sum_{\nu=0}^{p} \varepsilon^\nu \eta^\nu,$$

one obtains an approximate symmetry (up to the truncation error $o(\varepsilon^p)$) for Equation 2.48.

The situation is more interesting when the series 2.49 breaks off for a certain function H and certain symmetry X^0 of the unperturbed equation 2.47. If this occurs, the corresponding deformation X has a finite order and provides an *exact Lie-Bäcklund symmetry* for Equation 2.48 with an arbitrary constant ε, e.g., with $\varepsilon = 1$.

Example. Consider a nonlinear equation 2.47:

$$u_t = h(u)u_1, \qquad h'(u) \neq 0.$$

For this equation, let us take the simplest Lie-Bäcklund operator given by Formula 2.51, viz.

$$X^0 = \varphi(u)u_1\frac{\partial}{\partial u}.$$

Let us specify the perturbed equation 2.48 by letting $H = u_3$:

$$u_t = h(u)u_1 + \varepsilon u_3.$$

In this particular case, one can prove by inspecting the determining equations 2.52 that *for an arbitrary polynomial function $\varphi(u)$, the series 2.49 breaks off if and only if $h''' = 0$, i.e., if*

$$h(u) = au^2 + bu + c, \quad a, b, c = \text{const.}$$

The result of this example agrees with the well-known fact of the existence of Lie-Bäcklund symmetries for the modified Korteweg-de Vries equation

$$u_t = (au^2 + bu + c)u_1 + u_3.$$

Moreover, the new approach leads to a further understanding of the nature of Lie-Bäcklund symmetries. An algorithm for the calculation of exact symmetries provided by this approach is direct and simple. Namely, one takes any polynomial $\varphi(u)$ and solves Equations 2.52 recursively beginning with $\eta^0 = \varphi(u)u_1$. As a result, one obtains an exact Lie-Bäcklund symmetry of order $2n + 1$ provided that $\varphi(u)$ is a polynomial of degree n. Compare Section 9.4.3 of Chapter 9. For a more detailed discussion see Baikov, Gazizov, and Ibragimov [1989a], Section 6.

2.8.3. Formal Bäcklund transformations

Equation 2.48 with an arbitrary function $H(t, x, u, u_1, \ldots) \in \mathcal{A}$ can be mapped into the first-order equation 2.47:

$$v_t = h(v)v_1 \tag{2.53}$$

by a substitution $u \mapsto v$ given by a formal power series of the form

$$v = u + \sum_{i \geq 1} \varepsilon^i \Phi^i(t, x, u, u_1, \ldots), \qquad \Phi^i \in \mathcal{A}. \tag{2.54}$$

The coefficients Φ^i of the transformation 2.54 are determined recursively by the following infinite system of linear first-order partial differential equations:

$$\Phi_t^1 - h(u)\Phi_x^1 + \sum_{\alpha \geq 1}(D^\alpha(hu_1) - hu_{\alpha+1})\Phi_\alpha^1 - h'(u)u_1\Phi^1 = -H, \tag{2.55}$$

$$\Phi_t^i - h(u)\Phi_x^i + \sum_{\alpha \geq 1}(D^\alpha(hu_1) - hu_{\alpha+1})\Phi_\alpha^i - h'(u)u_1\Phi^i$$

$$= -\sum_{\alpha \geq 0} \Phi_\alpha^{i-1} D^\alpha(H) + u_1 \sum_{k=2}^{i} \frac{1}{k!} h^{(k)}(u) \sum_{i_1+\ldots+i_k=i} \Phi^{i_1} \cdots \Phi^{i_k}$$

$$+ \sum_{j+l=i} D(\Phi^j) \left(\sum_{k=1}^{l} \frac{1}{k!} h^{(k)}(u) \sum_{i_1+\ldots+i_k=l} \Phi^{i_1} \cdots \Phi^{i_k} \right), \qquad i \geq 2. \qquad (2.56)$$

Provided that $H = H(t, x, u, \ldots, u_n)$ is a differential function of the order n, the system 2.55 – 2.56 can be solved by differential functions Φ^k of the orders nk.

The transformation 2.54 is called a *formal Bäcklund transformation* relating Equations 2.48 and 2.53.

The point transformation

$$y = h(v), \quad w = x + t\, h(v)$$

reduces Equation 2.53 to the linear equation

$$w_t = 0 \qquad (2.57)$$

for the function $w = w(t, y)$. Hence, any equation 2.48 with a differential function $H \in \mathcal{A}$ of an arbitrary order can be reduced to the simplest first-order equation 2.57 via the formal Bäcklund transformation

$$y = h\left(u + \sum_{i \geq 1} \varepsilon^i \Phi^i\right), \quad w = x + t\, h\left(u + \sum_{i \geq 1} \varepsilon^i \Phi^i\right),$$

with the coefficients Φ^i determined by Equations 2.55 – 2.56. This is a generalization of the well-known fact of the existence of a Lie tangent transformation relating any two scalar partial differential equations of the first order.

2.8.4.   Bäcklund via formal Bäcklund transformations

Any finite sum of the series 2.54 gives what is called an *approximate Bäcklund transformation*.

Furthermore, if the series 2.54 breaks off, one gets an *exact Bäcklund transformation*.

If a formal Bäcklund transformation relating Equations 2.48 and 2.53 is known, one can transform infinitesimal symmetries of Equation 2.53 (e.g., the Lie-Bäcklund Symmetries 2.51) into formal and approximate symmetries (or exact Lie-Bäcklund symmetries, provided that the break-off conditions hold) of Equation 2.48. For this, one can use the following *transition formula*:

$$\eta_u = \left[1 + \sum_{i \geq 1} \varepsilon^i \Phi^i_* \right]^{-1} \eta_v, \qquad (2.58)$$

where η_u and η_v are symmetries of the equations 2.48 and 2.53, respectively, and Φ_*^i is the linear differential operator defined in the preamble to Section 2.8.

For applications of this approach, see Chapter 9 of this volume and Baikov, Gazizov, and Ibragimov [1989a].

2.8.5. Formal recursions

Symmetries of Equation 2.53 can be constructed recursively by means of the following *recursion operator*,

$$M = \frac{\alpha}{h'(v)} D_x \frac{1}{v_1} + \beta + \gamma v_1 D_x^{-1} h'(v) + \dots, \qquad (2.59)$$

previously discussed in this Handbook ([H1], Section 11.3). Here $\alpha, \beta, \gamma, \dots \in \mathcal{A}$ are arbitrary functions of $v, x + t\, h(v), t + 1/(h'(v)v_1), \dots$. Compare Formula 2.51. For details, see Baikov, Gazizov, and Ibragimov [1989a], Appendix.

The formal Bäcklund transformation 2.54 maps the operator 2.59 into a *formal recursion operator* L for Equation 2.48. This map is given by the following *transition formula*:

$$L = \left(1 + \sum_{i \geq 1} \varepsilon^i \Phi_*^i\right)^{-1} M \left(1 + \sum_{i \geq 1} \varepsilon^i \Phi_*^i\right). \qquad (2.60)$$

Any finite sum of the series 2.60 provides an approximate recursion operator. Furthermore, if break-off conditions for the series 2.60 hold, L is an *exact recursion operator* for Equation 2.48 with any ε, e.g., with $\varepsilon = 1$.

3

Differential Equations with Distributions: Group Theoretic Treatment of Fundamental Solutions

This chapter is integrated with Chapter 3 of [H2] and continues the development of Lie group methods for solving initial value problems. Here, the main emphasis is on the construction of fundamental solutions.

The majority of linear differential equations of physical relevance and of obvious mathematical importance have fundamental solutions in the space of distributions (generalized functions). This necessitates the extension of Lie group methods to differential equations in distributions. This natural path of development of Lie group analysis venturing into the space of distributions has been sketched in Ibragimov [1989] and then evolved in Berest [1991] and Ibragimov [1992a,b] (see also Berest [1993], Berest and Ibragimov [1994]).

The group theoretic derivation of fundamental solutions for the Cauchy problem rather than for differential operators has been presented in Ibragimov [1994d]. This recent presentation based on the *Invariance principle for boundary value problems* is adopted in what follows.

3.1. Generalities

3.1.1. Extension of group transformations to distributions

Let $f \in \mathcal{D}'$ be a distribution, i.e., a linear continuous functional over the space $\mathcal{D}$ of $\mathbf{C}^\infty$ functions with compact supports (test functions). The action of f on $\varphi(x) \in \mathcal{D}$, where $x = (x^1, ..., x^n) \in \mathbf{R}^n$, is denoted by $\langle f(x), \varphi(x) \rangle$.

Consider a one-parameter group of transformations in $\mathbf{R}^n$:

$$\bar{x} = g(x, a). \tag{3.1}$$

Its infinitesimal transformation is written in the form

$$\bar{x}^i \approx x^i + a\xi^i(x), \quad i = 1, ..., n. \tag{3.2}$$

The Jacobian of Transformations 3.1,

$$J = \det\left(\frac{\partial \bar{x}^i}{\partial x^j}\right),$$

is positive in a small neighborhood of $a = 0$. In what follows, the group parameter a is assumed to be sufficiently small, so that $J > 0$.

DEFINITION 3.1.   *Given a distribution f, its image $\bar{f}$ under Transformation 3.1 is defined by the following invariance condition for functionals:*

$$\langle f(x), \varphi(x) \rangle = \langle (\bar{f} \circ g^{-1})(\bar{x}), (\varphi \circ g^{-1})(\bar{x}) \rangle. \tag{3.3}$$

Here $\circ$ denotes the composition of transformations, g^{-1} is the inversion of g, and φ is an arbitrary test function.

Consider the regular case, when the distribution f is defined by a locally integrable function $f(x)$ as follows:

$$\langle f(x), \varphi(x) \rangle = \int_{\mathbf{R}^n} f(x)\varphi(x)dx.$$

According to the usual change of variables formula in the integral, we have

$$\int_{\mathbf{R}^n} f(x)\varphi(x)dx = \int_{\mathbf{R}^n} (f \circ g^{-1})(\bar{x})(\varphi \circ g^{-1})(\bar{x})J^{-1}d\bar{x},$$

that is,

$$\langle f(x), \varphi(x) \rangle = \langle (J^{-1}f \circ g^{-1})(\bar{x}), (\varphi \circ g^{-1})(\bar{x}) \rangle. \tag{3.4}$$

Comparison of Equations 3.3 and 3.4 suggests that $\bar{f} = J^{-1}f$.

Thus, we obtain the following *extension of the group transformations 3.1 to arbitrary distributions* (Ibragimov [1972]):

$$\bar{f} = \left[\det\left(\frac{\partial \bar{x}^i}{\partial x^j}\right)\right]^{-1}f. \tag{3.5}$$

In particular, for the Dirac distribution $f = \delta$, this formula together with the equation

$$\Phi(x)\delta(x) = \Phi(0)\delta(x)$$

where $\Phi(x)$ is an arbitrary C^∞ function, yield:

$$\bar{\delta} = \left[\det\left(\frac{\partial \bar{x}^i}{\partial x^j}\right)\right]^{-1}_{x=0} \delta. \tag{3.6}$$

For the infinitesimal transformation 3.2, the formulas 3.5 and 3.6 reduce, respectively, to the following forms:

$$\bar{f} \approx f - aD_i(\xi^i)f \tag{3.7}$$

and

$$\bar{\delta} \approx \delta - aD_i(\xi^i)\big|_{x=0}\delta. \tag{3.8}$$

Here

$$D_i(\xi^i) = \sum_{i=1}^{n} \frac{\partial \xi^i}{\partial x^i}.$$

The usual infinitesimal test for a group invariance of equations given by classical functions applies to equations with distributions as well.

THEOREM 3.1. *Let $F(x, f)$ be a linear function on a distribution f with smooth coefficients depending on the variables x. The equation*

$$F(x, f) = 0$$

is invariant under the group of Transformations 3.1 and 3.5 if and only if it is invariant infinitesimally, i.e., if

$$F(\bar{x}, \bar{f})\big|_{F(x,f)=0} = o(a).$$

REMARK. For an arbitrary transformation on $\mathbf{R}^n$ given by

$$\bar{x} = \Phi(x),$$

where Φ is a C^∞ diffeomorphism, the transformation formula 3.5 is written

$$\bar{f} = \left|\det\left(\frac{\partial \bar{x}^i}{\partial x^j}\right)\right|^{-1} f.$$

3.1.2. The Leray form and the Dirac measure on hypersurfaces

Consider a hypersurface in $\mathbf{R}^n$ defined by the equation

$$P(x) = 0,$$

where $P(x)$ is a continuously differentiable function such that $\nabla P \neq 0$ on the surface $P = 0$.

DEFINITION 3.2.   *The Leray form for this hypersurface is an $(n-1)$-differential form ω such that*

$$dP \wedge \omega = dx^1 \wedge \cdots \wedge dx^n.$$

It can be represented in the form (for any fixed i)

$$\omega = (-1)^{i-1} \frac{dx^1 \wedge \cdots \wedge dx^{i-1} \wedge dx^{i+1} \wedge \cdots \wedge dx^n}{P_i}$$

provided that $P_i \equiv D_i P \neq 0$ (see Leray [1953], Chapter IV, Section 1).

DEFINITION 3.3.   *Let $\theta(P)$ be the Heaviside function, or the characteristic function of the domain $P \geq 0$, that is,*

$$\theta(P) = \begin{cases} 1, & P \geq 0, \\ 0, & P < 0. \end{cases}$$

It defines the distribution

$$\langle \theta(P), \varphi \rangle = \int_{P \geq 0} \varphi(x) dx.$$

DEFINITION 3.4.   *The Dirac measure $\delta(P)$ on the surface $P(x) = 0$ is defined by*

$$\langle \delta(P), \varphi \rangle = \int_{P=0} \varphi \omega,$$

where ω is the Leray form.

3.1.3. Auxiliary equations

Let us begin with the simplest first-order ordinary differential equation,

$$xf' = 0$$

with one independent variable x. The only classical solution of this equation is f = const., while its general solution in distributions depends on two arbitrary constants and has the form

$$f = C_1\theta(x) + C_2.$$

We will use the following natural generalization of this equation. It is known that the distributions $\theta(P)$ and $\delta(P)$ given by Definitions 3.3 and 3.4 satisfy the equations (see, e.g., Gel'fand and Shilov [1959]):

$$\theta'(P) = \delta(P),$$

$$P\delta(P) = 0,$$

$$P\delta^{(m)}(P) + m\delta^{(m-1)}(P) = 0, \quad m = 1, 2, ...,$$

where $\delta^{(m)}$ is the mth derivative of $\delta(P)$ with respect to P. It follows that the first-order differential equation

$$Pf'(P) + mf(P) = 0 \tag{3.9}$$

has the general solution in distributions given by

$$f = C_1\theta(P) + C_2, \quad \text{for} \quad m = 0, \tag{3.10}$$

$$f = C_1\delta^{(m-1)}(P) + C_2 P^{-m}, \quad \text{for} \quad m = 1, 2, ..., \tag{3.11}$$

where C_1 and C_2 are arbitrary constants.

3.1.4. Invariance principle

Given a linear partial differential operator L, consider a boundary value (in particular, initial value) problem,

$$Lu = F(x), \tag{3.12}$$

$$u\big|_S = h(x) \tag{3.13}$$

with the data $h(x)$ defined on the manifold $S \subset \mathbf{R}^n$.

DEFINITION 3.5. The problem 3.12 – 3.13 is said to be invariant under a group G if the following hold:

1) the differential equation 3.12 admits G,
2) the manifold S together with Equation 3.13 are invariant under the group G.
The invariance principle. *Let the boundary value problem 3.12 – 3.13 be invariant under the group G. Then we should seek the solution among the functions invariant under G.*

Example 1. Consider the following boundary value problem in the circle $r \leq 1$:

$$u_{xx} + u_{yy} = e^u, \tag{3.14}$$

$$u\big|_{r=1} = 0, \tag{3.15}$$

where $r = \sqrt{x^2 + y^2}$.

The Liouville equation 3.14 admits the group of conformal transformations on the (x, y) plane properly extended to the variable u. The infinitesimal generator of this group is of the form

$$X = \xi \frac{\partial}{\partial x} + \eta \frac{\partial}{\partial y} - 2\xi_x \frac{\partial}{\partial u},$$

where the coefficients $\xi(x, y)$ and $\eta(x, y)$ are arbitrary analytic functions, i.e., defined by the Cauchy-Riemann equations:

$$\xi_x - \eta_y = 0, \quad \xi_y + \eta_x = 0.$$

It is convenient to investigate the invariance of the boundary condition 3.15 in the polar coordinates (r, ψ):

$$x = r \cos \psi, \quad y = r \sin \psi.$$

Then the symmetry operator for the Liouville equation

$$u_{rr} + \frac{1}{r} u_r + \frac{1}{r^2} u_{\psi\psi} = e^u$$

has the form

$$X = \alpha(r, \psi) \frac{\partial}{\partial r} + \beta(r, \psi) \frac{\partial}{\partial \psi} - 2\alpha_r \frac{\partial}{\partial u}.$$

Its coefficients are determined by the system of Cauchy-Riemann equations in polar coordinates:

$$\beta_r + \frac{1}{r^2} \alpha_\psi = 0, \quad \beta_\psi - \alpha_r + \frac{\alpha}{r} = 0.$$

One obtains from this system, by eliminating β, the second-order equation for $\alpha(r, \psi)$:

$$\alpha_{rr} + \frac{1}{r^2}\alpha_{\psi\psi} - \frac{1}{r}\alpha_r + \frac{1}{r^2}\alpha = 0.$$

The infinitesimal invariance condition of the boundary manifold $r = 1$ and of Equation 3.15 yield:

$$\alpha\big|_{r=1} = 0, \quad \alpha_r\big|_{r=1} = 0.$$

It follows from the Cauchy-Kovalevskaya theorem that $\alpha = 0$ in a neighborhood of $r = 1$. Then the Cauchy-Riemann system yields $\beta = \text{const}$.

Thus, the boundary value problem 3.14 – 3.15 admits the operator

$$X_1 = \frac{\partial}{\partial\psi},$$

i.e., the problem is invariant under the one-parameter group of rotations. According to the invariance principle, the solution is taken in the form $u = U(r)$. Then Equation 3.14 reduces to the ordinary differential equation

$$U'' + \frac{1}{r}U' - e^U = 0. \tag{3.16}$$

We will consider the solutions bounded at the "singular" point $r = 0$. Taking into account Equation 3.15, we have the following side conditions:

$$U(1) = 0, \quad |U(0)| < \infty.$$

Equation 3.16 admits the two-dimensional Lie algebra spanned by

$$Y_1 = r\frac{\partial}{\partial r} - 2\frac{\partial}{\partial U}, \quad Y_2 = r\ln r\frac{\partial}{\partial r} - 2(1 + \ln r)\frac{\partial}{\partial U}.$$

In accordance with Lie's integration algorithm (see [H1], Section 2.2.2), we find the transformation of the symmetry algebra to its canonical form. This transformation is given by the change of variables:

$$t = \ln r, \quad v = U + 2\ln r.$$

Equation 3.16 is rewritten, for the function $v(t)$, in the integrable form:

$$v'' = e^v.$$

An inspection of the general solution of this equation shows that the function

$$U(r) = \ln(8c) - \ln(1 - cr^2)^2$$

solves Equation 3.16 and satisfies the condition $|U(0)| < \infty$. Here, c is an arbitrary positive constant. The boundary condition $U(1) = 0$ yields $8c = (1 - c)^2$.

Hence, we have two solutions to the problem 3.14 – 3.15. One of them (corresponding to $c = 5 - 2\sqrt{6}$) is bounded everywhere in the circle $r \leq 1$, while the second solution (corresponding to $c = 5 + 2\sqrt{6}$) is singular at the circle $r = 1/\sqrt{c}$.

Example 2. Consider the two-body problem for a motion of a planet. The equation of motions,

$$\frac{d^2\mathbf{x}}{dt^2} = \alpha \frac{\mathbf{x}}{|\mathbf{x}|^3}, \quad \mathbf{x} \in \mathbf{R}^3, \ \alpha = \text{const.},$$

admits the group of rotations in $\mathbf{R}^3$. However, the orbits of planets are not invariant under this group: the orbit of any planet lies on a fixed plane determined by the Kepler laws. It may seem that this asymmetry and the invariance principle are contradictory. But there is no disagreement here. Indeed, we have to take into account an initial position and an initial velocity of the planet under consideration. Then the initial conditions break the rotational symmetry of the problem, and hence of the trajectory of the planet.

3.2. Heat equation

For a detailed discussion of heat diffusion problems from the standpoint of the Galilean invariance principle, the reader is referred to Chapter 6; see also [H2], Section 7.2.

3.2.1. Symmetry algebra of the equation

Consider the heat equation with n spatial variables,

$$u_t = \Delta u \tag{3.17}$$

where Δ is the n-dimensional Laplacian in the variables $x = (x^1, ..., x^n) \in \mathbf{R}^n$.

The maximal Lie algebra admitted by Equation 3.17 is composed of the finite-dimensional subalgebra spanned by

$$X_0 = \frac{\partial}{\partial t}, \quad X_i = \frac{\partial}{\partial x^i}, \quad X_{ij} = x^j \frac{\partial}{\partial x^i} - x^i \frac{\partial}{\partial x^j}, \quad X_{0i} = 2t \frac{\partial}{\partial x^i} - x^i u \frac{\partial}{\partial u},$$

$$Z_1 = 2t \frac{\partial}{\partial t} + x^i \frac{\partial}{\partial x^i}, \quad Z_2 = u \frac{\partial}{\partial u}, \quad Y = t^2 \frac{\partial}{\partial t} + t x^i \frac{\partial}{\partial x^i} - \frac{1}{4}(2nt + |x|^2) u \frac{\partial}{\partial u}$$

and the infinite-dimensional ideal consisting of the operators

$$X_\tau = \tau(t, x) \frac{\partial}{\partial u},$$

where $\tau(t, x)$ is an arbitrary solution of Equation 3.17.

In what follows, we let $\tau(t, x) = 0$ and use only the $[\frac{1}{2}(n + 1)(n + 2) + 3]$-dimensional symmetry algebra spanned by

$$X_0, \ X_i, \ X_{ij}, \ X_{0i}, \ Z_1, \ Z_2, \ Y, \quad i, j = 1, ..., n. \tag{3.18}$$

3.2.2. The symmetry algebra of the initial value problem for the fundamental solution

The theory of distributions reduces an arbitrary Cauchy problem for the heat equation to determining the following *fundamental solution.*

DEFINITION 3.6. *The distribution* $u = E(t, x)$ *is called the fundamental solution of the Cauchy problem for the heat equation if it solves Equation 3.17 for* $t > 0$ *and satisfies the initial condition*

$$u\big|_{t=0} = \delta(x), \tag{3.19}$$

where $u\big|_{t=0} \equiv \lim_{t \to +0} E(t, x)$, *and* $\delta(x)$ *is the Dirac distribution at* $x = 0$.

THEOREM 3.2. *Let L be the Lie algebra with the basis 3.18, and K its maximal subalgebra admitted by the initial value problem 3.17, 3.19. Then K is the linear span of the operators*

$$X_{ij}, \ X_{0i}, \ Z_1 - n Z_2, \ Y, \quad i, j = 1, ..., n. \tag{3.20}$$

PROOF. Since the algebra L is admitted by the differential equation 3.17, we shall consider only the invariance condition 2 of Definition 3.5.

In our case, the initial manifold S is given by $t = 0$. Further, the invariance of the initial data 3.19 requires, in particular, that the support of $\delta(x)$, i.e., the point $x = 0$, be unaltered. Thus, Definition 3.5 requires that the system of equations $t = 0$, $x = 0$ be invariant. This reduces the algebra L by the translation operators X_i, X_0.

Hence, the operators 3.18 are restricted to the following:

$$X_{ij},\ X_{0i},\ Z_1,\ Z_2,\ Y.$$

Equation 3.19 is invariant under the operators X_{ij}, X_{0i}, and Y. It is not invariant under the two-dimensional algebra spanned by Z_1, Z_2. Therefore, we inspect the infinitesimal invariance test for the linear combination

$$(Z_1 + kZ_2)\big|_{t=0} = x^i \frac{\partial}{\partial x^i} + ku \frac{\partial}{\partial u}, \quad k = \text{const.}$$

Under this operator, the variable u and the δ-function are subjected to the infinitesimal transformations (see Equation 3.8):

$$\bar{u} \approx u + aku, \quad \bar{\delta} \approx \delta - an\delta.$$

It follows that $\bar{u} - \bar{\delta} = u - \delta + a(ku + n\delta) + o(a)$ and that

$$(\bar{u} - \bar{\delta})\big|_{u=\delta} = a(k + n)\delta + o(a).$$

According to Theorem 3.1, the invariance condition of Equation 3.19 has the form $k + n = 0$.

Thus, we arrive at the algebra K spanned by the operators 3.20.

3.2.3. Derivation of the fundamental solution from the invariance principle

THEOREM 3.3. *The fundamental solution of the Cauchy problem for the heat equation,*

$$E(t, x) = (2\sqrt{\pi t})^{-n} \exp[-|x|^2/(4t)], \tag{3.21}$$

is uniquely determined by the invariance principle. Namely, it is the only function $u = \phi(t, x)$ which satisfies the initial condition 3.19 and is invariant under the group of rotations, Galilean transformations, and dilations with the infinitesimal generators

$$X_{ij}, \quad X_{0i}, \quad Z_1 - nZ_2, \quad i, j = 1, ..., n. \tag{3.22}$$

PROOF. We first notice that the functionally independent invariants of the rotations are t, r, u, where

$$r = |x| \equiv \sqrt{(x^1)^2 + \cdots + (x^n)^2}.$$

Then we write the restriction of the Galilean operators X_{0i} to functions of these invariants as follows:

$$X_{0i} = x^i (2\frac{t}{r} \frac{\partial}{\partial r} - u\frac{\partial}{\partial u}).$$

For these operators, the independent invariants are t and $p = u \exp[r^2/(4t)]$. The last operator 3.22 is written in these variables in the form:

$$Z_1 - nZ_2 = 2t\frac{\partial}{\partial t} - np\frac{\partial}{\partial p}.$$

It has the only independent invariant $J = t^{n/2} p$. Hence, the function

$$J = (\sqrt{t})^n u \exp[r^2/(4t)]$$

is the only common invariant for the operators 3.22.

It follows that the general form of the function $u = \phi(t, x)$ invariant under the operators 3.22 is given by $J = C$, or by

$$u = C(\sqrt{t})^{-n} \exp[-r^2/(4t)], \quad C = \text{const.}$$

In view of the known formula

$$\lim_{t \to +0} \left((\sqrt{t})^{-n} \exp[-|x|^2/(4t)] \right) = (2\sqrt{\pi})^n \delta(x),$$

the initial condition 3.19 yields $C = (2\sqrt{\pi})^{-n}$.

Thus, we have obtained (uniquely) the fundamental solution 3.21.

REMARK 1. One can readily verify that the function 3.21 is also invariant under the operator Y from Basis 3.20, i.e., admits the Lie algebra K. However, we do not need here this *excess symmetry* of the fundamental solution.

REMARK 2. The fundamental solution $\mathcal{E}$ of the heat operator, i.e., the solution of the equation

$$(\frac{\partial}{\partial t} - \Delta)\mathcal{E} = \delta(t, x),$$

is obtained by the formula $\mathcal{E} = \theta(t)E$, where $\theta(t)$ is the Heaviside function. Hence,

$$\mathcal{E} = \theta(t)(2\sqrt{\pi t})^{-n} \exp[-|x|^2/(4t)].$$

3.2.4. Solution of the Cauchy problem

The solution $u(t, x)$ of the Cauchy problem for Equation 3.17 with an arbitrary initial data,

$$u\big|_{t=0} = u_0(x)$$

is given by the convolution of the data and the fundamental solution 3.21:

$$u(t, x) = E * u_0 \equiv \int_{\mathbf{R}^n} u_0(y)E(t, x - y)dy.$$

Hence,

$$u(t, x) = (2\sqrt{\pi t})^{-n} \int_{\mathbf{R}^n} u_0(y) \exp(-|x - y|^2/(4t))dy, \quad t > 0.$$

3.3.   Wave equation

3.3.1. Symmetry algebra of the equation

Consider the wave equation with $n > 1$ spatial variables:

$$u_{tt} = \Delta u \qquad (3.23)$$

where Δ is the n-dimensional Laplacian in the variables $x = (x^1, ..., x^n) \in \mathbf{R}^n$.

The maximal Lie algebra admitted by Equation 3.23 is composed of the finite-dimensional subalgebra spanned by

$$X_0 = \frac{\partial}{\partial t}, \quad X_i = \frac{\partial}{\partial x^i}, \quad X_{ij} = x^j \frac{\partial}{\partial x^i} - x^i \frac{\partial}{\partial x^j}, \quad X_{0i} = t \frac{\partial}{\partial x^i} + x^i \frac{\partial}{\partial t},$$

$$Z_1 = t \frac{\partial}{\partial t} + x^i \frac{\partial}{\partial x^i}, \quad Z_2 = u \frac{\partial}{\partial u}, \quad Y_0 = (t^2 + |x|^2) \frac{\partial}{\partial t} + 2tx^i \frac{\partial}{\partial x^i} - (n-1)tu \frac{\partial}{\partial u},$$

$$Y_i = 2tx^i \frac{\partial}{\partial t} + (2x^i x^j + (t^2 - |x|^2)\delta^{ij}) \frac{\partial}{\partial x^j} - (n-1)x^i u \frac{\partial}{\partial u}, \quad i, j = 1, ..., n,$$

and the infinite-dimensional ideal consisting of the operators

$$X_\tau = \tau(t, x) \frac{\partial}{\partial u},$$

where $\tau(t, x)$ is an arbitrary solution of Equation 3.23.

In what follows, we let $\tau(t, x) = 0$ and use only the $[\frac{1}{2}(n + 2)(n + 3) + 1]$-dimensional symmetry algebra spanned by

$$X_0, \ X_i, \ X_{ij}, \ X_{0i}, \ Z_1, \ Z_2, \ Y_0, \ Y_i, \quad i, j = 1, ..., n. \tag{3.24}$$

3.3.2. Fundamental solution of the Cauchy problem

LEMMA 3.1. *The Cauchy problem with arbitrary initial conditions,*

$$u_{tt} - \Delta u = 0, \quad t > 0,$$

$$u\big|_{t=0} = u_0(x), \quad u_t\big|_{t=0} = u_1(x) \tag{3.25}$$

reduces to the following special Cauchy problem:

$$u_{tt} - \Delta u = 0, \quad u\big|_{t=0} = 0, \quad u_t\big|_{t=0} = h(x). \tag{3.26}$$

PROOF. Let $v(t, x)$ and $w(t, x)$ be the solutions of the problem 3.26 with $h(x) = u_0(x)$ and $h(x) = u_1(x)$, respectively. Then the solution $u(t, x)$ to the general problem 3.25 is given by

$$u(t, x) = w(t, x) + \frac{\partial v(t, x)}{\partial t}. \tag{3.27}$$

DEFINITION 3.7. *The distribution $u = E(t, x)$ is called the fundamental solution of the Cauchy problem for the wave equation if it solves Equation 3.23 for $t > 0$ and satisfies the initial conditions*

$$u\big|_{t=0} = 0 \tag{3.28}$$

and

$$u_t\big|_{t=0} = \delta(x), \tag{3.29}$$

where

$$u\big|_{t=0} \equiv \lim_{t \to +0} E(t, x), \quad u_t\big|_{t=0} \equiv \lim_{t \to +0} \frac{\partial E(t, x)}{\partial t}.$$

REMARK.   The solution $u(t, x)$ to the problem 3.26 is given by the convolution (in the spatial variables x) of the data with the fundamental solution:

$$u(t, x) = E * h(x).$$

3.3.3.   The symmetry of the initial data

THEOREM 3.4.   *Let L be the Lie algebra with the basis 3.24, and K its maximal subalgebra admitted by the initial conditions 3.28 and 3.29. Then K is the linear span of the operators*

$$X_{ij}, \ X_{0i}, \ Z_1 + (1 - n)Z_2, \ Y_0, \ Y_i, \quad i, j = 1, ..., n. \qquad (3.30)$$

PROOF.   In accordance with Definition 3.5, we require the invariance of the equations $t = 0$, $x = 0$ (see the proof of Theorem 3.2). Under this requirement, the operators 3.24 are restricted to the generators of rotations, Lorentz transformations, dilations, and conformal transformations:

$$X_{ij}, \ X_{0i}, \ Z_1, \ Z_2, \ Y_0, \ Y_i. \qquad (3.31)$$

It is easy to see that Equation 3.28 is invariant under the operators 3.31. Therefore we shall inspect the invariance of Equation 3.29 only. It is obvious that Equation 3.29 is invariant with respect to the rotations.

Turning to the Lorentz transformations, consider their generators X_{0i} in the prolonged form:

$$X_{0i} = t \frac{\partial}{\partial x^i} + x^i \frac{\partial}{\partial t} - u_i \frac{\partial}{\partial u_t}.$$

Under these operators, the variable u_t and the δ-function are subjected to the following infinitesimal transformations:

$$\bar{u}_t \approx u_t - a u_i, \quad \bar{\delta} \approx \delta.$$

In view of Equation 3.28, we have $u_i \big|_{t=0} = 0$. Hence, $\bar{u}_t \approx u_t$ at $t = 0$. Hence, Equation 3.29 is Lorentz invariant.

Likewise, one can verify that Equation 3.29 is invariant under the conformal transformations with the generators Y_0, Y_i.

Finally, consider the linear combination of the remaining operators 3.31 prolonged to u_t:

$$Z_1 + kZ_2 = t \frac{\partial}{\partial t} + x^i \frac{\partial}{\partial x^i} + ku \frac{\partial}{\partial u} + (k - 1)u_t \frac{\partial}{\partial u_t}.$$

It follows that

$$\overline{u}_t \approx u_t + a(k-1)u_t\,, \quad \overline{\delta} \approx \delta - an\delta.$$

Hence, Theorem 3.1 yields $k = 1 - n$.

3.3.4. Derivation of the fundamental solution from the invariance principle

We discuss here the case when n is odd. The fundamental solution for the wave equations with even n is easily obtained by Hadamard's method of descent (Hadamard [1923]) and has been presented in [H2], Chapter 7.

THEOREM 3.5. *The fundamental solution of the Cauchy problem for the wave equation 3.23 with an odd number n of spatial variables has the form:*

$$E(x,t) = \begin{cases} \frac{1}{2}\pi^{\frac{1-n}{2}}\delta^{(\frac{n-3}{2})}(\Gamma)\,, & n \geq 3, \\[2mm] \frac{1}{2}\theta(\Gamma)\,, & n = 1\,, \end{cases} \tag{3.32}$$

where

$$\Gamma = t^2 - |x|^2. \tag{3.33}$$

It is determined uniquely by the invariance principle. Namely, $E(t,x)$ given by Equation 3.32 is the only distribution which satisfies the initial conditions 3.28 and 3.29 and is invariant under the group of rotations, Lorentz transformations, and dilations with the infinitesimal generators

$$X_{ij}\,, \quad X_{0i}\,, \quad Z_1 + (1-n)Z_2\,, \quad i, j = 1, ..., n. \tag{3.34}$$

PROOF. (Cf. Proof of Theorem 3.3.) We first find a basis of invariants for the generators X_{ij}, X_{0i} of the isometric motions (rotations and Lorentz transformations). In the space of the variables (t, x, u), we have the following two independent invariants:

$$u, \quad \Gamma = t^2 - |x|^2.$$

Then we write the restriction of the last operator 3.34 to functions of these invariants as follows:

$$Z_1 + (1-n)Z_2 = 2\Gamma\frac{\partial}{\partial\Gamma} + (1-n)u\frac{\partial}{\partial u}. \tag{3.35}$$

Now, let us look for invariant distributions of the form

$$u = f(\Gamma).$$

The invariance under the dilation operator 3.35 yields the following first-order ordinary differential equation:

$$2\Gamma f'(\Gamma) + (n-1)f(\Gamma) = 0.$$

By setting $n = 2m + 1$, we rewrite it in the form of Equation 3.9:

$$\Gamma f'(\Gamma) + mf(\Gamma) = 0, \quad m = 0, 1, \dots.$$

Using the solution formulas 3.10 and 3.11 for this equation, we arrive at the following general form of distributions invariant under the operators 3.34:

$$u = C_1\theta(\Gamma) + C_2, \quad n = 1,$$

$$u = C_1\delta^{(\frac{n-3}{2})}(\Gamma) + C_2\Gamma^{\frac{1-n}{2}}, \quad n \geq 3.$$

The initial conditions 3.28 and 3.29, together with the known equations:

$$\lim_{t \to 0} \delta^{(\frac{n-3}{2})}(\Gamma) = 0, \qquad \lim_{t \to 0} \theta(\Gamma) = 0,$$

lead to

$$C_1 = \frac{1}{2}\pi^{\frac{1-n}{2}}, \quad C_2 = 0.$$

Thus, we have obtained (uniquely) the fundamental solution 3.32.

REMARK 1.   One can readily verify that the distribution 3.32 is also invariant under the operators Y from Basis 3.30, i.e., admits the Lie algebra K. However, we do not need here this *excess symmetry* of the fundamental solution.

REMARK 2.   The fundamental solution $\mathcal{E}$ of the wave operator, i.e., the solution of the equation

$$\left(\frac{\partial^2}{\partial t^2} - \Delta\right)\mathcal{E} = \delta(t, x),$$

is obtained by the formula

$$\mathcal{E} = \theta(t)E,$$

where $\theta(t)$ is the Heaviside function.

3.4. Wave equations with nontrivial conformal group

The classical wave equation 3.23 is an example of the equations mentioned in the title. We will show here that *the invariance principle* applies to all the wave equations with nontrivial conformal group.

The definition of the wave equation in a curved space-time as a conformally invariant equation is due to Penrose [1964] (a discussion of general properties) and to Ibragimov [1968], [1969a] (uniqueness of the conformally invariant equation in any space-time with nontrivial conformal group and non-uniqueness in all other space-times).

The solution of the Cauchy problem for the equations under consideration has been given in Ibragimov and Mamontov [1970], [1977] by adapting the Fourier transformation approach. The Lie group approach to the Cauchy problem and to the Huygens principle has been suggested in Ibragimov [1970]. The fundamental solution has been first constructed in Mamontov [1984] by an *ad hoc* method, and then discussed from the group theoretic viewpoint in Berest [1991] and Berest and Ibragimov [1994].

The reader interested in details of the conformal group analysis of the wave equation in Riemannian spaces is referred to Ibragimov [1983]. See also Chapters 4 and 7 of [H2], and Ibragimov [1992c]. For the notation and terminology used in this Section, see [H2], Section 3.1.

3.4.1. Riemannian spaces with nontrivial conformal group

Consider $(n + 1)$-dimensional Riemannian spaces $V_{(n+1)}$ of the hyperbolic signature $(+ - \cdots -)$ given by the fundamental metric forms

$$ds^2 = g_{ij}(x)dx^i dx^j.$$

Here $i, j = 0, 1, \cdots, n$. We will assume in what follows that $n \geq 3$.

DEFINITION 3.8. *A space $V_{(n+1)}$ is said to be a Riemannian space with nontrivial conformal group if, in $V_{(n+1)}$ and in every space conformal to $V_{(n+1)}$, the maximal group of conformal transformations does not reduce to the group of isometric motions.*

THEOREM 3.6. *$V_{(n+1)}$ is a space with nontrivial conformal group if and only* *if it is conformal to a space with the plane-wave metric:*

$$ds^2 = (dx^0)^2 - (dx^1)^2 - \sum_{i,j=2}^{n} a_{ij}(x^1 - x^0)dx^i dx^j. \tag{3.36}$$

Here, $\left[a_{ij}\right]$ *is an arbitrary positive definite matrix with entries depending on a single variable* $x^1 - x^0$.

This result is due to Bilyalov [1963] (for $n = 3$) and to Chupakhin [1979] (for $n > 3$). Detailed discussions are to be found in Petrov [1966], Chapter VII, and in Ibragimov [1983], Section 8.5.

3.4.2. The wave equation in $V_{(n+1)}$ and its symmetry algebra

THEOREM 3.7.   *Let* $V_{(n+1)}$ *be a space with nontrivial conformal group. Then any second-order linear differential equation in* $V_{(n+1)}$,

$$g^{ij}(x)u_{ij} + b^i(x)u_i + c(x)u = 0,$$

admitting the maximal group of conformal motions in $V_{(n+1)}$ *is equivalent to the following equation:*

$$\Delta_2 u + \frac{n-1}{4n} Ru = 0. \tag{3.37}$$

Here, $g_{ik}g^{kj} = \delta_i^j$, *$R$ is the scalar curvature of* $V_{(n+1)}$, *and* $\Delta_2 u$ *is Beltrami's second-order differential parameter defined by*

$$\Delta_2 u = g^{ij}(u_{ij} - \Gamma_{ij}^k u_k)$$

where u_k *and* u_{ij} *denote the partial derivatives of the first and the second order, respectively, and* Γ_{ij}^k *are the Christoffel symbols.*

Thus, in the spaces with nontrivial conformal group, Equation 3.37 is the only *conformally invariant* equation (up to equivalence transformations defined in [H2], Section 3.1.2); other equations admit only a subgroup of the conformal group. Another important property of this equation, related to the Huygens principle, underscores its physical significance.

In spaces with trivial conformal group, Equation 3.37 is also conformally invariant. However, it is not unique.

For the proof of Theorem 3.7 and for examples of conformally invariant equations different from Equation 3.37, in spaces with trivial conformal group, see Ibragimov [1983], Section 10.5.

DEFINITION 3.9. *The conformally invariant equation 3.37 is called the wave equation in* $V_{(n+1)}$.

THEOREM 3.8. *(Ibragimov [1983], Sections 12.2 and 13.1). Let the space* $V_{(n+1)}$ *with nontrivial conformal group be represented by a plane-wave metric 3.36. Then the wave equation 3.37 can be taken in the following form:*

$$u_{tt} - u_{x^1 x^1} - \sum_{i,j=2}^{n} a^{ij}(x^1 - t)u_{x^i x^j} = 0, \tag{3.38}$$

where $t = x^0$ *and* $a_{ik}a^{kj} = \delta_i^j$, $i, j = 2, ..., n$.

For Equation 3.38 with arbitrary coefficients $a^{ij}(x^1 - t)$, the maximal finite-dimensional symmetry algebra is spanned by the following $2n + 1$ operators (cf. Basis 3.24):

$$X_1 = \frac{\partial}{\partial t} + \frac{\partial}{\partial x^1}, \quad X_i = \frac{\partial}{\partial x^i},$$

$$Y_i = x^i\left(\frac{\partial}{\partial t} + \frac{\partial}{\partial x^1}\right) - \sum_{j=2}^{n} A^{ij}(x^1 - t)\frac{\partial}{\partial x^j}, \quad i = 2, ..., n,$$

$$Z_1 = (t + x^1)\left(\frac{\partial}{\partial t} + \frac{\partial}{\partial x^1}\right) + \sum_{i=2}^{n} x^i\frac{\partial}{\partial x^i}, \quad Z_2 = u\frac{\partial}{\partial u}. \tag{3.39}$$

Here, the functions $A^{ij}(x^1 - t)$ are defined by

$$A^{ij}(\sigma) = \int a^{ij}(\sigma)d\sigma. \tag{3.40}$$

3.4.3. Fundamental solution of the Cauchy problem

LEMMA 3.2. *For Equation 3.38, the Cauchy problem with arbitrary initial conditions*

$$u\big|_{t=0} = u_0(x), \quad u_t\big|_{t=0} = u_1(x) \tag{3.41}$$

reduces to the Cauchy problem with the following special initial data:

$$u\big|_{t=0} = 0, \quad u_t\big|_{t=0} = h(x). \tag{3.42}$$

Here,

$$x = (x^1, ..., x^n).$$

PROOF.   Let $v(t, x)$ and $w(t, x)$ be the solutions of the special Cauchy problem with

$$h(x) = u_0(x)$$

and

$$h(x) = \frac{\partial u_0(x)}{\partial x} - u_1(x),$$

respectively. Then the function

$$u(t, x) = \frac{\partial v(t, x)}{\partial t} + \frac{\partial v(t, x)}{\partial x} - w(t, x) \tag{3.43}$$

solves Equation 3.38 and satisfies the initial conditions 3.41.

DEFINITION 3.10.   *The distribution $u = E(t, x; t_0, x_0)$ is called the fundamental solution of the Cauchy problem for the wave equation 3.38, at the point (t_0, x_0), if it solves Equation 3.38 for $t > t_0$ and satisfies the initial conditions*

$$u\big|_{t=t_0} = 0, \quad u_t\big|_{t=t_0} = \delta(x - x_0). \tag{3.44}$$

3.4.4.   The symmetry of the initial data

THEOREM 3.9.   *Let L be the Lie algebra with the basis 3.39, and K its maximal subalgebra admitted by the initial conditions 3.44. Then K is the linear span of the following n operators:*

$$\tilde{Y}_i = (x^i - x_0^i)\left(\frac{\partial}{\partial t} + \frac{\partial}{\partial x^1}\right) - \sum_{j=2}^{n}\left(A^{ij}(x^1 - t) - A^{ij}(x_0^1 - t_0)\right)\frac{\partial}{\partial x^j}, i = 2, ..., n,$$

$$Z = (t - t_0 + x^1 - x_0^1)\left(\frac{\partial}{\partial t} + \frac{\partial}{\partial x^1}\right) + \sum_{k=2}^{n}(x^k - x_0^k)\frac{\partial}{\partial x^k} + (1 - n)u\frac{\partial}{\partial u}. \tag{3.45}$$

PROOF.   The natural adaptation of the proof of Theorem 3.4 applies here.

3.4.5. Derivation of the fundamental solution from the invariance principle

THEOREM 3.10. The fundamental solution of the Cauchy problem for the wave equation 3.38 with an odd $n \geq 3$ has the form:

$$E(t, x; t_0, x_0) = \frac{1}{2} \pi^{\frac{1-n}{2}} h(\sigma) \delta^{(\frac{n-3}{2})}(\Gamma). \tag{3.46}$$

Here,

$$\Gamma = (t - t_0)^2 - (x^1 - x_0^1)^2 - (t - t_0 - x^1 + x_0^1) \sum_{i,j=2}^{n} \tilde{A}_{ij}(x^i - x_0^i)(x^j - x_0^j) \tag{3.47}$$

is the geodesic distance between the points $(t, x) \in V_{(n+1)}$ and $(t_0, x_0) \in V_{(n+1)}$, and

$$h(\sigma) = \left[\frac{|\sigma - \sigma_0|^{n-1}}{|\det \|\tilde{A}^{ij}(\sigma)\||} \right]^{\frac{1}{2}}, \tag{3.48}$$

where

$$\|\tilde{A}_{ij}\| = \|A^{ij}(x^1 - t) - A^{ij}(x_0^1 - t_0)\|^{-1}.$$

It is determined uniquely by the invariance principle. Namely, $E(t, x)$ given by Equation 3.46 is the only distribution which satisfies the initial conditions 3.42 and is invariant under the n-parameter group with the infinitesimal generators 3.45.

PROOF. (See Proof of Theorem 3.5.) A basis of invariants for the generators $\tilde{Y}_i$ of isometric motions in $V_{(n+1)}$ consists of the following three independent invariants:

$$u, \quad \Gamma, \quad \sigma = x^1 - t,$$

where Γ is given by Equation 3.47 (its calculation is found in Ibragimov [1983], Section 12.1). Therefore, we look for the distributions, invariant under the operator Z of Basis 3.45, given in the general form:

$$u = f(\Gamma, \sigma).$$

We note that σ is an invariant for the operator Z and that $Z(\Gamma) = 2\Gamma$. It follows that the invariance condition under the operator Z is written as the first-order differential equation obtained in Section 3.3.4:

$$2\Gamma \frac{\partial f}{\partial \Gamma} + (n - 1)f(\Gamma) = 0. \tag{3.49}$$

In accordance with Section 3.3.4, the general solution of Equation 3.49 is given by

$$u = C_1(\sigma)\delta^{(\frac{n-3}{2})}(\Gamma) + C_2(\sigma)\Gamma^{\frac{1-n}{2}}$$

with arbitrary functions $C_1(\sigma)$ and $C_2(\sigma)$. The first initial condition 3.42 yields $C_2(\sigma) = 0$ (see Section 3.3.4), while the second condition 3.42 leads to

$$C_1(\sigma) = \frac{1}{2}\pi^{\frac{1-n}{2}}h(\sigma),$$

where $h(\sigma)$ is given by Equation 3.48. Thus, we have obtained (uniquely) the fundamental solution 3.46.

THEOREM 3.11.   *The fundamental solution of the Cauchy problem for the wave equation 3.38 with an even $n > 2$ is uniquely determined by the invariance principle and has the form:*

$$E(t, x; t_0, x_0) = \frac{1}{2}\pi^{-\frac{n}{2}}h(\sigma)\left(\frac{\theta(\Gamma)}{\sqrt{\Gamma}}\right)^{(\frac{n-2}{2})}. \tag{3.50}$$

PROOF.   The proof is similar to the case of odd n. The formula 3.49 is also obtained by the method of descent (for details of an application of this method to the solution of the Cauchy problem, see Ibragimov and Mamontov [1977] or Ibragimov [1983], Section 13.3).

4

Recursions

The object of this chapter[1] is a discussion of newly developed methods for obtaining infinitely many Lie-Bäcklund (see Chapter 1) and generalized symmetries of partial differential equations.

An algorithmic way to construct an infinite (countable) set of symmetries is to derive a recursion relation between symmetries. Usually, recursions are linear with respect to symmetries and hence, they are given by linear operators. These linear operators are known in the literature as *squared eigenfunction operators* (Ablowitz, Kaup, Newell, and Segur [1974]), *recursion operators* (Olver [1977], [1986], Ibragimov and Shabat [1979], [1980a], Ibragimov [1983]), *strong* or *hereditary symmetries* (Fuchssteiner [1979], Zakharov and Konopelchenko [1984]), *Kähler operators* (Magri [1978]), and *regular operators* (Gel'fand and Dorfman [1979], [1980]).

A method of finding linear and multilinear recursions is provided by the master symmetries approach (Fuchssteiner [1983]). This approach utilizes the structure of a Lie algebra of vector fields associated with partial differential equations.

A recursion operator plays at least two additional important roles. First, it generates infinite hierarchies of equations integrable by a spectral problem in the context of the inverse scattering transform method. Second, it generates, in the case of Hamiltonian equations, infinite sets of Hamiltonian structures associated with these equations.

4.1. Notation

(Cf. Chapter 1)
Consider evolution (system of) equations of the form

$$u_t = K(u, x, t). \tag{4.1}$$

---

[1]Research supported in part by International Science Foundation under Grant R5B000.

Here, $x = (x^1, ..., x^d)$, $u = (u^1, ..., u^n)$ is an element of the linear space S of n-dimensional (real) vector-functions $u(x, t)$ vanishing rapidly at $|x| \to \infty$, and $K : S \to S$ is a C^∞-map that may depend explicitly on x and t.

The derivative of K in the direction $v \in S$ is the Fréchet derivative:

$$K'(u, x, t)[v] = \frac{\partial}{\partial \epsilon} K(u + \epsilon v, x, t)\big|_{\epsilon=0}. \qquad (4.2)$$

Let $K = (K_1, \ldots, K_n)$ be a vector such that its components K_α are differential functions (see Chapter 1):

$$K_\alpha = K_\alpha(u, u_{(1)}, ..., u_{(m)}), \quad \alpha = 1, \ldots, n,$$

where $u_{(k)}$ denotes the set of k-order partial derivatives of u^α with respect to x^i. Then the Formula 4.2 coincides with the expressions $K_*(v, x, t)$ of Ibragimov [1983] and $D_k(v, x, t)$ of Olver [1986], viz.

$$(K')_{\alpha\beta} \equiv (D_k)_{\alpha\beta} \equiv (K_*)_{\alpha\beta} = \sum_J \left(\frac{\partial K_\alpha}{\partial u_J^\beta}\right) D_J, \quad \alpha, \beta = 1, \ldots, m. \qquad (4.3)$$

Here, $J = (j_1, ..., j_l)$ is a multi-index with $1 \le j_s \le n$,

$$u_J^\beta(x, t) = \frac{\partial^l u^\beta(x, t)}{\partial x^{j_1} ... \partial x^{j_l}},$$

and

$$D_J = D_{j_1} ... D_{j_l},$$

where D_{j_s} is the total derivative with respect to x^{j_s} (cf. Chapter 1, Formula 1.2). The summation in Equation 4.3 extends up to $l = k$.

Example 1. Korteweg-de Vries equation ($d = 1$):

$$u_t = u_{xxx} + 6uu_x. \qquad (4.4)$$

Here, $K(u) = u_{xxx} + 6uu_x$, and according to Formula 4.3,

$$K'(u) = D_x^3 + 6u D_x + 6u_x.$$

The following example illustrates the abstract form of a system 4.1 where $K(u)$ is not a differential function.

Example 2. Benjamin-Ono equation ($d = 1$):

$$u_t = Hu_{xx} + 2uu_x. \tag{4.5}$$

Here, H is the Hilbert transform given by

$$(Hf)(x) = \frac{1}{\pi} \int\limits_{-\infty}^{+\infty} \frac{f(\xi)}{\xi - x} d\xi, \tag{4.6}$$

where the principal value of the integral is implied. In this example, $K(u) = Hu_{xx} + 2uu_x$, and according to Formula 4.2,

$$K'[v] = Hv_{xx} + 2(uv)_x.$$

For functions $\gamma(u, x, t)$ and $\sigma(u, x, t)$ defined on S and vanishing sufficiently fast for $|x| \to \infty$, the scalar product is given by

$$\langle \gamma, \sigma \rangle = \int\limits_{\mathbf{R}^d} \gamma(u(x, t), x, t)\sigma(u(x, t), x, t)dx. \tag{4.7}$$

Let S^* be the dual space to S with respect to the bilinear form 4.7. Thus, for $\gamma \in S^*$ and $\sigma \in S$, the expression $\langle \gamma, \sigma \rangle$ is an application of the linear functional γ to σ.

Given an operator $T: S \to S$, the adjoint (transpose) operator $T^+ : S^* \to S^*$ is defined by the equation

$$\langle a, Tb \rangle = \langle T^+ a, b \rangle$$

for all $a \in S^*$, $b \in S$. An operator T is called symmetric if

$$\langle a, Tb \rangle = \langle Ta, b \rangle,$$

and skew-symmetric if

$$\langle a, Tb \rangle = -\langle Ta, b \rangle$$

for all $a \in S^*$, $b \in S$.

An operator $\theta: S^* \to S$ is called symmetric if

$$\langle a, \theta b \rangle = \langle b, \theta a \rangle,$$

and skew-symmetric if

$$\langle a, \theta b \rangle = -\langle b, \theta a \rangle$$

for all $a, b \in S^*$. For operators $J: S \to S^*$ the definitions are similar.

A function $\gamma: S \to S^*$ (it may depend explicitly on x, t) is said to be a *gradient* if it has a *potential* P, i.e., a map $P: S \to \mathbb{R}$ such that

$$\langle \gamma(u, x, t), v \rangle = P'(u, x, t)[v] \tag{4.8}$$

for all $u, v \in S$. The functional P may depend explicitly on x, t. We write $\gamma = \operatorname{grad} P$. Thus, the equation

$$\langle \operatorname{grad} P, v \rangle = P'[v] \tag{4.8'}$$

is a definition of a gradient of $P(u, x, t)$ with respect to u.

A function $\gamma(u, x, t)$ is a gradient, iff

$$\gamma'^{+} = \gamma',$$

i.e., if the linear operator $\gamma'(u, x, t)[\cdot]$ is symmetric. Then the potential P of γ is given by the homotopy formula

$$P(u, x, t) = \int_0^1 \langle \gamma(\lambda u, x, t), u \rangle d\lambda. \tag{4.9}$$

In the case of classical variational calculus, the functional P has the form

$$P = \int_{\mathbb{R}^d} \rho(u, u_{(1)}, ..., u_{(m)}, x, t) dx, \tag{4.10}$$

where ρ (a *Lagrangian*) is a differential function. In this particular case, the gradient coincides with the variational derivative of P:

$$\gamma = \operatorname{grad} P = \frac{\delta P}{\delta u^\alpha}, \tag{4.11}$$

where (in the notation used in Formula 4.3)

$$\frac{\delta}{\delta u^\alpha} = \sum_J (-D)_J \frac{\partial}{\partial u_J^\alpha} \tag{4.12}$$

is the Euler-Lagrange operator (see [H1], Section 6.2).

Specifically, if x and u are one-dimensional, and if

$$P = \int_{-\infty}^{+\infty} \rho(u, u_x, ..., u_{x...x}^{(M)}, x, t)dx \qquad (4.10')$$

then

$$\operatorname{grad} P = \frac{\partial \rho}{\partial u} - D_x \left(\frac{\partial \rho}{\partial u_x} \right) + D_x^2 \left(\frac{\partial \rho}{\partial u_{xx}} \right) - \cdots + (-1)^M D_x^M \left(\frac{\partial \rho}{\partial u_{x...x}^{(M)}} \right). \quad (4.13)$$

Example 3. Let

$$P = \int_{-\infty}^{+\infty} \left(-\frac{u_x^2}{2} + u^3 \right) dx. \qquad (4.14)$$

Then $P'[v] = \langle u_{xx} + 3u^2, v \rangle$. Hence

$$\operatorname{grad} P = u_{xx} + 3u^2. \qquad (4.15)$$

If $\gamma = u_{xx} + 3u^2$, then $\gamma' = D_x^2 + 6u = (\gamma')^+$. Hence the condition for γ to be a gradient function is satisfied.

Example 4. Let

$$P = \int_{-\infty}^{+\infty} \left(\frac{1}{2} u H u_x + \frac{u^3}{3} \right) dx, \qquad (4.16)$$

where H is the Hilbert transform 4.6. Here, P is not of the form 4.10, since the density ρ is not a differential function. Equation 4.8' yields:

$$\operatorname{grad} P = H u_x + u^2. \qquad (4.17)$$

4.2. Basic notions

Here we classify, as the basic notions, symmetries, conservation laws, recursion operators, and master symmetries. The notions, which refer to Hamiltonian structure, are treated in Section 4.3.

4.2.1. Symmetries

Consider a one-parameter Lie-Bäcklund transformation group G generated by a canonical Lie-Bäcklund operator (see Chapter 1):

$$\hat{X}_\sigma = \sum_{\alpha, J} D_J[\sigma_\alpha] \frac{\partial}{\partial u_J^\alpha}, \qquad (4.18)$$

where $\sigma = \{\sigma_\alpha\}$, $\sigma_\alpha = \sigma_\alpha(u, u_{(1)}, ..., u_{(m)}, x, t) \equiv \sigma_\alpha[u, x, t]$. We also say that the group G is generated by the Lie-Bäcklund equation:

$$u_\tau^\alpha = \sigma_\alpha(u, u_{(1)}, ..., u_{(m)}, x, t), \qquad (4.19)$$

where τ is the group parameter.

In particular, if x and u are the one-dimensional variables, then

$$\hat{X}_\sigma = \sigma \frac{\partial}{\partial u} + (D_t[\sigma]) \frac{\partial}{\partial u_t} + (D_x[\sigma]) \frac{\partial}{\partial u_x} + (D_x^2[\sigma]) \frac{\partial}{\partial u_{xx}} + ... \qquad (4.20)$$

The group G with the generator 4.18 is admitted by Equation 4.1 iff (see Chapter 1)

$$\hat{X}_\sigma (u_t - K(u, x, t))\big|_{u_t = K} = 0. \qquad (4.21)$$

If the infinitesimal criterion 4.21 is satisfied, then the infinitesimal transformation $u \to u + \epsilon \sigma$ leaves Equation 4.1 form invariant and $v(t) = \sigma[u, x, t]$ is a solution of the perturbation equation

$$v_t = K'(u, x, t)[v], \quad v \in S. \qquad (4.22)$$

The condition 4.21 provides the invariance of the solution manifold of Equation 4.1 with respect to G (cf. [H2], Chapter 1).

Equation 4.21 is equivalent to the commutability of the flows 4.1 and 4.19. Hence, one can define a symmetry of evolution equations, in a more general context, as follows (see, e.g., Fuchssteiner [1979]).

DEFINITION 4.1. *A function σ is called a symmetry of a system of evolution equations 4.1 if the flow*

$$u_\tau = \sigma(u, x, t) \qquad (4.19')$$

commutes with the flow 4.1, i.e., if σ satisfies the equation

$$\frac{\partial \sigma}{\partial t} + [\![\sigma, K]\!] = 0. \qquad (4.23)$$

Here, $\partial\sigma/\partial t$ denotes the partial derivative of σ with respect to its explicit dependence on t, and $[\![\sigma, K]\!]$ is the commutator defined by

$$[\![\sigma(u, x, t), K(u, x, t)]\!] = \sigma'[K] - K'[\sigma] \qquad (4.24)$$

$$\equiv \frac{\partial}{\partial\epsilon}[\sigma(u + \epsilon K(u, x, t), x, t) - K(u + \epsilon\sigma(u, x, t), x, t)]\big|_{\epsilon=0}.$$

In particular, if σ does not depend explicitly on t, Equation 4.23 has the form

$$[\![\sigma, K]\!] = 0. \qquad (4.25)$$

If σ and K are differential functions, then the obvious relation

$$\sigma'[K] = \hat{X}_K(\sigma) \qquad (4.26)$$

is valid and Equation 4.23 can be written in the form

$$\frac{\partial\hat{X}_\sigma}{\partial t} + [\hat{X}_K, \hat{X}_\sigma] = 0. \qquad (4.27)$$

In Equation 4.27, $\hat{X}_K$ and $\hat{X}_\sigma$ are canonical Lie-Bäcklund operators 4.18. They generate the flows 4.1 and 4.19, respectively. Furthermore, $[\hat{X}_K, \hat{X}_\sigma]$ is the usual Lie bracket of Lie-Bäcklund operators (see Chapter 1). Hence, the Lie bracket and the commutator 4.24 are connected by the equation

$$[\hat{X}_K, \hat{X}_\sigma] = \hat{X}_{[\![\sigma, K]\!]}. \qquad (4.28)$$

It follows that, for differential functions, $[\![\sigma, K]\!]$ coincides with the bracket $\{\sigma, K\} = \sigma_* K - K_* \sigma$ of Ibragimov [1981], [1983]. Here, the notation 4.24 is taken for this bracket in order to distinguish it from the Poisson bracket for Hamiltonian systems.

Equation 4.23 is known as *the determining equation* for symmetries. Given K, the determining equation is a system of linear homogeneous first-order partial differential equations for σ:

$$\left(\frac{\partial}{\partial t} + \hat{X}_K - K'\right)[\sigma] = 0, \qquad (4.23')$$

provided that K and σ are differential functions.

Example 1. For the Korteweg-de Vries equation 4.4, the function

$$\sigma = u_{xxxxx} + 10uu_{xxx} + 20u_x u_{xx} + 30u^2 u_x \qquad (4.29)$$

is a symmetry given by a differential function, and hence it is a Lie-Bäcklund symmetry (see, e.g., Ibragimov and Shabat [1979]).

Example 2. For the Benjamin-Ono Equation 4.5, the function

$$\sigma = (2u^3 + 3H(uu_x) + 3uHu_x - 2u_{xx})_x, \qquad (4.30)$$

is a symmetry (Bock and Kruskal [1979]). It follows from the definition of the Hilbert transform 4.6 that σ is not a differential function. Hence, it is not a Lie-Bäcklund symmetry.

4.2.2. Conservation laws

Let $P: S \to \mathbb{R}$ be a functional.

DEFINITION 4.2. *A functional $P(u, x, t)$ is said to be conserved by the flow 4.1, if*

$$D_t[P]\big|_{u_t=K} = 0,$$

or, equivalently, if

$$\frac{\partial P}{\partial t} + \langle \gamma, K \rangle = 0, \quad \gamma = grad P. \qquad (4.31)$$

The vector $\gamma \in S^*$ is called a conserved gradient for Equation 4.1.

Let the functional P be given by the integral 4.10. Then Equation 4.31 can be written in the differential form:

$$D_t[\rho]\big|_{u_t=K} + \mathrm{Div}[V] = 0. \qquad (4.32)$$

The differential function ρ is called a conserved density of order m, and $V = \{V_i\}$ is known as a flux. Here,

$$\mathrm{Div}[V] = \sum_{i=1}^{d} D_i[V_i].$$

Equation 4.31 is valid with γ defined by Equation 4.11:

$$\gamma = \frac{\delta P}{\delta u}. \tag{4.33}$$

Hence, the Euler-Lagrange operator maps a conserved density ρ into a conserved gradient γ.

The conserved gradient γ coincides with the characteristic of a conservation law (Olver [1986]), written in the following *characteristic form*:

$$D_t[\rho] + \text{Div}[V] = \gamma \cdot [u_t - K(u, x, t)]. \tag{4.34}$$

Thus, γ is an analog of an integrating factor well known for ordinary differential equations.

DEFINITION 4.3. *A function $\gamma: S \rightarrow S^*$ (it may depend explicitly on t, x) is called a conserved covariant for Equation 4.1 if it satisfies the equation (see Fuchssteiner and Fokas [1981]):*

$$\frac{\partial \gamma}{\partial t} + \gamma'[K] + (K')^+[\gamma] = 0. \tag{4.35}$$

THEOREM 4.1. *Let the function γ be a gradient, i.e.,*

$$(\gamma')^+ = \gamma'. \tag{4.36}$$

If $\gamma = \text{grad} P$ is a covariant of Equation 4.1, then it is a conserved gradient of Equation 4.1, and the potential functional P is conserved by Equation 4.1.

If P is given by the integral 4.10 and if K in Equation 4.1 is a differential function, then Equation 4.35 is equivalent to the corresponding equations of Ibragimov [1983] (with one-dimensional x and u):

$$D_t[\gamma]\big|_{u_t=K} + \sum_{j=0}^{m}(-D_x)^j\left[\frac{\partial K}{\partial u_{x...x}^{(j)}} \cdot \gamma\right] = 0, \tag{4.35'}$$

and of Olver [1986] (for $(\gamma')^+ = \gamma'$):

$$\frac{\partial \gamma}{\partial t} + D_K^+[\gamma] + D_\gamma^+[K] = 0. \tag{4.35''}$$

The determining equation (4.35″) for integrals can be written in the form

$$(\frac{\partial}{\partial t} + \hat{X}_K - K')^+[\gamma] = 0. \tag{4.35'''}$$

THEOREM 4.2. *(cf. Ibragimov [1983], Section 22.5). The determining equations 4.23′ and 4.35‴ for symmetries and integrals, respectively, are mutually adjoint.*

REMARK.   The function $W(t) = \gamma(u, x, t)$ is a conserved gradient for Equation 4.1 iff it is a solution of the adjoint perturbation equation (cf. Equation 4.22):

$$W_t = -(K')^+(u, x, t)[W], \quad W \in S^*. \tag{4.37}$$

Example 1. (see Fuchssteiner and Fokas [1981]). For the KdV equation 4.4, the quantity

$$P = \int\limits_{-\infty}^{+\infty} \left(\frac{u_{xx}^2}{2} + \frac{5}{2}u^2 u_{xx} + \frac{10}{4}u^4\right) dx \tag{4.38}$$

is a conserved functional. The corresponding conserved gradient of P is

$$\gamma = \operatorname{grad} P = u_{xxxx} + 10uu_{xx} + 5u_x^2 + 10u^3. \tag{4.39}$$

Example 2. (see Fuchssteiner [1983]). For the Benjamin-Ono equation 4.5, the functional

$$P = \int\limits_{-\infty}^{+\infty} \left(-uu_{xx} - \frac{3}{2}u_x Hu^2 + \frac{1}{2}u^4\right) dx, \tag{4.40}$$

where H is Hilbert transform 4.6, is conserved by the flow 4.5 and the conserved gradient of P is

$$\gamma = \operatorname{grad} P = 2u^3 + 3H(uu_x) + 3uHu_x - 2u_{xx}. \tag{4.41}$$

Note that the conserved density and gradient in Equations 4.38 and 4.39, respectively, are differential functions, while they are not differential functions for Equations 4.40 and 4.41.

4.2.3. Recursion operators

DEFINITION 4.4.   *A function R from S into the space of operators $S \to S$ is called a recursion operator or a strong symmetry for Equation 4.1 if it maps* *symmetries of Equation* 4.1 *into symmetries of the same equation (see Olver [1977], Ibragimov and Shabat [1979, 1980a], Fuchssteiner [1979]).*

THEOREM 4.3. *A necessary and sufficient condition for* $R = R(u, x, t)$ *to be a recursion operator of Equation 4.1 is that* R *commutes, when restricted on the solution manifold of Equation 4.23, with the operator of the determining equation 4.23 for symmetries. This condition is written:*

$$R_t + R'[K] - [K', R] = 0. \tag{4.42}$$

The zero in the right-hand side of Equation 4.42 means an operator, which annihilates any solution σ *of Equation 4.23.*

Example. For the KdV equation 4.4, a recursion operator due to A. Lenard (see Olver [1977], Ibragimov and Shabat [1979], Fuchssteiner and Fokas [1981]) is given by

$$R = D_x^2 + 4u + 2u_x D_x^{-1}. \tag{4.43}$$

Here and in what follows, D_x^{-1} is defined by

$$(D_x^{-1} f)(x) = \int_{-\infty}^{x} f(\xi) d\xi.$$

THEOREM 4.4. *Let* $R(u)$ *be a recursion operator. Then the adjoint operator* $R^+(u)$ *maps conserved covariants into conserved covariants.*

It follows that the operator R^+ generates solutions $\tilde{\gamma}$ of the determining equation 4.35 for integrals from a given solution γ of this equation:

$$\tilde{\gamma} = R^+(u)[\gamma].$$

However, $\tilde{\gamma}$ may be not a gradient function because the constraint $(\tilde{\gamma}')^+ = \tilde{\gamma}'$ may not be satisfied.

4.2.4. Lie-Bäcklund and generalized symmetries

A recursion operator R can generate an infinite set of symmetries. Indeed, if σ is a symmetry and R is a recursion operator for Equation 4.1, then $R^j[\sigma]$ ($j = 1, 2, ...$) are also symmetries of the same equation.

DEFINITION 4.5. *A symmetry σ of Equation 4.1 is called a Lie-Bäcklund symmetry if σ is a differential function of an arbitrary finite order (see Chapter 1). We call σ a generalized symmetry if it is not a differential function.*

Equation 4.30 gives an example of a generalized symmetry σ.

The following example provides first-order Lie-Bäcklund symmetries and recursion operators for semi-Hamiltonian systems.

Example. A first-order diagonal quasilinear system

$$u^i_t = v_i(u)u^i_x \quad (i = 1, 2, ..., n; \quad n \geq 3) \tag{4.44}$$

is called semi-Hamiltonian (Tsarev [1985]) if $v_i \neq v_j$ for $i \neq j$ and if

$$[v_{i,u^j}/(v_j - v_i)]_{,u^k} = [v_{i,u^k}/(v_k - v_i)]_{,u^j} \tag{4.45}$$

for $i \neq j \neq k \neq i$. Here, the subscripts t, x, u^j denote the partial derivations with respect to these variables. For a semi-Hamiltonian system (4.44) a continuum set of first-order symmetries is generated by Lie equations:

$$u^i_\tau = w_i(u)u^i_x \quad (i = 1, 2, ..., n), \tag{4.46}$$

where $w_i(u)$ satisfies the linear system

$$w_{i,u^j} = \Gamma^i_{ij}(u)(w_j - w_i) \quad (j \neq i) \tag{4.47}$$

with $\Gamma^i_{ij} = v_{i,u^j}/(v_j - v_i)$ $(j \neq i)$. Hence,

$$\sigma = \{w_i(u)u^i_x\}.$$

A recursion operator of the form

$$R = (AD_x + B)U_x^{-1}, \tag{4.48}$$

where $A(u)$, $B(u, u_x)$ are $n \times n$ matrices, and

$$U_x = \begin{pmatrix} u^1_x & & 0 \\ & \ddots & \\ 0 & & u^n_x \end{pmatrix},$$

is given in Teshukov [1989] for symmetries of Equation 4.44 under the following additional constraints:

$$S_{i,u^j} = \Gamma^i_{ij}(S_j - S_i) \quad (j \neq i) \tag{4.49}$$

where

$$S_i(u) = \sum_{k=1}^{n} c_k(u^k)\Gamma^i_{ik}(u) + d_i(u^i).$$

The recursion operator (4.48) generates the recursion formula

$$\bar{w}_i(u) = c_i(u^i)w_{i,u^i}(u) + d_i(u)w_i(u) + \sum_{k=1}^{n} c_k(u^k)\Gamma^i_{ik}(u)w_k(u). \tag{4.50}$$

Under the conditions 4.49, $\bar{\sigma}_i = \bar{w}_i(u)u^i_x$ is a first-order symmetry of Equations 4.44 if $\sigma_i = w_i(u)u^i_x$ is a symmetry.

In Sheftel' [1993], a second-order recursion operator is given in the form

$$R = (AD_x^2 + BD_x + C)U_x^{-1}, \tag{4.51}$$

where A, B, C are $n \times n$ matrices. Existence conditions of this recursion have the same form 4.49:

$$B_{i,u^j} = \Gamma^i_{ij}(B_j - B_i) \quad (j \neq i) \tag{4.52}$$

However, here $B_i(u)$ are different from $S_i(u)$. As a result, the constraints 4.52 are more liberal than Equation 4.49.

4.2.5. Infinite symmetry algebras and conservation laws

Here we consider the one-dimensional x and use the notation of Chapter 1.

Let the right-hand side of Equation 4.1 be a differential function of order m: $K = K(u, u_1, ..., u_m)$. Here K and u may be vectors. Then Equation 4.1 is a system of evolution equations.

DEFINITION 4.6. *Equation 4.1 is said to be formally integrable if a solution of Equation 4.42 for a recursion operator $R(u)$ exists and belongs to the class of formal operator power series*

$$\overset{(k)}{R} = P_k(u)D_x^k + ... + P_0(u) + P_{-1}(u)D_x^{-1} + P_{-2}(u)D_x^{-2} + ..., \tag{4.53}$$

where the coefficients $P_i(u)$ are differential functions. The integer k is termed the order of the operator $\overset{(k)}{R}$.

THEOREM 4.5. *(see Ibragimov and Shabat [1980a] through [1980c], and Ibragimov [1983], Section 19.3). The set of solutions 4.53 for Equation 4.42 is closed under the multiplication and extraction of a root.*

COROLLARY. *If Equation 4.1 is formally integrable, then there exists a first-order solution of Equation 4.42:*

$$L = \overset{(1)}{R} = r_1(u)D_x + r_0(u) + r_{-1}(u)D_x^{-1} + r_{-2}(u)D_x^{-2} + \cdots . \qquad (4.54)$$

DEFINITION 4.7. *The evolution equation 4.1 admits an infinite Lie-Bäcklund algebra if there exists a symmetry $\sigma(u, u_1, ..., u_N, x, t)$ of an arbitrarily high order N.*

THEOREM 4.6. *(Ibragimov and Shabat [1980a]-[1980c]). For any scalar equation 4.1, which admits an infinite Lie-Bäcklund algebra, Equation 4.42 for a recursion operator is solvable in the class of formal operator power series 4.53 of arbitrary order k.*

COROLLARY. *Under the conditions of Theorem 4.6, any solution of order m of Equation 4.42 has the form*

$$\overset{(m)}{R} = K' + pD_x + q + rD_x^{-1} + \cdots . \qquad (4.55)$$

For a motivation of the following definitions, see Mikhailov and Shabat [1985], [1986]; Mikhailov, Shabat, and Yamilov [1987]; Sokolov [1988]; and Shabat [1989].

DEFINITION 4.8. *The coefficient P_{-1} in the formal power series 4.53 (or r_{-1} in the series 4.54) is termed the residue of the operator R (or of L).*

DEFINITION 4.9. *The following differential functions of coefficients of the operator $L = \overset{(1)}{R}$ form the canonical series of residues of the operator $\overset{(k)}{R} = L^k$:*

$$\rho_{-1} = \frac{1}{r_1}, \quad \rho_0 = \frac{r_0}{r_1}, \quad \rho_i = res(L^i) \ (i = 1, 2, ...). \qquad (4.56)$$

In other words, the canonical series of residues of the operator $\overset{(k)}{R}$ is the set of *residues of its fractional powers:*

$$\rho_i = res(R^{\overset{(k)}{}\frac{i}{k}}), \quad i = -1, 1, 2, \dots.$$

(see Gel'fand and Dikii [1976] and Manin [1979]), supplemented by the logarithmic residue ρ_0.

THEOREM 4.7. *(Shabat and Sokolov [1982]). If Equation 4.1 possesses a conservation law of an arbitrarily high order, then Equation 4.42 for recursion operators is solvable in the class of the formal operator power series 4.53 of an arbitrary order k.*

THEOREM 4.8. *(Shabat [1989]). Let R be a recursion operator which satisfies Equation 4.42 and has the form 4.53. Then the traces of all canonical residues of a formal power series R are conserved densities for Equation 4.1:*

$$D_t[tr\ res\ln L] = D_x[V_0], \quad D_t[tr\ resL] = D_x[V_1],$$

$$D_t[tr\ resL^i] = D_x[V_i], \quad i = -1, 1, 2, \dots., \tag{4.57}$$

where D_t is calculated according to Equation 4.1 and the fluxes V_i are differential functions.

4.2.6. Hereditary recursion operators

DEFINITION 4.10. *(Fuchssteiner [1979]). A recursion operator is said to be hereditary (or Nijenhuis) if it generates an Abelian symmetry algebra out of commuting symmetries.*

Such recursion operators are also called Kähler (Magri [1978]) or regular (Gel'fand and Dorfman [1979, 1980]).

More specifically, let the flow $u_t = v$ commute with the flows $u_t = w, u_t = Rw$ and let the flow $u_t = w$ commute with the flow $u_t = Rv$, where v, w are arbitrary smooth functions. Then a recursion operator R is hereditary iff we require that the flows $u_t = Rv$ and $u_t = Rw$ also commute.

Assume for simplicity that $\partial R/\partial t = 0$, i.e., $R = R(u): S \to S$ for $u \in S$.

THEOREM 4.9. *(Fuchssteiner [1979]). A recursion operator $R(u)$ is hereditary iff the commutator $[R'(u), R(u)]$ is a symmetric bilinear operator for all $u \in S$,* *i.e., the map* $(v, w) \rightarrow R'[Rv]w - RR'[v]w$ *is symmetric with respect to* $v, w \in S$:

$$R'[Rv]w - RR'[v]w = R'[Rw]v - RR'[w]v. \tag{4.58}$$

THEOREM 4.10. *(Fuchssteiner [1980]). A recursion operator R is hereditary iff*

$$R^2(\llbracket f, g \rrbracket) + \llbracket Rf, Rg \rrbracket - R(\llbracket Rf, g \rrbracket) - \llbracket f, Rg \rrbracket) = 0 \tag{4.59}$$

for arbitrary functions $f(u)$ and $g(u)$ on S, where the commutator (4.24) is used.

The left-hand side of Equation 4.59 is known as the Nijenhuis torsion of a recursion operator R (see Kosmann-Schwarzbach [1986]).

THEOREM 4.11. *(Fuchssteiner [1979]). If $R(u)$ is a hereditary recursion operator for Equation 4.1 then $R(u)$ is also a recursion operator for the associated hierarchy of equations $u_t = R^j(u)K(u)$ $(j = 1, 2, ...)$, and these flows mutually commute.*

Since every translation-invariant operator $R(u)$ is a recursion operator for $u_t = u_x$, then $R(u)$ must be a recursion operator for any of the equations $u_t = R^j(u)u_x$ $(j = 1, 2, ...)$. All well-known soliton evolution equations are of this type.

THEOREM 4.12. *(see Fuchssteiner and Fokas [1981]). If a recursion operator R for Equation 4.1 is hereditary, then all the conserved covariants, generated by the adjoint R^+ of R, are conserved gradients, providing the starting one is a conserved gradient.*

THEOREM 4.13. *(see Fokas and Anderson [1982]). Let γ be the gradient of a conserved functional P of Equation 4.1 and let $R(u)$ be a recursion operator for Equation 4.1. Then P is conserved by the whole hierarchy of equations*

$$u_t = R^j(u)K(u) \quad (j = 1, 2, ...) \tag{4.60}$$

iff $(R^+)^j \gamma$ are gradient functions for $j = 1, 2, ...$

Example. The recursion operator R for KdV in Equation 4.43 is hereditary.

4.2.7. Master symmetries

(Fuchssteiner [1983], Fokas [1987]; see also Fokas and Fuchssteiner [1981a,b]).

One way to generate an infinite set of symmetries is to use a recursion operator. An alternative method is to use a master symmetry.

Let L^* be Lie algebra of C^∞-vector fields on S with a commutator $[\![,]\!]$, defined by Equation 4.24. Let L be some suitable subalgebra containing K from Equation 4.1. Given $\tau \in L^*$, the adjoint map $\hat{\tau}$ of the Lie algebra L^* is defined by

$$\hat{\tau}\sigma = [\![\tau, \sigma]\!], \quad \sigma \in L^*. \tag{4.61}$$

A concept of master symmetry is closely related to time-dependent symmetries, which have explicit polynomial dependence on t. A formal solution of Equation 4.23 for time-dependent symmetries can be defined by formal Taylor series

$$\sigma_\tau(t) = \exp(t\hat{K})\tau \equiv \sum_{i=0}^{\infty} \frac{t^i}{i!}(\hat{K})^i\tau \tag{4.62}$$

for any $\tau = \sigma_\tau(0) \in L^*$. The formula 4.62 makes sense for those τ, for which the series reduces to a finite sum:

$$\sigma_\tau(t) = \sum_{i=0}^{n} \frac{t^i}{i!}(\hat{K})^i\tau. \tag{4.62'}$$

This would be the case if τ satisfies the condition

$$(\hat{K})^{n+1}\tau = 0. \tag{4.63}$$

Such τ is called a K-generator of degree n. Hence K-generators of degree 0 satisfy Equation 4.25. Thus they are symmetries of Equation 4.1 and form a commutant K° of the element $K \in L^*$.

A commutant of K in L is denoted by $K^\perp$. It is a subalgebra of the time-independent symmetry algebra K°. Usually, these symmetries are x independent as well.

DEFINITION 4.11. *A function $\tau \in L^*$ is called a master symmetry (of degree 1) of Equation 4.1 if, for any symmetry $\sigma \in K^\perp$, the commutator $[\![\tau, \sigma]\!]$ defined by Equation 4.24 is also a symmetry of Equation 4.1 which belongs to $K^\perp$. Thus, the operator $\hat{\tau}$ maps $K^\perp$ into itself, i.e., $\hat{\tau}$ maps symmetries, which belong to L, into symmetries which also belong to L.*

DEFINITION 4.12. *A function $\tau \in L^*$ is called a master symmetry of degree $(n + 1)$ of Equation 4.1 if, for any symmetry $\sigma \in K^\perp$ of Equation 4.1, the commutator $[\![\tau, \sigma]\!]$ is a master symmetry of degree n of Equation 4.1.*

Hence by induction we define a master symmetry of any degree n ($n = 1, 2, ...$).

THEOREM 4.14.   *A function $\tau \in L^*$ is a master symmetry of Equation 4.1 of degree n iff, for any symmetries $\sigma_1, \sigma_2, ..., \sigma_n$ of Equation 4.1 which belong to L, the function $\hat{\sigma}_1 \cdot \hat{\sigma}_2 \cdot ... \cdot \hat{\sigma}_n \tau$ is also a symmetry of Equation 4.1, which belongs to L.*

It is clear that all master symmetries are K-generators (of the same degree), but the inverse is not true without additional conditions.

THEOREM 4.15.   *Let $K^\perp$ be abelian. Let τ be a K-generator of degree n such that $\hat{\sigma}_1 \cdot \hat{\sigma}_2 \cdot ... \cdot \hat{\sigma}_n \tau \in L$ for arbitrary $\sigma_1, ..., \sigma_n \in K^\perp$. Then τ is a master symmetry of degree n.*

Example.   For Equation 4.5 with $K(u) = Hu_{xx} + 2uu_x$, where H is the Hilbert transform 4.6, the function

$$\tau_1 = -6xK(u) - 6u^2 - 9Hu_x \tag{4.64}$$

is a master symmetry of degree 1. It generates an infinite sequence of commuting symmetries of Equation 4.5 by means of the relations

$$K_1 = K, \quad K_2 = \frac{1}{6}[\![\tau_1, K]\!], ..., K_{n+1} = \frac{1}{6}[\![\tau_1, K_n]\!]. \tag{4.65}$$

Infinitely many master symmetries of degree 1 are obtained by

$$\tau_n = [\![x, K_{n+1}]\!]. \tag{4.66}$$

They are related through Equation 4.62' to time-dependent symmetries of Equation 4.5 linear in t:

$$\sigma_{\tau_n}(t) = \tau_n + t[\![K_1, \tau_n]\!]. \tag{4.67}$$

Master symmetries $\tau_{m,n}$ of degree m are given by the recursion ($\tau_{1,n} = \tau_n$):

$$\tau_{2,n} = [\![x, \tau_n]\!], ..., \tau_{m+1,n} = [\![x, \tau_{m,n}]\!] \quad (m = 1, 2, ...; \ n = 1, 2, ...). \tag{4.68}$$

They are related through Equation 4.62' to time-dependent symmetries of Equation 4.5 which are polynomial of degree m in t:

$$\sigma_{\tau_{m,n}}(t) = \sum_{i=0}^{m} \frac{t^i}{i!} (\hat{K})^i \tau_{m,n}. \tag{4.69}$$

4.3. Recursion operators in Hamiltonian formalism

The major part of equations, which have infinitely many symmetries, possesses an additional structure which is a continual analog of Hamiltonian formalism of classical mechanics (see Manin [1979]). We proceed now to expose the related concepts.

4.3.1. Noether and inverse Noether operators

DEFINITION 4.13. *(see Fuchssteiner and Fokas [1981]). Let θ be a function from S into the space of operators $S^* \to S$, which may explicitly depend on x, t, i.e., $\theta = \theta(u, x, t)$ — for any $u \in S$. The function θ is said to be Noether operator for Equation 4.1 if, for any $u \in S$ satisfying Equation 4.1, $\theta(u, x, t)$ maps any conserved covariant γ of Equation 4.1 into a symmetry σ of Equation 4.1:*

$$\sigma = \theta\gamma \tag{4.70}$$

THEOREM 4.16. *(see Oevel and Fokas [1982]). $\theta(u, x, t)$ is a Noether operator for Equation 4.1 iff for any $u \in S$ it satisfies the equation:*

$$\frac{\partial\theta}{\partial t} + \theta'[K] - \theta(K')^+ - K'\theta = 0. \tag{4.71}$$

DEFINITION 4.14. *Let J be a function from S into the space of operators $S \to S^*$: $J = J(u, x, t)$ for any $u \in S$. The function J is said to be an inverse Noether operator for Equation 4.1 iff, for any $u \in S$ satisfying Equation 4.1, $J(u, x, t)$ maps any symmetry σ of Equation 4.1 into a conserved covariant γ of Equation 4.1:*

$$\gamma = J\sigma. \tag{4.72}$$

THEOREM 4.17. *(see Oevel and Fokas [1982]). $J(u, x, t)$ is an inverse Noether operator for Equation 4.1 iff for any $u \in S$ it satisfies the equation:*

$$\frac{\partial J}{\partial t} + J'[K] + JK' + (K')^+ J = 0. \tag{4.73}$$

If θ satisfies Equation 4.71 and θ^{-1} exists, then $J = \theta^{-1}$ satisfies Equation 4.73.

THEOREM 4.18.   *(see Magri [1978], Fuchssteiner and Fokas [1981]).   If* θ *and* J *are Noether and inverse Noether operators for Equation 4.1, then* $R = \theta J$ *is a recursion operator for Equation 4.1. Furthermore, the hierarchy* $\{R^j\theta | j = 1, 2, ...\}$ *is a hierarchy of Noether operators.*

THEOREM 4.19.   *If* θ *and* J *are Noether and inverse Noether operators and* R *is a recursion operator, then* $R\theta$ *is a Noether operator and* JR *is an inverse Noether operator.*

Example.   For Equation 4.4, $\theta_1 = D_x$ is a Noether operator and $J_1 = D_x^{-1}$ is an inverse Noether operator. If R is a recursion operator 4.43, then

$$\theta_2 = R\theta_1 = D_x^3 + 4uD_x + 2u_x \tag{4.74}$$

is also a Noether operator.

4.3.2.   Symplectic and Hamiltonian (implectic) operators

The following concepts and results are due to Magri [1978], Gel'fand and Dorfman [1979,1980], Fuchssteiner and Fokas [1981].

DEFINITION 4.15.   *An operator-valued function* $J(u, x, t)$: $S \to S^*$, $u \in S$, *which is skew-symmetric, is said to be a symplectic operator if the bracket, defined on* S^3 *by*

$$[\![a, b, c]\!] = \langle J'(u, x, t)[a]b, c\rangle, \tag{4.75}$$

satisfies the Jacobi identity:

$$[\![a, b, c]\!] + [\![b, c, a]\!] + [\![c, a, b]\!] = 0 \tag{4.76}$$

for all $a, b, c \in S$.

Definition 4.15 means that an exterior derivative of the two-form $\omega(u, x, t)$, defined by

$$\omega(u, x, t)(b, c) = \langle J(u, x, t)b, c\rangle \tag{4.77}$$

for all $b, c \in S$, vanishes, i.e., the two-form $\omega(u, x, t)$ is closed.

If J has an inverse $\theta = J^{-1}$, then the corresponding identity for θ is the Jacobi identity for the bracket $\{., ., .\}$ defined on S^{*3} by

$$\{a, b, c\} = \langle b, \theta'(u, x, t)[\theta(u, x, t)a]c\rangle \tag{4.78}$$

for arbitrary $a, b, c \in S^*$.

To account for the case when the operator θ is not invertible we accept the following definition.

DEFINITION 4.16. *An operator-valued function $\theta(u, x, t)$: $S^* \rightarrow S$, $u \in S$, which is skew-symmetric $\theta^+ = -\theta$, is called implectic (inverse-symplectic) or a Hamiltonian operator if the bracket defined by the formula 4.78 satisfies the Jacobi identity:*

$$\langle b, \theta'[\theta a]c \rangle + \langle c, \theta'[\theta b]a \rangle + \langle a, \theta'[\theta c]b \rangle = 0 \tag{4.79}$$

for all $a, b, c \in S^*$.

Here the existence of θ^{-1} is not assumed. The left-hand side of Equation 4.79 is called a Schouten bracket (see Lichnerowicz [1977], Olver [1984], Kosmann-Schwarzbach [1986], Kosmann-Schwarzbach and Magri [1988, 1989]).

Consider obvious examples of Hamiltonian operators.

1. A constant skew-symmetric operator.
2. The inverse of a symplectic operator, if it exists.

THEOREM 4.20. *Let $\theta(u)$, $u \in S$ be skew-symmetric operators $S^* \rightarrow S$. Then the following assertions are equivalent:*

1. θ is Hamiltonian;

2. θ is a Noether operator for every evolution equation of the form:

$$u_t = \theta(u)h(u) \tag{4.80}$$

where $h(u)$ is a gradient function: $h(u) = grad H(u)$;

3. for all gradient functions f and g we have:

$$(\theta f)'[\theta g] - (\theta g)'[\theta f] = \theta \langle f, \theta g \rangle'[\cdot]. \tag{4.81}$$

Example. $\theta_1 = D_x$ and θ_2 in Equation 4.74 are Hamiltonian operators. $J_1 = D_x^{-1}$ is a symplectic operator.

4.3.3. Hamiltonian systems

The following concepts and results are due to Magri [1978], Gel'fand and Dorfman [1979, 1980], Fuchssteiner and Fokas [1981], Oevel and Fokas [1982] (see also Olver [1980, 1986]).

DEFINITION 4.17.   *Evolution equations 4.1 are said to be a Hamiltonian system if they have the form*

$$u_t = \theta(u, x, t)h(u, x, t), \tag{4.82}$$

where θ is a Hamiltonian (implectic) operator and h is a gradient function: $h = gradH$, i.e., in Equation 4.1 $K = \theta gradH$. Then H is said to be a Hamiltonian of the system 4.82.

THEOREM 4.21.   *Let $\theta(u, x, t)$ be a Hamiltonian operator. Then θ is a Noether operator for Equation 4.82 iff the operator $\theta(u, x, t)$ and the function $h(u, x, t)$ satisfy the equation*

$$\frac{\partial \theta}{\partial t} - \theta(h' - (h')^+)\theta = 0. \tag{4.83}$$

THEOREM 4.22.   *If $J(u, x, t)K(u, x, t) = h(u, x, t)$ and J is a symplectic operator, then J is an inverse Noether operator for Equation 4.1 iff the operator $J(u, x, t)$ and the function $h(u, x, t)$ satisfy the equation*

$$\frac{\partial J}{\partial t} + h' - (h')^+ = 0. \tag{4.84}$$

DEFINITION 4.18.   *If f and g are gradients, $f = gradF$ and $g = gradG$, and θ is a Hamiltonian operator, then the expression*

$$\{F, G\} = \langle f, \theta g \rangle = \langle gradF, \theta gradG \rangle \tag{4.85}$$

is said to be Poisson bracket of functions F and G.

It follows from Equation 4.85, with the use of Equations 4.26 and 4.9, that

$$\{F, G\} = F'[\theta gradG] = \hat{X}_{\theta gradG}[F] \quad . \tag{4.86}$$

DEFINITION 4.19. $\hat{X}_{\theta gradG}$ *is called a Hamiltonian vector field corresponding to a functional G.*

THEOREM 4.23.   *A skew-symmetry of Hamiltonian operator θ is equivalent to a skew-symmetry of the Poisson bracket:*

$$\{G, F\} = -\{F, G\}. \tag{4.87}$$

The Jacobi identity 4.79 for Hamiltonian operator θ is equivalent to the Jacobi identity for the Poisson bracket:

$$\{\{F, G\}, H\} + \{\{G, H\}, F\} + \{\{H, F\}, G\} = 0. \tag{4.88}$$

DEFINITION 4.20. *We say that a functional $P(u, x, t)$ is conserved by a Hamiltonian flow*

$$u_t = \theta(u, x, t)grad H(u, x, t) \tag{4.89}$$

iff Equation 4.31 with $K = \theta\, grad\, H$ is satisfied:

$$\frac{\partial P}{\partial t} + \langle grad P, \theta\, grad H \rangle = 0, \tag{4.90}$$

which with the use of the expression 4.85 takes the form:

$$\frac{\partial P}{\partial t} + \{P, H\} = 0. \tag{4.91}$$

We also say that P is an integral (conservation law) for the system 4.89.

If P is independent of t, then the condition 4.91 for P to be conserved by a flow with a Hamiltonian H takes the form:

$$\{P, H\} = 0. \tag{4.92}$$

DEFINITION 4.21. *If the condition 4.92 is satisfied for P and H, which do not explicitly depend on t, then P and H are said to be in involution with respect to the Poisson bracket 4.85.*

THEOREM 4.24. *(Hamiltonian form of Noether theorem). Let Equation 4.89 be a Hamiltonian system with the Hamiltonian operator θ. Then θ maps the gradients of conservation laws of Equation 4.89 into the symmetries of this system:*

$$\sigma = \theta\, grad P. \tag{4.93}$$

More precisely, σ is a symmetry of Equation 4.89 if there exists such P, satisfying Equation 4.93, which is a conservation law for Equation 4.89.

4.3.4. Recursion operators and bi-Hamiltonian systems

(Magri [1978]; see also Fuchssteiner and Fokas [1981], Olver [1986], [1987], [1987]).

DEFINITION 4.22.   *A system 4.1 is said to be bi-Hamiltonian if it can be written in two different ways as a Hamiltonian system:*

$$u_t = K(u, x, t) = \theta_1(u, x, t) grad H_1(u, x, t) = \theta_2(u, x, t) grad H_2(u, x, t)$$
$$(4.94)$$

with Hamiltonian operators θ_1 and θ_2 and Hamiltonians H_1 and H_2.

THEOREM 4.25.   *Consider a bi-Hamiltonian system 4.94. Let θ_1 be invertible. Then*

$$R = \theta_2 \theta_1^{-1} \qquad (4.95)$$

is a recursion operator for Equation 4.94.

Example.   Equation 4.4 can be written as a bi-Hamiltonian system 4.94:

$$u_t = D_x[u_{xx} + 3u^2] = (D_x^3 + 4u D_x + 2u_x)[u] \qquad (4.96)$$

with two Hamiltonian operators $\theta_1 = D_x$ and $\theta_2 = D_x^3 + 4u D_x + 2u_x$ and Hamiltonians $H_1 = \int\limits_{-\infty}^{+\infty} \left(-\frac{u_x^2}{2} + u^3\right) dx$ and $H_2 = \int\limits_{-\infty}^{+\infty} \frac{u^2}{2} dx$. Hence $R = \theta_2 \theta_1^{-1} = D_x^2 + 4u + 2u_x D_x^{-1}$ is a recursion operator for Equation 4.4 which coincides with Equation 4.43.

4.3.5.   Hereditary (Nijenhuis) property of recursion operators and compatibility of Hamiltonian structures

(Fuchssteiner and Fokas [1981])

DEFINITION 4.23.   *We call two Hamiltonian operators θ_1 and θ_2 compatible if $\theta_1 + \theta_2$ is again a Hamiltonian operator.*
Define the following mixed brackets, involving θ_1 and θ_2:

$$[a, b, c]_1 = \langle b, \theta'_1[\theta_2 a]c \rangle, \quad [a, b, c]_2 = \langle b, \theta'_2[\theta_1 a]c \rangle. \qquad (4.97)$$

THEOREM 4.26.   *θ_1 and θ_2 are compatible iff the bracket*

$$[a, b, c] = [a, b, c]_1 + [a, b, c]_2 \qquad (4.98)$$

satisfies the Jacobi identity.

COROLLARY. *Hamiltonian operators θ_1 and θ_2 are compatible iff $\theta_1 + \alpha\theta_2$ is Hamiltonian for all constant $\alpha \in \mathbb{R}$.*

THEOREM 4.27. *Let θ_1 and θ_2 be compatible Hamiltonian operators and assume that θ_1 is invertible. Then a recursion operator $R = \theta_2\theta_1^{-1}$ is hereditary (Nijenhuis).*

COROLLARY. *If $(R^+)^{-1}$ exists, then θ_1 and θ_2 are compatible iff $R = \theta_2\theta_1^{-1}$ is hereditary.*

REMARK. If in addition to the assumptions of Theorem 4.27 θ_2 is also invertible, then R and R^{-1} are hereditary.

THEOREM 4.28. *Under the assumptions of Theorem 4.27 all the operators $R^n\theta_1$ ($n = 1, 2, ...$) are Hamiltonian.*

COROLLARY. *For invertible R the equation*

$$u_t = R^n u_x \quad (n = 1, 2, ...) \tag{4.99}$$

is N-Hamiltonian, i.e., it has N different Hamiltonian structures, where N is an arbitrary natural number.

DEFINITION 4.24. *If a recursion operator R can be written as $R = \theta J$, where J is a symplectic invertible operator and θ and J^{-1} are a compatible pair of Hamiltonian (implectic) operators, then we say that R admits a symplectic-implectic factorization.*

THEOREM 4.29. *If R is a recursion operator which admits a symplectic-implectic factorization, then all the conserved covariants γ_n, defined by*

$$\gamma_n = J K_n \tag{4.100}$$

with

$$K_n = R^n K \quad (n = 0, 1, 2, ...) \tag{4.101}$$

are gradients: $\gamma_n = \mathrm{grad}\,\Gamma_n$, if γ_0 is a gradient. Furthermore, all $\Gamma_n - s$ are in *involution with respect to the Poisson bracket*

$$\{\Gamma_n, \Gamma_m\} = \langle \gamma_n, \theta \gamma_m \rangle = 0. \tag{4.102}$$

THEOREM 4.30.   *(see Fokas and Anderson [1982], Shabat and Sokolov [1982], Boiti and Konopelchenko [1985]). Let R be a recursion operator for a Hamiltonian system 4.89 with a Hamiltonian operator θ. Then there exists the explicit relationship between θ and R:*

$$\theta R^+ = R\theta. \tag{4.103}$$

THEOREM 4.31.   *(see Ablowitz, et al. [1974]). In a context of the inverse scattering transform method a family of equations, which are integrable by a given spectral problem, is generated by the adjoint recursion operator $L = R^+$ in the form*

$$u_t = \theta f(R^+) grad H. \tag{4.104}$$

Here, $f(L)$ is an arbitrary entire (meromorphic for an invertible L) function of $L = R^+$, θ is a Hamiltonian operator, and H is a Hamiltonian of the "initial" member of the family.

The function f is determined by a dispersion relation of, corresponding to Equation 4.104, linearized equations and a recursion operator $L = R^+$ is determined by a given spectral problem (see Ablowitz and Segur [1981]).

COROLLARY.   *If, in particular, $f(L) = L^n$ $(n = 1, 2, ...)$, then the family of Equations 4.104, integrable by a given spectral problem, takes the form*

$$u_t = \theta (R^+)^n grad H \quad (n = 1, 2, ...). \tag{4.105}$$

Example.   The family of integrable equations, associated with Equation 4.4, has a recursion operator R (for symmetries) defined by Equation 4.43 and a Hamiltonian operator $\theta = D_x$. Hence $L = R^+ = \theta^{-1} R\theta = D_x^{-1}(D_x^2 + 4u + 2u_x D_x^{-1})D_x = D_x^2 + 2u + 2D_x^{-1}u D_x$ and L is the Lenard operator (see Section 4.2.3). According to Equation 4.105, the KdV hierarchy takes the form

$$u_t = D_x L^n u. \tag{4.106}$$

REMARK.   In Sections 4.3.6 to 4.3.10, we mean, for brevity, by a recursion operator R an adjoint recursion operator R^+.

4.3.6. Weak recursion operators

The results of Sections 4.3.6 to 4.3.9 are due to Zakharov and Konopelchenko [1984].

Let $u^1(x, t), ..., u^n(x, t)$ be a system of n fields defined in the d-dimensional space of variables $x = (x_1, ..., x_d)$.

DEFINITION 4.25. *A system of n equations is called Hamiltonian if it can be represented in the form*

$$\int dx' \Omega_{\alpha\beta}(x, x') \frac{\partial u^\beta(x', t)}{\partial t} = \frac{\delta H}{\delta u^\alpha(x, t)}, \tag{4.107}$$

which is a particular functional representation of a more abstract form

$$\Omega \frac{\partial u}{\partial t} = grad H. \tag{4.108}$$

Here, $\Omega_{\alpha\beta}(x, x')$ is a kernel of a nondegenerate linear operator, depending, in general, on $u^1, ..., u^n$, which is symplectic, i.e., it satisfies conditions of closeness:

$$\frac{\delta \Omega_{\alpha\beta}(x, x')}{\delta u^\gamma(x'')} + \frac{\delta \Omega_{\gamma\alpha}(x'', x)}{\delta u^\beta(x')} + \frac{\delta \Omega_{\beta\gamma}(x', x'')}{\delta u^\alpha(x)} = 0 \tag{4.109}$$

and of skew-symmetry:

$$\Omega_{\alpha\beta}(x, x') = -\Omega_{\beta\alpha}(x', x). \tag{4.110}$$

Such a matrix $\Omega_{\alpha\beta}(x, x')$ defines a symplectic form, i.e., a closed skew-symmetric form.

DEFINITION 4.26. *An operator R_H is called an H-weak recursion operator if any of its powers converts the gradient of a functional $H \neq const.$ into a gradient:*

$$\int dx' (R_H^N)_\alpha^\beta(x, x') \frac{\delta H}{\delta u^\beta(x')} = \frac{\delta H_N}{\delta u^\alpha(x)}, \tag{4.111}$$

or in the abstract form:

$$R_H^N grad H = grad H_N. \tag{4.112}$$

DEFINITION 4.27. *An operator R_Ω is called an Ω-weak recursion operator if any of its powers converts a symplectic form Ω into a symplectic form Ω_N:*

$$\int dx' (R_\Omega^N)_\alpha^\beta (x, x') \Omega_{\beta\gamma} (x', x'') = \Omega_{N\alpha\gamma} (x, x''), \qquad (4.113)$$

or in the abstract form

$$R_\Omega^N \Omega = \Omega_N. \qquad (4.114)$$

A Ω-weak recursion operator is called a regular operator by Gel'fand and Dorfman [1979,1980].

THEOREM 4.32. *An operator $R = R_H$ is an H-weak recursion operator if it satisfies the following system of equations:*

$$\int d\tilde{x} \left\{ \left(\frac{R_\alpha^\delta (x, \tilde{x})}{\delta u^\beta (x')} - \frac{\delta R_\beta^\delta (x', \tilde{x})}{\delta u^\alpha (x)} \right) R_\delta^\gamma (\tilde{x}, x'') \right.$$

$$\left. + R_\alpha^\delta (x, \tilde{x}) \frac{\delta R_\beta^\gamma (x', x'')}{\delta u^\delta (\tilde{x})} - R_\beta^\delta (x', \tilde{x}) \frac{\delta R_\alpha^\gamma (x, x'')}{\delta u^\delta (\tilde{x})} \right\} = 0. \qquad (4.115)$$

$$\int d\tilde{x} \left\{ \frac{\delta R_\alpha^\rho (x, \tilde{x})}{\delta u^\beta (x')} \frac{\delta H}{\delta u^\rho (\tilde{x})} + R_\alpha^\rho (x, \tilde{x}) \frac{\delta^2 H}{\delta u^\beta (x') \delta u^\rho (\tilde{x})} - \right.$$

$$\left. - \frac{\delta R_\beta^\rho (x', \tilde{x})}{\delta u^\alpha (x)} \frac{\delta H}{\delta u^\rho (\tilde{x})} - R_\beta^\rho (x', \tilde{x}) \frac{\delta^2 H}{\delta u^\alpha (x) \delta u^\rho (\tilde{x})} \right\} = 0. \qquad (4.116)$$

THEOREM 4.33. *An operator $R = R_\Omega$ is an Ω-weak recursion operator if it satisfies Equation 4.115 and the equations*

$$\frac{\delta (R\Omega)_{\alpha\beta} (x, x')}{\delta u^\gamma (x'')} + \frac{\delta (R\Omega)_{\gamma\alpha} (x'', x)}{\delta u^\beta (x')} + \frac{\delta (R\Omega)_{\beta\gamma} (x', x'')}{\delta u^\alpha (x)} = 0, \qquad (4.117)$$

$$(R\Omega)_{\beta\alpha} (x', x) = -(R\Omega)_{\alpha\beta} (x, x') \Leftrightarrow (R\Omega)^+ = -R\Omega, \qquad (4.118)$$

where

$$(R\Omega)_{\alpha\beta} (x, x') = \int dx'' R_\alpha^\gamma (x, x'') \Omega_{\gamma\beta} (x'', x). \qquad (4.119)$$

THEOREM 4.34.   *If, together with the form Ω, the forms $R_\Omega \Omega$ and $R_\Omega^2 \Omega$ are closed and skew-symmetric, then R_Ω is an Ω-weak recursion operator.*

THEOREM 4.35.   *In the case when R_Ω is invertible, the closeness and skew-symmetry of one of the two sets of forms, $R_\Omega^{-1}\Omega, \Omega, R_\Omega\Omega$ or $R_\Omega^{-2}\Omega, \Omega, R_\Omega^{-1}\Omega$, also provide sufficient conditions for R_Ω to be an Ω-weak recursion operator.*

THEOREM 4.36.   *If R_Ω is invertible, then the Conditions 4.115, 4.117, and 4.118 are necessary and sufficient for R_Ω to be an Ω-weak recursion operator.*

The application of weak recursion operators generates infinite families of Hamiltonian systems, starting from Equation 4.107, which have the following abstract form:

$$\Omega \frac{\partial u}{\partial t} = R_H^N \operatorname{grad} H \equiv \operatorname{grad} H_N, \tag{4.120}$$

$$R_\Omega^N \Omega \frac{\partial u}{\partial t} = \Omega_N \frac{\partial u}{\partial t} = \operatorname{grad} H \quad (N = 0, 1, 2, ...). \tag{4.121}$$

THEOREM 4.37.   *Equations 4.120 are Hamiltonian with respect to the same symplectic form Ω and different Hamiltonians H_N. Equations 4.121 are Hamiltonian with respect to the same Hamiltonian H and different symplectic forms Ω_N.*

COROLLARY 1.
Any entire function of recursion operator is a recursion operator of the same type.

COROLLARY 2.
The most general family of equations, which are generated by H-weak and Ω-weak recursion operators from Equation 4.107, has the form

$$\Phi(R_\Omega)\Omega \frac{\partial u}{\partial t} \equiv \Omega_\Phi \frac{\partial u}{\partial t} = \operatorname{grad} H_f \equiv f(R_H)\operatorname{grad} H, \tag{4.122}$$

where Φ and f are arbitrary entire functions.

If R_Ω and Ω are invertible, then Equation 4.122 can be represented in the form

$$\frac{\partial u}{\partial t} = \Omega_\Phi^{-1} \operatorname{grad} H_f \equiv \Omega^{-1}\Phi^{-1}(R_\Omega)f(R_H)\operatorname{grad} H, \tag{4.123}$$

which corresponds to Equation 4.82, Definition 4.17 of a Hamiltonian system.

4.3.7. Strong recursion operators

DEFINITION 4.28.   *Operator R is called a strong recursion operator if any of its powers transforms a gradient into gradients and a symplectic form into symplectic forms:*

$$R^N \operatorname{grad} H = \operatorname{grad} H_N, \quad R^N \Omega = \Omega_N. \tag{4.124}$$

COROLLARY.   *R is a strong recursion operator iff it is simultaneously H-weak and Ω-weak.*

Hence a strong recursion operator generates both an infinite family of Hamiltonians H_N and an infinite family of symplectic forms Ω_N.

Consider an equation $\partial u / \partial t = \Omega^{-1} \operatorname{grad} H$ from the family of Equations 4.123 and let $R = R_H = R_\Omega$ be an invertible strong recursion operator. Then this equation can be written as

$$\frac{\partial u}{\partial t} = \Omega^{-1} \operatorname{grad} H \equiv (R^N \Omega)^{-1} R^N \operatorname{grad} H \equiv \Omega_N^{-1} \operatorname{grad} H_N, \tag{4.125}$$

where N is any integer and Ω_N are closed symplectic forms.

THEOREM 4.38.   *Any Equation 4.125 generated by a strong recursion operator R is Hamiltonian with respect to an infinite set of Hamiltonian structures (i.e., pairs Ω_N, H_N).*

Let us denote Poisson brackets, which correspond to a form Ω_N, as $\{\cdot, \cdot\}_N$:

$$\{F, H\}_N = \langle \operatorname{grad} F, \Omega_N^{-1} \operatorname{grad} H \rangle. \tag{4.126}$$

THEOREM 4.39.   *All the Hamiltonians H_N, which are generated by a strong recursion operator R, are in involution with respect to any Poisson bracket $\{\cdot, \cdot\}_N$, generated by this operator, and form an infinite set of integrals for Equation 4.125.*

Let $R = R_H = R_\Omega$ be an invertible strong recursion operator for a family of Equations 4.123 and $f(R)$ be an arbitrary entire function of R. Then Equation 4.123 can be written as

$$\frac{\partial u}{\partial t} = \Omega^{-1} f(R) \operatorname{grad} H, \tag{4.127}$$

or more explicitly as

$$\frac{\partial u^{\alpha}(x, t)}{\partial t} = \int dx' (\Omega^{-1} f(R))^{\alpha\beta}(x, x') \frac{\delta H}{\delta u^{\beta}(x', t)}. \tag{4.128}$$

For any entire function $\Phi(R)$ introduce Poisson bracket

$$\{F, H\}_{\Phi} = \int dx dx' \frac{\delta F}{\delta u^{\alpha}(x)} (\Omega^{-1} \Phi(R))^{\alpha\beta}(x, x') \frac{\delta H}{\delta u^{\beta}(x')}. \tag{4.129}$$

THEOREM 4.40. *Each member of Equations 4.127 possesses an infinite set of integrals H_N and is a Hamiltonian system with respect to an infinite family of Poisson brackets 4.129 with arbitrary entire Φ.*

THEOREM 4.41. *If the initial equation 4.107 possesses a strong recursion operator R, then each member of Equations 4.127 possesses the same strong recursion operator.*

THEOREM 4.42. *If an operator R satisfies the whole system of Equations 4.115 to 4.118, then it is a strong recursion operator.*

THEOREM 4.43. *(see Gel'fand and Dorfman [1979], [1980]). Let Ω be an invertible symplectic form and R be a strong recursion operator. Then Ω^{-1} and $(R\Omega)^{-1}$ provide a compatible pair of Hamiltonian operators iff the forms Ω, $R\Omega$, $R^2\Omega$ are closed.*

4.3.8. Recursion operators in momentum representation

(Zakharov and Konopelchenko [1984])
Consider the Fourier transform

$$u^{\alpha}(x, t) = (2\pi)^{-\frac{d}{2}} \int dp a_p^{\alpha}(t) \exp(ipx), \tag{4.130}$$

$$\Omega_{\alpha\beta}(x, x') = (2\pi)^{-d} \int dp dp' \Omega_{\alpha\beta, pp'} \exp(ipx + ip'x'), \tag{4.131}$$

$$R_{\alpha}^{\beta}(x, x') = (2\pi)^{-d} \int dp dp' R_{\alpha, pp'}^{\beta} \exp(ipx + ip'x'), \tag{4.132}$$

where $p = (p_1, \ldots, p_d)$, $px = p_1 x^1 + \cdots + p_d x^d$.

Closeness and skew-symmetry conditions 4.109 and 4.110 take the following form in a momentum representation:

$$\frac{\delta \Omega_{\alpha\beta,pq}}{\delta a^{\gamma}_{-k}} + \frac{\delta \Omega_{\gamma\alpha,kp}}{\delta a^{\beta}_{-q}} + \frac{\delta \Omega_{\beta\gamma,qk}}{\delta a^{\alpha}_{-p}} = 0, \tag{4.133}$$

$$\Omega_{\alpha\beta,pq} = -\Omega_{\beta\alpha,qp} \tag{4.134}$$

Equations 4.115 to 4.118, defining a strong recursion operator, have the following momentum representation:

$$\int d\tilde{p} \left\{ R^{\rho}_{\alpha,p\tilde{p}} \frac{\delta R^{\gamma}_{\beta,qk}}{\delta a^{\rho}_{\tilde{p}}} - R^{\rho}_{\beta,q\tilde{p}} \frac{\delta R^{\gamma}_{\alpha,pk}}{\delta a^{\rho}_{\tilde{p}}} \right.$$

$$\left. + \left(\frac{\delta R^{\rho}_{\alpha,p\tilde{p}}}{\delta a^{\beta}_{-q}} - \frac{\delta R^{\rho}_{\beta,q\tilde{p}}}{\delta a^{\alpha}_{-p}} \right) R^{\gamma}_{\rho,-\tilde{p}k} \right\} = 0, \tag{4.135}$$

$$\int dk \left\{ \frac{\delta R^{\rho}_{\alpha,pk}}{\delta a^{\beta}_{-q}} \frac{\delta H}{\delta a^{\rho}_{k}} + R^{\rho}_{\alpha,pk} \frac{\delta^2 H}{\delta a^{\beta}_{-q} \delta a^{\rho}_{k}} \right. \tag{4.136}$$

$$\left. - \frac{\delta R^{\rho}_{\beta,qk}}{\delta a^{\alpha}_{-p}} \frac{\delta H}{\delta a^{\rho}_{k}} - R^{\rho}_{\beta,qk} \frac{\delta^2 H}{\delta a^{\alpha}_{-p} \delta a^{\rho}_{k}} \right\} = 0,$$

$$\frac{\delta (R\Omega)_{\alpha\beta,pq}}{\delta a^{\gamma}_{-k}} + \frac{\delta (R\Omega)_{\gamma\alpha,kp}}{\delta a^{\beta}_{-q}} + \frac{\delta (R\Omega)_{\beta\gamma,qk}}{\delta a^{\alpha}_{-p}} = 0, \tag{4.137}$$

$$(R\Omega)_{\beta\alpha,qp} = -(R\Omega)_{\alpha\beta,pq}, \tag{4.138}$$

where

$$(R\Omega)_{\alpha\beta,pq} = \int dk R^{\gamma}_{\alpha,pk} \Omega_{\gamma\beta,-kq}. \tag{4.139}$$

4.3.9. Nonsingular and regular recursion operators

(Zakharov and Konopelchenko [1984])

Consider nonlinear systems which are described by one real field, i.e., $n = 1$. Without loss of generality, one can choose a local u-independent symplectic form, such as

$$\Omega_{(0)pq} = -\frac{i}{p} \delta(p + q) \tag{4.140}$$

for one-dimensional space ($d = 1$), where $\delta(p)$ is Dirac-delta function and

$$\Omega_{(0)pq} = f_p \delta(p + q) \tag{4.141}$$

for $d > 1$, where $f_p = -f_{-p}$ is an antisymmetric function.

Consider translation-invariant systems which smoothly reduce to linear systems in the limit $a_p \to 0$ (i.e., for weak fields).

Thus assume the following forms of a Hamiltonian and of a symplectic form:

$$H = \sum_{n=2}^{\infty} \int dq_1 \dots dq_n \delta(q_1 + \dots + q_n) V_{(n)q_1 \dots q_n} a_{q_1} \dots a_{q_n}, \tag{4.142}$$

$$\Omega_{pq} = \sum_{n=0}^{\infty} \int dq_1 \dots dq_n \delta(p + q - q_1 - \dots - q_n) \Omega_{(n)pq(q_1 \dots q_n)} a_{q_1} \dots a_{q_n}, \tag{4.143}$$

where $V_{(n)q_1 \dots q_n}$ are some functions which are completely symmetric in their variables, and $\Omega_{(n)pq(q_1 \dots q_n)}$ are functions which are symmetric in $q_1, \dots, q_n$.

Assume a recursion operator R_{pq} in a momentum representation to be an "entire" function of a_p:

$$R_{pq} = \sum_{n=0}^{\infty} \int dq_1 \dots dq_n \tilde{R}_{(n)pq(q_1 \dots q_n)} a_{q_1} \dots a_{q_n}, \tag{4.144}$$

where $\tilde{R}_{(n)pq(q_1 \dots q_n)}$ are completely symmetric in $q_1, \dots, q_n$. A restriction on the form of functions $\tilde{R}_{(n)pq(q_1 \dots q_n)}$ follows from translation invariance:

$$\tilde{R}_{(n)pq(q_1 \dots q_n)} = \delta(p + q - q_1 - \dots - q_n) R_{(n)pq(q_1 \dots q_n)}. \tag{4.145}$$

THEOREM 4.44. *Any Hamiltonian system 4.107 with a Hamiltonian 4.142 possesses a formal strong recursion operator.*

REMARK. The word "formal" means that functions $R_{(n)pq(q_1 \dots q_n)}$ in the integral representation 4.144 of R_{pq}, and hence coefficients $V_{(n)q_1 \dots q_n}$ in the representations of the form 4.142 for higher integrals H_N ($R^N \operatorname{grad} H = \operatorname{grad} H_N$), may have nonintegrable singularities. Therefore in a general case the integrals H_N can not be defined properly.

DEFINITION 4.29. *Hamiltonians H_N of the form 4.142 are called well-defined functionals on a_p if their coefficients $V_{(n)q_1 \dots q_n}$ do not contain nonintegrable singularities.*

DEFINITION 4.30. *A recursion operator, which generates a family of well-defined Hamiltonians H_N out of a well-defined initial Hamiltonian H, is referred to as a nonsingular recursion operator.*

DEFINITION 4.31. *A recursion operator, which produces a family of Hamiltonians of the form 4.142 with such coefficients $V_{(n)q_1...q_n}$ that have no singularities at all, is referred to as a regular recursion operator.*

THEOREM 4.45. *In a one-dimensional space $(d = 1)$ there exist Hamiltonian systems which possess a regular recursion operator. For a certain subclass of these systems the recursion operator is a polynomial of a finite order in the field variables a_q.*

Denote by $\Gamma^{n,m}$ a manifold which is defined by

$$p_1 + ... + p_n = p_{n+1} + ... + p_{n+m},$$

$$\omega(p_1) + ... + \omega(p_n) = \omega(p_{n+1}) + ... + \omega(p_{n+m}). \qquad (4.146)$$

Here $\omega(p)$ determines a dispersion law for the equation in a momentum representation, corresponding to Equation 4.107:

$$\frac{\partial a_p}{\partial t} = \omega(p)a_p + ... , \qquad (4.147)$$

which is a "linear part" of this equation.

$\Gamma^{n,m}$ turn out to be singularity manifolds for the coefficients $V_{(n)q_1...q_n}$ of Hamiltonians H_N, which have the form 4.142.

THEOREM 4.46. *Nonlinear equations in two-dimensional and higher-dimensional spaces $(d \geq 2)$, which describe a nontrivial scattering of n waves into m waves $(n \neq m)$ with $\Gamma^{n,m}$ having maximum dimension, do not possess a nonsingular recursion operator of the form 4.132, 4.144.*

Example. Consider the Kadomtsev-Petviashvili (KP) equation

$$u_t = u_{xxx} + 6uu_x + 3\alpha^2 D_x^{-1}u_{yy}. \qquad (4.148)$$

Here, $u = u(x, y, t)$, α is a constant parameter (particular values $\alpha^2 = -1$ and $\alpha^2 = 1$ correspond to KPI and KPII). For Equation 4.148, $\Gamma^{1,2}$ has maximum dimension, i.e., $\dim\Gamma^{1,2} = 3$. The KP equation has no nonsingular recursion operator of the form 4.132, 4.144.

4.3.10. Nonlocal recursion operators

(Fokas and Santini [1986], [1988])

We shall refer to recursion operators, which were considered previously in Section 4.3, as local recursion operators.

The KP and Benjamin-Ono (BO) equations are typical examples of equations which do not possess a nonsingular local recursion operator. To bypass the "prohibition law" formulated in Theorem 4.46 (it states nonexistence of recursion operators and hence of a bi-Hamiltonian formulation) one introduces extended representations as reductions of nonlocal systems in higher dimensions. This implies the existence of nonlocal recursion operators and Hamiltonian structures.

Example 1.
Equation 4.148 can be presented in the extended form

$$u_{1,t} = K_{11} \equiv \int_{\mathbb{R}} dy_2 \delta_{12} \theta_{12}^{(1)} \gamma_{12}^{(1)} = \int_{\mathbb{R}} dy_2 \delta_{12} \theta_{12}^{(2)} \gamma_{12}^{(2)}, \tag{4.149}$$

where $u_i = u(x, y_i, t)$ $(i = 1, 2)$, $\delta_{12} = \delta(y_1 - y_2)$. $\theta_{12}^{(1)}$ and $\theta_{12}^{(2)}$ are two extended nonlocal Hamiltonian operators:

$$\theta_{12}^{(1)} = D_x, \quad \theta_{12}^{(2)} = D_x^3 + D_x u_{12}^+ + u_{12}^+ D_x + u_{12}^- D_x^{-1} u_{12}^-, \tag{4.150}$$

where

$$u_{12}^{\pm} = u_1 \pm u_2 + \alpha(D_{y^1} \mp D_{y^2}). \tag{4.151}$$

$\gamma_{12}^{(i)}$ $(i = 1, 2)$ are suitable extended gradients.

Equation 4.149 is an extended bi-Hamiltonian representation of the KP equation.

A nonlocal recursion operator, which generates extended symmetries σ_{12}, is calculated according to the formula 4.95:

$$R_{12} = \theta_{12}^{(2)} (\theta_{12}^{(1)})^{-1} = D_x^2 + D_x u_{12}^+ D_x^{-1} + u_{12}^+ + u_{12}^- D_x^{-1} u_{12}^- D_x^{-1}. \tag{4.152}$$

The adjoint recursion operator is determined by Equation 4.103:

$$\Delta_{12} = R_{12}^+ = (\theta_{12}^{(1)})^{-1} R \theta_{12}^{(1)} = (\theta_{12}^{(1)})^{-1} \theta_{12}^{(2)}$$

$$= D_x^2 + u_{12}^+ + D_x^{-1} u_{12}^+ D_x + D_x^{-1} u_{12}^- D_x^{-1} u_{12}^-. \tag{4.153}$$

It generates extended conserved gradients γ_{12}. For example, we have in Equation 4.149:

$$\gamma_{12}^{(1)} = \Delta_{12}^2(-\frac{1}{2}).\tag{4.154}$$

Example 2. The BO equation 4.5, defined with the use of the Hilbert transform 4.6, admits an extended representation of the form:

$$u_{1,t} = \int_{\mathbb{R}} dx_2 \delta(x_1 - x_2) K_{12},\tag{4.155}$$

where K_{12} is some function of $u_i = u(x_i, t)$ $(i = 1, 2)$.

We define the operators:

$$u_{12}^{\pm} = u_1 \pm u_2 + i(D_{x_1} \mp D_{x_2}).\tag{4.156}$$

The BO equation possesses two compatible extended nonlocal Hamiltonian operators:

$$\theta_{12}^{(1)} = u_{12}^-, \quad \theta_{12}^{(2)} = (u_{12}^+ - iu_{12}^- H_{12})u_{12}^-,\tag{4.157}$$

where H_{12} is an extended Hilbert transform operator:

$$(H_{12}f)(x_1, x_2) = \frac{1}{\pi}\int_{\mathbb{R}} \frac{F(\xi, x_1 - x_2)}{\xi - (x_1 + x_2)}d\xi,\tag{4.158}$$

where the principal value of integral is meant and $F(x_1 + x_2, x_1 - x_2) = f(x_1, x_2)$. Hence the BO equation is a bi-Hamiltonian system.

A nonlocal recursion operator for extended symmetries of the BO equation is again calculated according to the formula 4.95:

$$R_{12} = \theta_{12}^{(2)}(\theta_{12}^{(1)})^{-1} = u_{12}^+ - iu_{12}^- H_{12}.\tag{4.159}$$

The adjoint operator R_{12}^+ generates extended conserved gradients.

An explicit form of the extended representation for the BO equation is

$$u_{1,t} = (4i)^{-1}\int_{\mathbb{R}} dx_2 \delta(x_1 - x_2) R_{12}^2 u_{12}^- \cdot 1.\tag{4.160}$$

The adjoint recursion operator is given by

$$R_{12}^+ = u_{12}^+ - iH_{12}u_{12}^-\tag{4.161}$$

and generates extended conserved gradients $\gamma_{12}^{(n)}$:

$$\gamma_{12}^{(n)} = (R_{12}^+)^n \cdot 1 \quad n = 1, 2, \dots .$$ (4.162)

Then one can give an extended bi-Hamiltonian formulation of the BO equation:

$$u_{1,t} = (4i)^{-1} \int_{\mathbb{R}} dx_2 \delta_{12} \theta_{12}^{(1)} \gamma_{12}^{(2)} = (4i)^{-1} \int_{\mathbb{R}} dx_2 \delta_{12} \theta_{12}^{(2)} \gamma_{12}^{(1)},$$ (4.163)

where $\delta_{12} = \delta(x_1 - x_2)$.

4.4. Bäcklund transformations of recursion operators and of Hamiltonian structures

4.4.1. Bäcklund transformations of evolution equations

(Fokas and Fuchssteiner [1981a], Fuchssteiner and Fokas [1981], Fokas and Anderson [1982]. See also Section 1.5 of this volume.)

Let S_1, S_2, S_3 be linear vector spaces. Let $B(u, v)$ be a function of $u \in S_1$ and $v \in S_2$ with values in S_3. By B_u, B_v we denote Fréchet derivatives with respect to u, v.

DEFINITION 4.32. *A function $B(u, v)$ is called admissible if an implicit function, given by the equation $B(u, v) = 0$, gives rise to a one-to-one map between corresponding tangent spaces; i.e., we require, for $B = 0$, linear maps B_u and B_v, from S_1 to S_3 and from S_2 to S_3, respectively, be invertible.*

Consider two evolution equations:

$$u_t = K(u), \quad u(t) \in S_1$$ (4.164)

$$v_t = G(v), \quad v(t) \in S_2.$$ (4.165)

DEFINITION 4.33. *An admissible function $B(u, v)$ (or rather an implicit function between u and v, given by the equation $B(u, v) = 0$) is said to be a Bäcklund transformation between Equations 4.164 and 4.165 if, for all t*

$$B(u(t), v(t)) = 0 \quad whenever \quad B(u(0), v(0)) = 0.$$ (4.166)

The t-derivation of Equation 4.166 with the use of the evolution equations 4.164 and 4.165 yields:

$$B_u(u, v)[K(u)] + B_v(u, v)[G(v)] = 0 \quad \text{if} \quad B(u, v) = 0, \qquad (4.167')$$

or in short

$$B_u[K] + B_v[G] = 0 \quad \text{if} \quad B = 0. \qquad (4.167)$$

THEOREM 4.47. *An admissible function $B(u, v)$ is a Bäcklund transformation between evolution equations 4.164 and 4.165 iff it satisfies the condition 4.167.*

Example. (Miura [1968]). The KdV equation 4.4 is related to the modified KdV (MKdV) equation

$$v_t = v_{xxx} + 6v^2 v_x \qquad (4.168)$$

by the Bäcklund transformation

$$B(u, v) \equiv u - v^2 - iv_x = 0 \quad (i = \sqrt{-1}). \qquad (4.169)$$

It is easy to verify that the condition 4.167 is satisfied by Equation 4.169. That is, Equation 4.169 is indeed a transformation between Equations 4.4 and 4.168. In particular, $B_u = 1$, $B_v = -2v - i D_x$, and hence Equation 4.167, i.e., $u_{xxx} + 6uu_x - (2v + i D_x)(v_{xxx} + 6v^2 v_x) = 0$ becomes an identity in virtue of $u = v^2 + iv_x$.

4.4.2. Bäcklund transformations of symmetries and recursion operators

Let us introduce the operator $T: S_1 \rightarrow S_2$ by

$$T = (B_v(u, v))^{-1} B_u(u, v). \qquad (4.170)$$

We denote by D_u and D_v total Fréchet derivatives with respect to u and v, which act on functions $\Gamma(u, v): S_1 \times S_2 \rightarrow S_3$. They have to be distinguished from the Fréchet derivatives B_u, B_v, which are denoted by subscripts and treat u and v as independent variables. Total derivatives D_u and D_v take into account that u and v are related by an implicit function defined by $B(u, v) = 0$. Then $T = T(u)$ or $T = T(v)$ in Equation 4.170. We define a directional total derivative (in the "direction" z) of an operator-valued function $T(u)$ by

$$D_u T(u)[z]w = \frac{\partial}{\partial \epsilon} (T(u + \epsilon z)w) |_{\epsilon=0} . \qquad (4.171)$$

This notion and the following two theorems are due to Fokas and Fuchssteiner [1981a].

LEMMA. *If $B(u, v)$ is an admissible function and $B(u, v) = 0$, then*

$$D_v T[z]w = D_v T[Tw]T^{-1}z \qquad (4.172)$$

for all $z \in S_2$, $w \in S_1$. A similar property for $T^+: S_2^* \to S_1^*$ is

$$< D_v T^+[z]a, w > = < (D_v T^+)[Tw]a, T^{-1}z > = < a, (D_v T)[Tw]T^{-1}z >$$
$$(4.173)$$

for arbitrary $z \in S_2$, $w \in S_1$, $a \in S_2^*$.

THEOREM 4.48. *(Bäcklund transformation of recursion operators). Operator-valued function $R(u)$ is a recursion operator for Equation 4.164 iff*

$$L(v) = TRT^{-1}, \quad \text{where } B(u, v) = 0, \qquad (4.174)$$

is a recursion operator for Equation 4.165. Here the operator T is defined by Equation 4.170.

THEOREM 4.49. *Let B be an admissible implicit function of $u \in S_1$ and $v \in S_2$, and let $B(u, v) = 0$. Then $R(u)$ is a hereditary recursion operator iff the operator $L(v)$ defined by Equation 4.174 is hereditary.*

It follows that all the members of the hierarchy 4.60, generated by a hereditary recursion operator, possess the same Bäcklund transformation. In other words, if $B(u, v) = 0$ is a Bäcklund transformation between Equations 4.164 and 4.165, and if $R(u)$ is a hereditary recursion operator for Equation 4.164, then the equation $B(u, v) = 0$ defines, for every $n = 1, 2, ...$, a Bäcklund transformation between the equations

$$u_t = K_n(u) = R^n(u)K(u) \qquad (4.175)$$

and

$$v_t = G_n(v) = L^n(v)G(v), \qquad (4.176)$$

where $L(v)$ is the operator defined by Equation 4.174.

Example 1. For Equation 4.4, a hereditary recursion operator is given by Equation 4.43: $R = D_x^2 + 4u + 2u_x D_x^{-1}$ and is said to be hereditary (see Fuchssteiner [1979]). The operator T in Equation 4.170, which corresponds to the Bäcklund transformation 4.169 of Equation 4.4 into Equation 4.168, is $T = -(2v + iD_x)^{-1}$. In accordance with Equation 4.174, $L(v)$ is defined by the relation: $R(u)(2v + iD_x) = (2v + iD_x)L(v)$, where $u = v^2 + iv_x$. Hence

$$L(v) = D_x^2 + 4v^2 + 4v_x D_x^{-1} v \qquad (4.177)$$

is a hereditary recursion operator for Equation 4.168.

Ibragimov [1981], [1983] uses more general Bäcklund transformations of evolution equations, which involve transformations of independent variables x, t as well. We proceed now to the exposition of this theory.

DEFINITION 4.34. *A transformation of the form*

$$s = t, \quad y = \phi(x, u, u_1, ..., u_n), v = \Phi(x, u, u_1, ..., u_n), \qquad (4.178)$$

where ϕ, Φ are arbitrary differential functions, is called a restricted Bäcklund transformation, or a differential substitution.

The transformation 4.178 induces transformations $D_x \rightarrow D_y$, $u_i \rightarrow v_i = D_y^i[v]$, $u_t \rightarrow v_s$ according to the formulas:

$$D_x = D_x[\phi]D_y, \qquad (4.179)$$

$$D_t = D_s + D_t[\phi]D_y, \qquad (4.180)$$

$$v_y = v_1 = \frac{D_x[\Phi]}{D_x[\phi]}, \quad v_{i+1} = D_y[v_i], \qquad (4.181)$$

$$D_t[\Phi] = v_s + v_y D_t[\phi]. \qquad (4.182)$$

DEFINITION 4.35. *Evolution equations*

$$u_t = K(x, u, u_1, ..., u_m) \qquad (4.183)$$

and

$$v_s = G(y, v, v_1, ..., v_m) \qquad (4.184)$$

are said to be equivalent under restricted Bäcklund transformations if there exists a differential substitution 4.178 mapping Equation 4.183 into Equation 4.184.

The substitution 4.178 yields the operator

$$T = \Phi' - v_y \phi' \quad , \qquad (4.185)$$

which transforms symmetries and recursion operators. Here v_y is defined by Equation 4.181, and Φ', ϕ' are the Fréchet derivatives of Φ, ϕ with respect to u.

THEOREM 4.50. *(Ibragimov [1981,1983]). A function $\sigma(x, u)$ is a symmetry of Equation 4.183 iff the function $\rho(y, v)$ defined by*

$$\rho = T\sigma \mid_{(4.178)} \tag{4.186}$$

is a symmetry of Equation 4.184. Here we have used an abstract notion of symmetry (see Fuchssteiner [1979]) without the assumption that σ, ρ are differential functions.

DEFINITION 4.36. *(see [H1], Chapter 7). If a symmetry σ of Equation 4.183 is a differential function, then σ is called a local symmetry. If σ is a local symmetry of Equation 4.183, and ρ is a symmetry of Equation 4.184 related to σ by a transformation 4.186, then ρ is called a quasi-local symmetry (for Equation 4.184) associated with σ.*

Example 2. (For a more detailed discussion, see also [H1], Section 13.1.1.) Consider the equations of a one-dimensional adiabatic gas flow:

$$\rho_t + v\rho_x + \rho v_x = 0, \quad v_t + vv_x + \frac{1}{\rho}P_x = 0,$$

$$P_t + vP_x + B(P, \frac{1}{\rho})\frac{1}{\rho}v_x = 0, \quad B(P, \frac{1}{\rho}) = \frac{\partial S}{\partial(1/\rho)} \Big/ \frac{\partial S}{\partial P}. \tag{4.187}$$

Here ρ, v, P are the gas density, velocity, and pressure, respectively, and S is the entropy, which is a given function of the pressure P and of the specific volume $1/\rho$.

A transition from the Euler coordinates (t, x, ρ) to the Lagrangian mass variables (s, y, q) is given by the non-point transformation

$$s = t, \quad y = \int \rho dx, \quad q = 1/\rho. \tag{4.188}$$

Then the system (4.184) has the form

$$q_s - v_y = 0, \quad v_s + P_y = 0, \quad P_s + B(P, q)v_y = 0. \tag{4.189}$$

The transformation 4.188 takes the form 4.178 after introducing an intermediate system (Akhatov, Gazizov, and Ibragimov [1987a]):

$$R_t + vR_x = 0, \quad v_t + vv_x + P_x/R_x = 0,$$

$$P_t + vP_x + B(P, 1/R_x)v_x/R_x = 0. \tag{4.190}$$

The transition from Equations 4.190 to Equations 4.187 and 4.189 is performed by the differential substitutions $\rho = R_x$ and $s = t$, $y = R$, $q = 1/R_x$, respectively. Hence the Bäcklund transformation 4.188 as the composition of the second differential substitution with the inverse to the first differential substitution.

Consider the polytropic gas. Then $B(P, q) = \gamma P/q$. Here, we take $\gamma = 3$. Then the vector $\sigma = (\sigma_v, \sigma_p, \sigma_\rho)$ with

$$\sigma_v = -x + t(v + tv_t + xv_x), \quad \sigma_p = t(3P + tP_t + xP_x),$$

$$\sigma_\rho = t(\rho + t\rho_t + x\rho_x) \tag{4.191}$$

is a local symmetry of Equations 4.187 (Ovsiannikov [1962], [1978]). The Bäcklund transformation maps σ into the quasi-local symmetry $\rho = (\rho_v, \rho_p, \rho_q)$ for Equations 4.189, in accordance with Definition 4.36. Here, ρ is given by

$$\rho_v = sv + s^2 v_s - x, \quad \rho_p = 3sP + s^2 P_s, \quad \rho_q = -sq + s^2 q_s \tag{4.192}$$

where x a nonlocal variable defined by

$$x_y = q, \quad x_s = v. \tag{4.193}$$

THEOREM 4.51. *(Ibragimov [1981]). An operator-valued function $R(u, x)$ is a recursion operator for Equation 4.183 iff*

$$L(v, y) = TRT^{-1} \tag{4.194}$$

is a recursion operator for Equation 4.184. Here the operator T is defined by the formula 4.185.

This theorem gives a transformation law of recursion operators with respect to differential substitutions 4.178. It is in agreement with the transformation law 4.174 under Bäcklund transformations 4.166.

DEFINITION 4.37. *(Ibragimov [1983]). The evolution equations 4.183 and 4.184 are said to be equivalent with respect to a Bäcklund transformation*

$$y = \phi(t, x, u, u_1, ..., u_n), \quad s = \psi(t, x, u, u_1, ..., u_n),$$

$$v = \Phi(t, x, u, u_1, ..., u_n), \tag{4.195}$$

if this transformation maps Equations 4.183 and 4.184 one into another in accordance with the following formulas:

$$D_x = D_x[\phi]D_y + D_x[\psi]D_s, \quad D_t = D_t[\phi]D_y + D_t[\psi]D_s, \tag{4.196}$$

$$D_x[\phi]v_y + D_x[\psi]v_s = D_x[\Phi], \quad D_t[\phi]v_y + D_t[\psi]v_s = D_t[\Phi]. \tag{4.197}$$

Here, v_y and v_s are determined by

$$v_y = v_1 = D_y[v], \quad v_{i+1} = D_y[v_i].$$

The definition 4.185 of the transition operator T generalizes as follows:

$$T = \Phi' - v_y\phi' - v_s\psi'. \tag{4.198}$$

The corresponding transformation laws for symmetries and recursion operators are given by the following statements (Ibragimov [1983]).

THEOREM 4.52. *A function $\sigma(u, x, t)$ is a symmetry of Equation 4.183 if a function*

$$\rho = T\sigma \mid_{(4.195)} \tag{4.199}$$

is a symmetry of Equation 4.184.

THEOREM 4.53. *An operator $R(u, x, t)$ is a recursion operator for Equation 4.183 iff*

$$L(v, y, s) = TRT^{-1}, \tag{4.200}$$

is a recursion operator for Equation 4.184. Here, T is defined by the formula 4.198.

4.4.3. Bäcklund transformations of Hamiltonian structures

(Fuchssteiner [1980], Fuchssteiner and Fokas [1981], Anderson and Fokas [1982])

THEOREM 4.54. *For $v \in S_2$ define operators $\Omega(v): S_2^* \to S_2$ by*

$$\Omega(v) = T\theta(u)T^+, \tag{4.201}$$

where T is defined by a Bäcklund transformation 4.166 *via the formula* 4.170, *and* $\theta(u)$, $u \in S_1$ *are given operators* $S_1^* \to S_1$. *Then* Ω *is Hamiltonian iff* θ *is Hamiltonian.*

THEOREM 4.55.   *Let* $B(u, v) = 0$ *be a Bäcklund transformation between Equations* 4.164 *and* 4.165. *Let*

$$\Omega(v) = T\theta(u)T^+. \tag{4.202}$$

Then Ω *is a Noether operator for Equation* 4.165 *iff* θ *is a Noether operator for Equation* 4.164.

THEOREM 4.56.   *For* $v \in S_2$ *define operators* $\tilde{J}(v)$: $S_2 \to S_2^*$ *by*

$$\tilde{J}(v) = (T^+)^{-1} J(u) T^{-1} \tag{4.203}$$

where T is defined by the formula 4.170 *corresponding to a Bäcklund transformation* $B(u, v) = 0$. *Let* $J(u)$, $u \in S_1$ *be given operators* $S_1 \to S_1^*$. *Then* $\tilde{J}$ *is a symplectic (or an inverse Noether) operator for Equation* 4.165 *iff* J *is a symplectic (or an inverse Noether) operator for Equation* 4.164.

THEOREM 4.57.   *The transformation* 4.201 *preserves a compatibility of Hamiltonian operators.*

Consider two Hamiltonian systems:

$$u_t = \theta_1(u) \operatorname{grad} H_1(u) \tag{4.204}$$

and

$$v_t = \Omega_1(v) \operatorname{grad} \tilde{H}_1(v). \tag{4.205}$$

Here, θ_1 and Ω_1 are Hamiltonian operators, which are related through Bäcklund transformation $B(u, v) = 0$ and the corresponding operator T is defined by the formula 4.170. Assume that $\theta_1(u)$ is invertible. Then:
1) the operator

$$\theta_2(u) = T^{-1} \Omega_1(v)(T^{-1})^+, \text{ where } B = 0, \tag{4.206}$$

is a second Hamiltonian operator, and Equation 4.204 can be written

$$u_t = \theta_2(u) \operatorname{grad} H_2(u), \tag{4.207}$$

2) the operator

$$\tilde{J}_2(v) = (T^{-1})^+ \theta_1^{-1}(u) T^{-1}, \quad \text{where } B = 0, \tag{4.208}$$

is a symplectic operator, which defines a second Hamiltonian structure for Equation 4.205,

3) the operators

$$R(u) = \theta_2(u)\theta_1^{-1}(u), \quad L(v) = \Omega_1(v)\tilde{J}_2(v), \quad \text{where } B = 0, \tag{4.209}$$

are hereditary recursion operators for Equations 4.204 and 4.205, respectively.

Example. (see Kupershmidt and Wilson [1981]). The KdV equation 4.4 is related to the MKdV equation

$$v_t = v_{xxx} - 6v^2 v_x \tag{4.210}$$

by the Miura transformation

$$u = v_x - v^2. \tag{4.211}$$

The MKdV equation can be written in a Hamiltonian form, viz.

$$v_t = (-\frac{1}{2}D_x)\frac{\delta H_1}{\delta v} \tag{4.212}$$

with

$$H_1 = u^2 = (v_x - v^2)^2. \tag{4.213}$$

According to Equation 4.206, one obtains a second Hamiltonian operator for the KdV equation:

$$(D_x - 2v)(-\frac{1}{2}D_x)(D_x - 2v)^+ = \frac{1}{2}D_x^3 + uD_x + D_x u. \tag{4.214}$$

Hence the KdV equation admits a second Hamiltonian form (see Magri [1978]):

$$u_t = (\frac{1}{2}D_x^3 + uD_x + D_x u)\frac{\delta H_2}{\delta u} \tag{4.215}$$

with $H_2 = u^2$.

4.5. Inverse scattering transform and hereditary recursion operators

(Fuchssteiner [1980], Fokas and Anderson [1982], Oevel and Fokas [1982])

If there exists an isospectral eigenvalue problem for solving a given equation by the inverse scattering transform, then the knowledge of the eigenvalue problem provides one by a hereditary recursion operator for the whole hierarchy of equations solvable by this eigenvalue problem.

LEMMA 4.1. *Consider a linear eigenvalue problem*

$$\Delta \Psi = \mu \Psi, \tag{4.216}$$

where $\Delta(u, x, t)$: $S^* \to S^*$, $u \in S$ and $\Psi(u, x, t)$: $S \to S^*$. Assume that, if u satisfies Equation 4.1, then Equation 4.216 is isospectral, i.e., $D_t[\mu] = 0$. Let Ψ be a conserved covariant for Equation 4.1 (Definition 4.3). Then

$$(\Delta_t + \Delta'[K] + [(K')^+, \Delta])\Psi = 0. \tag{4.217}$$

THEOREM 4.58. *If Equation 4.216 is satisfied for a sufficiently large number of solutions Ψ of Equation 4.35, then $R = \Delta^+$ is a recursion operator for Equation 4.1.*

THEOREM 4.59. *Consider a linear isospectral eigenvalue problem for Equation 4.1*

$$L\Psi = \lambda \Psi \tag{4.218}$$

where $L = L(u, x, t)$. Let $D_t[\lambda]\,|_{u_t = K} = 0$ so that the eigenvalues λ, which depend on u, are conserved quantities for Equation 4.1. Denote by G_λ the gradient of the eigenvalue λ with respect to u. If Equation 4.218 transforms into an eigenvalue equation 4.216, satisfied by G_λ:

$$\Delta G_\lambda = \mu(\lambda)G_\lambda \tag{4.219}$$

then the operator

$$R = \Delta^+ \tag{4.220}$$

is a recursion operator for Equation 4.1.

LEMMA 4.2. *Assume that a recursion operator R for Equation 4.1 satisfies the isospectral eigenvalue problem*

$$R^+ G_\lambda = \mu G_\lambda \qquad (4.221)$$

where G_λ are gradients of conserved quantities λ of Equation 4.1. Let the conserved quantities λ be the eigenvalues of a linear scattering eigenvalue problem 4.218. Then λ and R are conserved quantities and a recursion operator, respectively, for the whole hierarchy 4.60 generated by R from Equation 4.1.

REMARK. The eigenvalue problem is isospectral for the whole hierarchy of Equations 4.60.

THEOREM 4.60. *Assume that a recursion operator R for Equation 4.1 satisfies the isospectral eigenvalue problem 4.221. Then all symmetries of Equation 4.1 generated by R commute and R is hereditary provided that these symmetries are dense.*

Example. (Oevel and Fokas [1982]). Consider the cylindrical Korteweg-de Vries equation

$$u_t + u_{xxx} + uu_x + \frac{u}{2t} = 0. \qquad (4.222)$$

The isospectral eigenvalue problem associated with this equation is given by (see Calogero and Degasperis [1978])

$$(t D_x^2 - \frac{x}{12} + \frac{1}{6}tu)\Psi = \lambda\Psi. \qquad (4.223)$$

Taking a directional derivative of Equation 4.223 and the scalar product with Ψ, one obtains

$$G_\lambda = \text{grad}\lambda = t\Psi^2. \qquad (4.224)$$

It follows:

$$(t D_x^2 + \frac{2}{3}tu - \frac{t}{3}D_x^{-1}u_x - \frac{x}{3} + \frac{1}{6}D_x^{-1})(t\Psi^2) = 4\lambda(t\Psi^2). \qquad (4.225)$$

Equations 4.219, 4.220, and 4.225 yield the hereditary recursion operator for Equation 4.223:

$$R(u, x, t) = t(D_x^2 + \frac{2}{3}u + \frac{1}{3}u_x D_x^{-1}) - \frac{x}{3} - \frac{1}{6}D_x^{-1}. \qquad (4.226)$$

5

Calculation of Symmetry Groups for Integro-Differential Equations

The main obstacle for the application of Lie's infinitesimal technique to integro-differential equations is their nonlocality. These equations do not have a frame locally defined in the space $\mathcal{A}$ of differential functions (see [H1], Section 1.3, and [H2], Section 1.3.1). Consequently, the algorithmic approach based on the definition of a symmetry group as a group of transformations leaving the frame unalterable (see the *second definition* of a symmetry group in [H2], Section 1.3.3) does not apply to integro-differential equations.

However, one can use the definition of a symmetry group as a group of transformations converting every solution of an equation into a solution of the same equation (see the *first definition* of a symmetry group in [H2], Section 1.3.2). In this way, one encounters no complexity in constructing determining equations for infinitesimal symmetries. The problem is, however, in solving the determining equations.

This chapter is aimed to illustrate, by an example, a practical method for solving the determining equations for integro-differential equations. Though the following calculations are pertinent to the particular illustrative example, it is hoped, however, that the reader can readily adapt them to his own problems.

Several *ad hoc* methods are also known in the literature. Namely, *an a priori choice* of a simple form of symmetries, a transition to an infinite system of differential equations by using the *method of moments*, a transformation of the integral equation under consideration into a differential equation by means of a *non-local transformation*, etc. For these simplified approaches, see, e.g., Nikol'skii [1963a], [1963b], [1965]; Bobylev [1975a] to [1984]; Taranov [1976]; Krook and Wu [1977]; Ernst and Hendriks [1979]; Muncaster [1979]; Bunimovich and Krasnoslobodtsev [1982], [1983]; Nonenmacher [1984]; Grigoriev and Meleshko [1986]; Bobylev and Ibragimov [1989]; and Fushchich, Shtelen and Serov [1989].

In this Handbook, symmetries of integro-differential equations were previously presented in [H1], Chapter 15, and [H2], Chapter 16.

5.1. Determining equation

Let

$$F(x, [u]) = 0 \qquad (5.1)$$

be an integro-differential equation. Here, $x = (x^1, \ldots, x^n)$ and $u = (u^1, \ldots, u^m)$ are the independent and dependent variables, respectively, $F(x, [u])$ is an integro-differential operator involving x, u and finite-order derivatives of u.

DEFINITION. *A one-parameter group G of transformations*

$$\overline{x} = f(x, u, a), \quad \overline{u} = h(x, u, a) \qquad (5.2)$$

is said to be a symmetry group for Equation 5.1 if G converts every solution $u(x)$ of Equation 5.1 into a solution $u_a(x)$ of the same equation.

The term "Equation 5.1 admits the group G" is used interchangeably in the literature (cf. [H1], Section 1.3, or [H2], Section 1.3.2).

THEOREM. *An integro-differential equation 5.1 admits the group G of Transformations 5.2 if and only if*

$$\left. \frac{\partial}{\partial a} F(x, [u_a(x)]) \right|_{a=0} = 0 \qquad (5.3)$$

for every solution $u(x)$ of Equation 5.1. Here $u_a(x)$ is obtained from the solution $u(x)$ by the transformation 5.2.

Equation 5.3 is termed *the determining equation* for the symmetry group G. In general, the determining equation 5.3 is written as a complicated equation, for the coordinates of the generator

$$X = \xi^i \frac{\partial}{\partial x^i} + \eta^\alpha \frac{\partial}{\partial u^\alpha} \qquad (5.4)$$

of the group G, involving non-local (in particular, integral) operators.

REMARK.   If $F \in \mathcal{A}$ is a differential function, i.e., if Equation 5.1 is a system of differential equations, then Equation 5.3 reduces to Lie's determining equation

$$\left. XF \right|_{(5.1)} = 0.$$

The practical construction of determining equations for integro-differential equations is performed by using the canonical Lie-Bäcklund's representation (see Chapter 1, Section 1.1.4) of the operator 5.4:

$$\tilde{X} = \tilde{\eta}^\alpha \frac{\partial}{\partial u^\alpha}, \quad \tilde{\eta}^\alpha = \eta^\alpha - \xi^i u_i^\alpha, \tag{5.5}$$

and by replacing the partial derivatives $\partial/\partial u^\alpha$ by the Fréchet derivative.

5.2. Illustrative example

(Grigoriev and Meleshko [1986], [1987], [1990])
Consider the integro-differential equation (Bobylev [1975a], [1984])

$$F(x, t, [u]) \equiv \frac{\partial u(x, t)}{\partial t} + u(x, t)u(0, t) - \int_0^1 u(xs, t)u(x(1-s), t)\, ds. \tag{5.6}$$

The group generators are written in the form

$$X = \xi(x, t, u)\frac{\partial}{\partial x} + \eta(x, t, u)\frac{\partial}{\partial t} + \zeta(x, t, u)\frac{\partial}{\partial u}. \tag{5.7}$$

In this example, the determining equation 5.3 has the following form:

$$D_t \psi(x, t) + \psi(0, t)u(x, t) + \psi(x, t)u(0, t) \tag{5.8}$$

$$-2 \int_0^1 u(x(1-s)s, t)\psi(xs, t)\, ds = 0,$$

where D_t is the total differentiation with respect to t, and the function ψ is determined by Operator 5.7 in accordance with the transition to the canonical operator 5.5:

$$\psi(x, t) = \zeta(x, t, u(x, t)) - \xi(x, t, u(x, t))u_x(x, t) - \eta(x, t, u(x, t))u_t(x, t).$$

In Equation 5.8, $u(x, t)$ is an arbitrary solution of Equation 5.6.

5.3. The approach

We restrict the determining equation 5.8 to the subset of solutions of Equation 5.6 determined by the initial conditions[1]

$$u(x, t_0) = bx^n, \quad b = \text{const.}, \tag{5.9}$$

where n is a positive integer. More precisely, we let $t = t_0$ in the determining equation 5.8, and substitute the function $u(x, t)$ and its derivatives obtained from Equations 5.9 and 5.6. We consider the resulting equation at an arbitrary initial time t_0 which is denoted again by t. Accordingly, Equation 5.8 is written in terms of the following functions:

$$\hat{\xi}(x, t) = \xi(bx^n, x, t), \quad \hat{\eta}(x, t) = \eta(bx^n, x, t),$$

$$\hat{\zeta}(x, t) = \zeta(bx^n, x, t), \quad \hat{\zeta}_u(x, t) = \zeta_u(bx^n, x, t),$$

$$\hat{\zeta}_t(x, t) = \zeta_t(bx^n, x, t), \quad \hat{\zeta}_x(x, t) = \zeta_x(bx^n, x, t), \dots \tag{5.10}$$

Then we solve Equation 5.8 by letting $n = 0, 1, 2, \dots$ in the initial condition 5.9, and simultaneously varying the parameter b.

In this particular example, these calculations are sufficient for obtaining operators 5.7 admitted by the integro-differential equation under consideration.

5.4. Solution of the determining equation

We proceed now to the calculations. The coefficients of Operator 5.7 are assumed to be locally analytic functions. Hence they can be represented by the Taylor series with respect to u:

$$\xi(x, t, u) = \sum_{l \geq 0} q_l(x, t)u^l,$$

$$\eta(x, t, u) = \sum_{l \geq 0} r_l(x, t)u^l,$$

---

[1] Solvability of the Cauchy problems 5.6 – 5.9 is proved, e.g., in Bobylev [1984].

$$\zeta(x, t, u) = \sum_{l \geq 0} p_l(x, t) u^l. \tag{5.11}$$

Let $n = 0$ in Equation 5.9. Then the determining equation 5.8 has the form

$$\hat{\zeta}_t(x, t) + b\hat{\zeta}(0, t) + b\hat{\zeta}(x, t) - 2b \int_0^1 \hat{\zeta}(xs, t) \, ds = 0.$$

It follows:

$$\frac{\partial p_0}{\partial t} = 0, \quad \frac{\partial p_{l+1}(x, t)}{\partial t} + p_l(x, t) + p_l(0, t) - 2 \int_0^1 p_l(xs, t) \, ds = 0 \tag{5.12}$$

with $l = 0, 1, \dots$.
Let $n \geq 1$ in Equation 5.9. Then

$$u_t = P_n b^2 x^{2n}, \quad u_x = nbx^{n-1}, \quad u_{tt} = Q_n b^3 x^{3n}, \quad u_{tx} = 2n P_n b^2 x^{2n-1},$$

and the determining equation 5.8 yields:

$$\hat{\zeta}_t + b[-nx^{n-1}\hat{\xi}_t + x^n\hat{\zeta}(0, t) - 2x^n \int_0^1 (1 - s)^n \hat{\zeta}(xs, t) \, ds]$$

$$+ b^2[-P_n x^{2n}\hat{\eta}_t + P_n x^{2n}\hat{\zeta}_u - 2n P_n x^{2n-1}\hat{\xi} - \delta_{n1}x\hat{\xi}(0, t)$$

$$+ 2nx^{2n-1} \int_0^1 (1 - s)^n s^{n-1}\hat{\xi}(xs, t) \, ds]$$

$$+ b^3[-n P_n x^{2n-1}\hat{\xi}_u - Q_n x^{3n}\hat{\eta} + 2P_n x^{3n} \int_0^1 (1 - s)^n s^{2n}\hat{\eta}(xs, t) \, ds]$$

$$- b^4[P_n^2 x^{4n}\hat{\eta}_\varphi] = 0 \tag{5.13}$$

where

$$P_n = \frac{(n!)^2}{(2n + 1)!}, \quad Q_n = 2P_n \frac{(2n)!n!}{(3n + 1)!}.$$

Now we treat b as an arbitrary parameter. Accordingly, we split Equation 5.13 into a series of equations by equating to zero the coefficients of b^k, $k = 0, 1, \dots$ in the left-hand side of Equation 5.13. The resulting system of equations, together with Equations 5.12, solve the determining equation 5.8.

It follows from the expansions 5.11 and the expression 5.10 of $\hat{\zeta}_t$, that for $k = 0$ the corresponding coefficient in the left-hand side of Equation 5.13 vanishes in virtue of the first equation 5.12.

For $k = 1$, Equation 5.13 yields:

$$x[-p_0(x,t) + 2\int_0^1 (1 - (1-s)^n)p_0(xs,t)\,ds] - n\frac{\partial q_0(x,t)}{\partial t} = 0.$$

Since n is arbitrary, it follows:

$$p_0(x,t) = 0, \qquad \frac{\partial q_0(x,t)}{\partial t} = 0.$$

Hence $\hat{\zeta}(0,t) = 0$.

Similarly, one obtains for $k = 2$:

$$x[-p_1(x,t) - p_1(0,t) + 2\int_0^1 (1 - (1-s)^n s^n)p_1(xs,t)\,ds - P_n\frac{\partial r_0(x,t)}{\partial t}$$

$$+ P_n p_1(x,t)] - n\frac{\partial q_1(x,t)}{\partial t} - 2nP_n q_0(x,t) + 2n\int_0^1 (1-s)^n s^{n-1} q_0(xs,t)\,ds = 0.$$

Whence

$$p_1(x,t) = c_0 + c_1 x, \quad \frac{\partial q_1(x,t)}{\partial t} = 0, \quad q_0(x,t) = c_2 x, \quad \frac{\partial r_0(x,t)}{\partial t} = -c_0$$

where c_0, c_1, c_2 are arbitrary constants.

For $k = 3$, one has

$$x^{n+1}[-p_2(x,t) - p_2(0,t) + 2\int_0^1 (1 - (1-s)^n s^{2n})p_2(xs,t)\,ds$$

$$-P_n\frac{\partial r_1(x,t)}{\partial t} + 2P_n p_2(x,t) + 2P_n\int_0^1 (1-s)^n s^{2n} r_0(xs,t)\,ds - Q_n r_0(x,t)]$$

$$+ x^n[-n\frac{\partial q_2(x,t)}{\partial t} - 2nP_n q_1(x,t) + 2n\int_0^1 (1-s)^n s^{2n-1} q_1(xs,t)\,ds]$$

$$-nP_n q_1(x,t) = 0.$$

Whence

$$q_1(x,t) = 0, \quad \frac{\partial q_2(x,t)}{\partial t} = 0, \quad p_2(x,t) = 0, \quad \frac{\partial r_1(x,t)}{\partial t} = 0, \quad r_0(x,t) = -c_0 t + c_3$$

where c_3 is an arbitrary constant.

For $k = 4 + l$, $l = 0, 1, \ldots$, Equation 5.13 yields

$$x^{n+1}[\frac{\partial p_{\alpha+4}(x, t)}{\partial t} - 2\int_0^1 (1 - s)^n s^{(3+\alpha)n} p_{3+\alpha}(xs, t) \, ds$$

$$+(3 + \alpha) P_n p_{3+\alpha}(x, t) - P_n \frac{\partial r_{2+\alpha}(x, t)}{\partial t} - (\alpha + 1) P_n^2 r_{\alpha+1}(x, t)$$

$$+2 P_n \int_0^1 (1 - s)^n s^{(3+\alpha)n} r_{\alpha+1}(xs, t) \, ds - Q_n r_{\alpha+1}]$$

$$+x^n[-n \frac{\partial q_{3+\alpha}(x, t)}{\partial t} - 2n P_n q_{\alpha+2}(x, t) + 2n \int_0^1 (1 - s)^n s^{(3+\alpha)n-1} q_{\alpha+2}(xs, t) \, ds]$$

$$-n(\alpha + 2) P_n q_{\alpha+2}(x, t) = 0.$$

Whence

$$p_{l+3}(x, t) = 0, \quad q_{l+2}(x, t) = 0, \quad r_{l+1}(x, t) = 0, \quad l = 0, 1, \ldots .$$

It follows from the above equations that

$$\xi = c_3 x, \quad \eta = c_1 - c_4 t, \quad \zeta = (c_2 x + c_4) u$$

with arbitrary constants c_1, c_2, c_3, c_4.

5.5. The symmetry Lie algebra

Thus, Equation 5.6 admits the four-dimensional Lie algebra spanned by the operators

$$X_1 = \frac{\partial}{\partial t}, \quad X_2 = xu \frac{\partial}{\partial u}, \quad X_3 = x \frac{\partial}{\partial x}, \quad X_4 = u \frac{\partial}{\partial u} - t \frac{\partial}{\partial t}. \tag{5.14}$$

Part II

Body of Results

6

Group Theoretic Modeling

The present chapter is designed to emphasize a fundamental significance of the Lie group philosophy in mathematical modeling.

Experience gained from the relativistic physics and from the group analysis of concrete problems has convinced us that numerous natural phenomena can be modeled directly in group theoretical terms. Differential equations, conservation laws, solutions to initial and/or boundary value problems, and so forth can be obtained as immediate consequences of the group invariance principle ([H2], Chapter 5, and Chapter 3 of this volume; see also Ibragimov [1993]).

Here, no adherence to general discussions is attempted; the emphasis is on physically relevant examples.

6.1. Galilean principle in diffusion problems

It is shown that the Galilean group determines the differential equations of linear diffusion. Further, it is demonstrated that the extension of this group by scaling transformations is unique. The extended Galilean group determines (uniquely) the linear heat equation. Consequently, this group allows one to construct the fundamental solution and hence, to solve the problem of the heat diffusion with an arbitrary initial distribution of the temperature field.

Nonlinear diffusion type equations can also be obtained from the Galilean principle. Here, this problem is considered in the one-dimensional case.

The material of this section is integrated with that of Chapter 3, Section 3.2, and is based on the recent works of Ibragimov [1989], [1994d].

6.1.1. Statement of the Galilean principle

The group analysis, due to Lie [1881], of the heat equation

$$u_t = u_{xx} \tag{6.1}$$

makes evident how the temperature behaves (in the linear approximation) under a change to different inertial coordinate frames. Namely, Equation 6.1 is invariant under the Galilean transformation

$$\bar{t} = t, \quad \bar{x} = x + 2at, \tag{6.2}$$

provided that the temperature u transforms as follows:

$$\bar{u} = ue^{-(ax+a^2t)}. \tag{6.3}$$

It follows that if an observer at rest in the coordinate frame x detects the temperature field

$$u = \tau(t, x),$$

an observer at rest in the inertial frame $\bar{x}$, moving with the velocity $V = 2a$ relative to the original frame, will detect the temperature field

$$\bar{u} = e^{(a^2t - a\bar{x})} \tau(t, \bar{x} - 2at).$$

Thus, if one acknowledges a physical significance of the linear heat equation 6.1, one has to accept the following as well:

Galilean principle in the linear heat diffusion. *The temperature is not a scalar under the Galilean transformation 6.1. It behaves, in different inertial frames with the relative velocity $V = 2a$, in accordance with the transformation law 6.3.*

In other words, the Galilean principle states that the linear heat equation is invariant under the one-parameter group of Transformations 6.2 and 6.3 with the generator

$$Y = 2t\frac{\partial}{\partial x} - xu\frac{\partial}{\partial u}. \tag{6.4}$$

The Galilean principle in the multi-dimensional case is formulated similarly with replacing Generator 6.4 by the corresponding Galilean boosts along the x^i axes.

6.1.2.   Semi-scalar linear representations of the Galilean group

Consider the Galilean group in the (t, x) space, where t is the time coordinate and $x = (x^1, ..., x^n) \in \mathbf{R}^n$. It comprises the isometric motions (i.e., the translations and rotations) in $\mathbf{R}^n$ with the infinitesimal generators

$$X_i = \frac{\partial}{\partial x^i}, \quad X_{ij} = x^j\frac{\partial}{\partial x^i} - x^i\frac{\partial}{\partial x^j}, \quad i, j = 1, ..., n, \tag{6.5}$$

and the time translations and the Galilean boost generated, respectively, by

$$X_0 = \frac{\partial}{\partial t}, \quad Y_i = 2t \frac{\partial}{\partial x^i}. \tag{6.6}$$

We treat here the case of one *differential* variable u (see Chapter 1, Section 1.1.1) with $n + 1$ independent variables t, x. The differential variable is termed also the *dependent* variable.

It is well known (Lie [1881], Ovsiannikov [1962]) that under the Lie point symmetry groups of linear partial differential equations, the dependent variable u undergoes linear transformations. Therefore, when dealing with linear equations, one considers only linear group representations in the variable u. Furthermore, we consider here homogeneous equations. Consequently, we deal with groups containing the dilation generator $T_1 = u\partial/\partial u$.

DEFINITION 6.1. *A linear representation of the Galilean group is said to be semi-scalar if its infinitesimal generators have the following form:*

$$X_0 = \frac{\partial}{\partial t}, \quad X_i = \frac{\partial}{\partial x^i}, \quad X_{ij} = x^j \frac{\partial}{\partial x^i} - x^i \frac{\partial}{\partial x^j},$$

$$T_1 = u \frac{\partial}{\partial u}, \quad Y_i = t \frac{\partial}{\partial x^i} + \mu^i(t, x) u \frac{\partial}{\partial u}, \quad i, j = 1, ..., n. \tag{6.7}$$

That is, u is invariant under the time translations and isometric motions in $\mathbf{R}^n$. In the particular case $\mu^i(t, x) \equiv 0$, $i = 1, \dots, n$, the operators 6.7 define the scalar representation of the Galilean group.

THEOREM 6.1. *There exist[1] two non-similar semi-scalar linear representations of the Galilean group. Namely, the scalar representation defined by Operators 6.7 with $\mu_i(t, x) \equiv 0$, $i = 1, \dots, n$, and the representation with the generators*

$$X_0 = \frac{\partial}{\partial t}, \quad X_i = \frac{\partial}{\partial x^i}, \quad X_{ij} = x^j \frac{\partial}{\partial x^i} - x^i \frac{\partial}{\partial x^j},$$

$$T_1 = u \frac{\partial}{\partial u}, \quad Y_i = 2t \frac{\partial}{\partial x^i} - x^i u \frac{\partial}{\partial u}, \quad i, j = 1, ..., n. \tag{6.8}$$

PROOF. One can verify by inspection that the linear span of Operators 6.7 is closed under the Lie bracket (commutator) if and only if $\mu^i(t, x) = kx^i$, $i =$

---

[1]N.H. Ibragimov, unpublished.

$1, \ldots, n$. If $k \neq 0$, one can set $k = -1/2$ by choosing an appropriate scaling of t or x. The operators 6.8 and the operators 6.7 with $\mu_i(t, x) \equiv 0$, $i = 1, \ldots, n$, span two non-isomorphic Lie algebras. Consequently, these Lie algebras define two non-similar semi-scalar linear representations of the Galilean group.

DEFINITION 6.2.   *The group* **H** *with the infinitesimal generators 6.8 is called the heat representation of the Galilean group.*

Thus, there exist two distinctly different semi-scalar linear representations of the Galilean group, namely, the scalar representation and the heat representation.

6.1.3.   Derivation of diffusion equations from the Galilean principle

THEOREM 6.2.   *A second-order linear differential equation is invariant under the heat representation* **H** *of the Galilean group if and only if it has the form*

$$u_t = \Delta u + ku, \quad k = \text{const.,} \tag{6.9}$$

where Δ is the n-dimensional Laplacian in $\mathbf{R}^n$.

PROOF.   We set $t = x^0$ and consider the equations

$$a^{\alpha\beta} u_{\alpha\beta} + b^\alpha u_\alpha + cu = 0, \tag{6.10}$$

with constant coefficients $a^{\alpha\beta}, b^\alpha, c$, where $\alpha, \beta = 0, \ldots, n$. The subscripts α, β denote the derivations with respect to x^α, x^β. This is the general form of the linear equations that are invariant under the translations and dilations with the generators $X_\alpha, \alpha = 0, \ldots, n$, and T_1.

Thus, the problem is to determine the coefficients of Equation 6.10 from the invariance conditions under the generators X_{ij} and Y_i of the rotations and the Galilean transformations.

Consider the operators X_{ij}. After the second prolongation, they have the form (there is no necessity to use a special notation to indicate the prolongation; see the footnote to Section 2.5.5 of Chapter 2):

$$X_{ij} = x^j \frac{\partial}{\partial x^i} - x^i \frac{\partial}{\partial x^j} + u_j \frac{\partial}{\partial u_i} - u_i \frac{\partial}{\partial u_j} + (\delta_k^i u_{\alpha j} + \delta_\alpha^i u_{kj} - \delta_k^j u_{\alpha i} - \delta_\alpha^j u_{ki}) \frac{\partial}{\partial u_{\alpha k}} \quad .$$

Here, $i, j, k = 1, \ldots, n$, $\alpha = 0, \ldots, n$. Hence:

$$X_{ij}(a^{\alpha\beta} u_{\alpha\beta} + b^\alpha u_\alpha + cu) = a^{\alpha i} u_{\alpha j} + a^{ik} u_{kj} - a^{\alpha j} u_{\alpha i} - a^{jk} u_{ki} + b^i u_j - b^j u_i \; .$$

By the infinitesimal invariance test, this expression vanishes on the frame of the differential equation 6.10. By inspecting this test, one arrives at the equation

$$Au_{tt} + a\Delta u + bu_t + cu = 0, \quad A, a, b, c = \text{const.} \tag{6.11}$$

Finally, consider the invariance conditions of Equation 6.11 under Y_i. After the second prolongation, these operators have the form:

$$Y_i = 2t\frac{\partial}{\partial x^i} - x^i u\frac{\partial}{\partial u} - (x^i u_t + 2u_i)\frac{\partial}{\partial u_t} - (\delta_k^i u + x^i u_k)\frac{\partial}{\partial u_k} -$$

$$(x^i u_{tt} + 4u_{ti})\frac{\partial}{\partial u_{tt}} - (\delta_k^i u_t + x^i u_{tk} + 2u_{ik})\frac{\partial}{\partial u_{tk}} - (\delta_k^i u_l + \delta_l^i u_k - x^i u_{kl})\frac{\partial}{\partial u_{kl}}$$

In virtue of this expression, the invariance conditions

$$Y_i(Au_{tt} + a\Delta u + bu_t + cu) = 0, \quad i = 1, ..., n,$$

yield:

$$A = 0, \quad a + b = 0.$$

Hence, by setting $k = c/a$, one obtains the equation 6.9.

The following statement can be proved in a similar way.

THEOREM 6.3. *A second-order linear differential equation is invariant under the scalar representation of the Galilean group (i.e., under the group with the infinitesimal generators 6.7 with $\mu^i(t, x) \equiv 0$) if and only if it is the Helmholtz equation*

$$\Delta u + ku = 0, \quad k = \text{const.}$$

6.1.4. Derivation of the heat equation

THEOREM 6.4. *The heat representation* **H** *of the Galilean group admits an extension by scaling transformations (dilations). This extension is unique and is given by the following generator:*

$$T_2 = 2t\frac{\partial}{\partial t} + x^i\frac{\partial}{\partial x^i}. \tag{6.12}$$

PROOF. We look for the most general infinitesimal dilation generator

$$T = at\frac{\partial}{\partial t} + b^1 x^1\frac{\partial}{\partial x^1} + ... + b^n x^n\frac{\partial}{\partial x^n} \tag{6.13}$$

such that the linear span of the operators 6.8 and 6.13 is a Lie algebra. Here, a and b^i are arbitrary constants.

The commutator of Operator 6.13 with X_{ij} is given by (no summation with respect to i, j):

$$[T, X_{ij}] = (b^j - b^i)(x^j \frac{\partial}{\partial x^i} + x^i \frac{\partial}{\partial x^j}), \quad i, j = 1, ..., n.$$

We require that these commutators belong to the linear span of the operators 6.8 and 6.13. It follows that $b^1 = \cdots = b^n$.

Hence, T has the form

$$T = at \frac{\partial}{\partial t} + bx^i \frac{\partial}{\partial x^i}.$$

The commutator of T with the last operator 6.8, i.e., with Y_i has the form

$$[T, Y_i] = 2(a - b)t \frac{\partial}{\partial x^i} - bx^i u \frac{\partial}{\partial u}.$$

It belongs to the linear span of the operators 6.8 and 6.13 if and only if $a = 2b$. Hence, we arrive at the operator 6.12 considered up to an incidental constant factor.

DEFINITION 6.3. *The group with the infinitesimal generators 6.8 and 6.12 is termed the extended heat representation of the Galilean group. This extension of the group* **H** *by scaling transformations is denoted by* **S.**

THEOREM 6.5. *The heat equation*

$$u_t = \Delta u \tag{6.14}$$

is an only linear second-order equation invariant under the group **S.**

PROOF. In virtue of Theorem 6.2, it suffices to identify those diffusion equations 6.9 invariant under the scaling generator 6.12. The invariance test has the form (here, the prolonged action of the operator T_2 is implied):

$$T_2(u_t - \Delta u - ku)\Big|_{(6.9)} \equiv 2ku = 0.$$

It follows that $k = 0$.

REMARK.   Equation 6.14 admits the extension of the group **S** by the projective transformations (see Section 3.2.1, where the generator of the projective transformations is denoted by Y). However, this extension is not necessary to obtain the heat equation and therefore it is considered as an *excess symmetry* of Equation 6.14. Cf. Remark 1 in Section 3.2.3.

6.1.5.   Determination of a linear heat diffusion

The Galilean principle can be used to solve an arbitrary Cauchy problem for the linear heat equation 6.14. Namely, it was observed in Ibragimov [1989] that the fundamental solution for the heat equation is determined up to a constant factor by the assumption of its Galilean invariance. The constant factor was found by substituting into Equation 6.14.

Subsequently, it was shown (Ibragimov [1994d]) that the extended heat representation **S** of the Galilean group determines the fundamental solution *uniquely*, without using the differential equation 6.14.

Thus, *given an initial distribution of the temperature, the linear heat diffusion is governed completely by the Galilean principle.*

For the group theoretic derivation of the fundamental solution, the reader is referred to Section 3.2, Theorem 3.3.

REMARK.   For the non-homogeneous equation

$$u_t - \Delta u = f(x),$$

the group **S** is not a symmetry group. However, **S** is a group of equivalence transformations for the family of non-homogeneous linear heat equations with arbitrary $f(x)$ (for the definition of an equivalence group, see, e.g., [H2], Chapter 2). This fact is a group theoretic background for solving the Cauchy problem for non-homogeneous heat equations.

6.1.6.   Nonlinear diffusion type equations

We will consider here the one-dimensional case ($n = 1$). The multi-dimensional case can be treated similarly.

DEFINITION 6.4.   *The evolution equation*

$$u_t = F(x, u, u_x, u_{xx})$$

is said to be a diffusion type if it is invariant under the heat representation **H** *of the Galilean group.*

THEOREM 6.6.   *The most general diffusion type equation has the form*

$$u_t = u_{xx} + f\left[\left(\frac{u_x}{u}\right)_x\right]. \tag{6.15}$$

PROOF.   The group **H** has the following two functionally independent invariants depending on u, u_t, u_x, u_{xx} :

$$J_1 = u_t - u_{xx}, \quad J_2 = \frac{u_{xx}}{u} - \frac{u_x^2}{u^2} \equiv \left(\frac{u_x}{u}\right)_x.$$

Hence, the most general regular invariant equation is given by (see [H2], Chapter 1)

$$J_1 = f(J_2).$$

After substituting the expressions of the invariants J_1 and J_2, we arrive at Equation 6.15.

THEOREM 6.7.   *The most general diffusion type equation invariant under the extended group **S** has the form*

$$u_t = ku_{xx} + (1 - k)\frac{u_x^2}{u}, \quad k = \text{const.} \tag{6.16}$$

PROOF.   The group **S** has one functionally independent invariant depending upon the variables u, u_t, u_x, u_{xx} :

$$J = \frac{uu_t - u_x^2}{uu_{xx} - u_x^2}.$$

Hence, the most general equation is given by $J = k$ with an arbitrary constant k. Thus, we arrive at Equation 6.16.

6.2.   Lie-Bäcklund group approach to Newton's gravitational field

Originally, the "law of inverse squares" of Newton's gravitation theory was derived by testing central force fields to satisfy Kepler's empirical laws. Recently, all three Kepler's laws have been derived from symmetry principles (Ibragimov [1993]; see also [H2], Section 5.2). This fact reveals a group theoretic background of Newton's gravitation theory.

Central to this section is a three-parameter Lie-Bäcklund symmetry group of Newton's gravitation field. Application of the Noether theorem to this group yields the Laplace vector and hence, the first Kepler law. Furthermore, the quantization shows that the well-known specific symmetry of the hydrogen atom is an immediate consequence of this Lie-Bäcklund group.

6.2.1. Historical note on Lie-Bäcklund transformation groups

This section is based on the work of Lie [1874], all published mathematical papers of Bäcklund [1874] to [1883], and an excellent biographical paper by Oseen [1929].

As a result of Sophus Lie's work, contact (i.e., first-order tangent) transformations occupied a special position in the theory of first-order partial differential equations with one dependent variable, because they can be used to transform first-order partial differential equations into one another. This is not true for systems of differential equations because, in the case of many dependent variables, contact transformations reduce to prolonged Lie point transformations (for the proof, see, e.g., Ibragimov [1983], Section 14.1).

The set of contact transformations is not sufficient, however, for tackling higher-order differential equations in a similar way. Therefore, Lie [1874] raised the problem of the existence of higher-order tangent transformations.

Independently, the young Swedish mathematician Albert Victor Bäcklund from Lund encountered the question of whether or not there are surface transformations for which the role of an invariant condition is played by the second-order tangency (osculation) rather than by the first-order tangency. He investigated this question in a more general context of arbitrary finite-order tangent transformations of plane curves and surfaces in three-dimensional spaces. He gave a negative answer to the above question and published the result in Bäcklund [1874]. When the paper was published, Bäcklund learned that the idea underlying the existence of higher-order tangent transformations and their importance for the theory of differential equations had been discussed by Lie[2] [1874]. This fact encouraged him to prepare a revised and enlarged version (Bäcklund [1876]).

The paper Bäcklund [1876] is important for elucidating the principal difference and the common roots of *Lie-Bäcklund transformation groups* and of what is called in the literature *Bäcklund transformations*. Bäcklund [1876], p. 305, begins with transformations of the most general form (the notation is taken from Chapter 1 of

---

[2]Bäcklund mentioned Sophus Lie's name for the first time in 1872 in one of his student works. Thereafter, his mathematical work was strongly influenced by Lie's ideas.

this volume):

$$\bar{x}^i = f^i(x, u, u_{(1)}, u_{(2)}, \ldots), \quad \bar{u} = \varphi(x, u, u_{(1)}, u_{(2)}, \ldots) \tag{6.17}$$

where $i = 1, \ldots, n$. He extends Transformations 6.17 to all derivatives through differentiation to obtain the following:

$$\bar{u}_i = \psi_i(x, u, u_{(1)}, u_{(2)}, \ldots), \quad u_{ij} = \psi_{ij}(x, u, u_{(1)}, u_{(2)}, \ldots), \ldots \tag{6.18}$$

so that the infinite system of equations

$$du - u_i dx^i = 0, \quad du_i - u_{ij} dx^j = 0, \ldots \quad \text{to inf.} \tag{6.19}$$

remains invariant. Thus, Equations 6.17 and 6.18 define an infinite-order tangent transformation in the infinite-dimensional space.

The emphasis of the paper is on the following two types of Transformations 6.17 (Bäcklund [1876], pp. 303 to 305). The first type comprises single valued (invertible) transformations involving a finite number of variables $x, u, u_{(1)}, \ldots$. More precisely, these transformations are closed (and invertible) in a finite-dimensional space of variables $(x, u, u_{(1)}, \ldots, u_{(s)})$ with $s < \infty$, i.e., they have the form:

$$\bar{x}^i = f^i(x, u, u_{(1)}, \ldots, u_{(s)}),$$

$$\bar{u} = \varphi(x, u, u_{(1)}, \ldots, u_{(s)}),$$

$$\bar{u}_i = \psi_i(x, u, u_{(1)}, \ldots, u_{(s)}),$$

$$\ldots\ldots\ldots\ldots\ldots\ldots\ldots\ldots\ldots\ldots$$

$$\bar{u}_{i_1 \cdots i_s} = \psi_{i_1 \cdots i_s}(x, u, u_{(1)}, \ldots, u_{(s)}). \tag{6.20}$$

Lie point and contact transformations belong to this type with $s = 0$ and $s = 1$, respectively. The second type consists of infinite valued transformations widely known today as *Bäcklund transformations*.

The major part of the paper is devoted to the proof (both geometric and analytic) of the following statement (Bäcklund's non-existence theorem):

If an invertible transformation 6.20 with a finite s leaves invariant the sth-order tangency conditions

$$du - u_i dx^i = 0, \quad du_i - u_{ij} dx^j = 0, \ldots, \quad du_{i_1 \cdots i_{s-1}} - u_{j i_1 \cdots i_{s-1}} dx^j = 0$$

or the infinite-order tangency conditions 6.19, then it is a Lie contact (in particular, Lie point) transformation.

Hence, the first type of Transformations 6.17 does not generalize Lie contact transformations.

The second type of transformations is discussed in the last section of the paper. In particular, it is shown there that the transformation

$$\overline{x} = x, \quad \overline{y} = y, \quad \overline{z} = q$$

is of the second type and maps any second-order partial differential equation of the form

$$r = F(x, y, q, s, t)$$

into a second-order *quasilinear* equation. Here, the classical notation is used: x, y are independent variables, z is a dependent variable, $p = z_x$, $q = z_y$, $r = z_{xx}$, $s = z_{xy}$, $t = z_{yy}$. This was the first example, given by Bäcklund, of a transformation of the second class[3]. Subsequently, he investigated them in more detail in Bäcklund [1877a], [1878a], [1880], [1882], [1883].

According to Bäcklund's theorem, a generalization of Lie contact transformations by invertible Transformations 6.17 and 6.18 requires an investigation of those transformations acting intrinsically in an infinite-dimensional space of variables $(x, u, u_{(1)}, u_{(2)}, \ldots)$. The problem was whether or not such transformations exist.

An attempt was undertaken by Ibragimov and Anderson [1977] (see also Anderson and Ibragimov [1979]) to tackle the problem by imposing the one-parameter Lie group structure on these transformations and hence, by using Lie's infinitesimal technique. A Lie group of invertible infinite-order tangent transformation of general Bäcklund's form 6.17 and 6.18 is termed briefly *Lie-Bäcklund transformation group*. Prolonged Lie point and contact transformation groups provide examples of Lie-Bäcklund transformation groups. In several works (for the references, see [H1], Chapter 5), Lie's infinitesimal method has been carried over to infinitesimal Lie-Bäcklund symmetries of differential equations. However, the fundamental question of solvability of the corresponding Lie-Bäcklund equations has remained open. Thus, in general, the existence of Lie-Bäcklund transformation groups has not been proved.

A solution to the problem was given in Ibragimov [1979], [1983]. The crucial idea was to introduce into consideration the space $\mathcal{A}$ of differential functions and to use the formal power series with coefficients from $\mathcal{A}$. This theory provides an existence and uniqueness of a one-parameter Lie-Bäcklund transformation group for any infinitesimal transformation

$$\overline{x}^i = x^i + a\xi^i(x, u, u_{(1)}, \ldots), \quad \overline{u}^\alpha = u^\alpha + a\eta^\alpha(x, u, u_{(1)}, \ldots)$$

---

[3]Laplace [1773] used transformations of linear hyperbolic second-order equations $s + a(x, y)p + b(x, y)q + c(x, y)z = 0$ given by the following relations: $\overline{x} = x$, $\overline{y} = y$, $\overline{z} = q + az$, $\overline{p} = (a_x - c)z - bq$. This is precisely what is called Bäcklund transformation.

with arbitrary ξ^i, $\eta^\alpha \in \mathcal{A}$. A detailed discussion is given in Chapter 1.

For his 1874 work, Bäcklund was awarded a Swedish state research grant for travelling abroad. In 1874, he spent six months in Germany where he met, in particular, Felix Klein. Klein clearly realized the significance of Bäcklund's results and subsequently popularized them in his lectures and books (see, e.g., Klein [1926]). Naturally, these results were well known to colleagues and students of F. Klein. For example, Noether [1918] used infinitesimal transformations involving higher-order derivatives in her conservation theorem. Therefore, one finds now and then in the literature credit for these transformations given erroneously to E. Noether.

We summarize: *Bäcklund transformations and Lie-Bäcklund transformation groups have common roots, but they are distinctly different in their main features. Bäcklund transformations are intrinsically infinite valued. Lie-Bäcklund transformation groups are given by single-valued Transformations 6.17 and 6.18, but they act in an infinite-dimensional space and are represented by specific formal series.*

For applications, the Lie-Bäcklund group approach is not as simple as the classical Lie theory. Therefore, it is mainly of theoretical value. However, in certain problems the use of Lie-Bäcklund symmetries is inevitable. The hidden symmetry of Newton's gravitational field provides one of these problems.

6.2.2. Lie-Bäcklund symmetries of Newton's field

For this section, see Ibragimov [1983], Section 25.1, and [H2], Section 5.2.

A motion of a particle with mass m in a central potential field $U(r)$ is described by the Lagrangian

$$L = \frac{m}{2}|\mathbf{v}|^2 - U(r).$$ (6.21)

Here, $r = |\mathbf{x}|$, $\mathbf{x} = (x^1, x^2, x^3)$, $\mathbf{v} = d\mathbf{x}/dt$. The equation of motion has the form

$$m\frac{d\mathbf{v}}{dt} = -\frac{U'(r)}{r}\mathbf{x}.$$ (6.22)

For Equation 6.22, the canonical Lie-Bäcklund symmetries 1.10 (Chapter 1) are written

$$X = \eta^i \frac{\partial}{\partial x^i}.$$ (6.23)

The vector $\eta = (\eta^1, \eta^2, \eta^3)$ depends on t, $\mathbf{x}$, $\mathbf{v}$, and is defined by the determining equation (cf. Section 1.3)

$$m\left(\frac{d}{dt}\right)^2 \eta^i + \frac{1}{r}\left[U'(r)\eta^i - \left(\frac{U'(r)}{r}\right)'(\mathbf{x} \cdot \eta)x^i\right] = 0,$$ (6.24)

where $(\mathbf{x} \cdot \eta)$ denotes the scalar product and

$$\frac{d}{dt} = \frac{\partial}{\partial t} + v^i\frac{\partial}{\partial x^i} - \frac{U'(r)}{mr}x^i\frac{\partial}{\partial v^i}.$$

is the differentiation along the trajectory of the particle.

Let's consider Newton's gravitational field

$$U(r) = \frac{\mu}{r}, \quad \mu = \text{const.} \tag{6.25}$$

The equation of motion

$$m\frac{d^2\mathbf{x}}{dt^2} = \mu\frac{\mathbf{x}}{r^3} \tag{6.26}$$

admits the five-parameter group of point transformations with the generators

$$X_0 = \frac{\partial}{\partial t}, \quad X_{ij} = x^j\frac{\partial}{\partial x^i} - x^i\frac{\partial}{\partial x^j}, \quad T = 3t\frac{\partial}{\partial t} + 2x^i\frac{\partial}{\partial x^i}. \tag{6.27}$$

The linear combination of these generators written in the canonical Lie-Bäcklund form 6.23 has the coordinates

$$\eta^i = (A + 3Ct)v^i + B^i_k x^k - 2Cx^i, \tag{6.28}$$

where A, B^i_k, C are constants such that $B^i_k + B^k_i = 0$. The differential functions 6.28 are linear in $\mathbf{v}$ and have the special form prescribed by Theorem 4 of Chapter 1. The following statement (Ibragimov [1983], p. 346) yields all symmetries 6.23 with coefficients linear in $\mathbf{v}$, and provides non-trivial Lie-Bäcklund symmetries for Equation 6.26.

THEOREM 6.8. *For the determining equations 6.24 with Newton's field 6.25, the general solution of the form*

$$\eta^i = \alpha^i_j(t, \mathbf{x})v^j + \beta^i(t, \mathbf{x})$$

is given by the functions 6.28 and by three vectors η_k, $k = 1, 2, 3$, with the coordinates

$$\eta^i_k = 2x^k v^i - x^i v^k - (\mathbf{x} \cdot \mathbf{v})\delta^i_k \tag{6.29}$$

where δ^i_k is the Kronecker symbol.

It is convenient to represent the symmetries of Newton's field by the *infinitesimal canonical Lie-Bäcklund transformations*. Namely, the infinitesimal group transformation $\bar{\mathbf{x}} \approx \mathbf{x} + \delta\mathbf{x}$, corresponding to the canonical Lie-Bäcklund operator 6.23, is defined by the increment $\delta\mathbf{x}$ with the coordinates $\delta x^i = a\eta^i$ where a is the group parameter. Consider the generators 6.27. The operators X_0 and

T generate one-parameter infinitesimal transformations with the increments (see Equation 6.27 with $B_k^i = 0$)

$$\delta \mathbf{x} = a \mathbf{v} \tag{6.30}$$

and

$$\delta \mathbf{x} = (3t\mathbf{v} - 2\mathbf{x})a, \tag{6.31}$$

respectively. The operators X_{ij} generate the three-parameter infinitesimal rotations with the increment

$$\delta \mathbf{x} = \mathbf{x} \times \mathbf{a} \tag{6.32}$$

where $\mathbf{a} = (a^1, a^2, a^3)$ is the vector parameter. Similarly, the symmetries 6.29 yield the infinitesimal three-parameter Lie-Bäcklund transformations with the increment

$$\delta \mathbf{x} = (\mathbf{x} \times \mathbf{v}) \times \mathbf{a} + \mathbf{x} \times (\mathbf{v} \times \mathbf{a}). \tag{6.33}$$

DEFINITION 6.5.   *Given an increment $\delta \mathbf{x}$, the corresponding increments of the velocity $\mathbf{v}$ and of the Lagrangian 6.21 are determined as follows:*

$$\delta \mathbf{v} = \frac{d}{dt} \delta \mathbf{x}, \quad \delta L = m(\mathbf{v} \cdot \delta \mathbf{v}) - \frac{U'(r)}{r}(\mathbf{x} \cdot \delta \mathbf{x}).$$

The infinitesimal canonical Lie-Bäcklund transformation $\overline{\mathbf{x}} \approx \mathbf{x} + \delta \mathbf{x}$ is said to be a Noether symmetry for Equation 6.22 if

$$\delta L = \frac{dF}{dt}, \quad F \in \mathcal{A}. \tag{6.34}$$

The Noether theorem associates with a Noether symmetry the following conserved quantity for Equation 6.22:

$$J = m(\mathbf{v} \cdot \delta \mathbf{v}) - F, \tag{6.35}$$

where F is determined by Equation 6.34.

The infinitesimal transformations given by the increments 6.30, 6.32, and 6.33 are Noether symmetries for Equation 6.26. Consequently, the formulae 6.34 and 6.35 yield the energy

$$E = \frac{1}{2m}|\mathbf{p}|^2 - \frac{\mu}{r}, \quad \text{where } \mathbf{p} = m\mathbf{v}, \tag{6.36}$$

the angular momentum

$$\mathbf{M} = \mathbf{x} \times \mathbf{p}, \tag{6.37}$$

and the Laplace vector

$$\mathbf{A} = \mathbf{v} \times \mathbf{M} + \mu \frac{\mathbf{x}}{r}. \tag{6.38}$$

The conservation of the vectors $\mathbf{A}$ and $\mathbf{M}$ is equivalent to the first and second Kepler laws, respectively. The scaling transformation defined by the increment 6.31 is not a Noether symmetry. However, it yields the third Kepler's law.

Thus, Theorem 6.7 provides a group theoretic background of the Kepler laws. Consequently, the Noether symmetries given by the increments 6.30, 6.31, and 6.33 determine uniquely Newton's gravitation field 6.25 (Ibragimov [1993]).

6.2.3. Derivation of Lie-Bäcklund symmetries for the hydrogen atom via quantization

The quantization is performed, in appropriate physical units, by the substitution

$$\mathbf{p} \mapsto -\sqrt{-1}\,\nabla. \tag{6.39}$$

After the substitution 6.39, Equation 6.36 yields

$$-\frac{1}{2m}\nabla^2 - \frac{\mu}{r} - E = 0. \tag{6.40}$$

Hence, we arrive at the stationary Schrödinger equation for the hydrogen atom:

$$\frac{1}{2m}\Delta\psi + \frac{\mu}{r} + E\psi = 0, \quad \Delta = \nabla^2. \tag{6.41}$$

We will consider the case of a negative energy $E < 0$.

The operator X_0 is commutative with X_{ij} and with the Lie-Bäcklund operators given by Equations 6.29. Therefore the second-order linear differential operator 6.40 commutes with the operators obtained from $\mathbf{M}$ and $\mathbf{A}$ by the substitution 6.39. It follows that Substitution 6.39 maps the angular momentum and the Laplace vector into Lie-Bäcklund symmetries of the Schrödinger equation 6.41.

The angular momentum 6.37 is linear in $\mathbf{p}$ and hence it is mapped into Lie point symmetries, namely into the generators of rotations X_{ij}.

The Laplace vector $\mathbf{A}$ is non-linear with respect to $\mathbf{p}$. Consequently, it is mapped into non-point symmetries. Indeed, the substitution 6.39 into the coordinates

$$A_k = \frac{1}{m}\left(|\mathbf{p}|^2 x^k - (\mathbf{x} \cdot \mathbf{p})p_k\right) + \frac{\mu}{r}x^k, \quad k = 1, 2, 3,$$

of the vector 6.38 yields:

$$A_k = \frac{1}{m}\left(-\nabla^2 x^k + (\mathbf{x}\cdot\nabla)\nabla_k\right) + \frac{\mu}{r}x^k \equiv \frac{1}{m}\left(\sum_{i\neq k}(x^i\nabla_k - x^k\nabla_i)\nabla_i - \nabla_k\right) + \frac{\mu}{r}x^k.$$

Hence, we arrive at the canonical Lie-Bäcklund operators $X_k = \eta_k\frac{\partial}{\partial\psi}$, $k = 1, 2, 3$, with the coordinates (cf. Equations 6.40 and 6.41)

$$\eta_k \equiv A_k(\psi) = \frac{1}{m}\left(\sum_{i\neq k}(x^i\psi_{ik} - x^k\psi_{ii}) - \psi_k\right) + \frac{\mu}{r}x^k\psi. \tag{6.42}$$

The Lie-Bäcklund symmetries 6.42 of Equation 6.41 together with the generators X_{ij} of rotations span the well-known $so(4)$ symmetry algebra of the hydrogen atom.

We summarize: *The Lie-Bäcklund symmetries 6.29 (or 6.33) of Newton's gravitational field account for both the Laplace vector 6.38 and the specific Symmetries 6.42 of Equation 6.41.*

6.3. Groups and relativistic conservation laws

Minkowski [1908] convinced the scientific community that "Henceforth space by itself, and time by itself are doomed to fade away into mere shadows, and only a kind of union of the two will preserve an independent reality." Subsequently, Klein [1910] has underscored that what is called in physics the special theory of relativity is, in fact, a theory of invariants of the Lorentz group. Consequently, he has suggested identifying a "theory of relativity" as a "theory of invariants" of a group of transformations in a space-time manifold. The present section pursues this philosophy.

For this section, see [H2], Section 5.4, and Ibragimov [1983]. For an introduction to principles of relativity, see, e.g., Landau and Lifshitz [1967], Chapters 1, 2, and 10.

6.3.1. Relativistic principle of least action

The geometry of a Riemannian space is determined by the fundamental metric form

$$ds^2 = g_{ij}(x)dx^i dx^j \tag{6.43}$$

and by the group of isometric motions in the space under consideration. Here, $x = (x^1, \ldots, x^n)$, $i, j = 1, \ldots, n$.

One can develop a theory of relativity in any four-dimensional Riemannian space with a metric 6.43 of the space-time signature $(- - -+)$. We discuss here the relativistic mechanics of particles, more specifically, the motion of a free material particle. In this case, the motion of a particle with mass m is determined by the principle of least action, according to which the particle moves so that its world line $x^i = x^i(s)$, $i = 1, \ldots, 4$, is a *geodesic*. That is, the motion of a particle in the space with the metric 6.43 is determined by the Lagrangian

$$L = -mc\sqrt{g_{ij}(x)\dot{x}^i\dot{x}^j} \tag{6.44}$$

Here, $x = (x^1, \ldots, x^4)$, and $\dot{x} = dx/ds$ is the derivative of the four-vector x with respect to the arc length s measured from a fixed point x_0. Hence, the equations of motion have the form

$$\frac{d^2x^i}{ds^2} + \Gamma^i_{jk}(x)\frac{dx^j}{ds}\frac{dx^k}{ds} = 0, \quad i = 1, \ldots, 4, \tag{6.45}$$

where the coefficients Γ^i_{jk} are given by

$$\Gamma^i_{jk} = \frac{1}{2}g^{il}\left(\frac{\partial g_{lj}}{\partial x^k} + \frac{\partial g_{lk}}{\partial x^j} - \frac{\partial g_{jk}}{\partial x^l}\right)$$

and are known as the *Christoffel symbols*. Here, $g^{il}g_{lk} = \delta^i_k$.

A subsequent development of the relativistic mechanics is based, in accordance with Klein [1910], on the group of isometric motions of the metric 6.43. The generators

$$X = \eta^i(x)\frac{\partial}{\partial x^i}$$

of the isometric motions are determined by the *Killing equations*

$$\eta^k\frac{\partial g_{ij}}{\partial x^k} + g_{ik}\frac{\partial \eta^k}{\partial x^j} + g_{jk}\frac{\partial \eta^k}{\partial x^i} = 0.$$

A solution $\eta = (\eta^1, \ldots, \eta^4)$ of the Killing equations is termed also a *Killing vector*. By using the Noether theorem, one associates (see, e.g., Ibragimov [1983], Section 23.1) with any Killing vector the following conserved quantity (integral of motion) for Equations 6.45:

$$J = mcg_{ij}\eta^i d\dot{x}^j, \quad \text{where } \dot{x}^j = dx^j/ds. \tag{6.46}$$

6.3.2.   Conservation laws in special relativity

The Minkowski space-time is a flat Riemannian space of signature $(- - -+)$ defined by the fundamental metric form

$$ds^2 = c^2 dt^2 - |d\mathbf{x}|^2, \tag{6.47}$$

where c is the velocity of light in vacuum.

The group of isometric motions in the Minkowski space-time is the Lorentz group with the basic infinitesimal generators

$$X_0 = \frac{\partial}{\partial t}, \; X_i = \frac{\partial}{\partial x^i}, \; X_{ij} = x^j \frac{\partial}{\partial x^i} - x^i \frac{\partial}{\partial x^j}, \; X_{0i} = t \frac{\partial}{\partial x^i} + \frac{1}{c^2} x^i \frac{\partial}{\partial t}$$

where $i, j = 1, 2, 3$.

Let us apply the conservation formula 6.46 to the generators of the Lorentz group. In what follows, the spatial vector is denoted by $\mathbf{x} = (x^1, x^2, x^3)$, hence the physical velocity $\mathbf{v} = d\mathbf{x}/dt$ is a three-dimensional vector $\mathbf{v} = (v^1, v^2.v^3)$. Their scalar and vector products are denoted by $(\mathbf{x} \cdot \mathbf{v})$ and $\mathbf{x} \times \mathbf{v}$, respectively. The equation 6.47 written in the form

$$ds = c\sqrt{1 - \beta^2}\, dt, \quad \beta^2 = |\mathbf{v}|^2/c^2 ,$$

yields:

$$\frac{d\mathbf{x}}{ds} = \frac{\mathbf{v}}{c\sqrt{1 - \beta^2}}, \quad \frac{dt}{ds} = \frac{1}{c\sqrt{1 - \beta^2}} .$$

Using these equations and substituting the coordinates of X_0 into the formula 6.46, one obtains the relativistic energy:

$$E = \frac{mc^2}{\sqrt{1 - \beta^2}} .$$

Similarly, by using the operators X_i and X_{ij}, one easily obtains the relativistic momentum

$$\mathbf{p} = \frac{m\mathbf{v}}{\sqrt{1 - \beta^2}}$$

and the angular momentum

$$\mathbf{M} = \mathbf{x} \times \mathbf{p}$$

respectively.

The Lorentz boosts X_{0i} give rise to the vector

$$Q = \frac{m(\mathbf{x} - t\mathbf{v})}{\sqrt{1 - \beta^2}}.$$

The conservation of this vector, in the case of an N-body problem, is the relativistic center-of-mass theorem.

6.3.3. Conservation laws in the de Sitter space-time

The de Sitter space-time is a Riemannian space of a constant curvature K. It is defined by the fundamental metric form

$$ds^2 = \theta^{-2}(c^2 dt^2 - |d\mathbf{x}|^2), \quad K = \text{const.}, \tag{6.48}$$

where

$$\theta = 1 + \frac{K}{4}(|\mathbf{x}|^2 - c^2 t^2).$$

The de Sitter space-time has a 10-parameter group of isometric motions known as the de Sitter group. Its basic infinitesimal generators are

$$X_0 = \frac{1}{c^2}\left[1 - \frac{K}{4}(c^2 t^2 + |\mathbf{x}|^2)\right]\frac{\partial}{\partial t} - \frac{K}{2}t\sum_{i=1}^{3}x^i\frac{\partial}{\partial x^i},$$

$$X_1 = \left(1 + \frac{K}{4}[(x^1)^2 - (x^2)2 - (x^3)2 + c^2 t^2]\right)\frac{\partial}{\partial x^1} + \frac{K}{2}x^1\left(x^2\frac{\partial}{\partial x^2} + x^3\frac{\partial}{\partial x^3} + t\frac{\partial}{\partial t}\right),$$

$$X_2 = \left(1 + \frac{K}{4}[(x^2)^2 - (x^1)2 - (x^3)2 + c^2 t^2]\right)\frac{\partial}{\partial x^2} + \frac{K}{2}x^2\left(x^1\frac{\partial}{\partial x^1} + x^3\frac{\partial}{\partial x^3} + t\frac{\partial}{\partial t}\right),$$

$$X_3 = \left(1 + \frac{K}{4}[(x^3)^2 - (x^1)2 - (x^2)2 + c^2 t^2]\right)\frac{\partial}{\partial x^3} + \frac{K}{2}x^3\left(x^1\frac{\partial}{\partial x^1} + x^2\frac{\partial}{\partial x^2} + t\frac{\partial}{\partial t}\right),$$

$$X_{ij} = x^j\frac{\partial}{\partial x^i} - x^i\frac{\partial}{\partial x^j}, \quad X_{0i} = t\frac{\partial}{\partial x^i} + \frac{1}{c^2}x^i\frac{\partial}{\partial t}, \quad i, j = 1, 2, 3.$$

Let us apply the conservation formula 6.46 to the generators of the de Sitter group. In what follows, the notation of the preceding section is used: $\mathbf{x} = (x^1, x^2, x^3)$, $\mathbf{v} = d\mathbf{x}/dt$. Equation 6.48 written in the form

$$ds = \frac{c}{\theta}\sqrt{1 - \beta^2}\,dt, \quad \beta^2 = |\mathbf{v}|^2/c^2,$$

yields:

$$\frac{d\mathbf{x}}{ds} = \frac{\theta \mathbf{v}}{c\sqrt{1-\beta^2}}, \quad \frac{dt}{ds} = \frac{\theta}{c\sqrt{1-\beta^2}}.$$

Using these equations and substituting the coordinates of X_0 into the formula 6.46, one obtains; the energy of a particle moving freely in the de Sitter space-time with curvature K:

$$\mathcal{E} = \frac{mc^2}{\sqrt{1-\beta^2}}\left[1 - \frac{K}{2\theta}(\mathbf{x} - t\mathbf{v})\cdot\mathbf{x}\right].$$

In the linear approximation with respect to K, this formula yields:

$$\mathcal{E} \approx E\left[1 - \frac{K}{2}(\mathbf{x} - t\mathbf{v})\cdot\mathbf{x}\right]$$

where E is the relativistic energy given in Section 6.3.3.
    Similarly, by using the operators X_i, one obtains the momentum

$$\mathcal{P} = \frac{m}{\theta\sqrt{1-\beta^2}}\left[(2-\theta)\mathbf{v} - \frac{K}{2}(c^2 t - \mathbf{x}\cdot\mathbf{v})\mathbf{x}\right].$$

In the linear approximation with respect to K, this formula yields:

$$\mathcal{P} \approx \mathbf{p} - \frac{K}{2}E\left[t(\mathbf{x} - t\mathbf{v}) + \frac{1}{c^2}(\mathbf{x}\times\mathbf{v})\times\mathbf{x}\right]$$

where E and $\mathbf{p}$ are the relativistic energy and momentum, respectively, given in Section 6.3.3.
    By using the infinitesimal rotations X_{ij} and Lorentz boosts X_{0i}, one obtains the angular momentum

$$\mathcal{M} = \frac{1}{2-\theta}(\mathbf{x}\times\mathcal{P})$$

and the vector

$$\mathcal{Q} = \frac{m(\mathbf{x} - t\mathbf{v})}{\theta\sqrt{1-\beta^2}},$$

respectively.
    Certain peculiarities of the theory of relativity in the de Sitter universe (an approximate representation of the de Sitter group, the Kepler problem, splitting of a neutrino into two neutrinos by the curvature) are discussed in [H2], Section 5.4.

7

Generalized Hydrodynamic-Type Systems

This chapter[1] summarizes results on Lie and Lie-Bäcklund symmetries for generalizations of hydrodynamic-type systems,[2] one-dimensional gas dynamics, and semi-Hamiltonian equations.

Hydrodynamic-type systems were introduced by Dubrovin and Novikov [1983], [1989] as quasi-linear systems of first-order partial differential equations possessing a Hamiltonian structure.

Here, we identify as generalized hydrodynamic-type systems more general systems of equations which generalize one-dimensional gas dynamics equations and comprise, e.g., semi-Hamiltonian equations investigated by Tsarev [1985], [1991]. These equations can depend explicitly on t. They are rich in symmetries and therefore can be linearized. Furthermore, an infinite set of their exact solutions can be obtained. Thus, they are similar to Hamiltonian equations in their integrability properties.

Hydrodynamic-type systems describe various physical phenomena: gas dynamics and hydrodynamics, magnetic hydrodynamics (see Rozhdestvenskii and Yanenko [1978]), nonlinear elasticity and phase-transition models (see Olver and Nutku [1988]), chromatography and electrophoresis equations from physical chemistry and biology (see Pavlov [1987], Ferapontov and Tsarev [1991]).

Applications of another kind are obtained by a representation of physically interesting higher-order equations as integrability conditions of hydrodynamic-type systems: the Euler and Poisson equations of nonlinear acoustics (see Gümral and Nutku [1990]), the Born-Infeld equation of nonlinear electrodynamics (see Arik et al. [1989]), etc. Modern applications of hydrodynamic-type systems arise in the theory of averaging nonlinear soliton equations (see Dubrovin and Novikov [1983], Tsarev [1985]).

An attractive feature of hydrodynamic-type systems is the possibility of a geometric formulation of their properties by close analogy with the Hamilton-Jacoby and eikonal equations of mechanics and optics.

---

[1]Research supported in part by International Science Foundation under Grant R5B000.
[2]Editor's note: In a concise form, the group analysis of hydrodynamic-type systems is considered in [H1], Chapter 14.

Group analysis of these systems leads naturally and algorithmically to associated differential-geometric structures: metrics, connections, curvatures, curvilinear orthogonal coordinate systems, and their transformations. If a Hamiltonian structure exists, then it turns out to be merely an aspect of this geometrical theory (see Dubrovin and Novikov [1983], Tsarev [1985],[1991]).

A purely differential-geometric approach does not solve the problem of integrating hydrodynamic-type equations. An alternative approach to the problem is based on a systematic study of Lie-Bäcklund symmetries and recursion operators (see Sheftel' [1993], [1994a,b]). The essence of this approach is to pick out a class of those hydrodynamic-type systems which possesses an infinite set of symmetries depending on arbitrary solutions of linear differential equations. Then, assuming an additional constraint, one can construct recursion operators mapping symmetries into symmetries.

7.1. Hydrodynamic-type systems

Here we consider two-component diagonal hydrodynamic-type systems depending explicitly on t, which are linear and homogeneous in derivatives of the unknowns $s(x, t), r(x, t)$:

$$s_t = \phi(s, r, t)s_x, \qquad r_t = \psi(s, r, t)r_x \qquad (7.1)$$

where the subscripts t and x denote the partial differentiations with respect to t and x, respectively. The functions ϕ and ψ satisfy the nondegeneracy conditions:

$$\phi \neq \psi, \qquad \phi_r(s, r, t)\psi_s(s, r, t) \neq 0, \qquad (7.2)$$

where the subscripts r and s denote the partial differentiations with respect to r and s, respectively.

Symmetries of System 7.1 are generated by canonical Lie-Bäcklund operators

$$X = f\frac{\partial}{\partial s} + g\frac{\partial}{\partial r}, \qquad f, g \in \mathcal{A}, \qquad (7.3)$$

where $\mathcal{A}$ is the space of differential functions (see Chapter 1).

REMARK.   A system

$$s_t = \phi^*(s, r, x)s_x, \qquad r_t = \psi^*(s, r, x)r_x \qquad (7.4)$$

depending explicitly on x reduces to Equation 7.1 by the transformation $t \mapsto x$, $x \mapsto t$, $\phi = \phi^{*-1}$, $\psi = \psi^{*-1}$.

The determining equations for Symmetries 7.3 have the form:

$$D_t[f]\big|_{(7.1)} - \phi D_x[f] - s_x(\phi_s f + \phi_r g) = 0,$$
$$D_t[g]\big|_{(7.1)} - \psi D_x[g] - r_x(\psi_s f + \psi_r g) = 0. \tag{7.5}$$

Here, D_t and D_x are the total derivatives:

$$D_x = \frac{\partial}{\partial x} + s_x\frac{\partial}{\partial s} + r_x\frac{\partial}{\partial r} + \sum_{k=1}^{\infty}\left(s_x^{(k+1)}\frac{\partial}{\partial s_x^{(k)}} + r_x^{(k+1)}\frac{\partial}{\partial r_x^{(k)}}\right),$$
$$D_t\big|_{(7.1)} = \frac{\partial}{\partial t} + \phi s_x\frac{\partial}{\partial s} + \psi r_x\frac{\partial}{\partial r} \tag{7.6}$$
$$+ \sum_{k=1}^{\infty}\left(D_x^k(\phi s_x)\frac{\partial}{\partial s_x^{(k)}} + D_x^k(\psi r_x)\frac{\partial}{\partial r_x^{(k)}}\right)$$

where D_t is calculated with the use of System 7.1.

7.1.1. First-order Lie-Bäcklund symmetries

(see Sheftel' [1994b])

First-order symmetries are given by Operators 7.3 with the coefficients of the form

$$f = f(x, t, s, r, s_x, r_x), \quad g = g(x, t, s, r, s_x, r_x). \tag{7.7}$$

Let us define the functions $\Phi(s, r, t)$ and $\Theta(s, r, t)$ by the first-order differential equations

$$\Phi_r(s, r, t) = \phi_r(s, r, t)/(\phi - \psi), \quad \Theta_s(s, r, t) = \psi_r(s, r, t)/(\psi - \phi), \tag{7.8}$$

Let

$$\hat{\Phi}(s, r, t) = b(s)\Phi_s(s, r, t) + d(r)\Phi_r(s, r, t) + \Phi_0(s),$$
$$\hat{\Theta}(s, r, t) = b(s)\Theta_s(s, r, t) + d(r)\Theta_r(s, r, t) + \Theta_0(r), \tag{7.9}$$

$$\hat{\phi}(s, r, t) = b(s)\phi_s(s, r, t) + d(r)\phi_r(s, r, t),$$
$$\hat{\psi}(s, r, t) = b(s)\psi_s(s, r, t) + d(r)\psi_r(s, r, t). \tag{7.10}$$

System 7.1 is said to be generic with respect to first-order symmetries if its coefficients ϕ, ψ do not satisfy the constraints:

$$\phi_t = \mu(t)\phi^2 + \varepsilon(t)\phi + \lambda(t), \quad \psi_t = \mu(t)\psi^2 + \varepsilon(t)\psi + \lambda(t) \tag{7.11}$$

with arbitrary functions $\mu(t)$, $\varepsilon(t)$, $\lambda(t)$.

INFINITE SET OF FIRST-ORDER SYMMETRIES

A diagonal two-component generic hydrodynamic-type system 7.1 possesses an infinite set of first-order Lie-Bäcklund symmetries with a functional arbitrariness iff

$$\Phi_{rt} = \beta\phi_r, \quad \Theta_{st} = \beta\psi_s, \quad \beta = \text{const.} \tag{7.12}$$

and

$$\hat{\Phi}_r = \Phi_r(\hat{\Phi} - \hat{\Theta}), \quad \hat{\Theta}_s = \Theta_s(\hat{\Theta} - \hat{\Phi}), \tag{7.13}$$

where the functions $\Phi(s, r, t)$, $\Theta(s, r, t)$ and $\hat{\Phi}(s, r, t)$, $\hat{\Theta}(s, r, t)$ are defined by Equations 7.8 and 7.9, respectively, and the partial derivatives with respect to t are taken at constant values of s and r.

The symmetries 7.3 have the coordinates

$$\begin{aligned} f &= \tilde{\phi}(x, t, s, r)s_x + b(s), \\ g &= \tilde{\psi}(x, t, s, r)r_x + d(r), \end{aligned} \tag{7.14}$$

where the functions $\tilde{\phi}$, $\tilde{\psi}$ are given as follows:

if $\beta \neq 0$ then

$$\begin{aligned} \tilde{\phi}(x, t, s, r) &= a(s, r)\exp\left\{\beta\left[x + \int_0^t \phi(s, r, t)dt\right]\right\} + \frac{1}{\beta}\hat{\Phi}(s, r, t), \\ \tilde{\psi}(x, t, s, r) &= c(s, r)\exp\left\{\beta\left[x + \int_0^t \psi(s, r, t)dt\right]\right\} + \frac{1}{\beta}\hat{\Theta}(s, r, t); \end{aligned} \tag{7.15}$$

if $\beta = 0$ then

$$\begin{aligned} \tilde{\phi}(x, t, s, r) &= a(s, r) + \int_0^t \hat{\phi}(s, r, t)dt \\ &\quad - \hat{\Phi}(s, r)\left[x + \int_0^t \phi(s, r, t)dt\right], \\ \tilde{\psi}(x, t, s, r) &= c(s, r) + \int_0^t \hat{\psi}(s, r, t)dt \\ &\quad - \hat{\Theta}(s, r)\left[x + \int_0^t \psi(s, r, t)dt\right]. \end{aligned} \tag{7.16}$$

Here, the integrals with respect to t are taken at constant values of s and r and the functions $\hat{\phi}$, $\hat{\psi}$ are defined by Formulas 7.10. The functions $a(s, r)$, $c(s, r)$ satisfy the linear first-order differential equations

$$a_r(s, r) = \Phi_r(s, r, 0)(a - c),$$
$$c_s(s, r) = \Theta_s(s, r, 0)(c - a). \tag{7.17}$$

This system has, e.g., the trivial solution $a(s, r) = c(s, r) = c_0 = \text{const.}$

One can use the freedom in Definition 7.8 of the functions Φ, Θ to simplify Equations 7.12 to the following form:

$$\Phi_t(s, r, t) = \beta\phi(s, r, t), \quad \Theta_t(s, r, t) = \beta\psi(s, r, t). \tag{7.18}$$

The solution manifold of the linear system 7.17 is locally parameterized by two arbitrary functions $c_1(s), c_2(r)$ of one variable. They determine a functional arbitrariness of the symmetries 7.14.

Equation 7.13 has the trivial solution

$$b(s) = d(r) = 0, \quad \Phi_0(s) = \Theta_0(r) = c_0 = \text{const.} \tag{7.19}$$

Then $\hat{\Phi} = \hat{\Theta} = c_0$, $\hat{\phi} = \hat{\psi} = 0$. It follows that Equation 7.12 is necessary and sufficient for System 7.1 to possess an infinite set of symmetries

$$f = \tilde{\phi}(x, t, s, r)s_x, \quad g = \tilde{\psi}(x, t, s, r)r_x. \tag{7.20}$$

The coefficients $\tilde{\phi}, \tilde{\psi}$ are determined by the following formulas:
if $\beta \neq 0$ then

$$\tilde{\phi}(x, t, s, r) = a(s, r)\exp\left\{\beta\left[x + \int_0^t \phi(s, r, t)dt\right]\right\} + c_0,$$

$$\tilde{\psi}(x, t, s, r) = c(s, r)\exp\left\{\beta\left[x + \int_0^t \psi(s, r, t)dt\right]\right\} + c_0; \tag{7.21}$$

if $\beta = 0$ then

$$\tilde{\phi}(x, t, s, r) = a(s, r) + c_0\left[x + \int_0^t \phi(s, r, t)dt\right],$$

$$\tilde{\psi}(x, t, s, r) = c(s, r) + c_0\left[x + \int_0^t \psi(s, r, t)dt\right]. \tag{7.22}$$

The condition 7.12 with $\beta = 0$ is satisfied, e.g., for Systems 7.1 with the coefficients $\phi(s, r), \psi(s, r)$ independent of t. Hence these systems always have an infinite set of first-order Lie-Bäcklund symmetries 7.22 with the coefficients

$$\tilde{\phi}(x, t, s, r) = a(s, r) + c_0 [x + t\phi(s, r)],$$
$$\tilde{\psi}(x, t, s, r) = c(s, r) + c_0 [x + t\psi(s, r)].$$

(7.23)

INVARIANT SOLUTIONS

Formulas 7.21, 7.22 for $\tilde{\phi}$, $\tilde{\psi}$ (with $c_0 = 1$) yield the following conditions for invariant solutions:

if $\beta \neq 0$ then

$$
\left.
\begin{array}{l}
a(s, r) + \exp\left\{-\beta\left[x + \int_0^t \phi(s, r, t)dt\right]\right\} = 0, \\[3ex]
c(s, r) + \exp\left\{-\beta\left[x + \int_0^t \psi(s, r, t)dt\right]\right\} = 0;
\end{array}
\right.
$$

(7.24)

if $\beta = 0$ then

$$
\begin{array}{l}
a(s, r) + x + \int_0^t \phi(s, r, t)dt = 0, \\[2ex]
c(s, r) + x + \int_0^t \psi(s, r, t)dt = 0.
\end{array}
$$

(7.25)

Furthermore, any solution of System 7.24 (for $\beta \neq 0$) or of System 7.25 (for $\beta = 0$) is also a solution of System 7.1 provided that the following conditions are satisfied :

for $\beta \neq 0$

$$
\left\{\left[\ln|a(s, r)| + \beta \int_0^t \phi(s, r, t)dt\right]_s \left[\ln|c(s, r)| + \beta \int_0^t \psi(s, r, t)dt\right]_r\right\}\Bigg|_{(7.24)} \neq 0,
$$

for $\beta = 0$

$$
\left\{\left[a(s, r) + \int_0^t \phi(s, r, t)dt\right]_s \left[c(s, r) + \int_0^t \psi(s, r, t)dt\right]_r\right\}\Bigg|_{(7.25)} \neq 0.
$$

Equation 7.25 coincides with the hodograph transformation (see Rozhdestven-skii and Yanenko [1978]):

$$a(s, r) + x + t\phi(s, r) = 0, \quad c(s, r) + x + t\psi(s, r) = 0.$$

7.1.2. Second-order Lie-Bäcklund symmetries

Here we let

$$f = f(x, t, s, r, s_x, r_x, s_{xx}, r_{xx}), \quad g = g(x, t, s, r, s_x, r_x, s_{xx}, r_{xx})$$

in the operators 7.3 and in the determining equations 7.5. The functions Φ and Θ are again defined by Equation 7.8.

System 7.1 is said to be generic with respect to second-order symmetries if its coefficients ϕ, ψ do not satisfy the constraints:

$$\Phi(s, r, t) = \ln \left| \frac{\phi_s}{c(s, t)\phi + d(s, t)} \right|, \quad \Theta(s, r, t) = \ln \left| \frac{\psi_r}{G(r, t)\psi + H(r, t)} \right|,$$

$$\Phi_t(s, r, t) = A(s, t)\phi + B(s, t), \quad \Theta_t(s, r, t) = E(r, t)\psi + F(r, t)$$

with arbitrary functions A, B, c, d, E, F, G, H.

For a generic system 7.1, a necessary condition for the existence of second-order symmetries is provided by Equation 7.12.

Let us define the function $\Lambda(s, r)$ by the equation

$$\Lambda_{sr}(s, r) = -\Phi_r(s, r, t)\Theta_s(s, r, t),$$

and let

$$\hat{\Phi}(s, r, t) = A(s)(\Phi_s^2 - \Phi_{ss} + 2\Lambda_{ss}) + A'(s)\Lambda_s$$

$$+ C(r)(2\Phi_r\Theta_r + \Phi_{rr} - \Phi_r^2) + C'(r)\Phi_r + b(s)\Phi_s + d(r)\Phi_r + \Phi_0(s),$$

$$\hat{\Theta}(s, r, t) = A'(s)\Theta_s + A(s)(2\Theta_s\Phi_s + \Theta_{ss} - \Theta_s^2)$$

$$+ C(r)(\Theta_r^2 - \Theta_{rr} + 2\Lambda_{rr}) + C'(r)\Lambda_r(s, r) + b(s)\Theta_s + d(r)\Theta_r + \Theta_0(r) \quad (7.26)$$

where $A(s), C(r), b(s), d(r), \Phi_0(s), \Theta_0(r)$ are arbitrary functions.

If in Equation 7.12 $\beta = 0$, it follows from Equation 7.18 that the functions Φ, Θ do not depend on t. In this case, we define the functions

$$\hat{\phi}(s, r, t) = A(s)(2\Phi_s\phi_s - \phi_{ss}) + C(r)[2\Theta_r\phi_r + \phi_{rr} - 2\Phi_r(\phi_r - \psi_r)]$$
$$+ C'(r)\phi_r + b(s)\phi_s + d(r)\phi_r,$$

$$\hat{\psi}(s, r, t) = A(s)[2\Phi_s\psi_s + \psi_{ss} - 2\Theta_s(\psi_s - \phi_s)] + A'(s)\psi_s$$
$$+ C(r)(2\Theta_r\psi_r - \psi_{rr}) + b(s)\psi_s + d(r)\psi_r.$$

INFINITE SET OF SECOND-ORDER SYMMETRIES
(Sheftel' [1994b])

A diagonal two-component generic hydrodynamic-type system 7.1 possesses an infinite set of second-order Lie-Bäcklund symmetries (depending on two arbitrary functions $c_1(s)$, $c_2(r)$) iff the coefficients ϕ, ψ of System 7.1 satisfy Equations 7.12 and 7.13, where the functions Φ, Θ are defined by Equation 7.8 and the functions $\hat{\Phi}(s, r, t)$, $\hat{\Theta}(s, r, t)$ are given by Formula 7.26.

The symmetries 7.3 have the coordinates

$$f = A(s)\frac{s_{xx}}{s_x^2} + \Phi_r\left(A(s)\frac{r_x}{s_x} + C(r)\frac{s_x}{r_x}\right) + \beta\frac{A(s)}{s_x} + s_x v(x, t, s, r) + 2A(s)\Phi_s + b(s),$$

$$g = C(r)\frac{r_{xx}}{r_x^2} + \Theta_s\left(A(s)\frac{r_x}{s_x} + C(r)\frac{s_x}{r_x}\right) + \beta\frac{C(r)}{r_x} + r_x \rho(x, t, s, r) + 2C(r)\Theta_r + d(r)$$

where the functions v, ρ are determined as follows:
if $\beta \neq 0$ then

$$v(x, t, s, r) = a(s, r)\exp\left\{\beta\left[x + \int_0^t \phi(s, r, t)dt\right]\right\} + \frac{1}{\beta}\hat{\Phi}(s, r, t),$$

$$\rho(x, t, s, r) = c(s, r)\exp\left\{\beta\left[x + \int_0^t \psi(s, r, t)dt\right]\right\} + \frac{1}{\beta}\hat{\Theta}(s, r, t);$$

if $\beta = 0$ then

$$v(x, t, s, r) = a(s, r) + \int_0^t \hat{\phi}(s, r, t)dt - \hat{\Phi}(s, r)(x + \int_0^t \phi(s, r, t)dt),$$

$$\rho(x, t, s, r) = c(s, r) + \int_0^t \hat{\psi}(s, r, t)dt - \hat{\Theta}(s, r)(x + \int_0^t \psi(s, r, t)dt).$$

Here, the integrals with respect to t are taken at constant values of s, r. The functions $a(s, r)$, $c(s, r)$ constitute an arbitrary smooth solution of the linear system 7.17.

7.1.3. First-order recursion operator

(Teshukov [1989], Sheftel' [1994b])

In Sections 7.1.3 and 7.1.4, the recursion operators for Equation 7.1 are considered in the form of matrix differential operators

$$R = A_N D_x^N + A_{N-1} D_x^{N-1} + \ldots + A_1 D_x + A_0$$

where A_i are 2×2 matrices. Their entries are differential functions depending on s, r and their finite order derivatives with respect to x. If $A_N \neq 0$, then N is called the order of R. In this section, the case $N = 1$ is considered.

Let

$$S(s, r) = \hat{\Phi}(s, r, 0) = A(s)\Phi_s(s, r, 0) + C(r)\Phi_r(s, r, 0),$$
$$T(s, r) = \hat{\Theta}(s, r, 0) = A(s)\Theta_s(s, r, 0) + C(r)\Theta_r(s, r, 0)$$

where $A(s)$ and $C(r)$ are arbitrary functions.

Let a system 7.1 satisfy the condition 7.12. Then it has a first-order recursion operator iff there exist functions $A(s)$ and $C(r)$ such that

$$S_r(s, r) = \Phi_r(s, r, 0)(S - T), \quad T_s(s, r) = \Theta_s(s, r, 0)(T - S).$$

The recursion operator is given by

$$R = \left[\begin{pmatrix} A(s) & 0 \\ 0 & C(r) \end{pmatrix} (D_x - \beta) \right.$$
$$\left. - \begin{pmatrix} A(s)D_x(\Phi(s, r, t)), & -\Phi_r(s, r, t)[A(s)r_x - C(r)s_x] \\ \Theta_s(s, r, t)[A(s)r_x - C(r)s_x] & C(r)D_x(\Theta(s, r, t)) \end{pmatrix} \right].$$
$$\cdot \begin{pmatrix} 1/s_x & 0 \\ 0 & 1/r_x \end{pmatrix}$$

7.1.4. Second-order recursion operator

(Sheftel' [1994b])

System 7.1 is said to be generic with respect to second-order recursion operators if its coefficients do not satisfy the constraints

$$\frac{\Phi_{rs}}{\Phi_r} + \Theta_s = c_1(s)e^\Phi, \quad \frac{\Theta_{sr}}{\Theta_s} + \Phi_r = c_2(r)e^\Theta$$

with arbitrary functions $c_1(s), c_2(r)$.

Let

$$S(s, r) = A(s)(\Phi_s^2 - \Phi_{ss} + 2\Lambda_{ss}) + A'(s)\Lambda_s + C(r)(2\Phi_r\Theta_r + \Phi_{rr} - \Phi_r^2)$$
$$+ C'(r)\Phi_r + b(s)\Phi_s + d(r)\Phi_r + \Phi_0(s),$$
$$T(s, r) = A'(s)\Theta_s + A(s)(2\Theta_s\Phi_s + \Theta_{ss} - \Theta_s^2) + C(r)(\Theta_r^2 - \Theta_{rr} + 2\Lambda_{rr})$$
$$+ C'(r)\Lambda_r + b(s)\Theta_s + d(r)\Theta_r + \Theta_0(r).$$

where $A(s)$, $C(r)$, $b(s)$, $d(r)$, $\Phi_0(s)$, $\Theta_0(r)$ are arbitrary functions. The functions $\Phi(s, r, t)$, $\Theta(s, r, t)$ are given by Equations 7.8, and the function $\Lambda(s, r)$ is defined in Section 7.1.2.

A generic system 7.1 satisfying the condition 7.12 has a second-order recursion operator iff there exist functions $A(s)$, $C(r)$, $b(s)$, $d(r)$, $\Phi_0(s)$, $\Theta_0(r)$ such that

$$S_r(s, r) = \Phi_r(s, r, 0)(S - T), \quad T_r(s, r) = \Theta_s(s, r, 0)(T - S).$$

The recursion operator is given by

$$R = (AD_x^2 + BD_x + F) \begin{pmatrix} \dfrac{1}{s_x} & 0 \\ 0 & \dfrac{1}{r_x} \end{pmatrix}.$$

Here,

$$A = \begin{pmatrix} \dfrac{A(s)}{s_x} & 0 \\ 0 & \dfrac{C(r)}{r_x} \end{pmatrix},$$

and

$$B = \begin{pmatrix} b_{11} & b_{12} \\ b_{21} & b_{22} \end{pmatrix}, \quad F = \begin{pmatrix} f_{11} & f_{12} \\ f_{21} & f_{22} \end{pmatrix}$$

are the matrices with the following elements:

$$b_{11} = -[A(s)\frac{s_{xx}}{s_x^2} + 2A(s)\left(\Phi_s + \Phi_r \frac{r_x}{s_x} + \frac{\beta}{s_x}\right) + b(s)],$$

$$b_{12} = \Phi_r(s, r, t)\left[A(s)\frac{r_x}{s_x} - C(r)\frac{s_x}{r_x}\right],$$

$$b_{21} = -\Theta_s(s, r, t)\left[A(s)\frac{r_x}{s_x} - C(r)\frac{s_x}{r_x}\right],$$

$$b_{22} = -[C(r)\frac{r_{xx}}{r_x^2} + 2C(r)\left(\Theta_s\frac{s_x}{r_x} + \Theta_r + \frac{\beta}{r_x}\right) + d(r)]$$

and

$$f_{12} = A(s)\left[-\Phi_r \frac{r_x}{s_x}\left(\frac{s_{xx}}{s_x} - \frac{r_{xx}}{r_x}\right) + (\Phi_{rr} - \Phi_r^2)\frac{r_x^2}{s_x}\right]$$
$$-C(r)\Lambda_{sr}\frac{s_x^2}{r_x} + \{A(s)[2(\Phi_{rs} - \Phi_r\Phi_s) - \Lambda_{sr}] - b(s)\Phi_r\}r_x$$
$$+ \{C(r)(2\Phi_r\Theta_r + \Phi_{rr} - \Phi_r^2) + [C'(r) + d(r)]\Phi_r\}s_x$$
$$-\beta\Phi_r\left[A(s)\frac{r_x}{s_x} - C(r)\frac{s_x}{r_x}\right],$$

$$f_{21} = C(r)\left[\Theta_s \frac{s_x}{r_x}\left(\frac{s_{xx}}{s_x} - \frac{r_{xx}}{r_x}\right) + (\Theta_{ss} - \Theta_s^2)\frac{s_x^2}{r_x}\right]$$
$$-A(s)\Lambda_{sr}\frac{r_x^2}{s_x} + \{C(r)[2(\Theta_{sr} - \Theta_s\Theta_r) - \Lambda_{sr}] - d(r)\Theta_s\}s_x$$
$$+ \{A(s)(2\Theta_s\Phi_s + \Theta_{ss} - \Theta_s^2) + [A'(s) + b(s)]\Theta_s\}r_x$$
$$+\beta\Theta_s\left[A(s)\frac{r_x}{s_x} - C(r)\frac{s_x}{r_x}\right],$$

$$f_{11} + f_{12} = \hat{\Phi}(s, r, t)s_x + \beta\left\{A(s)\left[\frac{s_{xx}}{s_x^2} + 2\Phi_s(s, r, t) + \frac{\beta}{s_x}\right]\right.$$
$$\left. +\Phi_r(s, r, t)\left[A(s)\frac{r_x}{s_x} + C(r)\frac{s_x}{r_x}\right] + b(s)\right\},$$

$$f_{21} + f_{22} = \hat{\Theta}(s, r, t)r_x + \beta\left\{C(r)\left[\frac{r_{xx}}{r_x^2} + 2\Theta_r(s, r, t) + \frac{\beta}{r_x}\right]\right.$$
$$\left. +\Theta_s(s, r, t)\left[A(s)\frac{r_x}{s_x} + C(r)\frac{s_x}{r_x}\right] + d(r)\right\}.$$

The functions $\hat{\Phi}(s, r, t)$ and $\hat{\Theta}(s, r, t)$ are defined by Formulas 7.26.

7.2. One-dimensional isoentropic gas flows

7.2.1. Lie-Bäcklund symmetries

Consider one-dimensional equations for isoentropic gas flows:

$$u_t + uu_x + \alpha^2(\rho)\rho\rho_x = 0, \quad \rho_t + \rho u_x + u\rho_x = 0.$$

Here, $u(x, t)$ and $\rho(x, t)$ are gas velocity and density, respectively, and the function $\alpha(\rho)$ is connected with the speed of sound $c = \rho\alpha(\rho)$ in the gas. In practice $\alpha(\rho)$ is determined by a gas-state equation $p = P(\rho)$, where p is the gas pressure, by the formula $\alpha(\rho) = (1/\rho)\sqrt{P'(\rho)}$.

The gas dynamics equations can be written in the diagonal form 7.1 (see, e.g., Rozhdestvenskii and Yanenko [1978]):

$$s_t = \phi(s, r)s_x, \quad r_t = \psi(s, r)r_x \tag{7.27}$$

where

$$\phi(s, r) = -\left[\frac{s + r}{2} - \rho\alpha(\rho)\right], \quad \psi(s, r) = -\left[\frac{s + r}{2} + \rho\alpha(\rho)\right],$$

by using the transformation to the Riemann invariants s, r:

$$s = u - \int_{\rho_0}^{\rho} \alpha(\rho)d\rho, \quad r = u + \int_{\rho_0}^{\rho} \alpha(\rho)d\rho, \quad \rho_0 = \text{const.}$$

The inverse transformation is given by

$$u = \frac{r + s}{2}, \quad \int_{\rho_0}^{\rho} \alpha(\rho)d\rho = \frac{r - s}{2}.$$

PARTIAL CLASSIFICATION (LIE-BÄCKLUND SYMMETRIES)

(Sheftel' [1983])

Lie-Bäcklund symmetries of the system 7.27 are written in the canonical form 7.3. Here, the symmetries are presented with coordinates f_N, g_N of order $N = 1, 2, 3$ (that is, f_N and g_N are differential functions of order N; see Chapter 1, Section 1.1).

For the arbitrary function $\alpha(\rho)$, the principal Lie algebra $L_{\mathcal{P}}$ is determined by the following coordinates.

$N = 1$:

$$f_1 = a(s, r)s_x, \quad g_1 = c(s, r)r_x$$

where $a(s, r), c(s, r)$ satisfy the linear equations:

$$a_r(s, r) = \frac{\alpha'(\rho)}{4\alpha^2}(a - c), \quad c_s(s, r) = \frac{\alpha'(\rho)}{4\alpha^2}(a - c).$$

$N = 2$:

$$f_2 = -\frac{s_{xx}}{s_x^2} - \frac{\alpha'(\rho)}{4\alpha^2}\frac{(s_x - r_x)^2}{s_x r_x},$$

$$g_2 = -\frac{r_{xx}}{r_x^2} + \frac{\alpha'(\rho)}{4\alpha^2}\frac{(s_x - r_x)^2}{s_x r_x}.$$

$N = 3$:

$$f_3 = -\left(\frac{s_{xxx}}{s_x^3} - \frac{3s_{xx}^2}{s_x^4}\right) - \frac{\alpha'(\rho)}{4\alpha^2}\left(\frac{1}{s_x^3} - \frac{1}{r_x^3}\right)s_x r_{xx}$$
$$-\frac{3\alpha'(\rho)}{4\alpha^2}\frac{s_{xx}}{s_x^3}(s_x - r_x) - \left[\frac{1}{2}\left(\frac{\alpha'}{\alpha^2}\right)^2\left(\frac{1}{s_x} + \frac{1}{r_x}\right) - \left(\frac{\alpha'}{\alpha^2}\right)'\frac{1}{\alpha s_x}\right]\frac{(s_x - r_x)^3}{8s_x r_x},$$

$$g_3 = -\left(\frac{r_{xxx}}{r_x^3} - \frac{3r_{xx}^2}{r_x^4}\right) - \frac{\alpha'(\rho)}{4\alpha^2}\left(\frac{1}{s_x^3} - \frac{1}{r_x^3}\right)r_x s_{xx}$$
$$-\frac{3\alpha'(\rho)}{4\alpha^2}\frac{r_{xx}}{r_x^3}(s_x - r_x) + \left[\frac{1}{2}\left(\frac{\alpha'}{\alpha^2}\right)^2\left(\frac{1}{s_x} + \frac{1}{r_x}\right) - \left(\frac{\alpha'}{\alpha^2}\right)'\frac{1}{\alpha r_x}\right]\frac{(s_x - r_x)^3}{8s_x r_x}.$$

The algebra L_P, with $N = 2$, extends in the following cases:
1. The function $\alpha(\rho)$ satisfies the differential equation

$$\left[(l\rho + b)\alpha^2(\rho)/\alpha'(\rho)\right]'_\rho = -\frac{A}{2}\alpha(\rho)$$

with arbitrary constants A, b, l. The second-order symmetries are determined by the formula

$$f_2 = (As + \bar{A})\frac{s_{xx}}{s_x^2} + \frac{\alpha'(\rho)}{4\alpha^2}\left[(As + \bar{A})\left(\frac{r_x}{s_x} - 2\right) + (Ar + \bar{C})\frac{s_x}{r_x}\right] + b + a(s, r)s_x,$$

$$g_2 = (Ar + \bar{C})\frac{r_{xx}}{r_x^2} - \frac{\alpha'(\rho)}{4\alpha^2}\left[(Ar + \bar{C})\left(\frac{s_x}{r_x} - 2\right) + (As + \bar{A})\frac{r_x}{s_x}\right] + b + c(s, r)r_x,$$

where $\bar{A}, \bar{C}$ are arbitrary constants, and $a(s, r), c(s, r)$ are the functions defined in L_P with $N = 1$.

In particular, if $l = 0$, one obtains a physically interesting state equation describing a polytropic gas:

$$P(\rho) = k^2 \rho^\gamma, \quad \alpha(\rho) = k\sqrt{\gamma}\rho^{(\gamma-3)/2}, \quad k, \gamma = \text{const.}$$

2. $\alpha = \text{const.}$ This is the polytropic gas with $\gamma = 3$. Then Equation 7.27 admits second-order symmetries

$$f_2 = s_x\psi_1\left(\frac{s_{xx}}{s_x^3}, s\right), \quad g_2 = r_x\psi_2\left(\frac{r_{xx}}{r_x^3}, r\right)$$

depending on two arbitrary functions ψ_1, ψ_2.

In this case, the system 7.27 is written

$$s_t = -ss_x, \qquad r_t = -rr_x.$$

Its general solution is given by

$$x - st = F(s), \qquad x - rt = G(r)$$

and depends upon two arbitrary functions F and G. It follows that the solution manifold consists of invariant solutions only.

3. $\alpha(\rho) = k\rho^{-2}$. This is the Chaplygin gas (see, e.g., Ovsiannikov [1981]) with the state equation

$$P(\rho) = P_0 - \frac{k^2}{\rho}, \qquad k = \text{const.}, \quad P_0 = \text{const.} > 0.$$

In this case, the second-order symmetries again depend upon two arbitrary functions of two variables.

7.2.2. Recursion operator

For the gas dynamics equation 7.27 with arbitrary function $\alpha(\rho)$ one has the following first-order recursion operator (cf. Section 7.1.3):

$$R = \left(I D_x - \frac{\alpha_x(\rho)}{2\alpha} \begin{pmatrix} 1 & -1 \\ -1 & 1 \end{pmatrix} \right) \begin{pmatrix} \dfrac{1}{s_x} & 0 \\ 0 & \dfrac{1}{r_x} \end{pmatrix}.$$

7.3.   Separable Hamiltonian systems

7.3.1. Hydrodynamic-type Hamiltonian systems

Consider two-component systems of the form:

$$D_t \begin{pmatrix} u \\ \rho \end{pmatrix} = \sigma_1 D_x \begin{pmatrix} H_u(u, \rho) \\ H_\rho(u, \rho) \end{pmatrix}, \tag{7.28}$$

where $\sigma_1 = \begin{pmatrix} 0 & 1 \\ 1 & 0 \end{pmatrix}$ is the Pauli matrix. The function $H(u, \rho)$ is called a hydrodynamic-type Hamiltonian density corresponding to the Hamiltonian $\mathcal{H} = \int\limits_{-\infty}^{\infty} H(u, \rho)dx$. For brevity, $H(u, \rho)$ is also called a Hamiltonian.

Equations 7.28 can be written as Hamilton equations:

$$D_t \begin{pmatrix} u \\ \rho \end{pmatrix} = \{ \begin{pmatrix} u \\ \rho \end{pmatrix}, H \}$$

with the hydrodynamic-type Poisson bracket (see Dubrovin and Novikov [1983]):

$$\{H, h\} = (h_u, h_\rho) D_x \sigma_1 (H_u, H_\rho)^T \qquad (7.29)$$

By using the Hamiltonian matrix

$$\hat{H} = \begin{pmatrix} H_{\rho u} & H_{\rho\rho} \\ H_{uu} & H_{u\rho} \end{pmatrix} \qquad (7.30)$$

one can rewrite Equation 7.28 in the form

$$\begin{pmatrix} u \\ \rho \end{pmatrix}_t = \hat{H} \begin{pmatrix} u \\ \rho \end{pmatrix}_x \iff (I D_t - \hat{H} D_x) \begin{pmatrix} u \\ \rho \end{pmatrix} = \begin{pmatrix} 0 \\ 0 \end{pmatrix}.$$

Gas dynamics equations (see Section 7.2.1) have the Hamiltonian form 7.28 with the Hamiltonian

$$H(u, \rho) = -(\rho u^2/2 + \int\limits_0^\rho d\rho \int\limits_0^\rho \alpha^2(\rho)\rho d\rho).$$

7.3.2. Separable systems

We say that Hamiltonians $\int\limits_{-\infty}^{\infty} H \, dx$ and $\int\limits_{-\infty}^{\infty} h \, dx$ (or the densities H and h) commute if the Poisson bracket 7.29 of their densities is a total derivative with respect to x:

$$\{H(u, \rho, t), h(u, \rho, t)\} = D_x[Q(u, \rho, t)].$$

For any hydrodynamic-type Hamiltonian $H(u, \rho)$, there exists an infinite set of Hamiltonians $h(u, \rho)$ commuting with H and with each other. They are arbitrary smooth solutions of the wave equation

$$H_{\rho\rho} h_{uu} - h_{\rho\rho} H_{uu} = 0.$$

We assume that $H_{uu}H_{\rho\rho} \neq 0$ and write the wave equation in the form:

$$h_{uu} - V(u, \rho)h_{\rho\rho} = 0, \quad \text{where } V(u, \rho) = H_{uu}/H_{\rho\rho}.$$

The system 7.28 is called *a separable Hamiltonian system* if $V(u, \rho) = \beta^2(u)/\alpha^2(\rho)$. That is, the wave equation admits a separation of variables:

$$(1/\beta^2(u))h_{uu} = (1/\alpha^2(\rho))h_{\rho\rho} . \tag{7.31}$$

Gas dynamics equations provide an example of a separable system with $\beta^2(u) = 1$. More generally, if $H_{uu}/H_{\rho\rho}$ depends only on ρ, then $H(u, \rho)$ is called a generalized gas dynamics Hamiltonian density.

A discussion of physical applications of separable Hamiltonian systems are to be found in Olver and Nutku [1988].

7.3.3. Second-order recursion operator for separable Hamiltonian systems

Notation:

$$\partial_u = \partial/\partial u, \quad \partial_\rho = \partial/\partial\rho, \quad \partial_u^{-1} = \int_0^u du, \quad \partial_\rho^{-1} = \int_0^\rho d\rho.$$

Given a separable Hamiltonian system with the wave equation 7.31, define the matrices

$$U_1 = \begin{pmatrix} u & \partial_\rho^{-1}\alpha^2(\rho) \\ \rho & \partial_u^{-1}\beta^2(u) \end{pmatrix}, \quad U_2 = \begin{pmatrix} \partial_u^{-1}\beta^2(u) & \partial_\rho^{-1}\alpha^2(\rho) \\ \rho & u \end{pmatrix}$$

Lie-Bäcklund transformation groups admitted by a Hamiltonian system 7.28 are written in the form

$$\begin{pmatrix} u \\ \rho \end{pmatrix}_\tau = \begin{pmatrix} f_n \\ g_n \end{pmatrix} \tag{7.32}$$

where f_n, $g_n \in \mathcal{A}$ are differential functions of order n.

Any *separable* Hamiltonian system 7.28 possesses a second-order matrix recursion operator. Namely,

$$L = D_x(U_{2x})^{-1}D_x(U_{1x})^{-1}, \tag{7.33}$$

where

$$(U_{1x})^{-1} = \frac{1}{\beta^2 u_x^2 - \alpha^2 \rho_x^2} \begin{pmatrix} \beta^2 u_x & -\alpha^2 \rho_x \\ -\rho_x & u_x \end{pmatrix},$$

$$(U_{2x})^{-1} = \frac{1}{\beta^2 u_x^2 - \alpha^2 \rho_x^2} \begin{pmatrix} u_x & -\alpha^2 \rho_x \\ -\rho_x & \beta^2 u_x \end{pmatrix}.$$

The operator 7.33 satisfies the recursion relation

$$L \begin{pmatrix} f_n \\ g_n \end{pmatrix} = \begin{pmatrix} f_{n+2} \\ g_{n+2} \end{pmatrix}, \qquad n = 2, 3, \ldots \quad .$$

7.3.4. Lie-Bäcklund symmetries

Separable systems 7.28 admit an infinite set of first-order ($n = 1$) symmetries 7.32 with the coordinates f_1, g_1 homogeneous in derivatives. The symmetries depending upon x, t and those independent of x, t are given by the formulas

$$\begin{pmatrix} f_1 \\ g_1 \end{pmatrix} = (xI + t\hat{H} + \hat{h}) \begin{pmatrix} u \\ \rho \end{pmatrix}_x \tag{7.34}$$

and

$$\begin{pmatrix} f_1 \\ g_1 \end{pmatrix} = \hat{h} \begin{pmatrix} u \\ \rho \end{pmatrix}_x \equiv \sigma_1 D_x \begin{pmatrix} h_u(u, \rho) \\ h_\rho(u, \rho) \end{pmatrix}, \tag{7.35}$$

respectively. Here, $h(u, \rho)$ is an arbitrary solution of Equation 7.31 and the matrices $\hat{H}$ and $\hat{h}$ are defined by the formula 7.30.

The space spanned by the vectors 7.35 is invariant under the operator 7.33. Namely,

$$L\hat{h} \begin{pmatrix} u \\ \rho \end{pmatrix}_x = \sigma_1 D_x \begin{pmatrix} h_{1u}(u, \rho) \\ h_{1\rho}(u, \rho) \end{pmatrix} \equiv \hat{h}_1 \begin{pmatrix} u \\ \rho \end{pmatrix}_x .$$

Hence, L generates the recursion of Hamiltonians $h(u, \rho)$:

$$h_1(u, \rho) = \frac{1}{\beta^2(u)} h_{uu}(u, \rho) = \frac{1}{\alpha^2(\rho)} h_{\rho\rho}(u, \rho),$$

$$h_m(u, \rho) = (\beta^{-2}(u)\partial^2/\partial u^2)^m h(u, \rho) = (\alpha^{-2}(\rho)\partial^2/\partial \rho^2)^m h(u, \rho). \tag{7.36}$$

It follows that the first-order symmetries 7.35 span an infinite-dimensional commutative Lie-Bäcklund algebra.

For any separable Hamiltonian system 7.28, higher-order Lie-Bäcklund symmetries of an even order $n = 2m$ are generated by the iterated action L^m of the recursion operator 7.33 upon the first-order symmetries 7.34:

$$\begin{pmatrix} f_{2m} \\ g_{2m} \end{pmatrix} = L^m x \begin{pmatrix} u \\ \rho \end{pmatrix}_x + (t\hat{H}_m + \hat{h}_m) \begin{pmatrix} u \\ \rho \end{pmatrix}_x , \qquad m = 1, 2, \ldots . \tag{7.37}$$

Here, $H_m(u, \rho)$ and $h_m(u, \rho)$ are obtained from $H(u, \rho)$ and $hH(u, \rho)$, respectively, by Transformation 7.36. By letting $m = 1$ in Formula 7.37, one obtains all the second-order symmetries.

7.4. Semi-Hamiltonian equations

7.4.1. Geometry and first-order symmetries

(Tsarev [1985], [1991])

Consider diagonal systems of quasi-linear equations

$$u_t^i = v_i(u)u_x^i, \qquad i = 1, 2, \ldots, n. \tag{7.38}$$

Here $u = (u^1, u^2, \ldots, u^n)$ is an n-vector. Henceforth no summation on repeated indices is assumed.

REMARK.   A group analysis of time-dependent diagonal systems

$$u_t^i = v_i(t, u)u_x^i, \qquad i = 1, 2, \ldots, n,$$

is to be found in Sheftel' [1994b].

Canonical Lie-Bäcklund symmetries (see Chapter 1)

$$X = \eta^i \frac{\partial}{\partial u^i}$$

of Equations 7.38 are given by the coordinates

$$\eta^i = \eta^i(t, x, u, u_x, u_{xx}, \ldots), \qquad \eta^i \in \mathcal{A}, \tag{7.39}$$

determined by the equations

$$\left(I D_t - V D_x - U_x(\frac{\partial V}{\partial U}) \right) \eta = 0. \tag{7.40}$$

Here, η is the n-vector with components η^i, D_x is the total x-differentiation, D_t is the total t-differentiation along the "trajectories" of the system 7.38, I is the

$n \times n$ unit matrix, $U_x = \mathrm{diag}(u_x^i)$, $V(u) = \mathrm{diag}(v_i(u))$, and $\partial V/\partial U = (v_{i,u^j}) \equiv (\partial v_i/\partial u^j)$ is the Jacobian matrix of the mapping $v = v(u)$.

If

$$v_i(u) \neq v_j(u) \qquad (i \neq j). \tag{7.41}$$

one can define the symmetric connection (the Christoffel symbols)

$$\Gamma_{ij}^i(u) = \Gamma_{ji}^i(u) = v_{i,u^j}(u)/(v_j - v_i) \qquad (i \neq j)$$

$$\Gamma_{ii}^j = -(g_{ii}/g_{jj})\Gamma_{ij}^i \quad (i \neq j), \qquad \Gamma_{jk}^i = 0 \quad (i \neq j \neq k \neq i).$$

This connection is compatible with the nondegenerate diagonal metric

$$g_{ii}(u) = H_i^2(u) = e^{2\Phi_i(u)}, g_{ij} = 0 \quad (i \neq j), \qquad \det(g^{ij}) \neq 0,$$

so that

$$\Gamma_{ij}^i = (\ln \sqrt{g_{ii}})_{u^j} = (\ln H_i)_{u^j} = \Phi_{i,u^j}(u). \tag{7.42}$$

The functions $H_i(u)$ are known as the Lamé coefficients. The integrability conditions $\Gamma_{ij,u^k}^i = \Gamma_{ik,u^j}^i$ of Equations 7.42 have the form:

$$[v_{i,u^j}/(v_j - v_i)]_{u^k} = [v_{i,u^k}/(v_k - v_i)]_{u^j} \quad (i \neq j \neq k \neq i). \tag{7.43}$$

A diagonal system 7.38 is said to be *semi-Hamiltonian* if its coefficients $v_i(u)$ satisfy the conditions 7.41 and Equation 7.43. A semi-Hamiltonian system is a *Hamiltonian system* iff, in the above metric, Riemann's curvature tensor R_{jkl}^i satisfies the condition $R_{jji}^i = 0$ ($i \neq j$). Then the curvature tensor vanishes identically and the variables u^i provide an orthogonal curvilinear coordinate system on a flat (pseudo-Euclidean) space.

Any semi-Hamiltonian system possesses an infinite set of first-order symmetries 7.39 of the form

$$\eta^i = w_i(u)u_x^i, \qquad i = 1, 2, \ldots, n. \tag{7.44}$$

The functions $w_i(u)$ are determined by the first-order linear differential equations

$$w_{i,u^j} = \Gamma_{ij}^i(u)(w_j - w_i), \qquad i \neq j. \tag{7.45}$$

Furthermore, it is evident that any system 7.38 admits the group of homogeneous dilations of variables t, x with the following coordinates of the canonical Lie-Bäcklund generator 7.39:

$$\eta^i = (x + tv_i(u))u_x^i.$$

The linear combination with the symmetries 7.44,

$$\eta^i = [w_i(u) + c(x + t v_i(u))]u_x^i, \quad i = 1, 2, \ldots, n, \ c = \text{const.} \quad (7.46)$$

provides the most general first-order symmetry known for semi-Hamiltonian systems 7.38 (see Sheftel' [1993]).

7.4.2. Representation of the general solution via invariant solutions

Given a semi-Hamiltonian system 7.38, consider the symmetries 7.46 with $c = -1$. For these symmetries, the regular invariant solutions (see [H2], Section 1.5.11) are given by the equations

$$w_i(u) = t v_i(u) + x, \quad i = 1, 2, \ldots, n, \quad (7.47)$$

provided that $u_x^i \neq 0, \ i = 1, 2, \ldots, n$.

The following result due to Tsarev [1991] linearizes the semi-Hamiltonian systems and generalizes the classical hodograph transformation to the multicomponent case.

Given a solution $w_i(u)$ of Equations 7.45, any solution $u^i(x, t), \ i = 1, \ldots, n$, of the algebraic equations 7.47 is a solution of the semi-Hamiltonian system 7.38. Conversely, any solution $u^i(x, t), \ i = 1, \ldots, n$, of the system 7.38 such that $u_x^i(x_0, t_0) \neq 0, \ i = 1, \ldots, n$, at a point (x_0, t_0), satisfies, in a neighborhood of the point (x_0, t_0), the invariance conditions 7.47 with an appropriately chosen solution $w_i(u), \ i = 1, \ldots, n$, of Equations 7.45.

7.4.3. Recursion operator

(Teshukov [1989]; see also Sheftel' [1994a])

A recursion operator R for a system 7.38 commutes with the operator of determining equation 7.40 (see Chapter 4, Section 4.2.3):

$$\left[I D_t - V D_x - U_x \left(\frac{\partial V}{\partial U} \right), R \right] = 0. \quad (7.48)$$

For any semi-Hamiltonian system 7.38, Equation 7.48 written for a first-order operator R of the form

$$R = (A D_x + B) U_x^{-1},$$

where $A = A(u), B = B(u, u_x)$ are $n \times n$ matrices, yields the matrix-operator R

with the elements:

$$R_{ij} = [\delta_{ij}c_j(u^j)(D_x + \sum_{k=1}^{n} \Gamma_{ik}^i(u)u_x^k)$$
$$+\Gamma_{ij}^i(u)(c_j(u^j)u_x^i - c_i(u^i)u_x^j)](1/u_x^j) + d_j(u^j)\delta_{ij}. \tag{7.49}$$

Here, δ_{ij} is the Kronecker symbol, and the functions $c_i(u^i)$, $d_i(u^i)$ are defined by the equations

$$S_{i,u^j}(u) = \Gamma_{ij}^i(u)(S_j - S_i), \qquad i \neq j, \tag{7.50}$$

where

$$S_i(u) = \sum_{k=1}^{n} \Gamma_{ik}^i(u)c_k(u^k) + d_i(u^i), \qquad i = 1, 2, \ldots, n.$$

Thus, any semi-Hamiltonian system 7.38 possesses the matrix-valued first-order recursion operator R given by the formula 7.49.

8

Ordinary Differential Equations

Since the fundamental work of Sophus Lie on ordinary differential equations started approximately 120 years ago, numerous important results on group analysis of ordinary differential equations have been obtained by Lie himself and various researchers using the Lie approach. In this chapter we collect and present the majority of the results dealing with integration algorithms. The reader interested in conceptual discussions is referred to the relevant chapters of [H1].

One of Lie's most remarkable results was the complete classification (in the complex domain) of all possible continuous groups acting in the plane (see Lie [1880a], [1883], [1891], [1893]) and Tschebotaröw [1940]). We give here *in extenso* his own words taken from Lie [1883]: 'In a short communication to the Scientific Society of Göttingen (3 December 1874), I gave, among other things, a listing of all continuous transformation groups in two variables x, y, and specially emphasized that this might be made the basis of a classification and rational integration theory of all differential equations $f(x, y, y', \ldots, y^{(m)}) = 0$ admitting a continuous group of transformations. The great program sketched there I have subsequently carried out in detail.'

8.1. Linear nth-order equations

$$y^{(n)} + \sum_{i=0}^{n-1} a_i(x) y^{(i)} = 0, \quad n \geq 3. \tag{8.1}$$

8.1.1. Equivalent equations[1]

Two equations are equivalent by means of a point transformation if one can

---

[1] We remind the reader that all considerations in this Handbook are local. Thus, here we mean local equivalence.

be transformed into the other by the said transformation. The following are well known.

FIRST-ORDER EQUATIONS

All first-order equations $y' = f(x, y)$ are equivalent to each other. In particular, an equation of the first order can be transformed into the simplest one, viz., $y' = 0$.

LINEAR SECOND-ORDER EQUATIONS

All linear second-order equations are equivalent to each other and can, for example, be reduced to the simplest equation $y'' = 0$.

However, a linear equation of order $n \geq 3$ need not be transformable into its simplest form. Therefore, in Section 8.1, we consider $n \geq 3$.

8.1.2.　Laguerre-Forsyth canonical form

(Laguerre [1879], Forsyth [1899])

The transformation

$$y \mapsto y \exp(\frac{1}{n} \int_{x_0}^{x} a_{n-1}(s)\, ds)$$

reduces Equation 8.1 to

$$y^{(n)} + \sum_{i=0}^{n-2} a_i(x) y^{(i)} = 0. \tag{8.2}$$

Furthermore, one can set $a_{n-2} = 0$ by the transformation

$$x \mapsto f(x), \quad y \mapsto yg(x),$$

with $f(x)$ and $g(x)$ defined by $f' = h^{-2}$, $g = h^{1-n}$, where $h(x)$ is a solution of

$$\frac{(n+1)!}{(n-2)!3!} h'' + a_{n-2}h = 0.$$

The corresponding differential equation

$$y^{(n)} + \sum_{i=0}^{n-3} a_i(x) y^{(i)} = 0 \tag{8.3}$$

is called the *Laguerre-Forsyth canonical form* (see, e.g., Neuman [1991]) and is characterized by the vanishing of a_{n-1} and a_{n-2} in Equation 8.1. The transformations that reduce Equation 8.1 to Equation 8.3 are known as *Laguerre-Forsyth transformations*. In our discussions, we invoke both forms of Equations 8.2 and 8.3.

8.1.3. Determining equation

(Mahomed [1989], Mahomed and Leach [1990])

THEOREM 8.1. *The most general form of the Lie point symmetry operator admitted by Equation 8.2 for* $n \geq 3$ *is*

$$X = \xi(x)\frac{\partial}{\partial x} + [(\frac{n-1}{2}\xi' + c_0)y + \eta(x)]\frac{\partial}{\partial y}, \qquad (8.4)$$

where c_0 is a constant, $\eta(x)$ satisfies Equation 8.2, and $\xi(x)$ is determined by the relations

$$\frac{(n+1)!(i-1)}{(n-i)!(i+1)!}\xi^{(i+1)} + 2i\xi'a_{n-i} + 2\xi a'_{n-i}$$

$$+ \sum_{j=2}^{i-1} a_{n-j}\frac{(n-j)![n(i-j-1)+i+j-1]}{(n-i)!(i-j+1)!}\xi^{(i-j+1)} = 0, \ i = 1, \dots, n.$$
$$(8.5)$$

It is widely known from Lie's works that the maximal point symmetry Lie algebra of an nth-order ($n \geq 3$) ordinary differential equation has dimension $r \leq n + 4$. Hence Equations 8.5 have at most three independent solutions.

8.1.4. Principal Lie algebra L_p

(Mahomed and Leach [1990], Krause and Michel [1991])

The Lie algebra admitted by Equation 8.2 for arbitrary coefficients $a_i(x)$ is called the principal Lie algebra of this equation and is denoted by L_p (see [H2] for details on principal Lie algebras).

For arbitrary $a_i(x)$, the determining equation for ξ implies that $\xi = 0$. Thus, the principal Lie algebra of Equation 8.2 is, invoking Operator 8.4, spanned by the operators

$$X_0 = y\frac{\partial}{\partial y}, \qquad (8.6)$$

$$X_i = \eta_i(x)\frac{\partial}{\partial y}, \qquad i = 1, \dots, n, \qquad (8.7)$$

where $\eta_i(x)$ are n linearly independent solutions of Equation 8.2.

8.1.5. Example of an extension of L_P

(Mahomed and Leach [1990])

Consider the simplest equation, viz.,

$$y^{(n)} = 0, \tag{8.8}$$

for an arbitrary $n \geq 3$. The $n + 1$-dimensional principal Lie algebra is spanned by Operators 8.6 and 8.7 with $\eta_i(x)$ given by

$$\eta_i(x) = C_i x^{i-1}, \quad i = 1, \ldots, n,$$

where the C_i,s are n arbitrary constants. Moreover, using Equation 8.5 we have, since $a_i = 0$,

$$\xi = A_0 + A_1 x + A_2 x^2, \quad A_i \text{ constants}.$$

The extension is maximal, i.e., three dimensional. Thus, the maximal symmetry algebra of Equation 8.8 is spanned by Operators 8.6, 8.7, and

$$X_{n+2} = \frac{\partial}{\partial x}, \quad X_{n+3} = x\frac{\partial}{\partial x}, \quad X_{n+4} = x^2\frac{\partial}{\partial x} + (n-1)xy\frac{\partial}{\partial y}. \tag{8.9}$$

8.1.6. Iterative equations

(Mahomed [1989], Krause and Michel [1991])

Let y_1 and y_2 be two linearly independent solutions of a linear homogeneous second-order equation. Define the following n functions:

$$u_k = y_1^{n-k} y_2^{k-1}, \quad k = 1, \ldots, n. \tag{8.10}$$

Then $u_1, \ldots, u_n$ are n linearly independent solutions of a linear nth-order equation of the Form 8.2. This equation is referred to as the *iterative equation* (see, e.g., Neuman [1991]) and its entire solution set is generated by just two solutions y_1 and y_2 of a linear homogeneous second-order equation.

Example 1. Suppose

$$y'' + a_0(x)y = 0$$

has solutions y_1 and y_2. Then the iterative third-order linear equation is

$$u''' + 4a_0 u' + 2a'_0 u = 0. \tag{8.11}$$

Using Equation 8.10, one obtains the following three solutions:

$$u_1 = y_1^2, \quad u_2 = y_1 y_2, \quad u_3 = y_2^2 .$$

The following theorem provides a Lie algebraic criterion for a linear nth-order equation of the Form 8.2 to be iterative.

THEOREM 8.2. *A linear nth-order Equation 8.2 is iterative if and only if*
(a) it possesses the maximal Lie algebra of dimension $n + 4$ or
(b) it is reducible by the Laguerre-Forsyth transformations (see Section 8.1.2.) to $y^{(n)} = 0.$

Example 2. Equation 8.11 has $n + 4 = 7$ point symmetries and it is iterative.

8.1.7. Equations that have a symmetry algebra of maximal order $n+4$

(Mahomed and Leach [1990])

Here, we give a list of linear equations of orders $n = 3, \ldots, 8$, that admit $n + 4$ point symmetries. In the case $n \geq 9$, the general form of the equation admitting $n + 4$ point symmetries is not as yet known.

$$y^{(3)} + a_1 y^{(1)} + \frac{1}{2} a_1^{(1)} y = 0,$$

$$y^{(4)} + a_2 y^{(2)} + a_2^{(1)} y^{(1)} + \left(\frac{3}{10} a_2^{(2)} + \frac{9}{100} a_2^2\right) y = 0,$$

$$y^{(5)} + a_3 y^{(3)} + \frac{3}{2} a_3^{(1)} y^{(2)} + \left(\frac{9}{10} a_3^{(2)} + \frac{16}{100} a_3^2\right) y^{(1)}$$

$$+ \left(\frac{1}{5} a_3^{(3)} + \frac{16}{100} a_3 a_3^{(1)}\right) y = 0,$$

$$y^{(6)} + a_4 y^{(4)} + 2 a_4^{(1)} y^{(3)} + \left(\frac{63}{35} a_4^{(2)} + \frac{259}{1225} a_4^2\right) y^{(2)}$$

$$+ \left(\frac{28}{35} a_4^{(3)} + \frac{518}{1225} a_4 a_4^{(1)}\right) y^{(1)}$$

$$+ \left(\frac{5}{35} a_4^{(4)} + \frac{130}{1225} a_4^{(1)2} + \frac{155}{1225} a_4 a_4^{(2)} + \frac{225}{42875} a_4^3\right) y = 0,$$

$$y^{(7)} + a_5 y^{(5)} + \frac{5}{2} a_5^{(1)} y^{(4)} + (3a_5^{(2)} + \frac{1}{4} a_5^2) y^{(3)} + (2a_5^{(3)} + \frac{3}{4} a_5^{(1)} a_5) y^{(2)}$$

$$+ (\frac{5}{7} a_5^{(4)} + \frac{295}{784} a_5^{(1)^2} + \frac{88}{196} a_5^{(2)} a_5 + \frac{36}{2744} a_5^3) y^{(1)}$$

$$+ (\frac{3}{28} a_5^{(5)} + \frac{177}{784} a_5^{(1)} a_5^{(2)} + \frac{39}{392} a_5 a_5^{(3)} + \frac{54}{2744} a_5^2 a_5^{(1)}) y = 0,$$

$$y^{(8)} + a_6 y^{(6)} + 3a_6^{(1)} y^{(5)} + (\frac{9}{2} a_6^{(2)} + \frac{47}{168} a_6^2) y^{(4)}$$

$$+ (4a_6^{(3)} + \frac{47}{42} a_6 a_6^{(1)}) y^{(3)} + (\frac{15}{7} a_6^{(4)} + \frac{1773}{1764} a_6^{(2)} a_6 + \frac{1485}{1764} a_6^{(1)^2}$$

$$+ \frac{3229}{148176} a_6^3) y^{(2)} + (\frac{9}{14} a_6^{(5)} + \frac{393}{882} a_6 a_6^{(3)} + \frac{891}{882} a_6^{(1)} a_6^{(2)}$$

$$+ \frac{9687}{148176} a_6^2 a_6^{(1)}) y^{(1)} + (\frac{1}{12} a_6^{(6)} + \frac{40}{441} a_6^{(4)} a_6 + \frac{51}{1680} a_6^{(2)^2} + \frac{113}{504} a_6^{(1)} a_6^{(3)}$$

$$+ \frac{25}{112896} a_6^4 + \frac{4135}{42336} a_6^{(2)} a_6^2 + \frac{347}{10584} a_6 a_6^{(1)^2}) y = 0.$$

If the coefficients a_i are constants ($a_i^{(j)} = 0$, for $j = 1, 2, \ldots$), the equations of odd order contain only odd derivatives of y and, similarly, equations of even order contain only even derivatives of y. In both these cases one can write the above equations in their factored form. In the general case we obtain the following. For n odd, $n + 4$ point symmetries occur if and only if the equation has the form

$$\frac{d}{dx} \left\{ \prod_{i=1}^{(n-1)/2} \left(\frac{(n+1)!}{(n-2)!3!} \frac{d^2}{dx^2} + (2i)^2 a_{n-2} \right) \right\} y = 0. \qquad (8.12)$$

For n even we have the form

$$\left\{ \prod_{i=1}^{n/2} \left(\frac{(n+1)!}{(n-2)!3!} \frac{d^2}{dx^2} + (2i-1)^2 a_{n-2} \right) \right\} y = 0 \qquad (8.13)$$

which has $n + 4$ point symmetries. Both equations (odd and even) are derivable from a second-order equation and are hence iterative.

8.1.8.   Nonexistence of $n + 3$-dimensional symmetry algebra

(Mahomed and Leach [1990])

THEOREM 8.3.   *A linear nth-order equation of the Form 8.2 does not admit a maximal symmetry algebra of dimension $n + 3$.*

8.1.9.   Equations with $n + 2$-dimensional symmetry algebra

(Mahomed [1989], Krause and Michel [1991])

THEOREM 8.4.   *A linear nth-order equation of the Form 8.2 has a maximal symmetry algebra of dimension $n + 2$ if and only if it is transformable into a constant coefficient linear equation which is not of the form of Equations 8.12 or 8.13.*

Example.   The linear equation of the Form 8.2,

$$y^{(n)} + \sum_{i=0}^{n-2} A_i y^{(i)} = 0, \tag{8.14}$$

where the A_is are n independent constants, has $n + 2$ point symmetries. Thus, the principal Lie algebra $L_{\mathcal{P}}$ extends by one, and the symmetry algebra of Equation 8.14 is spanned by the Operators 8.6 and 8.7 with $\eta_i(x)$ being n independent solutions of Equation 8.14, and

$$X_{n+2} = \frac{\partial}{\partial x}.$$

8.2.   Second-order equations

$$y'' = F(x, y, y') . \tag{8.15}$$

8.2.1.   Lie's classification in the complex domain

(Lie [1889], [1891]; see also Ibragimov [1992b])

Lie gave a complete classification of second-order equations admitting non-similar complex Lie algebras L_r of dimension r, where $0 \leq r \leq 8$.

Two complex Lie algebras of operators acting on the plane are *similar* if they are isomorphic and one can be transformed into the other by means of a complex change of variables. Otherwise the algebras are *non-similar*. Equations that admit similar complex algebras are said to be similar (equivalent) to each other as they can be transformed from one to the other by a complex change of variables. Lie showed that the maximal complex Lie algebra admitted by a given second-order equation is one of dimension 0, 1, 2, 3, or 8. Dimension 0 here means that the equation admits no point symmetry. It should be noted that an equation cannot admit a maximal 4-, 5-, 6-, or 7-dimensional Lie algebra. If it admits an 8-dimensional algebra, the equation is linearizable and equivalent to the simplest equation $y'' = 0$. The reader is referred to [H1] (Chapters 2, 3, and 8) for Lie's complex classification and linearizability (see also Sections 8.2.2 and 8.2.3 of this chapter) of second-order equations.

8.2.2. Classification in the real domain

(Mahomed and Leach [1989a], Ibragimov [1992b])

In applications, one mainly encounters real second-order differential equations. Therefore, it is beneficial to consider the Lie classification for these equations over the reals.

One- and two-dimensional algebras both have the same structure over the reals and over the complex numbers. In suitable bases, these can be taken to be $[X_1, X_2] = 0$ and $[X_1, X_2] = X_1$ for two-dimensional algebras, and $[X, X] = 0$ for any X for a one-dimensional algebra. Hence, the classification of second-order equations over the reals is the same as that over the complex numbers for one- and two-dimensional algebras. For a detailed discussion of the Lie classification of second-order equations admitting a two-dimensional algebra, see [H1] (Chapter 2). However, the situation is somewhat different for three-dimensional algebras. This is also true for higher-dimensional algebras.

For our investigation here, we focus attention on three-dimensional algebras. In the real domain, two of the complex Lie algebras of dimension three each splits up into two real non-isomorphic algebras (see Table 1). Thus, there are two more non-isomorphic real three-dimensional algebras than complex algebras. Consequently, there are more three-dimensional algebra of operators acting on the real plane than the complex plane. These result in additional non-similar second-order equations that admit real Lie algebras.

Table 1

Non-isomorphic structures of Real three-dimensional Lie Algebras

(by Bianchi [1918])

Algebra	Nonzero commutation relations
$L_{3;1}$	
$L_{3;2}$	$[X_2, X_3] = X_1$
$L_{3;3}$	$[X_1, X_3] = X_1,\ [X_2, X_3] = X_1 + X_2$
$L_{3;4}$	$[X_1, X_3] = X_1$
$L_{3;5}$	$[X_1, X_3] = X_1,\ [X_2, X_3] = X_2$
$L_{3;6}$	$[X_1, X_3] = X_1,\ [X_2, X_3] = aX_2 \quad a \neq 0, 1$
$L_{3;7}$	$[X_1, X_3] = bX_1 - X_2,\ [X_2, X_3] = X_1 + bX_2$
$L_{3;8}$	$[X_1, X_2] = X_1,\ [X_2, X_3] = X_3,\ [X_3, X_1] = -2X_2$
$L_{3;9}$	$[X_1, X_2] = X_3,\ [X_2, X_3] = X_1,\ [X_3, X_1] = X_2$

Labeling. In the above as well as the following, when dealing with more than one Lie algebra of the same dimension, we distinguish one from another by two indices, the first referring to the dimension and the second to the number of the algebra in an arbitrary chosen ordering. Thus, $L_{3;2}$ denotes the second algebra of dimension three.

REMARK. Lie [1883] gave his classification in the complex domain. In the complex plane, there are seven non-isomorphic structures: if one changes to the complex basis $\bar{X}_1 = X_2 + iX_1$, $\bar{X}_2 = X_1 + iX_2$, $\bar{X}_3 = (i + b)X_3$ with $a = (b - i)/(b + i)$, algebra $L_{3;6}$ takes the form $L_{3;7}$ while $L_{3;8}$ becomes $L_{3;9}$ in the basis $\bar{X}_1 = \frac{1}{2}X_1 + \frac{1}{2}X_3$, $\bar{X}_2 = iX_2$, $\bar{X}_3 = \frac{1}{2}X_1 - \frac{i}{2}X_3$. Thus, Lie's complex classification of non-isomorphic algebras L_3 is obtained by excluding $L_{3;7}$ and $L_{3;9}$.

Table 2

Realizations of 1, 2, and 3-dimensional Lie Algebras in the Real domain

Let $p = \partial/\partial x$ and $q = \partial/\partial y$.

Algebra	Realizations in (x, y) plane

L_1 $\quad X_1 = p$

$L_{2;1}^{I}$ $\quad X_1 = p, X_2 = q$

$L_{2;1}^{II}$ $\quad X_1 = q, X_2 = xq$

$L_{2;2}^{I}$ $\quad X_1 = q, X_2 = xp + yq$

$L_{2;2}^{II}$ $\quad X_1 = q, X_2 = yq$

$L_{3;1}$ $\quad X_1 = q, X_2 = xq, X_3 = h(x)q$

$L_{3;2}$ $\quad X_1 = q, X_2 = p, X_3 = xq$

$L_{3;3}^{I}$ $\quad X_1 = q, X_2 = p, X_3 = xp + (x + y)q$

$L_{3;3}^{II}$ $\quad X_1 = q, X_2 = xq, X_3 = p + yq$

$L_{3;4}^{I}$ $\quad X_1 = p, X_2 = q, X_3 = xp$

$L_{3;4}^{II}$ $\quad X_1 = q, X_2 = xq, X_3 = xp + yq$

$L_{3;5}^{I}$ $\quad X_1 = p, X_2 = q, X_3 = xp + yq$

$L_{3;5}^{II}$ $\quad X_1 = q, X_2 = xq, X_3 = yq$

$L_{3;6}^{I}$ $\quad X_1 = p, X_2 = q, X_3 = xp + ayq, a \neq 0, 1$

$L_{3;6}^{II}$ $\quad X_1 = q, X_2 = xq, X_3 = (1 - a)xp + yq, a \neq 0, 1$

$L_{3;7}^{I}$ $\quad X_1 = p, X_2 = q, X_3 = (bx + y)p + (by - x)q$

$L_{3;7}^{II}$ $\quad X_1 = xq, X_2 = q, X_3 = (1 + x^2)p + (x + b)yq$

$L_{3;8}^{I}$ $\quad X_1 = q, X_2 = xp + yq, X_3 = 2xyp + y^2 q$

$L_{3;8}^{II}$ $\quad X_1 = q, X_2 = xp + yq, X_3 = 2xyp + (y^2 - x^2)q$

$L_{3;8}^{III}$ $\quad X_1 = q, X_2 = xp + yq, X_3 = 2xyp + (y^2 + x^2)q$

$L_{3;8}^{IV}$ $\quad X_1 = q, X_2 = yq, X_3 = y^2 q$

$L_{3;9}$ $\quad X_1 = (1 + x^2)p + xyq, X_2 = xyp + (1 + y^2)q,$
$\quad\quad X_3 = yp - xq$

REMARK 1.   In Lie's realizations (up to similarity) in the complex domain, Lie algebras $L_{3;7}^{I}$, $L_{3;7}^{II}$, $L_{3;8}^{II}$, and $L_{3;9}$ do not arise. The realizations in the real domain given here are taken from Mahomed and Leach [1989a].

REMARK 2.   In Table 2, a and b are real parameters. Their values can be restricted to $-1 \leq a < 1$ and $b \geq 0$ by choosing a suitable basis of the algebras.

Table 3
Lie group classification of second-order equations in the real domain

Let $p = \partial/\partial x$ and $q = \partial/\partial y$.

Algebra	Basis operators	Representative Equations
L_1	$X_1 = p$	$y'' = f(y, y')$
$L_{2;1}^I$	$X_1 = p, X_2 = q.$	$y'' = f(y')$
$L_{2;2}^I$	$X_1 = q, X_2 = xp + yq.$	$xy'' = f(y')$
$L_{3;3}^I$	$X_1 = p, X_2 = q,$ $X_3 = xp + (x+y)q.$	$y'' = Ce^{-y'}$
$L_{3;6}^I$	$X_1 = p, X_2 = q,$ $X_3 = xp + ayq.$	$y'' = Cy'^{\frac{a-2}{a-1}}, \ a \neq 0, \frac{1}{2}, 2$
$L_{3;7}^I$	$X_1 = p, X_2 = q,$ $X_3 = (bx + y)p + (by - x)q.$	$y'' = C(1 + y'^2)^{\frac{3}{2}} e^{b \arctan y'}$
$L_{3;8}^I$	$X_1 = q, X_2 = xp + yq,$ $X_3 = 2xyp + y^2 q.$	$xy'' = Cy'^3 - \frac{1}{2}y'$
$L_{3;8}^{II}$	$X_1 = q, X_2 = xp + yq,$ $X_3 = 2xyp + (y^2 - x^2)q.$	$xy'' = y' + y'^3 + C(1 + y'^2)^{3/2}$
$L_{3;8}^{III}$	$X_1 = q, X_2 = xp + yq,$ $X_3 = 2xyp + (y^2 + x^2)q.$	$xy'' = y' - y'^3 + C(1 - y'^2)^{3/2}$
$L_{3;9}$	$X_1 = (1 + x^2)p + xyq,$ $X_2 = xyp + (1 + y^2)q,$ $X_3 = yp - xq.$	$y'' = C\left[\dfrac{1 + y'^2 + (y - xy')^2}{1 + x^2 + y^2}\right]^{3/2}$
L_8	$X_1 = p, X_2 = q \ X_3 = xq,$ $X_4 = xp, X_8 = yp, X_6 = yq,$ $X_7 = x^2 p + xyq,$ $X_8 = xyp + y^2 q.$	$y'' = 0$

In Table 3, f is an arbitrary function of its argument(s) and C is an arbitrary constant.

REMARK.   A second-order equation does not admit the Abelian Lie algebra $L_{3;1}$ or the algebra $L_{3;8}^{IV}$. Furthermore, a second-order equation does not admit as a maximal Lie algebra $L_{3;2}$, $L_{3;4}$ or $L_{3;5}$. In the real classification, algebra $L_{3;8}$ has three non-similar representations $L_{3;8}^I$, $L_{3;8}^{II}$, and $L_{3;8}^{III}$ that give rise to three non-equivalent equations. In the complex classification, $L_{3;7}^I$, $L_{3;8}^{II}$, and $L_{3;9}$, and their corresponding equations do not arise.

THE REPRESENTATION DUE TO LIE
   (Lie [1891])

The difference in Lie's classification occurs only in the three-dimensional symmetry Lie algebra $L_{3;8}$. Only the difference is given here.

$$L_{3;8}^I \quad X_1 = q, X_2 = 2xp + yq, \qquad y'' = Cy^{-3}$$
$$X_3 = x^2 p + xyq.$$

$$L_{3;8}^{III} \quad X_1 = p + q, X_2 = xp + yq, \quad y'' + 2\frac{y' + Cy'^{3/2} + y'^2}{x - y} = 0$$
$$X_3 = x^2 p + y^2 q.$$

REMARK.   The algebras $L_{3;7}$, $L_{3;8}^{II}$, and $L_{3;9}$ do not arise in Lie's complex classification.

ONE MORE REPRESENTATION
   (Mahomed [1989])

For the integration procedure (follow [H1], Section 2.2.2), one may find this additional form useful:

$$L_{3;3}^I \quad X_1 = q, X_2 = p - (\ln x)q, \quad xy'' = -1 + Ce^{-y'}$$
$$X_3 = xp + yq.$$

$$L_{3;6}^I \quad X_1 = q, X_2 = x^{1-a}p, \quad xy'' = (a - 1)y' + Cy'^{\frac{2a-1}{a-1}}, \ a \neq 0, \tfrac{1}{2}, 2$$
$$X_3 = xp + yq.$$

8.2.3.   Linearization

(Lie [1883], [1891], Tresse [1896], Sarlet, Mahomed, and Leach [1987], Mahomed and Leach [1989b], Grissom, Thompson, and Wilkens [1989], and Ibragimov [1991], [1992b])

The following linearization result supplements Theorem 3.1 of [H1]:

THEOREM 8.5. *The following are equivalent:*
i. A second-order equation

$$y'' = f(x, y, y')$$ (8.16)

is linearizable by a point change of variables;

ii. The quantities (called relative invariants)

$$I_1 = f_{y'y'y'y'}$$

and

$$I_2 = \frac{d^2}{dx^2} f_{y'y'} - 4\frac{d}{dx} f_{y'y} - 3f_y f_{y'y'} + 6f_{yy} + f_{y'}(4f_{y'y} - \frac{d}{dx} f_{y'y'})$$

both vanish identically for Equation 8.16;

iii. Equation 8.16 is cubic in the first derivatives, i.e., has the form

$$y'' = A(x, y)y'^3 + B(x, y)y'^2 + C(x, y)y' + D(x, y)$$

and the coefficients A to D satisfy[2]

$$3A_{xx} + 3A_x C - 3A_y D + 3AC_x + C_{yy}$$

$$- 6AD_y + BC_y - 2BB_x - 2B_{xy} = 0,$$

$$6A_x D - 3B_y D + 3AD_x + B_{xx} - 2C_{xy}$$

$$- 3BD_y + 3D_{yy} + 2CC_y - CB_x = 0;$$ (8.17)

iv. Equation 8.16 has two commuting symmetries X_1, X_2, with $X_1 \vee X_2 \neq 0$ such that a change of variables which brings X_1 and X_2 to their canonical form

$$X_1 = \frac{\partial}{\partial x}, \quad X_2 = \frac{\partial}{\partial y}$$

reduces the equation to one which is at most cubic in the first derivative;

---

[2]The Equations 8.17 turn out to be the compatibility conditions for the auxiliary overdetermined System 3.8 of [H1].

v. Equation 8.16 has two noncommuting symmetries X_1, X_2, *in a suitable basis* $[X_1, X_2] = X_1$, *with* $X_1 \vee X_2 \neq 0$ *such that a change of variables which brings* X_1 *and* X_2 *to their canonical form*

$$X_1 = \frac{\partial}{\partial y}, \quad X_2 = x\frac{\partial}{\partial x} + y\frac{\partial}{\partial y}$$

reduces the equation to

$$xy'' = ay'^3 + by'^2 + (1 + \frac{b^2}{3a})y' + \frac{b}{3a} + \frac{b^3}{27a^2}, \tag{8.18}$$

where $a(\neq 0)$ *and* b *are constants.*[3]

REMARK.   The change of variables

$$X = y + \frac{b}{3a}x, \quad Y = y^2/2 + \frac{b}{3a}xy + \frac{b^2}{18a^2}x^2 + \frac{x^2}{2a} \tag{8.19}$$

transforms Equation 8.18 of Condition v to $Y'' = 0$. Similarly, one can obtain changes of variables (see Sarlet, Mahomed, and Leach [1987]) that reduce the cubic in first derivative equations of Condition iv to linear equations.

Example. The well-known second-order equation

$$y'' + 3yy' + y^3 = 0 \tag{8.20}$$

is linearizable since Conditions 8.17 are satisfied. One can easily deduce the following two noncommuting symmetries

$$X_1 = \frac{\partial}{\partial x}, \quad X_2 = x\frac{\partial}{\partial x} - y\frac{\partial}{\partial y} \tag{8.21}$$

for Equation 8.20. Note that $X_1 \vee X_2 \neq 0$. So we invoke Condition v to find a linearizing transformation. The transformation that reduces the Symmetries 8.21 to their canonical form

$$X_1 = \frac{\partial}{\partial Y}, \quad X_2 = X\frac{\partial}{\partial X} + Y\frac{\partial}{\partial Y}$$

---

[3]Cf. Conditions iv and v of this Theorem with Condition iv of Theorem 3.1 of [H1].

is

$$X = \frac{1}{y}, \quad Y = x + \frac{1}{y}. \tag{8.22}$$

By Transformation 8.22, Equation 8.20 becomes

$$XY'' = -Y'^3 + 6Y'^2 - 11Y' + 6.$$

This is reducible to a linear equation via Transformation 8.19. Combining Transformations 8.19 and 8.22, Equation 8.20 linearizes to $\bar{Y}'' = 0$ via

$$\bar{Y} = x^2/2 - x/y, \quad \bar{X} = x - 1/y.$$

8.2.4. Ermakov's equation

Ermakov [1880] associated the nonlinear equation

$$y'' + b(x)y = ky^{-3}, \quad k = \quad \text{const.} \quad \neq 0 \tag{8.23}$$

with the time-dependent harmonic oscillator equation

$$y'' + b(x)y = 0. \tag{8.24}$$

Namely, he first showed that if two independent solutions of Equation 8.24 are known, one can find the complete integral of Equation 8.23 (see Ermakov [1880], Section 20).

It is well known that Equation 8.24 can be transformed by an equivalence transformation to $y'' = 0$. The same transformation maps Equation 8.23 to the form $y'' = ky^{-3}$ (see Section 8.2.2). The latter was integrated by Lie (see, e.g., Lie [1891]). The inversion of the equivalence transformation provides the solution of the original Equation 8.23 via solutions of the corresponding Equation 8.24. Lie's group theoretic method gives another way of solving Equation 8.23. Later, the solution was re-discovered by Pinney [1950]. Pinney showed that the solution of Equation 8.23 with arbitrary initial data $y(x_0) = y_0 \neq 0$, $y'(x_0) = y_1$, has the form

$$y(x) = [u^2(x) + kW^{-2}v^2(x)]^{\frac{1}{2}},$$

where u, v are linearly independent solutions of Equation 8.24 such that $u(x_0) = y_0$, $u'(x_0) = y_1$, $v(x_0) = 0$, $v'(x_0) \neq 0$, while W is the Wronskian: $W = uv' - vu' = \text{const.} \neq 0$.

The interest in Equation 8.23 was revived by Lewis [1967].

We refer to Equation 8.23 as the Ermakov equation. However, it should be mentioned that Equation 8.23 is also referred to in the literature by other names such as the Pinney equation or the Lewis-Pinney equation. After the discovery of Ermakov's paper, this equation is now called the Ermakov equation. Today, Equation 8.23 is important not only because of its connection with the harmonic oscillator Equation 8.24, but also of its occurrence in applications, e.g., see Lewis [1967], Boiti, Laddomana and Pempinelli [1981], Rogers and Ames [1989], and Vawda and Mahomed [1994b].

Ray and Reid [1979] presented the system

$$y'' + b(x)y = f(z/y)/y^3 \,,$$

$$z'' + b(x)z = g(z/y)/z^3$$

as a natural generalization of the Ermakov Equation 8.23. This system and its extensions to higher dimensions are known as Ermakov systems. Nowadays, there is considerable interest in Ermakov systems. Hereafter, we discuss only the scalar Ermakov equation and some of its variants from the group standpoint.

One should bear in mind that from the Lie group theoretic point of view, the linear Equation 8.24 is equivalent to the general homogeneous linear equation

$$y'' + a(x)y' + b(x)y = 0 \,. \qquad\qquad (8.25)$$

The maximal equivalence transformation group for Equation 8.25 is given by

$$\bar{x} = f(x), \quad \bar{y} = h(x)y \,. \qquad\qquad (8.26)$$

Just as Ermakov connected Equations 8.24 and 8.23, we associate to Equation 8.25 the nonlinear equation

$$y'' + a(x)y' + b(x)y = c(x)y^{-3} \,. \qquad\qquad (8.27)$$

THEOREM 8.6. *(Ibragimov [1992c]). Transformations 8.26 provide the most general equivalence transformation group for Equation 8.27, i.e., the family of Equations 8.27 is invariant under transformations $\bar{x} = F(x, y)$, $\bar{y} = H(x, y)$ if and only if this transformation has the Form 8.26.*

A group classification of Equation 8.27 is given as follows.

THEOREM 8.7. *(Ibragimov [1992b]). Equation 8.27 admits a Lie algebra*
i. L_8 if $c(x) = 0$,
ii. L_3 if $c(x) = k \exp(-2 \int a(x)\,dx)$, $k = $ const. $\neq 0$,
iii. L_0 in all other cases.

Naturally we call the equation

$$y'' + a(x)y' + b(x)y = ky^{-3}\exp\left(-2\int a(x)\,dx\right) \tag{8.28}$$

the Ermakov equation associated with Equation 8.25.

It should be noted that the algebra L_3 which appears in Theorem 8.7 is in fact $L_{3;8}^I$ of Table 1.

Also, group properties of Equation 8.28 were found simultaneously and independently by Kara and Mahomed [1992] using the method of equivalent Lagrangians for nonlinear equations of the form

$$y'' + a(x)y' + b(x)y = \mu y^{-1}y'^2 + c(x)y^n, \quad \mu = \text{const.} \tag{8.29}$$

Equation 8.29 admits L_8 (and is hence linearizable) iff $n = 1$ or μ and $L_{3;8}^I$ iff $n = 4\mu - 3$, $\mu \neq 1$ and $c(x)$ is as given in Theorem 8.7 (ii).

Equation 8.28 reduces to the Lie canonical form (see Section 8.2.2 under Lie Representation)

$$y'' = Cy^{-3} \tag{8.30}$$

by the equivalence transformation

$$\bar{x} = \int \frac{\exp(-\int a(x)\,dx)}{f^2(x)}\,dx, \quad \bar{y} = \frac{y}{f(x)},$$

where $f(x)$ is any non-zero solution of the linear Equation 8.25. It follows that Equation 8.28 can be integrated via solutions of the associated linear Equation 8.25. This integration leads to a nonlinear superposition principle for Equation 8.28. For this reason Equation 8.28 is termed the Ermakov equation.

REMARK. All solutions of Equation 8.30 are invariant solutions (Ibragimov [1991]; see also [H1], Chapter 4). It follows from the equivalence of Equations 8.28 and 8.30 that all solutions of the Ermakov Equation are invariant solutions. The same applies to Equation 8.29 when $n = 4\mu - 3$, $\mu \neq 1$ and c is given in Theorem 8.7 (ii), since in this case Equation 8.29 is equivalent to Equation 8.30.

8.3. Third-order equations

According to Lie (see Section 8.1.3), third-order equations of the form

$$y''' = f(x, y, y', y'') \tag{8.31}$$

have at most a seven-dimensional point symmetry algebra.

8.3.1. Historical introduction

Lie [1883] performed a complete group classification of all scalar ordinary differential equations of arbitrary order in the complex domain. This is sufficient for integration problems. However, to make this result easily accessible it is necessary to present the result on classification in a simple explicit form. Lie himself did it for second-order equations as we have explained in the previous section. In this section, we present the complete group classification of third-order equations in explicit form both in the complex and real domains. The real classification of operators in the plane is in agreement with the results of González-López, Kamran, and Olver [1992]. The classification given in Gat [1992] is incomplete. Namely, two equations are missing.

As far as integration of second-order equations is concerned, this was done by Lie and is given in [H1] (see Chapter 2). Integration of third-order equations based on the complex classification contained here is to be found in Ibragimov and Nucci [1994].

8.3.2. 4-dimensional Lie algebras in the (x, y) plane

Table 4

Some Realizations of four-dimensional Lie Algebras in the Real domain

Let $p = \partial/\partial x$ and $q = \partial/\partial y$.		

Algebra	Realizations in (x, y) plane	
$L_{4;1}$	$X_1 = q, X_2 = xq, X_3 = h(x)q, X_4 = yq$	
$L_{4;2}$	$X_1 = p, X_2 = q, X_3 = yq, X_4 = xp$	
$L_{4;3}$	$X_1 = q, X_2 = p, X_3 = xq, X_4 = xp + (1 + b)yq$	
$L_{4;4}$	$X_1 = q, X_2 = p, X_3 = xp + yq, X_4 = yp - xq$	
$L_{4;5}^{I}$	$X_1 = p, X_2 = q, X_3 = yq, X_4 = y^2q$	
$L_{4;5}^{II}$	$X_1 = q, X_2 = xp + yq, X_3 = 2xyp + y^2q, X_4 = xp$	
$L_{4;6}$	$X_1 = q, X_2 = xq, X_3 = p, X_4 = xp + (2y + \frac{1}{2}x^2)q$	

Third-order equations admitting the algebras given in Table 2 (excluding $L_{3;1}$ and $L_{3;5}^{II}$) and in the above Table 4 are presented in Table 5.

8.3.3. Group classification of third-order equations in the real domain

EQUATIONS ADMITTING $L_r, r \leq 4$

In Table 5, f is an arbitrary function of its argument(s) and C is an arbitrary constant.

REMARK. The admissible algebras up to dimension *three* and corresponding equations contained in Table 5 are taken from Mahomed and Leach [1988]. These algebras and equations are given in Gat [1992], except for $L_{3;7}^{II}$. The algebras $L_{4;1}$, $L_{4;2}$, $L_{4;3}$, $L_{4;4}$, $L_{4;5}^{I}$, and $L_{4;5}^{II}$ and their equations are given in Gat [1992]. The algebra $L_{4;6}$ was missed[4] in Gat [1992]. The algebras $L_{3;7}^{I}$, $L_{3;7}^{II}$, $L_{3;8}^{II}$, $L_{3;9}$, and $L_{4;4}$ and their corresponding equations do not arise in the complex classification.

---

[4]Algebra $L_{4;6}$ is not isomorphic and hence not similar to any one of the algebras $L_{4;1}$ to $L_{4;5}^{II}$ neither on the real nor on the complex domains; see, e.g., Petrov [1966] (Chapter 1, Section 10).

Table 5

Equations Admitting 1-, 2-, 3-, and 4-dimensional real Lie Algebras

Algebra	Representative Equations
L_1	$y''' = f(y, y', y'')$
$L_{2;1}^I$	$y''' = f(y', y'')$
$L_{2;1}^{II}$	$y''' = f(x, y'')$
$L_{2;2}^I$	$y''' = y''^2 f(y', xy'')$
$L_{2;2}^{II}$	$y''' = y' f(x, y''/y')$
$L_{3;2}$	$y''' = f(y'')$
$L_{3;3}^I$	$y''' = y''^2 f(y'' \exp y')$
$L_{3;3}^{II}$	$y''' = y'' f(\exp x/y'')$
$L_{3;4}^I$	$y''' = y''^{\frac{3}{2}} f(y'' y'^{-2})$
$L_{3;4}^{II}$	$y''' = y''^2 f(xy'')$
$L_{3;5}^I$	$y''' = y''^2 f(y')$
$L_{3;6}^I$	$y''' = y''^{\frac{a-3}{a-2}} f(y''^{1-a} y'^{a-2}),\ a \neq 0, 1, 2;\quad y''' = y'^{-1} f(y''),\ a = 2$
$L_{3;6}^{II}$	$y''' = y''^{\frac{2-3a}{1-2a}} f(x^{1-2a} y''^{1-a}),\ a \neq 0, 1, \tfrac{1}{2};\quad y''' = x^{-1} f(y''),\ a = \tfrac{1}{2}$
$L_{3;7}^I$	$y''' = \dfrac{y''^2}{1+y'^2}\left(3y' + f\left(y''(1+y'^2)^{-\frac{3}{2}} \exp(-b \arctan y')\right)\right)$
$L_{3;7}^{II}$	$y''' = \dfrac{y''}{1+x^2}\left(3x + f\left(y''(1+x^2)^{\frac{3}{2}} \exp(-b \arctan x)\right)\right)$
$L_{3;8}^I$	$y''' = 3\dfrac{y''^2}{y'} + \dfrac{y'^4}{x^2} f\left(\dfrac{2xy''+y'}{y'^3}\right)$
$L_{3;8}^{II}$	$y''' = \dfrac{3y'y''^2}{1+y'^2} + \dfrac{(1+y'^2)^2}{x^2} f\left(\dfrac{xy''-y'-y'^3}{(1+y'^2)^{\frac{3}{2}}}\right)$
$L_{3;8}^{III}$	$y''' = \dfrac{3y'y''^2}{y'^2-1} + \dfrac{(y'^2-1)^2}{x^2} f\left(\dfrac{xy''-y'+y'^3}{(1-y'^2)^{\frac{3}{2}}}\right)$
$L_{3;8}^{IV}$	$y''' = \dfrac{3}{2}\dfrac{y''^2}{y'} + y' f(x)$
$L_{3;9}$	$y''' = \dfrac{-3xy''}{1+x^2} + \dfrac{3y''[y'(1+x^2)-xy]}{1+x^2}\left[\dfrac{y''(1+x^2)}{1+y'^2+(y-xy')^2}\right.$ $\left. - \dfrac{y}{1+x^2+y^2}\right]$ $+ \dfrac{[1+y'^2+(y-xy')^2]^2}{(1+x^2+y^2)^{\frac{5}{2}}} f\left(y''\left(\dfrac{1+y'^2+(y-xy')^2}{1+x^2+y^2}\right)^{\frac{-3}{2}}\right)$

Table 5 – cont.

Equations Admitting 1-, 2-, 3-, and 4-dimensional real Lie Algebras

Algebra	Representative Equations
$L_{4;1}$	$y''' = \left(h'''(x)/h''(x)\right) y'', \quad h''' \neq 0$
$L_{4;2}$	$y''' = C(y''^2/y')$
$L_{4;3}$	$y''' = C y''^{\frac{b-2}{b-1}}, \quad b \neq 1, 2$
$L_{4;4}$	$y''' = \dfrac{3y'y''^2}{1+y'^2} + C \dfrac{y''^2}{1+y'^2}$
$L_{4;5}^I$	$y''' = \dfrac{3}{2}\dfrac{y''^2}{y'} + Cy'$
$L_{4;5}^{II}$	$y''' = 3\dfrac{y''^2}{y'} + C\dfrac{(2xy''+y')^{\frac{3}{2}}}{x^2\sqrt{y'}}$
$L_{4;6}$	$y''' = C\exp(-y'')$

EQUATIONS ADMITTING L_5
(Mahomed and Leach [1990]

Third-order equations with five-dimensional symmetry algebras are linearizable and equivalent to the linear equation

$$y''' + ky' + ly = 0, \qquad k, l(\neq 0) = \text{const.}$$

For this equation, the symmetry algebra L_5 is spanned by (see Section 8.1.9)

$$X_1 = \frac{\partial}{\partial x}, \quad X_2 = y\frac{\partial}{\partial y}, \quad X_3 = \eta_1\frac{\partial}{\partial y}, \quad X_4 = \eta_2\frac{\partial}{\partial y}, \quad X_5 = \eta_3\frac{\partial}{\partial y},$$

where η_i are independent solutions of the linear equation $\eta''' + k\eta' + l\eta = 0$.

EQUATIONS ADMITTING L_6
(Lie [1896], Campbell [1903], Mahomed and Leach [1990], Gat [1992])

There are two classes of real third-order equations that admit a six-dimensional symmetry algebra. They are

$$y''' = \frac{3}{2}\frac{y''^2}{y'} \tag{8.32}$$

and

$$y''' = \frac{3y'y''^2}{1+y'^2} \tag{8.33}$$

and their symmetry algebras respectively are

$$L_{6;1} \quad X_1 = \frac{\partial}{\partial x}, \; X_2 = x\frac{\partial}{\partial x}, \; X_3 = x^2\frac{\partial}{\partial x},$$

$$X_4 = \frac{\partial}{\partial y}, \; X_5 = y\frac{\partial}{\partial y}, \; X_6 = y^2\frac{\partial}{\partial y} \qquad (8.34)$$

and

$$L_{6;2} \quad X_1 = \frac{\partial}{\partial x}, \; X_2 = \frac{\partial}{\partial y}, \; X_3 = x\frac{\partial}{\partial x} + y\frac{\partial}{\partial y}, \; X_4 = y\frac{\partial}{\partial x} - x\frac{\partial}{\partial y},$$

$$X_5 = (x^2 - y^2)\frac{\partial}{\partial x} + 2xy\frac{\partial}{\partial y}, \; X_6 = 2xy\frac{\partial}{\partial x} + (y^2 - x^2)\frac{\partial}{\partial y}. \quad (8.35)$$

In the complex plane $L_{6;1}$ is isomorphic and similar to $L_{6;2}$. Hence there is only one equation that admits L_6 in the complex plane. Indeed, the complex transformation

$$x = z + it, \quad y = t + iz \qquad (8.36)$$

maps Equation 8.32 to the Equation 8.33, viz.,

$$z''' = \frac{3z'z''^2}{1 + z'^2},$$

where $z' = dz/dt$.

EQUATIONS ADMITTING L_7
  (Mahomed and Leach [1990], Krause and Michel [1991])

A third-order equation that admits a seven-dimensional symmetry algebra is linearizable and equivalent to the simplest equation

$$y''' = 0. \qquad (8.37)$$

The symmetry algebra L_7 is spanned by (see Section 8.1.5)

$$X_1 = y\frac{\partial}{\partial y}, \; X_2 = \frac{\partial}{\partial y}, \; X_3 = x\frac{\partial}{\partial y}, \; X_4 = x^2\frac{\partial}{\partial y},$$

$$X_5 = \frac{\partial}{\partial x}, \ X_6 = x\frac{\partial}{\partial x}, \ X_7 = x^2\frac{\partial}{\partial x} + 2xy\frac{\partial}{\partial y} \qquad (8.38)$$

and is maximal.

8.3.4. Utilization of contact transformations

(Lie [1896]; see also Campbell [1903] (Sections 238 and 239))

It is well known that Lie's infinitesimal approach for point transformations is not efficient for first-order ordinary differential equations as there results no overdetermined system of determining equations. However, Lie's method for point symmetries applies systematically to second- and higher-order equations, the reason being that the determining equations are overdetermined and in practice can easily be solved for the Lie group generators. In the same way, Lie's infinitesimal approach for contact (first-order tangent) symmetries is not efficient for second-order equations (see Mahomed and Leach [1991] for an approach in dealing with contact symmetries of second-order equations) but apply efficiently for third-order and higher-order equations.

We present here the results on contact symmetries for Equations 8.32, 8.33, and 8.37. Furthermore, we discuss their geometrical and contact transformation relations.

CONTACT SYMMETRIES FOR EQUATION 8.37

The Lie algebra of contact symmetries for Equation 8.37 is spanned by the Operators 8.38 together with

$$X_8 = 2y'\frac{\partial}{\partial x} + y'^2\frac{\partial}{\partial y},$$

$$X_9 = (y - xy')\frac{\partial}{\partial x} - \frac{1}{2}xy'^2\frac{\partial}{\partial y} - \frac{1}{2}y'^2\frac{\partial}{\partial y'}, \qquad (8.39)$$

$$X_{10} = (xy - \frac{1}{2}x^2y')\frac{\partial}{\partial x} + (y^2 - \frac{1}{4}x^2y'^2)\frac{\partial}{\partial y} + (yy' - \frac{1}{2}xy'^2)\frac{\partial}{\partial y'}.$$

CONTACT SYMMETRIES FOR EQUATION 8.32

The Lie algebra of contact symmetries for Equation 8.32 is spanned by the Operators 8.34 together with

$$X_7 = \frac{1}{\sqrt{y'}}\frac{\partial}{\partial x} - \sqrt{y'}\frac{\partial}{\partial y},$$

$$X_8 = \frac{y}{y'}\frac{\partial}{\partial x} - y\sqrt{y'}\frac{\partial}{\partial y} - 2y'\sqrt{y'}\frac{\partial}{\partial y'},$$

$$X_9 = \frac{x}{\sqrt{y'}}\frac{\partial}{\partial x} - x\sqrt{y'}\frac{\partial}{\partial y} - 2\sqrt{y'}\frac{\partial}{\partial y'},$$

$$X_{10} = \frac{xy}{\sqrt{y'}}\frac{\partial}{\partial x} - xy\sqrt{y'}\frac{\partial}{\partial y} - 2(y\sqrt{y'} + xy'\sqrt{y'})\frac{\partial}{\partial y'}. \tag{8.40}$$

CONTACT SYMMETRIES FOR EQUATION 8.33

The Lie algebra of contact symmetries for Equation 8.33 is spanned by the Operators 8.35 together with

$$X_7 = \frac{y'}{\sqrt{1 + y'^2}}\frac{\partial}{\partial x} - \frac{1}{\sqrt{1 + y'^2}}\frac{\partial}{\partial y},$$

$$X_8 = \frac{xy'}{\sqrt{1 + y'^2}}\frac{\partial}{\partial x} - \frac{x}{\sqrt{1 + y'^2}}\frac{\partial}{\partial y} - \sqrt{1 + y'^2}\frac{\partial}{\partial y'},$$

$$X_9 = \frac{yy'}{\sqrt{1 + y'^2}}\frac{\partial}{\partial x} - \frac{y}{\sqrt{1 + y'^2}}\frac{\partial}{\partial y} - y'\sqrt{1 + y'^2}\frac{\partial}{\partial y'}, \tag{8.41}$$

$$X_{10} = \frac{(x^2 + y^2)y'}{\sqrt{1 + y'^2}}\frac{\partial}{\partial x} - \frac{x^2 + y^2}{\sqrt{1 + y'^2}}\frac{\partial}{\partial y} - 2(x + yy')\sqrt{1 + y'^2}\frac{\partial}{\partial y'}.$$

GEOMETRICAL RELATIONS

Equations 8.32, 8.33, and 8.37 have evident geometrical representations. That is, Equation 8.32 describes a family of hyperbolas, Equation 8.33 a family of circles, and Equation 8.37 a family of parabolas in the (x, y) plane. Lie showed that all three families are connected by contact transformations. These transformations are expressed analytically as follows.

CONTACT TRANSFORMATION RELATIONS

Equation 8.32 after the Legendre transformation

$$t = y', \quad u = xy' - y, \quad u' = x \tag{8.42}$$

reduces to the linear equation

$$u''' + \frac{3}{2}\frac{u''}{t} = 0.$$

The latter, by means of the substitution

$$\tau = \sqrt{t}, \tag{8.43}$$

becomes

$$\frac{d^3 u}{d\tau^3} = 0.$$

Thus, the composition of Transformations 8.42 and 8.43 is a real contact transformation which maps Equation 8.32 into the Equation 8.37.

It follows that the composition of Transformations 8.36, 8.42, and 8.43 provides a complex contact transformation that relates Equations 8.33 and 8.37.

REMARK. Lie [1896] (Chapter 10, Section 2) established a remarkable connection between the group of contact symmetries of the linear Equation 8.37 and conformal mappings in three-dimensional space. Also, it is worth noting that higher-order linear Equations 8.1 ($n > 3$) do not admit nontrivial (i.e., non-point) contact symmetries.

9

Differential Equations with a Small Parameter: Exact and Approximate Symmetries

This chapter is integrated with Chapter 2.

Approximate symmetries are presented here in the first order of precision. Section 9.4 is an exception and contains examples of second-order approximate symmetries.

Approximate Lie algebras are given by their *essential operators* (see Chapter 2, Section 2.3.7) rather than by basic operators. According to Section 2.3.7, a basis of any approximate Lie algebra (in the first order of approximation) is obtained by multiplying essential operators by ε and neglecting the terms of order $o(\varepsilon)$.

Notation: X_α^0 denotes symmetries of an unperturbed equation, whereas X_α and Y_α denote, respectively, approximate and exact symmetries of a perturbed equation. In approximate group classifications, some of the symmetries X_α^0 are not stable when coefficients (functions or parameters) of the equations under consideration are arbitrary. However, it may happen that X_α^0 is stable for particular values of the coefficients. In these cases, the notation $\tilde{X}_\alpha$ is used for the deformation of X_α^0 caused by the particular perturbation.

9.1. Second-order ordinary differential equations

9.1.1. $\mathbf{u}'' + \mathbf{u} = \varepsilon \mathbf{F}(\mathbf{u})$

Here, $u = u(\theta)$ and $u'' = d^2u/d\theta^2$. An equation of this form is used, e.g., in the two-body (central force) problem. Then

$$F(u) = -\frac{mf(1/u)}{L^2 u^2}$$

where u denotes the inverse radial distance (i.e., $u = 1/r$ where r is the distance from the center of a force), m is the mass of a particle, $L = $ const. is the angular momentum of a particle, f is a central force, and θ is the angular distance from a fixed axis.

9.1.1.1. Unperturbed equation

$$u'' + u = 0$$

LIE POINT SYMMETRIES
The symmetry Lie algebra is eight-dimensional and is spanned by

$$X_1^0 = \frac{\partial}{\partial \theta}, \quad X_2^0 = u\frac{\partial}{\partial u}, \quad X_3^0 = \cos\theta\frac{\partial}{\partial u}, \quad X_4^0 = \sin\theta\frac{\partial}{\partial u},$$

$$X_5^0 = u\sin\theta\frac{\partial}{\partial \theta} + u^2\cos\theta\frac{\partial}{\partial u}, \quad X_6^0 = u\cos\theta\frac{\partial}{\partial \theta} - u^2\sin\theta\frac{\partial}{\partial u},$$

$$X_7^0 = \sin 2\theta\frac{\partial}{\partial \theta} + u\cos 2\theta\frac{\partial}{\partial u}, \quad X_8^0 = \cos 2\theta\frac{\partial}{\partial \theta} - u\sin 2\theta\frac{\partial}{\partial u} \quad .$$

9.1.1.2. Perturbed equation: Exact symmetries
In discussions of exact symmetries, ε is considered as an arbitrary constant. Therefore one can take εF as new F and write the equation in the form

$$u'' + u = F(u)$$

CLASSIFICATION (LIE POINT SYMMETRIES)
(Vawda and Mahomed [1994a])

Equivalence Transformations

$$\overline{\theta} = a_1\theta + b_1, \quad \overline{u} = a_2 u + b_2, \quad \overline{F} = \frac{a_2}{a_1}F + (1 - \frac{1}{a_1^2})a_2 u + b_2.$$

Here, a_1, a_2, b_1, b_2 are arbitrary constants such that $a_1, a_2 \neq 0$.

Classification Result
    The principal Lie algebra $L_{\mathcal{P}}$ is one-dimensional and is spanned by

$$Y_1 = \frac{\partial}{\partial \theta}.$$

The algebra $L_{\mathcal{P}}$ extends in the following cases:

1. $F(u) = u + Ae^u$, $A \neq 0$ is a constant.

$$Y_2 = \theta \frac{\partial}{\partial \theta} - 2\frac{\partial}{\partial u}$$

2. $F(u) = u + Au^\alpha$, $\alpha \neq 0, 1, -3$, and $A \neq 0$.

$$Y_2 = (1 - \alpha)\theta \frac{\partial}{\partial \theta} + 2u\frac{\partial}{\partial u}$$

3. $F(u) = \mu u + Au^{-3}$, $A \neq 0$.

(a) $\mu < 1$

$$Y_2 = \cos(2\theta\sqrt{1 - \mu})\frac{\partial}{\partial \theta} - \sqrt{1 - \mu}\sin(2\theta\sqrt{1 - \mu})u\frac{\partial}{\partial u},$$

$$Y_3 = \sin(2\theta\sqrt{1 - \mu})\frac{\partial}{\partial \theta} + \sqrt{1 - \mu}\cos(2\theta\sqrt{1 - \mu})u\frac{\partial}{\partial u}$$

(b) $\mu > 1$

$$Y_2 = \exp(2\theta\sqrt{\mu - 1})\frac{\partial}{\partial \theta} + \sqrt{\mu - 1}\exp(2\theta\sqrt{\mu - 1})u\frac{\partial}{\partial u},$$

$$Y_3 = \exp(-2\theta\sqrt{\mu - 1})\frac{\partial}{\partial \theta} - \sqrt{\mu - 1}\exp(-2\theta\sqrt{\mu - 1})u\frac{\partial}{\partial u}$$

(c) $\mu = 1$

$$Y_2 = \theta^2 \frac{\partial}{\partial \theta} + \theta u\frac{\partial}{\partial u}, \quad Y_3 = 2\theta \frac{\partial}{\partial \theta} + u\frac{\partial}{\partial u}$$

4. $F(u) = Au$. This case reduces to Section 9.1.1.1.

9.1.1.3. Perturbed equation: Approximate symmetries

$$u'' + u = \varepsilon F(u)$$

CLASSIFICATION (APPROXIMATE POINT SYMMETRIES)
(Baikov, Gazizov, Ibragimov, and Mahomed [1994])

Approximate Equivalence Transformations

$$\bar{\theta} \approx (1 + \varepsilon a_2)\theta + a_4 , \quad \bar{u} \approx (u + \varepsilon a_3)a_1 , \quad \bar{F} = (F + 2a_2 u + a_3)a_1 + O(\varepsilon),$$

where $O(\varepsilon)$ denotes terms of order ε. Here, $a_1, \dots, a_4$ are arbitrary constants such that $a_1 \neq 0$.

Classification Result
For arbitrary $F(u)$, the approximate symmetry algebra is given by the following essential operators:

$$X_1 = \frac{\partial}{\partial \theta}, \quad X_2 = \varepsilon u \frac{\partial}{\partial u}, \quad X_3 = \varepsilon \cos \theta \frac{\partial}{\partial \theta}, \quad X_4 = \varepsilon \sin \theta \frac{\partial}{\partial u},$$

$$X_5 = \varepsilon \left(u \sin \theta \frac{\partial}{\partial \theta} + u^2 \cos \theta \frac{\partial}{\partial u} \right), \quad X_6 = \varepsilon \left(u \cos \theta \frac{\partial}{\partial \theta} - u^2 \sin \theta \frac{\partial}{\partial u} \right),$$

$$X_7 = \varepsilon \left(\sin 2\theta \frac{\partial}{\partial \theta} + u \cos 2\theta \frac{\partial}{\partial u} \right), \quad X_8 = \varepsilon \left(\cos 2\theta \frac{\partial}{\partial \theta} - u \sin 2\theta \frac{\partial}{\partial u} \right).$$

This algebra extends in the following cases:
1. $F(u) = Au \ln u, \quad A \neq 0.$

$$\tilde{X}_2 = 2\varepsilon A\theta \frac{\partial}{\partial \theta} + (4u + \varepsilon Au)\frac{\partial}{\partial u}$$

2. $F(u) = A(u + \delta)^2, \quad \delta = 0, 1, \quad A \neq 0.$

$$\tilde{X}_3 = 4A\varepsilon \sin \theta \frac{\partial}{\partial \theta} + [3 \cos \theta + \varepsilon(2Au \cos \theta + 3\delta A\theta \sin \theta)]\frac{\partial}{\partial u},$$

$$\tilde{X}_4 = -4A\varepsilon \cos \theta \frac{\partial}{\partial \theta} + [3 \sin \theta + \varepsilon(2Au \sin \theta - 3\delta A\theta \cos \theta)]\frac{\partial}{\partial u}$$

3. $F(u) = Au^{-3}, \quad A \neq 0.$

$$\tilde{X}_7 = \sin 2\theta \frac{\partial}{\partial \theta} + u \cos 2\theta \frac{\partial}{\partial u}, \quad \tilde{X}_8 = \cos 2\theta \frac{\partial}{\partial \theta} - u \sin 2\theta \frac{\partial}{\partial u}$$

4. $F(u) = 0.$ See Section 9.1.1.1.

9.1.2. $\mathbf{u''} + \mathbf{u} - \mathbf{u}^{-3} = \varepsilon\mathbf{F(u)}$

Here, $u = u(\theta)$ and $u'' = d^2u/d\theta^2$.

Symmetries of the unperturbed equation $u'' + u - u^{-3} = 0$ are contained in Section 9.1.1.2 (see case 3 with $\mu = 0$). Furthermore, the classification of the perturbed equation with respect to exact symmetries also can be obtained from Section 9.1.1.2.

CLASSIFICATION (APPROXIMATE POINT SYMMETRIES)
(Baikov, Gazizov, Ibragimov, and Mahomed [1994])

Approximate Equivalence Transformations

$$\overline{\theta} \approx (1 + \varepsilon a_2)\theta + a_1, \quad \overline{u} \approx u + \varepsilon\left(\frac{a_2}{2}u + a_3u + a_4\right),$$

$$\overline{F} = F + (1 + 3u^{-4})a_4 + 4a_3u^{-3} + 2a_2u + O(\varepsilon)$$

where $a_1, \ldots, a_4$ are arbitrary constants, $O(\varepsilon)$ denotes terms of order ε.

Classification Result
For arbitrary $F(u)$, the approximate symmetry algebra is given by

$$X_1 = \frac{\partial}{\partial\theta}, \quad X_2 = \varepsilon\left(\cos 2\theta \frac{\partial}{\partial\theta} - u\sin 2\theta \frac{\partial}{\partial u}\right), \quad X_3 = \varepsilon\left(\sin 2\theta \frac{\partial}{\partial\theta} + u\cos 2\theta \frac{\partial}{\partial u}\right).$$

This algebra extends only in the case $F(u) = 0$, the additional operators being

$$\tilde{X}_2 = \cos 2\theta \frac{\partial}{\partial\theta} - u\sin 2\theta \frac{\partial}{\partial u},$$

$$\tilde{X}_3 = \sin 2\theta \frac{\partial}{\partial\theta} + u\cos 2\theta \frac{\partial}{\partial u}.$$

9.1.3. Van der Pol equation

$$u'' + u = \varepsilon(u' - \frac{1}{3}u'^3)$$

Here, $u = u(x)$ and $u' = du/dx$. The symmetries are written in the form of canonical Lie-Bäcklund operators 1.10 (see Chapter 1). They are taken from Baikov [1994].

9.1.3.1.   Unperturbed equation

$$u'' + u = 0$$

LIE POINT SYMMETRIES
The symmetry Lie algebra is eight-dimensional and is spanned by (cf. Section 9.1.1.1)

$$X_1^0 = u_x \frac{\partial}{\partial u} + \dots, \quad X_2^0 = u \frac{\partial}{\partial u}, \quad X_3^0 = \cos x \frac{\partial}{\partial u}, \quad X_4^0 = \sin x \frac{\partial}{\partial u},$$

$$X_5^0 = (u^2 \cos x - u_x u \sin x) \frac{\partial}{\partial u} + \dots, \quad X_6^0 = (u^2 \sin x + u_x u \cos x) \frac{\partial}{\partial u} + \dots,$$

$$X_7^0 = (u \cos 2x - u_x \sin 2x) \frac{\partial}{\partial u} + \dots, \quad X_8^0 = (u \sin 2x + u_x \cos 2x) \frac{\partial}{\partial u} + \dots$$

9.1.3.2.   Perturbed equation: Exact symmetries

$$u'' + u = \varepsilon(u' - \frac{1}{3}u'^3)$$

LIE POINT SYMMETRIES
   The symmetry Lie algebra is one-dimensional and is spanned by

$$Y_1 = u_x \frac{\partial}{\partial u} + \dots$$

9.1.3.3.   Perturbed equation: Approximate symmetries

$$u'' + u = \varepsilon(u' - \frac{1}{3}u'^3)$$

APPROXIMATE SYMMETRIES
   Here, approximate symmetries

$$X = (f_0 + \varepsilon f_1) \frac{\partial}{\partial u} + \dots$$

are given such that

$$X^0 = f_0 \frac{\partial}{\partial u} + \dots$$

is the canonical Lie-Bäcklund representation of point symmetries of the unperturbed equation, and f_1 is a function of x, u, and u_x only.

$$X_1 = u_x \frac{\partial}{\partial u} + \dots ,$$

$$X_2 = \left\{ u - \varepsilon \left[\frac{1}{4} u_x^2 ux + \frac{1}{12} u_x u^2 + \frac{1}{4} u^3 x \right] \right\} \frac{\partial}{\partial u} + \dots ,$$

$$X_3 = \left\{ \cos x + \varepsilon \left[-\frac{1}{8} u_x^2 x \cos x + \frac{1}{4} u_x ux \sin x \right. \right.$$
$$\left. \left. + \frac{1}{8} u^2 (\sin x - 3x \cos x) + \frac{1}{2} x \cos x \right] \right\} \frac{\partial}{\partial u} + \dots ,$$

$$X_4 = \left\{ \sin x + \varepsilon \left[-\frac{1}{8} u_x^2 x \sin x - \frac{1}{4} u_x ux \cos x \right. \right.$$
$$\left. \left. - \frac{1}{8} u^2 (\cos x + 3x \sin x) + \frac{1}{2} x \sin x \right] \right\} \frac{\partial}{\partial u} + \dots ,$$

$$X_5 = \left\{ u^2 \cos x - u_x u \sin x + \varepsilon \left[\frac{1}{8} u_x^3 ux \sin x - \frac{1}{8} u_x^2 u^2 (\sin x + x \cos x) \right. \right.$$
$$+ u_x u^3 \left(-\frac{1}{24} \cos x + \frac{1}{8} x \sin x \right) + \frac{1}{2} u_x ux \sin x$$
$$\left. \left. - \frac{u^4}{12} (\sin x + \frac{3}{2} x \cos x) - \frac{u^2}{2} (- \sin x + x \cos x) \right] \right\} \frac{\partial}{\partial u} + \dots ,$$

$$X_6 = \left\{ u^2 \sin x + u_x u \cos x - \varepsilon \left[\frac{1}{8} u_x^3 ux \cos x + \frac{1}{8} u_x^2 u^2 (- \cos x + x \sin x) \right. \right.$$
$$+ u_x u^3 \left(\frac{1}{24} \sin x + \frac{1}{8} x \cos x \right) + \frac{1}{2} u_x ux \cos x$$
$$\left. \left. - \frac{u^4}{12} (\cos x - \frac{3}{2} x \sin x) + \frac{u^2}{2} (\cos x + x \sin x) \right] \right\} \frac{\partial}{\partial u} + \dots ,$$

$$X_7 = \left\{ u\cos 2x - u_x\sin 2x + \varepsilon\left[\frac{1}{4}u_x^2 ux\cos 2x - \frac{1}{4}u_x u^2(\cos 2x - 2x\sin 2x)\right.\right.$$

$$\left.\left. -\frac{1}{6}u^3\sin 2x - \frac{1}{4}u^3 x\cos 2x + \frac{1}{2}u\sin 2x\right]\right\}\frac{\partial}{\partial u} + \dots ,$$

$$X_8 = \left\{ u\sin 2x + u_x\cos 2x - \varepsilon\left[-\frac{1}{4}u_x^2 ux\sin 2x + \frac{1}{4}u_x u^2(\sin 2x + 2x\cos 2x)\right.\right.$$

$$\left.\left. -\frac{1}{6}u^3\cos 2x + \frac{1}{4}u^3 x\sin 2x + \frac{1}{2}u\cos 2x\right]\right\}\frac{\partial}{\partial u} + \dots$$

APPROXIMATE BÄCKLUND TRANSFORMATION
Van der Pol equation

$$u'' + u = \varepsilon\left(u' - \frac{1}{3}u'^3\right)$$

is mapped to the linear equation

$$v'' + v = 0$$

by the approximate transformation

$$u = v + \varepsilon\left[\frac{1}{24}v_x v^2 + \frac{1}{8}(v_x^2 + v^2)vx - \frac{vx}{2}\right] + o(\varepsilon).$$

APPROXIMATE SOLUTION

$$u \approx A\cos x - \frac{\varepsilon}{24}A^3\sin x\cos^2 x,$$

where $A = A(x)$ is given by

$$A^2 = \frac{4}{1 + \left(\dfrac{4}{A_0^2} - 1\right)e^{-\varepsilon x}}, \qquad A_0 = \text{const.}$$

9.2. Wave equations with a small dissipation

This section contains results on approximate symmetries of nonlinear wave equations of the form

$$u_{tt} + \varepsilon u_t = \nabla \cdot (p(u) \cdot \nabla u).$$

Some generalizations are considered as well.

9.2.1. One-dimensional waves

In Sections 9.2.1.1 to 9.2.1.3 the sequence of the equations

$$w_{tt} + \varepsilon w_t = F(w_{xx}) \xrightarrow{w_x = v} v_{tt} + \varepsilon v_t = f(v_x) v_{xx} \xrightarrow{v_x = u} u_{tt} + \varepsilon u_t = (f(u)u_x)_x ,$$

connected by the Bäcklund transformations, are considered. Here, $f = F'$. For each equation of this sequence, exact and approximate point symmetries are given.

Furthermore, in the non-linear case ($f' \neq 0$) quasi-local symmetries of these equations are presented (see [H1], Section 7, or Akhatov, Gazizov, and Ibragimov [1987a], [1989].)

In Section 9.2.1.4, the wave equations with a non-linear dissipation,

$$u_{tt} + \varepsilon \varphi(u) u_t = (f(u)u_x)_x, \quad \varphi(u) \neq \text{const.},$$

are considered.

9.2.1.1. $u_{tt} + \varepsilon u_t = (f(u)u_x)_x$

9.2.1.1.1. Unperturbed Equation

$$u_{tt} = (f(u)u_x)_x$$

CLASSIFICATION (LIE POINT SYMMETRIES)
(Ames, Lohner, and Adams [1981])

Equivalence Transformations

$$\bar{t} = \alpha_1 t + \alpha_2, \quad \bar{x} = \alpha_3 x + \alpha_4, \quad \bar{u} = \alpha_5 u + \alpha_6,$$

where α_i, $i = 1, \dots, 6$, are arbitrary constants such that $\alpha_1 \alpha_3 \alpha_5 \neq 0$.

Classification Result

The principal Lie algebra L_P is three-dimensional and is spanned by

$$X_1^0 = \frac{\partial}{\partial t}, \quad X_2^0 = \frac{\partial}{\partial x}, \quad X_3^0 = t\frac{\partial}{\partial t} + x\frac{\partial}{\partial x}.$$

The algebra L_P extends in the following cases:

1. $f(u) = \delta e^u$, $\delta = \pm 1$.

$$X_4^0 = x\frac{\partial}{\partial x} + 2\frac{\partial}{\partial u}$$

2. $f(u) = \delta u^\sigma$, $\delta = \pm 1$, $\sigma \neq -4, -\frac{4}{3}, 0$ is an arbitrary constant.

$$X_4^0 = \sigma x\frac{\partial}{\partial x} + 2u\frac{\partial}{\partial u}$$

3. $f(u) = \delta u^{-4}$, $\delta = \pm 1$.

$$X_4^0 = 2x\frac{\partial}{\partial x} - u\frac{\partial}{\partial u}, \quad X_5^0 = t^2\frac{\partial}{\partial t} + tu\frac{\partial}{\partial u}$$

4. $f(u) = \delta u^{-4/3}$, $\delta = \pm 1$.

$$X_4^0 = 2x\frac{\partial}{\partial x} - 3u\frac{\partial}{\partial u}, \quad X_5^0 = x^2\frac{\partial}{\partial x} - 3xu\frac{\partial}{\partial u}$$

5. $f(u) = 1$.

$$X_\infty^0 = (\alpha_1 + \alpha_2)\frac{\partial}{\partial t} + (\alpha_1 - \alpha_2)\frac{\partial}{\partial x} + (Cu + \beta_1 + \beta_2)\frac{\partial}{\partial u},$$

where α_1, β_1 and α_2, β_2 are arbitrary functions of the arguments $t + x$ and $t - x$, respectively, and C is an arbitrary constant.

9.2.1.1.2. Perturbed equation: Exact symmetries

$$u_{tt} + \varepsilon u_t = (f(u)u_x)_x$$

CLASSIFICATION (LIE POINT SYMMETRIES)
(Oron and Rosenau [1986])

Equivalence Transformations are taken from Section 9.2.1.1.1.

Classification Result

The principal Lie algebra $L_\mathcal{P}$ is two-dimensional and is spanned by

$$Y_1 = \frac{\partial}{\partial t}, \quad Y_2 = \frac{\partial}{\partial x}.$$

The algebra $L_\mathcal{P}$ extends in the following cases:

1. $f(u) = \delta e^u$, $\delta = \pm 1$.

$$Y_3 = x\frac{\partial}{\partial x} + 2\frac{\partial}{\partial u}$$

2. $f(u) = \delta u^\sigma$, $\delta = \pm 1$, $\sigma \neq -2, -\frac{4}{3}, 0$.

$$Y_3 = \sigma x\frac{\partial}{\partial x} + 2u\frac{\partial}{\partial u}$$

3. $f(u) = \delta u^{-2}$, $\delta = \pm 1$.

$$Y_3 = x\frac{\partial}{\partial x} - u\frac{\partial}{\partial u}, \quad Y_4 = e^{-\varepsilon t}\frac{\partial}{\partial t} - \varepsilon e^{-\varepsilon t}u\frac{\partial}{\partial u}$$

4. $f(u) = \delta u^{-4/3}$, $\delta = \pm 1$.

$$Y_3 = 2x\frac{\partial}{\partial x} - 3u\frac{\partial}{\partial u}, \quad Y_4 = x^2\frac{\partial}{\partial x} - 3xu\frac{\partial}{\partial u}$$

5. $f(u) = 1$.

$$Y_3 = x\frac{\partial}{\partial t} + t\frac{\partial}{\partial x} - \frac{\varepsilon}{2}xu\frac{\partial}{\partial u}, \quad Y_4 = u\frac{\partial}{\partial u},$$

$$Y_\infty = \Theta(t, x)\frac{\partial}{\partial u},$$

where $\Theta(t, x)$ is an arbitrary solution of the equation $\Theta_{tt} + \varepsilon\Theta_t = \Theta_{xx}$.

9.2.1.1.3. Perturbed equation: Approximate symmetries

$$u_{tt} + \varepsilon u_t = (f(u)u_x)_x$$

CLASSIFICATION (APPROXIMATE POINT SYMMETRIES)
(Baikov, Gazizov, and Ibragimov [1987a], [1988a], [1988b], [1989a])

Equivalence Transformations are taken from Section 9.2.1.1.1.

Classification Result

For arbitrary $f(u)$, the approximate symmetry algebra is given by

$$X_1 = \frac{\partial}{\partial t}, \quad X_2 = \frac{\partial}{\partial x}, \quad X_3 = \varepsilon\left(t\frac{\partial}{\partial t} + x\frac{\partial}{\partial x}\right).$$

This algebra extends in the following cases:

1. $f(u) = \delta e^u$, $\delta = \pm 1$.

$$\tilde{X}_3 = \left(t + \frac{\varepsilon}{2}t^2\right)\frac{\partial}{\partial t} + x\frac{\partial}{\partial x} - 2\varepsilon t\frac{\partial}{\partial u},$$

$$X_4 = x\frac{\partial}{\partial x} + 2\frac{\partial}{\partial u}$$

2. $f(u) = \delta u^\sigma$, $\delta = \pm 1$, $\sigma \neq -4, -\frac{4}{3}, 0.$

$$\tilde{X}_3 = \left(t + \frac{\varepsilon\sigma t^2}{2\sigma + 8}\right)\frac{\partial}{\partial t} + x\frac{\partial}{\partial x} - \frac{2\varepsilon tu}{\sigma + 4}\frac{\partial}{\partial u},$$

$$X_4 = \sigma x\frac{\partial}{\partial x} + 2u\frac{\partial}{\partial u}$$

3. $f(u) = \delta u^{-4}$, $\delta = \pm 1$.

$$X_4 = 2x\frac{\partial}{\partial x} - u\frac{\partial}{\partial u}, \quad X_5 = \varepsilon\left(t^2\frac{\partial}{\partial t} + tu\frac{\partial}{\partial u}\right)$$

4. $f(u) = \delta u^{-4/3}$, $\delta = \pm 1$.

$$\tilde{X}_3 = \left(t - \frac{\varepsilon}{4}t^2\right)\frac{\partial}{\partial t} + x\frac{\partial}{\partial x} - \frac{3}{4}\varepsilon tu\frac{\partial}{\partial u},$$

$$X_4 = 2x\frac{\partial}{\partial x} - 3u\frac{\partial}{\partial u}, \quad X_5 = x^2\frac{\partial}{\partial x} - 3xu\frac{\partial}{\partial u}$$

5. $f(u) = 1.$

$$X_\infty = (\alpha_1 + \alpha_2)\frac{\partial}{\partial t} + (\alpha_1 - \alpha_2)\frac{\partial}{\partial x}$$

$$+ \left[Cu + \beta_1 + \beta_2 - \varepsilon\left(\frac{1}{2}(\alpha_1 + \alpha_2)u + \frac{1}{4}(t - x)\beta_1 + \frac{1}{4}(t + x)\beta_2\right)\right]\frac{\partial}{\partial u},$$

where α_1, β_1 and α_2, β_2 are arbitrary functions of the arguments $t + x$ and $t - x$, respectively, and C is an arbitrary constant.

APPROXIMATE TRANSFORMATIONS
(Baikov, Gazizov, and Ibragimov [1988b], [1988c], [1989a])

1. The equations

$$v_{ss} + \varepsilon v_s = \left(\delta \frac{v_y}{v}\right)_y$$

and

$$u_{tt} + \varepsilon u_t = (\delta e^u u_x)_x, \quad \delta = \pm 1$$

are connected by the approximate transformation

$$s \approx x\left(1 - \frac{\varepsilon}{6}x\right), \quad y \approx t\left(1 - \frac{\varepsilon}{2}t\right), \quad v \approx e^u\left[1 + 2\varepsilon\left(t - \frac{x}{3}\right)\right].$$

2. The equations

$$v_{ss} + \varepsilon v_s = (\delta v^\gamma v_y)_y$$

and

$$u_{tt} + \varepsilon u_t = (\delta u^\sigma u_x)_x, \quad \delta = \pm 1,$$

where $\gamma = -\sigma/(\sigma + 1)$ and $\gamma, \sigma \neq 1, -4/3, -4$, are connected by the approximate transformation

$$s \approx x\left(1 + \varepsilon \frac{\gamma}{\gamma + 4} \frac{x}{2}\right), \quad y \approx t\left(1 + \varepsilon \frac{\gamma}{3\gamma + 4} \frac{t}{2}\right),$$

$$v \approx u^{1/(\gamma+1)}\left[1 + 2\varepsilon\left(\frac{t}{3\gamma + 4} - \frac{x}{\gamma + 4}\right)\right].$$

3. The equations

$$v_{ss} + \varepsilon v_s = (\delta v^{-4} v_y)_y$$

and

$$u_{tt} + \varepsilon u_t = (\delta u^{-\frac{4}{3}} u_x)_x, \quad \delta = \pm 1$$

are connected by the approximate transformation

$$s \approx \varepsilon^{2\kappa} x, \quad y \approx t + \frac{\varepsilon}{4}t^2, \quad v \approx \varepsilon^\kappa \left(1 - \frac{\varepsilon}{4}t\right)u^{-\frac{1}{3}}$$

with a small positive κ such that $0 < \kappa \leq \frac{1}{2}$.

4. The equations

$$v_{ss} + \varepsilon v_s = (\delta e^v v_y)_y$$

and
$$u_{tt} = (\delta e^u u_x)_x, \quad \delta = \pm 1$$

are connected by the approximate transformation

$$s \approx t\left(1 + \frac{\varepsilon}{2}t\right), \quad y \approx x, \quad v \approx u - 2\varepsilon t.$$

5. The equations
$$v_{ss} + \varepsilon v_s = (\delta v^\sigma v_y)_y$$

and
$$u_{tt} = (\delta u^\sigma u_x)_x, \quad \sigma \neq -4, \quad \delta = \pm 1$$

are connected by the approximate transformation

$$s \approx t\left(1 + \frac{\varepsilon}{2}\frac{\sigma}{\sigma+4}t\right), \quad y \approx x, \quad v \approx u\left(1 - \frac{2\varepsilon t}{\sigma+4}\right).$$

In the case $\sigma = -4$, the connection is given by the transformation

$$s \approx \varepsilon^{2\kappa} t, \quad y \approx x, \quad v \approx \varepsilon^\kappa u, \quad 0 < \kappa \leq \frac{1}{2},$$

with the singularity at $\varepsilon = 0$.

APPROXIMATE SOLUTIONS
(Baikov, Gazizov, and Ibragimov [1988b], [1988c], [1989a])

1. The equation
$$u_{tt} + \varepsilon u_t = (u^\sigma u_x)_x$$

is approximately invariant under the operator

$$\tilde{X}_3 - \frac{1}{\sigma}X_4 = \left(t + \frac{\varepsilon\sigma}{2\sigma+8}t^2\right)\frac{\partial}{\partial t} - \frac{2}{\sigma}\left(1 + \frac{\varepsilon\sigma t}{\sigma+4}\right)u\frac{\partial}{\partial u}.$$

This operator yields the approximately invariant solution

$$u \approx t^{-2/\sigma}\left(1 - \frac{\varepsilon t}{\sigma+4}\right)\varphi(x),$$

where $\varphi(x)$ is given by the equation

$$x + x_0 = \int \frac{d\varphi}{\sqrt{C\varphi^{-2\sigma} + \dfrac{4}{\sigma^2}\varphi^{2-\sigma}}}$$

with arbitrary constants x_0 and C. If $C = 0$, then

$$u \approx t^{-2/\sigma} \left(1 - \frac{\varepsilon t}{\sigma + 4}\right)(x + x_0)^{2/\sigma}.$$

2. The equation

$$u_{tt} + \varepsilon u_t = (e^u u_x)_x$$

is approximately invariant under the operator

$$\tilde{X}_3 - X_4 = \left(t + \frac{\varepsilon}{2}t^2\right)\frac{\partial}{\partial t} - 2(1 + \varepsilon t)\frac{\partial}{\partial u}.$$

This operator yields the approximately invariant solution

$$u \approx -2\ln t - \varepsilon t + \varphi(x),$$

where $\varphi(x)$ is given by the equation

$$x + x_0 = \int \frac{d\varphi}{\sqrt{Ce^{-2\varphi} + (2\varphi - 1)e^{-\varphi}}}$$

with arbitrary constants x_0 and C.

9.2.1.2. $\mathbf{v}_{tt} + \varepsilon \mathbf{v}_t = \mathbf{f}(\mathbf{v}_x)\mathbf{v}_{xx}, \quad \mathbf{f}'(\mathbf{v}_x) \neq 0$

9.2.1.2.1. Unperturbed equation

$$v_{tt} = f(v_x)v_{xx}$$

CLASSIFICATION (LIE POINT AND QUASI-LOCAL SYMMETRIES)
(Oron and Rosenau [1986], Baikov and Gazizov [1994])

Equivalence Transformations

$$\bar{t} = \alpha_1 t + \alpha_2, \quad \bar{x} = \alpha_3 x + \alpha_4, \quad \bar{v} = \alpha_5 v + \alpha_6 x + \alpha_7 t + \alpha_8,$$

where α_i, $i = 1, \dots, 8$, are arbitrary constants such that $\alpha_1 \alpha_3 \alpha_5 \neq 0$.

Classification Result
   The principal Lie algebra $L_\mathcal{P}$ is five-dimensional and is spanned by

$$X_1^0 = \frac{\partial}{\partial t}, \quad X_2^0 = \frac{\partial}{\partial x}, \quad X_3^0 = \frac{\partial}{\partial v},$$

$$X_4^0 = t\frac{\partial}{\partial v}, \quad X_5^0 = t\frac{\partial}{\partial t} + x\frac{\partial}{\partial x} + v\frac{\partial}{\partial v}.$$

The algebra L_p extends in the following cases:

1. $f(v_x) = \delta e^{v_x}$, $\delta = \pm 1$.

$$X_6^0 = x\frac{\partial}{\partial x} + (v + 2x)\frac{\partial}{\partial v}$$

2. $f(v_x) = \delta v_x^\sigma$, $\delta = \pm 1$, $\sigma \neq -4, -\frac{4}{3}, 0$.

$$X_6^0 = \sigma x\frac{\partial}{\partial x} + (\sigma + 2)v\frac{\partial}{\partial v}$$

3. $f(v_x) = \delta v_x^{-4}$, $\delta = \pm 1$.

$$X_6^0 = 2x\frac{\partial}{\partial x} + v\frac{\partial}{\partial v}, \quad X_7^0 = t^2\frac{\partial}{\partial t} + tv\frac{\partial}{\partial v}$$

4. $f(v_x) = \delta v_x^{-\frac{4}{3}}$, $\delta = \pm 1$.

$$X_6^0 = 2x\frac{\partial}{\partial x} - v\frac{\partial}{\partial v}, \quad X_7^0 = x^2\frac{\partial}{\partial x} + (w - xv)\frac{\partial}{\partial v}$$

Here, X_7^0 is a quasi-local symmetry with the nonlocal variable w defined by the equations

$$w_x = v, \quad w_{tt} = -3\delta v_x^{-1/3}.$$

9.2.1.2.2. Perturbed equation: Exact symmetries

$$v_{tt} + \varepsilon v_t = f(v_x)v_{xx}$$

CLASSIFICATION (LIE POINT AND QUASI-LOCAL SYMMETRIES)
(Oron and Rosenau [1986])[1]

Equivalence Transformations

$$\bar{t} = t + \beta_1, \quad \bar{x} = \beta_2 x + \beta_3, \quad \bar{v} = \beta_4 v + \beta_5 x + \beta_6 e^{-\varepsilon t} + \beta_7,$$

where $\beta_i, i = 1, \dots, 7$, are arbitrary constants such that $\beta_2\beta_4 \neq 0$.

---

[1] The quasi-local symmetry is found by T.V. Avzjanov, unpublished.

Classification Result

The principal Lie algebra $L_\mathcal{P}$ is four-dimensional and is spanned by

$$Y_1 = \frac{\partial}{\partial t}, \quad Y_2 = \frac{\partial}{\partial x}, \quad Y_3 = \frac{\partial}{\partial v}, \quad Y_4 = e^{-\varepsilon t}\frac{\partial}{\partial v}.$$

The algebra $L_\mathcal{P}$ extends in the following cases:

1. $f(v_x) = \delta e^{v_x}$, $\delta = \pm 1$.

$$Y_5 = x\frac{\partial}{\partial x} + (v + 2x)\frac{\partial}{\partial v}$$

2. $f(v_x) = \delta v_x^\sigma$, $\delta = \pm 1$, $\sigma \neq -2, -\frac{4}{3}, 0$.

$$Y_5 = \sigma x\frac{\partial}{\partial x} + (\sigma + 2)v\frac{\partial}{\partial v}$$

3. $f(v_x) = \delta v_x^{-2}$, $\delta = \pm 1$.

$$Y_5 = x\frac{\partial}{\partial x}, \quad Y_6 = e^{-\varepsilon t}\frac{\partial}{\partial t} - \varepsilon e^{-\varepsilon t}v\frac{\partial}{\partial v}$$

4. $f(v_x) = \delta v_x^{-4/3}$, $\delta = \pm 1$.

$$Y_5 = 2x\frac{\partial}{\partial x} - v\frac{\partial}{\partial v}, \quad Y_6 = x^2\frac{\partial}{\partial x} + (w - xv)\frac{\partial}{\partial v}$$

Here, Y_6 is a quasi-local symmetry with the nonlocal variable w defined by the equations

$$w_x = v, \quad w_{tt} + \varepsilon w_t = -3\delta v_x^{-1/3}.$$

9.2.1.2.3. Perturbed equation: Approximate symmetries

$$v_{tt} + \varepsilon v_t = f(v_x)v_{xx}$$

CLASSIFICATION (APPROXIMATE POINT AND QUASI-LOCAL SYMMETRIES)[2]

Approximate Equivalence Transformations

---

[2]T.V. Avzjanov, unpublished.

$$\bar{t} = t + \gamma_1, \quad \bar{x} = \gamma_2 x + \gamma_3, \quad \bar{v} \approx \gamma_4 v + \gamma_5 x + \gamma_6 \left(t - \varepsilon \frac{t^2}{2}\right) + \gamma_7,$$

where $\gamma_i, i = 1, \ldots, 7$, are arbitrary constants such that $\gamma_2 \gamma_4 \neq 0$.

Classification Result

For arbitrary $f(v_x)$, the approximate symmetry algebra is given by

$$X_1 = \frac{\partial}{\partial t}, \quad X_2 = \frac{\partial}{\partial x}, \quad X_3 = \frac{\partial}{\partial v}, \quad X_4 = \left(t - \frac{\varepsilon}{2}t^2\right)\frac{\partial}{\partial v},$$

$$X_5 = \varepsilon \left(t\frac{\partial}{\partial t} + x\frac{\partial}{\partial x} + v\frac{\partial}{\partial v}\right).$$

This algebra extends in the following cases:

1. $f(v_x) = \delta e^{v_x}, \ \delta = \pm 1.$

$$\tilde{X}_5 = \left(t + \frac{\varepsilon}{2}t^2\right)\frac{\partial}{\partial t} + x\frac{\partial}{\partial x} + (v - 2\varepsilon tx)\frac{\partial}{\partial v},$$

$$X_6 = x\frac{\partial}{\partial x} + (v + 2x)\frac{\partial}{\partial v}$$

2. $f(v_x) = \delta v_x^\sigma, \ \delta = \pm 1, \ \sigma \neq -4, -\frac{4}{3}, 0.$

$$\tilde{X}_5 = \left(t + \frac{\varepsilon\sigma t^2}{2\sigma + 8}\right)\frac{\partial}{\partial t} + x\frac{\partial}{\partial x} + \left(v - \frac{2\varepsilon t v}{\sigma + 4}\right)\frac{\partial}{\partial v},$$

$$X_6 = \sigma x\frac{\partial}{\partial x} + (\sigma + 2)v\frac{\partial}{\partial v}$$

3. $f(v_x) = \delta v_x^{-4}, \ \delta = \pm 1.$

$$X_6 = 2x\frac{\partial}{\partial x} + v\frac{\partial}{\partial v}, \quad X_7 = \varepsilon \left(t^2\frac{\partial}{\partial t} + tv\frac{\partial}{\partial v}\right)$$

4. $f(v_x) = \delta v_x^{-4/3}, \ \delta = \pm 1.$

$$\tilde{X}_5 = \left(t - \frac{\varepsilon}{4}t^2\right)\frac{\partial}{\partial t} + x\frac{\partial}{\partial x} + (v - \frac{3}{4}\varepsilon t v)\frac{\partial}{\partial v},$$

$$X_6 = 2x\frac{\partial}{\partial x} - v\frac{\partial}{\partial v}, \quad X_7 = x^2\frac{\partial}{\partial x} + (w - xv)\frac{\partial}{\partial v}$$

Here, X_7 is a quasi-local symmetry with the nonlocal variable w defined by the equations

$$w_x = v, \quad w_{tt} + \varepsilon w_t = -3\delta v_x^{-1/3}$$

9.2.1.3. $w_{tt} + \varepsilon w_t = F(w_{xx})$, $F''(w_{xx}) \neq 0$

9.2.1.3.1. Unperturbed equation

$$w_{tt} = F(w_{xx})$$

CLASSIFICATION (LIE POINT SYMMETRIES)
(Oron and Rosenau [1986])

Equivalence Transformations

$$\bar{t} = \alpha_1 t + \alpha_2, \quad \bar{x} = \alpha_3 x + \alpha_4,$$

$$\bar{w} = \alpha_5 w + \alpha_6 t^2 + \alpha_7 tx + \alpha_8 t + \alpha_9 x^2 + \alpha_{10} x + \alpha_{11},$$

where α_i, $i = 1, \ldots, 11$, are arbitrary constants such that $\alpha_1 \alpha_3 \alpha_5 \neq 0$.

Classification Result
The principal Lie algebra $L_{\mathcal{P}}$ is seven-dimensional and is spanned by

$$X_1^0 = \frac{\partial}{\partial t}, \quad X_2^0 = \frac{\partial}{\partial x}, \quad X_3^0 = \frac{\partial}{\partial w}, \quad X_4^0 = x \frac{\partial}{\partial w},$$

$$X_5^0 = t \frac{\partial}{\partial w}, \quad X_6^0 = tx \frac{\partial}{\partial w}, \quad X_7^0 = t \frac{\partial}{\partial t} + x \frac{\partial}{\partial x} + 2w \frac{\partial}{\partial w}.$$

The algebra $L_{\mathcal{P}}$ extends in the following cases:
1. $F(w_{xx}) = \delta e^{w_{xx}}$, $\delta = \pm 1$.

$$X_8^0 = x \frac{\partial}{\partial x} + (2w + x^2) \frac{\partial}{\partial w}$$

2. $F(w_{xx}) = \delta w_{xx}^{\sigma+1}$, $\delta = \pm 1$, $\sigma \neq -4, -\frac{4}{3}, -1, 0$.

$$X_8^0 = \sigma x \frac{\partial}{\partial x} + 2(\sigma + 1)w \frac{\partial}{\partial w}$$

3. $F(w_{xx}) = \delta w_{xx}^{-3}$, $\delta = \pm 1$.

$$X_8^0 = 2x \frac{\partial}{\partial x} + 3w \frac{\partial}{\partial w}, \quad X_9^0 = t^2 \frac{\partial}{\partial t} + tw \frac{\partial}{\partial w}$$

4. $F(w_{xx}) = \delta w_{xx}^{-\frac{1}{3}}$, $\delta = \pm 1$.

$$X_8^0 = 2x\frac{\partial}{\partial x} + w\frac{\partial}{\partial w}, \quad X_9^0 = x^2\frac{\partial}{\partial x} + xw\frac{\partial}{\partial w}$$

5. $F(w_{xx}) = \delta \ln w_{xx}$, $\delta = \pm 1$.

$$X_8^0 = x\frac{\partial}{\partial x} - \delta t^2\frac{\partial}{\partial w}$$

9.2.1.3.2. Perturbed equation: Exact symmetries

$$w_{tt} + \varepsilon w_t = F(w_{xx})$$

CLASSIFICATION (LIE POINT SYMMETRIES)[3]

Equivalence Transformations

$$\bar{t} = t + \beta_1, \quad \bar{x} = \beta_2 x + \beta_3,$$

$$\overline{w} = \beta_4 w + \beta_5\frac{t}{\varepsilon} + \beta_6 x e^{-\varepsilon t} + \beta_7 e^{-\varepsilon t} + \beta_8 x^2 + \beta_9 x + \beta_{10},$$

where β_i, $i = 1, \ldots, 10$, are arbitrary constants such that $\beta_2\beta_4 \neq 0$.

Classification Result
The principal Lie algebra $L_{\mathcal{P}}$ is six-dimensional and is spanned by

$$Y_1 = \frac{\partial}{\partial t}, \quad Y_2 = \frac{\partial}{\partial x}, \quad Y_3 = \frac{\partial}{\partial w},$$

$$Y_4 = x\frac{\partial}{\partial w}, \quad Y_5 = e^{-\varepsilon t}\frac{\partial}{\partial w}, \quad Y_6 = xe^{-\varepsilon t}\frac{\partial}{\partial w}.$$

The algebra $L_{\mathcal{P}}$ extends in the following cases:
1. $F(w_{xx}) = \delta e^{w_{xx}}$, $\delta = \pm 1$.

$$Y_7 = x\frac{\partial}{\partial x} + (2w + x^2)\frac{\partial}{\partial w}$$

---

[3]T.V. Avzjanov, unpublished.

2. $F(w_{xx}) = \delta w_{xx}^{\sigma+1}, \; \delta = \pm 1, \; \sigma \neq -2, -\dfrac{4}{3}, -1, 0.$

$$Y_7 = \sigma x \frac{\partial}{\partial x} + 2(\sigma + 1) w \frac{\partial}{\partial w}$$

3. $F(w_{xx}) = \delta w_{xx}^{-1}, \; \delta = \pm 1.$

$$Y_7 = x \frac{\partial}{\partial x} + w \frac{\partial}{\partial w}, \quad Y_8 = e^{-\varepsilon t} \frac{\partial}{\partial t} - \varepsilon e^{-\varepsilon t} w \frac{\partial}{\partial w}$$

4. $F(w_{xx}) = \delta w_{xx}^{-1/3}, \delta = \pm 1.$

$$Y_7 = 2x \frac{\partial}{\partial x} + w \frac{\partial}{\partial w}, \quad Y_8 = x^2 \frac{\partial}{\partial x} + xw \frac{\partial}{\partial w}$$

5. $F(w_{xx}) = \delta \ln w_{xx}, \; \delta = \pm 1.$

$$Y_7 = x \frac{\partial}{\partial x} - \frac{2\delta t}{\varepsilon} \frac{\partial}{\partial w}$$

9.2.1.3.3. Perturbed equation: Approximate symmetries

$$w_{tt} + \varepsilon w_t = F(w_{xx})$$

CLASSIFICATION (APPROXIMATE POINT SYMMETRIES)[4]

Approximate Equivalence Transformations

$$\bar{t} = t + \gamma_1, \quad \bar{x} = \gamma_2 x + \gamma_3,$$

$$\bar{w} \approx \gamma_4 w + \gamma_5 \left(\frac{t^2}{2} - \varepsilon \frac{t^3}{6} \right) + \gamma_6 \left(tx - \varepsilon \frac{xt^2}{2} \right) + \gamma_7 \left(t - \varepsilon \frac{t^2}{2} \right) + \gamma_8 x^2 + \gamma_9 x + \gamma_{10},$$

where $\gamma_i, \; i = 1, \ldots, 10,$ are arbitrary constants such that $\gamma_2 \gamma_4 \neq 0.$

Classification Result
For arbitrary $F(w_{xx})$, the approximate symmetry algebra is given by

$$X_1 = \frac{\partial}{\partial t}, \quad X_2 = \frac{\partial}{\partial x}, \quad X_3 = \frac{\partial}{\partial w}, \quad X_4 = x \frac{\partial}{\partial w}, \quad X_5 = \left(t - \varepsilon \frac{t^2}{2} \right) \frac{\partial}{\partial w},$$

---

[4]T.V. Avzjanov, unpublished.

$$X_6 = (tx - \varepsilon \frac{t^2 x}{2}) \frac{\partial}{\partial w}, \quad X_7 = \varepsilon \left(t \frac{\partial}{\partial t} + x \frac{\partial}{\partial x} + 2w \frac{\partial}{\partial w} \right).$$

This algebra extends in the following cases:

1. $F(w_{xx}) = \delta e^{w_{xx}}$, $\delta = \pm 1$.

$$\tilde{X}_7 = \left(t + \frac{\varepsilon}{2} t^2 \right) \frac{\partial}{\partial t} + x \frac{\partial}{\partial x} + (2w - \varepsilon t x^2) \frac{\partial}{\partial w},$$

$$X_8 = x \frac{\partial}{\partial x} + (2w + x^2) \frac{\partial}{\partial w}$$

2. $F(w_{xx}) = \delta w_{xx}^{\sigma+1}$, $\delta = \pm 1$, $\sigma \neq -4, -\frac{4}{3}, -1, 0$.

$$\tilde{X}_7 = \left(t + \frac{\varepsilon \sigma t^2}{2\sigma + 8} \right) \frac{\partial}{\partial t} + x \frac{\partial}{\partial x} + \left(2w - \frac{2\varepsilon t w}{\sigma + 4} \right) \frac{\partial}{\partial w},$$

$$X_8 = \sigma x \frac{\partial}{\partial x} + 2(\sigma + 1)w \frac{\partial}{\partial w}$$

3. $F(w_{xx}) = \delta w_{xx}^{-3}$, $\delta = \pm 1$.

$$X_8 = 2x \frac{\partial}{\partial x} + 3w \frac{\partial}{\partial w}, \quad X_9 = \varepsilon \left(t^2 \frac{\partial}{\partial t} + tw \frac{\partial}{\partial w} \right)$$

4. $F(w_{xx}) = \delta w_{xx}^{-1/3}$, $\delta = \pm 1$.

$$\tilde{X}_7 = (t - \frac{\varepsilon}{4} t^2) \frac{\partial}{\partial t} + x \frac{\partial}{\partial x} + (2w - \frac{3}{4} \varepsilon t w) \frac{\partial}{\partial w},$$

$$X_8 = 2x \frac{\partial}{\partial x} + w \frac{\partial}{\partial w}, \quad X_9 = x^2 \frac{\partial}{\partial x} + xw \frac{\partial}{\partial w}$$

5. $F(w_{xx}) = \delta \ln w_{xx}$, $\delta = \pm 1$.

$$\tilde{X}_7 = \left(t - \frac{\varepsilon}{6} t^2 \right) \frac{\partial}{\partial t} + x \frac{\partial}{\partial x} + (2w - \frac{2}{3} \varepsilon t w - \frac{\delta}{9} \varepsilon t^3) \frac{\partial}{\partial w},$$

$$X_8 = x \frac{\partial}{\partial x} - \delta(t^2 - \frac{t^3}{3}) \frac{\partial}{\partial w}$$

9.2.1.4. $\mathbf{u}_{tt} + \varepsilon\varphi(\mathbf{u})\mathbf{u}_t = (\mathbf{f}(\mathbf{u})\mathbf{u}_x)_x$, $\varphi \neq$ const.
For the unperturbed equation, see Section 9.2.1.1.1.

9.2.1.4.1. Perturbed equation: Exact symmetries

$$u_{tt} + \varepsilon\varphi(u)u_t = (f(u)u_x)_x$$

CLASSIFICATION (LIE POINT SYMMETRIES)
(Baikov, Gazizov, and Ibragimov [1988b], [1988c], [1989a])

  Equivalence Transformations are taken from Section 9.2.1.1.1.

Classification Result
The principal Lie algebra $L_\mathcal{P}$ is two-dimensional and is spanned by

$$Y_1 = \frac{\partial}{\partial t}, \quad Y_2 = \frac{\partial}{\partial x}.$$

  The algebra $L_\mathcal{P}$ extends in the following cases:
I. $f(u) = \delta e^u$, $\varphi(u) = e^{vu}$, $\delta = \pm 1$, $v \neq 0$.

$$Y_3 = t\frac{\partial}{\partial t} + \left(1 - \frac{1}{2v}\right)x\frac{\partial}{\partial x} - \frac{1}{v}\frac{\partial}{\partial u}$$

II. $f(u) = \delta u^\sigma$, $\varphi(u) = u^v$, $\delta = \pm 1$, $\sigma \neq 0$ and $v \neq 0$.

$$Y_3 = t\frac{\partial}{\partial t} + \left(1 - \frac{\sigma}{2v}\right)x\frac{\partial}{\partial x} - \frac{u}{v}\frac{\partial}{\partial u}$$

III. $f(u) = \delta u^{-4}$, $\varphi(u) = u^v$, $\delta = \pm 1$, $v \neq 0$.

$$Y_3 = vt\frac{\partial}{\partial t} + (v+2)x\frac{\partial}{\partial x} - u\frac{\partial}{\partial u}$$

IV. $f(u) = \delta u^{-4/3}$, $\varphi(u) = u^v$, $\delta = \pm 1$, $v \neq 0$.

$$Y_3 = t\frac{\partial}{\partial t} + \left(1 + \frac{2}{3v}\right)x\frac{\partial}{\partial x} - \frac{u}{v}\frac{\partial}{\partial u}$$

V. $f(u) = 1$, $\varphi(u) = e^u$.

$$Y_3 = t\frac{\partial}{\partial t} + x\frac{\partial}{\partial x} - \frac{\partial}{\partial u}$$

VI. $f(u) = 1$, $\varphi(u) = u^\sigma$.

$$Y_3 = t\frac{\partial}{\partial t} + x\frac{\partial}{\partial x} - \frac{u}{\sigma}\frac{\partial}{\partial u}$$

9.2.1.4.2. Perturbed equation: Approximate symmetries

$$u_{tt} + \varepsilon\varphi(u)u_t = (f(u)u_x)_x$$

CLASSIFICATION (APPROXIMATE POINT SYMMETRIES)
(Baikov, Gazizov, and Ibragimov [1988b], [1988c], [1989a])

Equivalence Transformations are taken from Section 9.2.1.1.1.

Classification Result
For arbitrary $f(u)$ and $\varphi(u)$, the approximate symmetry algebra is given by

$$X_1 = \frac{\partial}{\partial t}, \quad X_2 = \frac{\partial}{\partial x}, \quad X_3 = \varepsilon\left(t\frac{\partial}{\partial t} + x\frac{\partial}{\partial x}\right).$$

This algebra extends in the following cases:
I. $f(u) = \delta e^u$, $\delta = \pm 1$, and $\varphi(u)$ is an arbitrary function.

$$X_4 = \varepsilon\left(x\frac{\partial}{\partial x} + 2\frac{\partial}{\partial u}\right)$$

For each of the following functions $\varphi(u)$ there is an additional extension:
I.1. $\varphi(u) = e^{\nu u} + k$, $\nu \neq 0$, and k is an arbitrary constant.

$$\tilde{X}_4 = \left(t + \frac{\varepsilon}{2}kt^2\right)\frac{\partial}{\partial t} + \left(1 - \frac{1}{2\nu}\right)x\frac{\partial}{\partial x} - \left(\frac{1}{\nu} + 2\varepsilon kt\right)\frac{\partial}{\partial u}$$

I.2. $\varphi(u) = u + k$, k is an arbitrary constant.

$$\tilde{X}_4 = \varepsilon t^2\frac{\partial}{\partial t} + x\frac{\partial}{\partial x} + (2 - 4\varepsilon t)\frac{\partial}{\partial u}$$

II. $f(u) = \delta u^\sigma$, $\delta = \pm 1$, and $\varphi(u)$ is an arbitrary function.

$$X_4 = \varepsilon\left(\sigma x\frac{\partial}{\partial x} + 2u\frac{\partial}{\partial u}\right)$$

For each of the following functions $\varphi(u)$ there is an additional extension:

II.1. $\varphi(u) = \ln u$.

$$\tilde{X}_4 = \frac{\varepsilon t^2}{\sigma + 4} \frac{\partial}{\partial t} + x \frac{\partial}{\partial x} + \frac{2}{\sigma} u \left(1 - \frac{2\varepsilon t}{\sigma + 4}\right) \frac{\partial}{\partial u}$$

II.2. $\varphi(u) = u^\nu + k$, $\nu \neq 0$, and k is an arbitrary constant.

$$\tilde{X}_4 = \left(t + \frac{\varepsilon \sigma k}{2\sigma + 8} t^2\right) \frac{\partial}{\partial t} + \left(1 - \frac{\sigma}{2\nu}\right) x \frac{\partial}{\partial x} - \left(\frac{1}{\nu} + \frac{2\varepsilon k}{\sigma + 4} t\right) u \frac{\partial}{\partial u}$$

III. $f(u) = \delta u^{-4}$, $\delta = \pm 1$, and $\varphi(u)$ is an arbitrary function.

$$X_4 = \varepsilon \left(2x \frac{\partial}{\partial x} - u \frac{\partial}{\partial u}\right), \quad X_5 = \varepsilon \left(t^2 \frac{\partial}{\partial t} + tu \frac{\partial}{\partial u}\right)$$

For each of the following functions $\varphi(u)$ there is an additional extension:

III.1. $\varphi(u) = u^\nu$, $\nu \neq 0$.

$$\tilde{X}_4 = \nu t \frac{\partial}{\partial t} + (\nu + 2) x \frac{\partial}{\partial x} - u \frac{\partial}{\partial u}$$

IV. $f(u) = \delta u^{-4/3}$, $\delta = \pm 1$, and $\varphi(u)$ is an arbitrary function.

$$X_4 = \varepsilon \left(x \frac{\partial}{\partial x} - \frac{3}{2} u \frac{\partial}{\partial u}\right), \quad X_5 = \varepsilon \left(x^2 \frac{\partial}{\partial x} - 3xu \frac{\partial}{\partial u}\right)$$

For each of the following functions $\varphi(u)$ there is an additional extension:

IV.1. $\varphi(u) = \ln u$.

$$\tilde{X}_4 = \frac{3}{8} \varepsilon t^2 \frac{\partial}{\partial t} + x \frac{\partial}{\partial x} + \left(-\frac{3}{2} u + \frac{9}{8} \varepsilon tu\right) \frac{\partial}{\partial u}$$

IV.2. $\varphi(u) = u^\nu + k$, $\nu \neq 0$, and k is an arbitrary constant.

$$\tilde{X}_3 = \left(t - \varepsilon \frac{k}{4} t^2\right) \frac{\partial}{\partial t} + \left(1 + \frac{2}{3\nu}\right) x \frac{\partial}{\partial x} - \left(\frac{u}{\nu} + \varepsilon \frac{3k}{4} tu\right) \frac{\partial}{\partial u}$$

V. $f(u) = 1$ and $\varphi(u)$ is an arbitrary function.

$$X_\infty = \varepsilon \left[(\alpha_1 + \alpha_2) \frac{\partial}{\partial t} + (\alpha_1 - \alpha_2) \frac{\partial}{\partial x} + (Cu + \beta_1 + \beta_2) \frac{\partial}{\partial u}\right]$$

Here, α_1, β_1 and α_2, β_2 are arbitrary functions of the arguments $t + x$ and $t - x$, respectively, and C is an arbitrary constant.

For each of the following functions $\varphi(u)$ there is an additional extension:

V.1. $\varphi(u) = \ln u$.

$$X_4 = \left(u - \frac{\varepsilon}{2} tu\right) \frac{\partial}{\partial u}$$

V.2. $\varphi(u) = e^u + k$, k is an arbitrary constant.

$$X_4 = t\frac{\partial}{\partial t} + x\frac{\partial}{\partial x} - \left(1 + \frac{\varepsilon}{2} ktu\right) \frac{\partial}{\partial u}$$

V.3. $\varphi(u) = u^\sigma + k$, $\sigma \neq 0$, and k is an arbitrary constant.

$$X_4 = t\frac{\partial}{\partial t} + x\frac{\partial}{\partial x} - \left(\frac{u}{\sigma} + \frac{\varepsilon}{2} ktu\right) \frac{\partial}{\partial u}$$

V.4. $\varphi(u) = u$.

$$X_4 = t\frac{\partial}{\partial t} + x\frac{\partial}{\partial x} - u\frac{\partial}{\partial u}, \quad X_5 = \left(1 - \frac{\varepsilon}{2} tu\right) \frac{\partial}{\partial u}$$

9.2.2.   Plane waves

9.2.2.1.   Isotropic Case

$$u_{tt} + \varepsilon u_t = \delta_1(f(u)u_x)_x + \delta_2(f(u)u_y)_y, \quad \delta_1, \delta_2 = \pm 1$$

9.2.2.1.1. Unperturbed Equation

$$u_{tt} = \delta_1(f(u)u_x)_x + \delta_2(f(u)u_y)_y$$

CLASSIFICATION (LIE POINT SYMMETRIES)
(Baikov, Gazizov, and Ibragimov [1990], [1991a])

Equivalence Transformations

$$\bar{t} = at + b, \quad \bar{x} = cx + d, \quad \bar{y} = ey + l, \quad \bar{u} = mu + n,$$

where a, b, c, d, e, l, m, and n are arbitrary constants such that $acem \neq 0$.

Classification Result
The principal Lie algebra $L_{\mathcal{P}}$ is five-dimensional and is spanned by

$$X_1^0 = \frac{\partial}{\partial t}, \qquad X_2^0 = \frac{\partial}{\partial x}, \qquad X_3^0 = \frac{\partial}{\partial y},$$

$$X_4^0 = \delta_2 y \frac{\partial}{\partial x} - \delta_1 x \frac{\partial}{\partial y}, \qquad X_5^0 = t \frac{\partial}{\partial t} + x \frac{\partial}{\partial x} + y \frac{\partial}{\partial y}.$$

The algebra $L_{\mathcal{P}}$ extends in the following cases:

1. $f(u) = e^u$.

$$X_6^0 = x \frac{\partial}{\partial x} + y \frac{\partial}{\partial y} + 2 \frac{\partial}{\partial u}$$

2. $f(u) = u^\sigma, \quad \sigma \neq -4, -1, 0$.

$$X_6^0 = x \frac{\partial}{\partial x} + y \frac{\partial}{\partial y} + \frac{2u}{\sigma} \frac{\partial}{\partial u}$$

3. $f(u) = u^{-4}$.

$$X_6^0 = x \frac{\partial}{\partial x} + y \frac{\partial}{\partial y} - \frac{u}{2} \frac{\partial}{\partial u}, \qquad X_7^0 = t^2 \frac{\partial}{\partial t} + tu \frac{\partial}{\partial u}$$

4. $f(u) = u^{-1}$.

$$X_\infty^0 = \alpha^1(x, y) \frac{\partial}{\partial x} + \alpha^2(x, y) \frac{\partial}{\partial y} - 2\alpha_x^1(x, y)u \frac{\partial}{\partial u}$$

Here, $\alpha^1(x, y)$ and $\alpha^2(x, y)$ are determined by the equations

$$\alpha_x^1 = \alpha_y^2, \quad \alpha_y^1 = -\delta_1 \delta_2 \alpha_x^2.$$

5. $f(u) = 1$.

$$X_6^0 = \delta_1 x \frac{\partial}{\partial t} + t \frac{\partial}{\partial x}, \qquad X_7^0 = \delta_2 y \frac{\partial}{\partial t} + t \frac{\partial}{\partial y},$$

$$X_8^0 = (t^2 + \delta_1 x^2 + \delta_2 y^2) \frac{\partial}{\partial t} + 2tx \frac{\partial}{\partial x} + 2ty \frac{\partial}{\partial y} - tu \frac{\partial}{\partial u},$$

$$X_9^0 = tx \frac{\partial}{\partial t} + \frac{1}{2}(\delta_1 t^2 + x^2 - \delta_1 \delta_2 y^2) \frac{\partial}{\partial x} + xy \frac{\partial}{\partial y} - xu \frac{\partial}{\partial u},$$

$$X_{10}^0 = ty \frac{\partial}{\partial t} + xy \frac{\partial}{\partial x} + \frac{1}{2}(\delta_2 t^2 - \delta_1 \delta_2 x^2 + y^2) \frac{\partial}{\partial y} - yu \frac{\partial}{\partial u},$$

$$X_{11}^0 = u\frac{\partial}{\partial u}, \quad X_\infty^0 = \theta^0(t, x, y)\frac{\partial}{\partial u}$$

Here, $\theta^0(t, x, y)$ is an arbitrary solution of the equation $\theta_{tt}^0 = \delta_1\theta_{xx}^0 + \delta_2\theta_{yy}^0$.

9.2.2.1.2. Perturbed equation: Exact symmetries

$$u_{tt} + \varepsilon u_t = \delta_1(f(u)u_x)_x + \delta_2(f(u)u_y)_y$$

CLASSIFICATION (LIE POINT SYMMETRIES)
(Baikov, Gazizov, and Ibragimov [1990], [1991a])

   Equivalence Transformations are taken from Section 9.2.2.1.1.

Classification Result
The principal Lie algebra L_P is four-dimensional and is spanned by

$$Y_1 = \frac{\partial}{\partial t}, \quad Y_2 = \frac{\partial}{\partial x}, \quad Y_3 = \frac{\partial}{\partial y}, \quad Y_4 = \delta_2 y\frac{\partial}{\partial x} - \delta_1 x\frac{\partial}{\partial y}.$$

   The algebra L_P extends in the following cases:
1. $f(u) = e^u$.

$$Y_5 = x\frac{\partial}{\partial x} + y\frac{\partial}{\partial y} + 2\frac{\partial}{\partial u}$$

2. $f(u) = u^\sigma$, $\sigma \neq -2, -1, 0$.

$$Y_5 = x\frac{\partial}{\partial x} + y\frac{\partial}{\partial y} + \frac{2u}{\sigma}\frac{\partial}{\partial u}$$

3. $f(u) = u^{-2}$.

$$Y_5 = x\frac{\partial}{\partial x} + y\frac{\partial}{\partial y} - u\frac{\partial}{\partial u}, \quad Y_6 = e^{-\varepsilon t}\frac{\partial}{\partial t} - \varepsilon e^{-\varepsilon t}u\frac{\partial}{\partial u}$$

4. $f(u) = u^{-1}$.

$$Y_\infty = \alpha^1(x, y)\frac{\partial}{\partial x} + \alpha^2(x, y)\frac{\partial}{\partial y} - 2\alpha_x^1(x, y)u\frac{\partial}{\partial u}$$

The functions $\alpha^1(x, y)$ and $\alpha^2(x, y)$ are determined by the equations

$$\alpha_x^1 = \alpha_y^2, \quad \alpha_y^1 = -\delta_1\delta_2\alpha_x^2.$$

5. $f(u) = 1.$

$$Y_5 = x\frac{\partial}{\partial t} + \delta_1 t\frac{\partial}{\partial x} - \frac{\varepsilon}{2}xu\frac{\partial}{\partial u}, \quad Y_6 = y\frac{\partial}{\partial t} + \delta_2 t\frac{\partial}{\partial y} - \frac{\varepsilon}{2}yu\frac{\partial}{\partial u},$$

$$Y_7 = u\frac{\partial}{\partial u}, \quad Y_\infty = \theta(t, x, y)\frac{\partial}{\partial u}$$

Here, $\theta(t, x, y)$ is an arbitrary solution of the equation $\theta_{tt} + \varepsilon\theta_t = \delta_1\theta_{xx} + \delta_2\theta_{yy}$.

9.2.2.1.3. Perturbed equation: Approximate symmetries

$$u_{tt} + \varepsilon u_t = \delta_1(f(u)u_x)_x + \delta_2(f(u)u_y)_y$$

CLASSIFICATION (APPROXIMATE POINT SYMMETRIES)
(Baikov, Gazizov, and Ibragimov [1990], [1991a])

Equivalence Transformations are taken from Section 9.2.2.1.1.

Classification Result
For arbitrary $f(u)$, the approximate symmetry algebra is given by

$$X_1 = \frac{\partial}{\partial t}, \quad X_2 = \frac{\partial}{\partial x}, \quad X_3 = \frac{\partial}{\partial y},$$

$$X_4 = \delta_2 y\frac{\partial}{\partial x} - \delta_1 x\frac{\partial}{\partial y}, \quad X_5 = \varepsilon\left(t\frac{\partial}{\partial t} + x\frac{\partial}{\partial x} + y\frac{\partial}{\partial y}\right).$$

This algebra extends in the following cases:
1. $f(u) = e^u.$

$$\tilde{X}_5 = \left(t + \frac{\varepsilon}{2}t^2\right) + x\frac{\partial}{\partial x} + y\frac{\partial}{\partial y} - 2\varepsilon t\frac{\partial}{\partial u},$$

$$X_6 = x\frac{\partial}{\partial x} + y\frac{\partial}{\partial y} + 2\frac{\partial}{\partial u}$$

2. $f(u) = u^\sigma, \quad \sigma \neq -4, -1, 0.$

$$\tilde{X}_5 = \left(t + \frac{\varepsilon\sigma}{2\sigma + 8}t^2\right)\frac{\partial}{\partial t} + x\frac{\partial}{\partial x} + y\frac{\partial}{\partial y} - \frac{2\varepsilon}{\sigma + 4}tu\frac{\partial}{\partial u},$$

$$X_6 = x\frac{\partial}{\partial x} + y\frac{\partial}{\partial y} + \frac{2u}{\sigma}\frac{\partial}{\partial u}$$

3. $f(u) = u^{-4}$.

$$X_6 = x\frac{\partial}{\partial x} + y\frac{\partial}{\partial y} - \frac{u}{2}\frac{\partial}{\partial u}, \quad X_7 = \varepsilon\left(t^2\frac{\partial}{\partial t} + tu\frac{\partial}{\partial u}\right)$$

4. $f(u) = u^{-1}$.

$$\tilde{X}_5 = \left(t - \frac{\varepsilon}{6}t^2\right)\frac{\partial}{\partial t} + x\frac{\partial}{\partial x} + y\frac{\partial}{\partial y} - \frac{2}{3}\varepsilon tu\frac{\partial}{\partial u},$$

$$X_\infty = \alpha^1(x, y)\frac{\partial}{\partial x} + \alpha^2(x, y)\frac{\partial}{\partial y} - 2\alpha_x^1(x, y)u\frac{\partial}{\partial u}$$

The functions $\alpha^1(x, y)$ and $\alpha^2(x, y)$ are determined by the equations

$$\alpha_x^1 = \alpha_y^2, \quad \alpha_y^1 = -\delta_1\delta_2\alpha_x^2.$$

5. $f(u) = 1$.

$$\tilde{X}_5 = t\frac{\partial}{\partial t} + x\frac{\partial}{\partial x} + y\frac{\partial}{\partial y} - \frac{\varepsilon}{2}tu\frac{\partial}{\partial u},$$

$$X_6 = x\frac{\partial}{\partial t} + \delta_1 t\frac{\partial}{\partial x} - \frac{\varepsilon}{2}xu\frac{\partial}{\partial u}, \quad X_7 = y\frac{\partial}{\partial t} + \delta_2 t\frac{\partial}{\partial y} - \frac{\varepsilon}{2}yu\frac{\partial}{\partial u},$$

$$X_8 = (t^2 + \delta_1 x^2 + \delta_2 y^2)\frac{\partial}{\partial t} + 2tx\frac{\partial}{\partial x} + 2ty\frac{\partial}{\partial y} - \left(t + \frac{\varepsilon}{2}\left(t^2 + \delta_1 x^2 + \delta_2 y^2\right)\right)u\frac{\partial}{\partial u},$$

$$X_9 = tx\frac{\partial}{\partial t} + \frac{1}{2}(\delta_1 t^2 + x^2 - \delta_1\delta_2 y^2)\frac{\partial}{\partial x} + xy\frac{\partial}{\partial y} - (1 + \frac{1}{2}\varepsilon t)xu\frac{\partial}{\partial u},$$

$$X_{10} = ty\frac{\partial}{\partial t} + xy\frac{\partial}{\partial x} + \frac{1}{2}(\delta_2 t^2 - \delta_1\delta_2 x^2 + y^2)\frac{\partial}{\partial y} - (1 + \frac{1}{2}\varepsilon t)yu\frac{\partial}{\partial u},$$

$$X_{11} = u\frac{\partial}{\partial u}, \quad X_\infty = \theta(t, x, y, \varepsilon)\frac{\partial}{\partial u}$$

Here, $\theta(t, x, y, \varepsilon)$ solves the approximate equation $\theta_{tt} + \varepsilon\theta_t \approx \delta_1\theta_{xx} + \delta_2\theta_{yy}$.

AN APPROXIMATE CONSERVATION LAW
(Baikov, Gazizov, and Ibragimov [1990], [1991a])

Consider the equation

$$u_{tt} + \varepsilon u_t = u_{xx} + u_{yy}.$$

It is the approximate Euler-Lagrange equation 2.34 (see Chapter 2) with the Lagrangian

$$L = \frac{1}{2}e^{\varepsilon t}(u_t^2 - u_x^2 - u_y^2).$$

The operator

$$X = (t^2 + x^2 + y^2)\frac{\partial}{\partial t} + 2tx\frac{\partial}{\partial x} + 2ty\frac{\partial}{\partial y} - \left[t + \frac{\varepsilon}{2}(t^2 + x^2 + y^2)\right]u\frac{\partial}{\partial u}$$

is an approximate Noether symmetry for the equation under consideration. Hence, Theorem 2.8 (see Section 2.6.1) applies and yields the following approximate conservation law:

$$D_t(C^t) + D_x(C^x) + D_y(C^y) = o(\varepsilon)$$

where

$$C^t \approx e^{\varepsilon t}\left[-\frac{s^2}{2}(u_t^2 + u_x^2 + u_y^2) - u_t(tu + \frac{\varepsilon}{2}s^2 u + 2txu_x + 2tyu_y) + \frac{1+\varepsilon t}{2}u^2\right],$$

$$C^x \approx e^{\varepsilon t}\left[tx(u_t^2 + u_x^2 - u_y^2) + u_x(tu + \frac{\varepsilon}{2}s^2 u \mid s^2 u_t - \frac{\varepsilon}{2}xu^2 + 2tyu_y)\right],$$

$$C^y \approx e^{\varepsilon t}\left[ty(u_t^2 - u_x^2 + u_y^2) + u_y(tu + \frac{\varepsilon}{2}s^2 u + s^2 u_t - \frac{\varepsilon}{2}yu^2 + 2txu_x)\right].$$

Here, $s^2 = t^2 + x^2 + y^2$.

APPROXIMATE SOLUTIONS
(Baikov [1990])

Consider the case 4 of the approximate group classification. Let us take, for the equation

$$u_{tt} + \varepsilon u_t = \left(\frac{u_x}{u}\right)_x + \left(\frac{u_y}{u}\right)_y,$$

the two-dimensional subalgebra generated by the operators

$$X = \left(t - \frac{\varepsilon}{6}t^2\right)\frac{\partial}{\partial t} + 2u\left(1 - \frac{\varepsilon}{3}t\right)\frac{\partial}{\partial u},$$

$$X_\infty = \alpha^1(x, y)\frac{\partial}{\partial x} + \alpha^2(x, y)\frac{\partial}{\partial y} - 2\alpha_x^1(x, y)u\frac{\partial}{\partial u}$$

with $\alpha^1 = e^x \cos y, \alpha^2 = e^x \sin y$. The corresponding approximately invariant solution has the form

$$u \approx t^2\left(1 - \frac{\varepsilon}{3}t\right)v(\lambda)e^{-2x}, \quad \lambda = \frac{e^x}{\sin y},$$

where the function $v(\lambda)$ is determined by the second-order ordinary differential equation

$$\lambda^4 \left(\frac{v'}{v}\right)' + 2\frac{v'}{v}\lambda^3 - 2v = 0.$$

After setting $\lambda = 1/\mu$, $2v(\lambda) = e^{V(\mu)}$, this equation becomes

$$V_{\mu\mu} = e^V.$$

It follows that the approximately invariant solution splits into the following three families:

$$u = \frac{t^2 \left(1 - \frac{\varepsilon}{3}t\right)}{(\sin y + C_2 e^x)^2} + o(\varepsilon),$$

$$u = \frac{C_1^2 t^2 \left(1 - \frac{\varepsilon}{3}t\right) e^{C_1(e^{-x} \sin y + C_2)}}{e^{2x} \left[1 - e^{C_1(e^{-x} \sin y + C_2)}\right]^2} + o(\varepsilon),$$

$$u = \frac{C_1^2 t^2 \left(1 - \frac{\varepsilon}{3}t\right)}{4e^{2x}} \left\{\tan^2\left[\frac{C_1}{2}\left(\frac{\sin y}{e^x} + C_2\right)\right] + 1\right\} + o(\varepsilon),$$

where C_1 and C_2 are arbitrary constants.

9.2.2.2. Anisotropic case

$$u_{tt} + \varepsilon u_t = (f(u)u_x)_x + (g(u)u_y)_y$$

where f and g are linearly independent functions.

9.2.2.2.1. Unperturbed equation

$$u_{tt} = (f(u)u_x)_x + (g(u)u_y)_y$$

CLASSIFICATION (LIE POINT SYMMETRIES)
(Baikov, Gazizov, and Ibragimov [1990], [1991a])

Equivalence Transformations are taken from Section 9.2.2.1.1.

Classification Result
The principal Lie algebra L_P is four-dimensional and is spanned by

$$X_1^0 = \frac{\partial}{\partial t}, \quad X_2^0 = \frac{\partial}{\partial x}, \quad X_3^0 = \frac{\partial}{\partial y}, \quad X_4^0 = t\frac{\partial}{\partial t} + x\frac{\partial}{\partial x} + y\frac{\partial}{\partial y}.$$

The algebra L_P extends in the following cases:

I. $f(u) = \delta_1$, $\delta_1 = \pm 1$, and $g(u)$ is an arbitrary function.

$$X_5^0 = \delta_1 x \frac{\partial}{\partial t} + t \frac{\partial}{\partial x}$$

For each of the following functions $g(u)$ there is an additional extension:

I.1. $g(u) = \delta_2 e^u$, $\delta_2 = \pm 1$.

$$X_6^0 = y \frac{\partial}{\partial y} + 2 \frac{\partial}{\partial u},$$

$$X_\infty^0 = \alpha^1(t, x) \frac{\partial}{\partial t} + \alpha^2(t, x) \frac{\partial}{\partial x} - 2\alpha_t^1(t, x) \frac{\partial}{\partial u}$$

where the functions $\alpha^1(t, x)$ and $\alpha^2(t, x)$ solve the system

$$\alpha_t^1 = \alpha_x^2, \qquad \alpha_x^1 = \delta_1 \alpha_t^2.$$

I.2. $g(u) = \delta_2 u^\gamma$, $\delta_2 = \pm 1$, $\gamma \neq -\frac{4}{3}$, 0.

$$X_6^0 = y \frac{\partial}{\partial y} + \frac{2u}{\gamma} \frac{\partial}{\partial u}$$

I.3. $g(u) = \delta_2 u^{-4/3}$, $\delta_2 = \pm 1$.

$$X_6^0 = y \frac{\partial}{\partial y} - \frac{3}{2} u \frac{\partial}{\partial u}, \quad X_7^0 = y^2 \frac{\partial}{\partial y} - 3yu \frac{\partial}{\partial u}$$

II. $f(u) = \delta_1 e^u$, $g(u) = \delta_2 e^{ku}$, $\delta_1, \delta_2 = \pm 1$, $k \neq 0, 1$.

$$X_5^0 = x \frac{\partial}{\partial x} + ky \frac{\partial}{\partial y} + 2 \frac{\partial}{\partial u}$$

III. $f(u) = \delta_1 u^\sigma$, $g(u) = \delta_2 u^\gamma$, $\delta_1, \delta_2 = \pm 1$, $\sigma \neq 0$, $\gamma \neq 0$, and $\sigma \neq \gamma$.

$$X_5^0 = \sigma x \frac{\partial}{\partial x} + \gamma y \frac{\partial}{\partial y} + 2u \frac{\partial}{\partial u}$$

9.2.2.2.2. Perturbed equation: Exact symmetries

$$u_{tt} + \varepsilon u_t = (f(u)u_x)_x + (g(u)u_y)_y$$

CLASSIFICATION (LIE POINT SYMMETRIES)
(Baikov, Gazizov, and Ibragimov [1990], [1991a])

Equivalence Transformations are taken from Section 9.2.2.1.1.

Classification Result
The principal Lie algebra L_P is three-dimensional and is spanned by

$$Y_1 = \frac{\partial}{\partial t}, \quad Y_2 = \frac{\partial}{\partial x}, \quad Y_3 = \frac{\partial}{\partial y}.$$

The algebra L_P extends in the following cases:

I. $f(u) = \delta_1, \quad \delta_1 = \pm 1.$
For each of the following functions $g(u)$ there is an extension:

I.1. $g(u) = \delta_2 e^u, \quad \delta_2 = \pm 1.$

$$Y_4 = y\frac{\partial}{\partial y} + 2\frac{\partial}{\partial u}$$

I.2. $g(u) = \delta_2 u^\gamma, \quad \delta_2 = \pm 1, \quad \gamma \neq -\frac{4}{3}, \quad 0.$

$$Y_4 = y\frac{\partial}{\partial y} + \frac{2u}{\gamma}\frac{\partial}{\partial u}$$

I.3. $g(u) = \delta_2 u^{-4/3}, \quad \delta_2 = \pm 1.$

$$Y_4 = y\frac{\partial}{\partial y} - \frac{3}{2}u\frac{\partial}{\partial u}, \quad Y_5 = y^2\frac{\partial}{\partial y} - 3yu\frac{\partial}{\partial u}$$

II. $f(u) = \delta_1 e^u, \quad g(u) = \delta_2 e^{ku}, \quad \delta_1, \delta_2 = \pm 1, \quad k \neq 0, 1.$

$$Y_4 = x\frac{\partial}{\partial x} + ky\frac{\partial}{\partial y} + 2\frac{\partial}{\partial u}$$

III. $f(u) = \delta_1 u^\sigma, \quad g(u) = \delta_2 u^\gamma, \quad \delta_1, \delta_2 = \pm 1, \quad \sigma \neq 0, \quad \gamma \neq 0, \text{ and } \sigma \neq \gamma.$

$$Y_4 = \sigma x\frac{\partial}{\partial x} + \gamma y\frac{\partial}{\partial y} + 2u\frac{\partial}{\partial u}.$$

9.2.2.2.3. Perturbed equation: Approximate symmetries

$$u_{tt} + \varepsilon u_t = (f(u)u_x)_x + (g(u)u_y)_y$$

CLASSIFICATION (APPROXIMATE POINT SYMMETRIES)
(Baikov, Gazizov, and Ibragimov [1990], [1991a])

Equivalence Transformations are taken from Section 9.2.2.1.1.

Classification Result
For arbitrary $f(u)$, the approximate symmetry algebra is given by

$$X_1 = \frac{\partial}{\partial t}, \quad X_2 = \frac{\partial}{\partial x}, \quad X_3 = \frac{\partial}{\partial y},$$

$$X_4 = \varepsilon \left(t\frac{\partial}{\partial t} + x\frac{\partial}{\partial x} + y\frac{\partial}{\partial y} \right).$$

This algebra extends in the following cases:
I. $f(u) = \delta_1$, $\delta_1 = \pm 1$, and $g(u)$ is an arbitrary function.

$$X_5 = \varepsilon \left(\delta_1 x\frac{\partial}{\partial t} + t\frac{\partial}{\partial x} \right)$$

For each of the following functions $g(u)$ there is an additional extension:
I.1. $g(u) = \delta_2 e^u$, $\delta_2 = \pm 1$.

$$X_6 = y\frac{\partial}{\partial y} + 2\frac{\partial}{\partial u},$$

$$X_\infty = \varepsilon \left(\alpha^1(t, x)\frac{\partial}{\partial t} + \alpha^2(t, x)\frac{\partial}{\partial x} - 2\alpha_t^1(t, x)\frac{\partial}{\partial u} \right)$$

where the functions $\alpha^1(t, x)$ and $\alpha^2(t, x)$ solve the system

$$\alpha_t^1 = \alpha_x^2, \qquad \alpha_x^1 = \delta_1\alpha_t^2.$$

I.2. $g(u) = \delta_2 u^\gamma$, $\delta_2 = \pm 1$, $\gamma \neq -\frac{4}{3}, 0$.

$$\tilde{X}_4 = \left(t + \frac{\varepsilon\gamma}{8}(t^2 + \delta_1 x^2) \right)\frac{\partial}{\partial t} + \left(x + \frac{\varepsilon\gamma}{4}tx \right)\frac{\partial}{\partial x} + y\frac{\partial}{\partial y} - \frac{\varepsilon}{2}tu\frac{\partial}{\partial u},$$

$$\tilde{X}_5 = (\delta_1 x + \frac{\varepsilon\gamma}{4}tx)\frac{\partial}{\partial t} + (t + \frac{\varepsilon\gamma}{8}(\delta_1 t^2 + x^2))\frac{\partial}{\partial x} - \frac{\varepsilon}{2}\delta_1 x u\frac{\partial}{\partial u},$$

$$X_6 = y\frac{\partial}{\partial y} + \frac{2u}{\gamma}\frac{\partial}{\partial u}$$

I.3. $g(u) = \delta_2 u^{-4/3}$, $\delta_2 = \pm 1$.

$$\tilde{X}_4 = (t - \frac{\varepsilon}{6}(t^2 + \delta_1 x^2))\frac{\partial}{\partial t} + (x - \frac{\varepsilon}{3}tx)\frac{\partial}{\partial x} + y\frac{\partial}{\partial y} - \frac{\varepsilon}{2}tu\frac{\partial}{\partial u},$$

$$\tilde{X}_5 = \left(\delta_1 x - \frac{\varepsilon}{3}tx\right)\frac{\partial}{\partial t} + (t - \frac{\varepsilon}{6}(\delta_1 t^2 + x^2))\frac{\partial}{\partial x} - \frac{\varepsilon}{2}\delta_1 x u\frac{\partial}{\partial u},$$

$$X_6 = y\frac{\partial}{\partial y} - \frac{3}{2}u\frac{\partial}{\partial u}, \quad X_7 = y^2\frac{\partial}{\partial y} - 3yu\frac{\partial}{\partial u}$$

II. $f(u) = \delta_1 e^u$, $g(u) = \delta_2 e^{ku}$, $\delta_1, \delta_2 = \pm 1$, $k \neq 0, 1$.

$$X_5 = x\frac{\partial}{\partial x} + ky\frac{\partial}{\partial y} + 2\frac{\partial}{\partial u}$$

III. $f(u) = \delta_1 u^\sigma$, $g(u) = \delta_2 u^\gamma$, $\delta_1, \delta_2 = \pm 1$, $\sigma \neq 0$, $\gamma \neq 0$, and $\sigma \neq \gamma$.

$$X_5 = \sigma x\frac{\partial}{\partial x} + \gamma y\frac{\partial}{\partial y} + 2u\frac{\partial}{\partial u}$$

9.2.3.   Three-dimensional waves

9.2.3.1.   Isotropic case

$$u_{tt} + \varepsilon u_t = \delta_1(f(u)u_x)_x + \delta_2(f(u)u_y)_y + \delta_3(f(u)u_z)_z, \quad \delta_1, \delta_2, \delta_3 = \pm 1.$$

9.2.3.1.1. Unperturbed equation

$$u_{tt} = \delta_1(f(u)u_x)_x + \delta_2(f(u)u_y)_y + \delta_3(f(u)u_z)_z$$

CLASSIFICATION (LIE POINT SYMMETRIES)
(Baikov, Gazizov, and Ibragimov [1990], [1991a])

Equivalence Transformations

$$\bar{t} = at + b, \quad \bar{x} = cx + d, \quad \bar{y} = ey + l, \quad \bar{z} = mz + n, \quad \bar{u} = pu + q,$$

where $a, b, c, d, e, l, m, n, p, q$ are arbitrary constants such that $acemp \neq 0$.

Classification Result

The principal Lie algebra $L_{\mathcal{P}}$ is eight-dimensional and is spanned by

$$X_1^0 = \frac{\partial}{\partial t}, \quad X_2^0 = \frac{\partial}{\partial x}, \quad X_3^0 = \frac{\partial}{\partial y}, \quad X_4^0 = \frac{\partial}{\partial z},$$

$$X_5^0 = \delta_2 y \frac{\partial}{\partial x} - \delta_1 x \frac{\partial}{\partial y}, \quad X_6^0 = \delta_3 z \frac{\partial}{\partial x} - \delta_1 x \frac{\partial}{\partial z},$$

$$X_7^0 = \delta_3 z \frac{\partial}{\partial y} - \delta_2 y \frac{\partial}{\partial z}, \quad X_8^0 = t \frac{\partial}{\partial t} + x \frac{\partial}{\partial x} + y \frac{\partial}{\partial y} + z \frac{\partial}{\partial z}.$$

The algebra $L_{\mathcal{P}}$ extends in the following cases:

1. $f(u) = e^u$.

$$X_9^0 = x \frac{\partial}{\partial x} + y \frac{\partial}{\partial y} + z \frac{\partial}{\partial z} + 2 \frac{\partial}{\partial u}$$

2. $f(u) = u^\sigma$, $\sigma \neq -4, -\frac{4}{5}, 0$.

$$X_9^0 = \sigma x \frac{\partial}{\partial x} + \sigma y \frac{\partial}{\partial y} + \sigma z \frac{\partial}{\partial z} + 2u \frac{\partial}{\partial u}$$

3. $f(u) = u^{-4}$.

$$X_9^0 = x \frac{\partial}{\partial x} + y \frac{\partial}{\partial y} + z \frac{\partial}{\partial z} - \frac{1}{2} u \frac{\partial}{\partial u}, \quad X_{10}^0 = t^2 \frac{\partial}{\partial t} + t u \frac{\partial}{\partial u}$$

4. $f(u) = u^{-4/5}$.

$$X_9^0 = x \frac{\partial}{\partial x} + y \frac{\partial}{\partial y} + z \frac{\partial}{\partial z} - \frac{5}{2} u \frac{\partial}{\partial u},$$

$$X_{10}^0 = \frac{1}{2}(x^2 - \delta_1 \delta_2 y^2 - \delta_1 \delta_3 z^2) \frac{\partial}{\partial x} + xy \frac{\partial}{\partial y} + xz \frac{\partial}{\partial z} - \frac{5}{2} xu \frac{\partial}{\partial u},$$

$$X_{11}^0 = xy \frac{\partial}{\partial x} + \frac{1}{2}(y^2 - \delta_1 \delta_2 x^2 - \delta_2 \delta_3 z^2) \frac{\partial}{\partial y} + yz \frac{\partial}{\partial z} - \frac{5}{2} yu \frac{\partial}{\partial u},$$

$$X_{12}^0 = xz \frac{\partial}{\partial x} + yz \frac{\partial}{\partial y} + (z^2 - \delta_1 \delta_3 x^2 - \delta_2 \delta_3 y^2) \frac{\partial}{\partial z} - \frac{5}{2} zu \frac{\partial}{\partial u}$$

5. $f(u) = 1$.

$$X_9^0 = \delta_1 x \frac{\partial}{\partial t} + t \frac{\partial}{\partial x}, \quad X_{10}^0 = \delta_2 y \frac{\partial}{\partial t} + t \frac{\partial}{\partial y}, \quad X_{11}^0 = \delta_3 z \frac{\partial}{\partial t} + t \frac{\partial}{\partial z},$$

$$X_{12}^0 = \frac{1}{2}(t^2 + \delta_1 x^2 + \delta_2 y^2 + \delta_3 z^2)\frac{\partial}{\partial t} + tx\frac{\partial}{\partial x} + ty\frac{\partial}{\partial y} + tz\frac{\partial}{\partial z} - tu\frac{\partial}{\partial u},$$

$$X_{13}^0 = tx\frac{\partial}{\partial t} + \frac{1}{2}(\delta_1 t^2 + x^2 - \delta_1\delta_2 y^2 - \delta_1\delta_3 z^2)\frac{\partial}{\partial x} + xy\frac{\partial}{\partial y} + xz\frac{\partial}{\partial z} - xu\frac{\partial}{\partial u},$$

$$X_{14}^0 = ty\frac{\partial}{\partial t} + xy\frac{\partial}{\partial x} + \frac{1}{2}(\delta_2 t^2 - \delta_1\delta_2 x^2 + y^2 - \delta_2\delta_3 z^2)\frac{\partial}{\partial y} + yz\frac{\partial}{\partial z} - yu\frac{\partial}{\partial u},$$

$$X_{15}^0 = tz\frac{\partial}{\partial t} + xz\frac{\partial}{\partial x} + yz\frac{\partial}{\partial y} + \frac{1}{2}(\delta_3 t^2 - \delta_1\delta_3 x^2 - \delta_2\delta_3 y^2 + z^2)\frac{\partial}{\partial z} - zu\frac{\partial}{\partial u},$$

$$X_{16}^0 = u\frac{\partial}{\partial u}, \quad X_\infty^0 = \theta^0(t, x, y, z)\frac{\partial}{\partial u}$$

where $\theta^0(t, x, y, z)$ solves the equation $\theta_{tt}^0 = \delta_1\theta_{xx}^0 + \delta_2\theta_{yy}^0 + \delta_3\theta_{zz}^0$.

9.2.3.1.2. Perturbed equation: Exact symmetries

$$u_{tt} + \varepsilon u_t = \delta_1(f(u)u_x)_x + \delta_2(f(u)u_y)_y + \delta_3(f(u)u_z)_z$$

CLASSIFICATION (LIE POINT SYMMETRIES)
(Baikov, Gazizov, and Ibragimov [1990], [1991a])

Equivalence Transformations are taken from Section 9.2.3.1.1.

Classification Result
The principal Lie algebra $L_\mathcal{P}$ is seven-dimensional and is spanned by

$$Y_1 = \frac{\partial}{\partial t}, \quad Y_2 = \frac{\partial}{\partial x}, \quad Y_3 = \frac{\partial}{\partial y}, \quad Y_4 = \frac{\partial}{\partial z},$$

$$Y_5 = \delta_2 y\frac{\partial}{\partial x} - \delta_1 x\frac{\partial}{\partial y}, \quad Y_6 = \delta_3 z\frac{\partial}{\partial x} - \delta_1 x\frac{\partial}{\partial z},$$

$$Y_7 = \delta_3 z\frac{\partial}{\partial y} - \delta_2 y\frac{\partial}{\partial z}.$$

The algebra $L_\mathcal{P}$ extends in the following cases:
1. $f(u) = e^u$.

$$Y_8 = x\frac{\partial}{\partial x} + y\frac{\partial}{\partial y} + z\frac{\partial}{\partial z} + 2\frac{\partial}{\partial u}$$

2. $f(u) = u^\sigma$, $\sigma \neq -2, -\frac{4}{5}, 0$.

$$Y_8 = \sigma x\frac{\partial}{\partial x} + \sigma y\frac{\partial}{\partial y} + \sigma z\frac{\partial}{\partial z} + 2u\frac{\partial}{\partial u}$$

3. $f(u) = u^{-2}$.

$$Y_8 = x\frac{\partial}{\partial x} + y\frac{\partial}{\partial y} + z\frac{\partial}{\partial z} - u\frac{\partial}{\partial u}, \quad Y_9 = e^{-\varepsilon t}\frac{\partial}{\partial t} - \varepsilon e^{-\varepsilon t}u\frac{\partial}{\partial u}$$

4. $f(u) = u^{-4/5}$.

$$Y_8 = x\frac{\partial}{\partial x} + y\frac{\partial}{\partial y} + z\frac{\partial}{\partial z} - \frac{5}{2}u\frac{\partial}{\partial u},$$

$$Y_9 = \frac{1}{2}(x^2 - \delta_1\delta_2 y^2 - \delta_1\delta_3 z^2)\frac{\partial}{\partial x} + xy\frac{\partial}{\partial y} + xz\frac{\partial}{\partial z} - \frac{5}{2}xu\frac{\partial}{\partial u},$$

$$Y_{10} = xy\frac{\partial}{\partial x} + \frac{1}{2}(y^2 - \delta_1\delta_2 x^2 - \delta_2\delta_3 z^2)\frac{\partial}{\partial y} + yz\frac{\partial}{\partial z} - \frac{5}{2}yu\frac{\partial}{\partial u},$$

$$Y_{11} = xz\frac{\partial}{\partial x} + yz\frac{\partial}{\partial y} + (z^2 - \delta_1\delta_3 x^2 - \delta_2\delta_3 y^2)\frac{\partial}{\partial z} - \frac{5}{2}zu\frac{\partial}{\partial u}$$

5. $f(u) = 1$.

$$Y_8 = x\frac{\partial}{\partial t} + \delta_1 t\frac{\partial}{\partial x} - \frac{\varepsilon}{2}xu\frac{\partial}{\partial u}, \quad Y_9 = y\frac{\partial}{\partial t} + \delta_2 t\frac{\partial}{\partial y} - \frac{\varepsilon}{2}yu\frac{\partial}{\partial u},$$

$$Y_{10} = z\frac{\partial}{\partial t} + \delta_3 t\frac{\partial}{\partial z} - \frac{\varepsilon}{2}zu\frac{\partial}{\partial u}, \quad Y_{11} = u\frac{\partial}{\partial u}, \quad Y_\infty = \theta(t, x, y, z)\frac{\partial}{\partial u}$$

where $\theta(t, x, y, z)$ solves the equation $\theta_{tt} + \varepsilon\theta_t = \delta_1\theta_{xx} + \delta_2\theta_{yy} + \delta_3\theta_{zz}$.

9.2.3.1.3. Perturbed equation: Approximate symmetries

$$u_{tt} + \varepsilon u_t = \delta_1(f(u)u_x)_x + \delta_2(f(u)u_y)_y + \delta_3(f(u)u_z)_z$$

CLASSIFICATION (APPROXIMATE POINT SYMMETRIES)
(Baikov, Gazizov, and Ibragimov [1990], [1991a])

Equivalence Transformations are taken from Section 9.2.3.1.1.

Classification Result
For arbitrary $f(u)$, the approximate symmetry algebra is given by

$$X_1 = \frac{\partial}{\partial t}, \quad X_2 = \frac{\partial}{\partial x}, \quad X_3 = \frac{\partial}{\partial y}, \quad X_4 = \frac{\partial}{\partial z},$$

$$X_5 = \delta_2 y\frac{\partial}{\partial x} - \delta_1 x\frac{\partial}{\partial y}, \quad X_6 = \delta_3 z\frac{\partial}{\partial x} - \delta_1 x\frac{\partial}{\partial z}, \quad X_7 = \delta_3 z\frac{\partial}{\partial y} - \delta_2 y\frac{\partial}{\partial z},$$

$$X_8 = \varepsilon \left(t\frac{\partial}{\partial t} + x\frac{\partial}{\partial x} + y\frac{\partial}{\partial y} + z\frac{\partial}{\partial z} \right).$$

This algebra extends in the following cases:

1. $f(u) = e^u$.

$$\tilde{X}_8 = \left(t + \frac{\varepsilon}{2}t^2 \right)\frac{\partial}{\partial t} + x\frac{\partial}{\partial x} + y\frac{\partial}{\partial y} + z\frac{\partial}{\partial z} - 2\varepsilon t\frac{\partial}{\partial u},$$

$$X_9 = x\frac{\partial}{\partial x} + y\frac{\partial}{\partial y} + z\frac{\partial}{\partial z} + 2\frac{\partial}{\partial u}$$

2. $f(u) = u^\sigma$, $\sigma \neq -4, -\frac{4}{5}, 0$.

$$\tilde{X}_8 = \left(t + \frac{\varepsilon\sigma}{2\sigma+8}t^2 \right)\frac{\partial}{\partial t} + x\frac{\partial}{\partial x} + y\frac{\partial}{\partial y} + z\frac{\partial}{\partial z} - \frac{2\varepsilon}{\sigma+4}tu\frac{\partial}{\partial u},$$

$$X_9 = \sigma x\frac{\partial}{\partial x} + \sigma y\frac{\partial}{\partial y} + \sigma z\frac{\partial}{\partial z} + 2u\frac{\partial}{\partial u}$$

3. $f(u) = u^{-4}$.

$$X_9 = x\frac{\partial}{\partial x} + y\frac{\partial}{\partial y} + z\frac{\partial}{\partial z} - \frac{1}{2}u\frac{\partial}{\partial u},$$

$$X_{10} = \varepsilon \left(t^2\frac{\partial}{\partial t} + tu\frac{\partial}{\partial u} \right)$$

4. $f(u) = u^{-4/5}$.

$$\tilde{X}_8 = \left(t - \frac{\varepsilon}{8}t^2 \right)\frac{\partial}{\partial t} + x\frac{\partial}{\partial x} + y\frac{\partial}{\partial y} + z\frac{\partial}{\partial z} - \frac{5}{8}\varepsilon tu\frac{\partial}{\partial u},$$

$$X_9 = x\frac{\partial}{\partial x} + y\frac{\partial}{\partial y} + z\frac{\partial}{\partial z} - \frac{5}{2}u\frac{\partial}{\partial u},$$

$$X_{10} = \frac{1}{2}(x^2 - \delta_1\delta_2 y^2 - \delta_1\delta_3 z^2)\frac{\partial}{\partial x} + xy\frac{\partial}{\partial y} + xz\frac{\partial}{\partial z} - \frac{5}{2}xu\frac{\partial}{\partial u},$$

$$X_{11} = xy\frac{\partial}{\partial x} + \frac{1}{2}(y^2 - \delta_1\delta_2 x^2 - \delta_2\delta_3 z^2)\frac{\partial}{\partial y} + yz\frac{\partial}{\partial z} - \frac{5}{2}yu\frac{\partial}{\partial u},$$

$$X_{12} = xz\frac{\partial}{\partial x} + yz\frac{\partial}{\partial y} + (z^2 - \delta_1\delta_3 x^2 - \delta_2\delta_3 y^2)\frac{\partial}{\partial z} - \frac{5}{2}zu\frac{\partial}{\partial u}$$

5. $f(u) = 1$.

$$\tilde{X}_8 = t\frac{\partial}{\partial t} + x\frac{\partial}{\partial x} + y\frac{\partial}{\partial y} + z\frac{\partial}{\partial z} - \frac{\varepsilon}{2}tu\frac{\partial}{\partial u},$$

$$X_9 = x\frac{\partial}{\partial t} + \delta_1 t\frac{\partial}{\partial x} - \frac{\varepsilon}{2}xu\frac{\partial}{\partial u},$$

$$X_{10} = y\frac{\partial}{\partial t} + \delta_2 t\frac{\partial}{\partial y} - \frac{\varepsilon}{2}yu\frac{\partial}{\partial u},$$

$$X_{11} = z\frac{\partial}{\partial t} + \delta_3 t\frac{\partial}{\partial z} - \frac{\varepsilon}{2}zu\frac{\partial}{\partial u},$$

$$X_{12} = \frac{1}{2}(t^2 + \delta_1 x^2 + \delta_2 y^2 + \delta_3 z^2)\frac{\partial}{\partial t} + tx\frac{\partial}{\partial x} + ty\frac{\partial}{\partial y} + tz\frac{\partial}{\partial z}$$

$$- \left(t + \frac{\varepsilon}{4}(t^2 + \delta_1 x^2 + \delta_2 y^2 + \delta_3 z^2)\right)u\frac{\partial}{\partial u},$$

$$X_{13} = tx\frac{\partial}{\partial t} + \frac{1}{2}(\delta_1 t^2 + x^2 - \delta_1\delta_2 y^2 - \delta_1\delta_3 z^2)\frac{\partial}{\partial x} + xy\frac{\partial}{\partial y} + xz\frac{\partial}{\partial z} - \left(1 + \frac{\varepsilon}{2}t\right)xu\frac{\partial}{\partial u},$$

$$X_{14} = ty\frac{\partial}{\partial t} + xy\frac{\partial}{\partial x} + \frac{1}{2}(\delta_2 t^2 - \delta_1\delta_2 x^2 + y^2 - \delta_2\delta_3 z^2)\frac{\partial}{\partial y} + yz\frac{\partial}{\partial z} - \left(1 + \frac{\varepsilon}{2}t\right)yu\frac{\partial}{\partial u},$$

$$X_{15} = tz\frac{\partial}{\partial t} + xz\frac{\partial}{\partial x} + yz\frac{\partial}{\partial y} + \frac{1}{2}(\delta_3 t^2 - \delta_1\delta_3 x^2 - \delta_2\delta_3 y^2 + z^2)\frac{\partial}{\partial z} - \left(1 + \frac{\varepsilon}{2}t\right)zu\frac{\partial}{\partial u},$$

$$X_{16} = u\frac{\partial}{\partial u}, \quad X_\infty = \theta(t, x, y, z, \varepsilon)\frac{\partial}{\partial u},$$

where $\theta(t, x, y, z, \varepsilon)$ is an arbitrary approximate solution of the equation $\theta_{tt} + \varepsilon\theta_t \approx \delta_1\theta_{xx} + \delta_2\theta_{yy} + \delta_3\theta_{zz}$.

9.2.3.2. Semi-isotropic case

$$u_{tt} + \varepsilon u_t = \delta_1(f(u)u_x)_x + \delta_2(f(u)u_y)_y + (h(u)u_z)_z, \quad \delta_1, \delta_2 = \pm 1,$$

where f and h are linearly independent functions.

9.2.3.2.1. Unperturbed equation

$$u_{tt} = \delta_1(f(u)u_x)_x + \delta_2(f(u)u_y)_y + (h(u)u_z)_z$$

CLASSIFICATION (LIE POINT SYMMETRIES)
(Baikov, Gazizov, and Ibragimov [1990], [1991a])

Equivalence Transformations are taken from Section 9.2.3.1.1.

Classification Result
The principal Lie algebra L_P is six-dimensional and is spanned by

$$X_1^0 = \frac{\partial}{\partial t}, \quad X_2^0 = \frac{\partial}{\partial x}, \quad X_3^0 = \frac{\partial}{\partial y}, \quad X_4^0 = \frac{\partial}{\partial z},$$

$$X_5^0 = \delta_2 y \frac{\partial}{\partial x} - \delta_1 x \frac{\partial}{\partial y}, \quad X_6^0 = t\frac{\partial}{\partial t} + x\frac{\partial}{\partial x} + y\frac{\partial}{\partial y} + z\frac{\partial}{\partial z}.$$

The algebra L_P extends in the following cases:

I. $f(u) = e^u$ and $h(u) = \delta_3 e^{ku}, \quad \delta_3 = \pm 1, \quad k \neq 0, 1.$

$$X_7^0 = x\frac{\partial}{\partial x} + y\frac{\partial}{\partial y} + kz\frac{\partial}{\partial z} + 2\frac{\partial}{\partial u}$$

II. $f(u) = u^\nu, \quad h(u) = \delta_3 u^\gamma, \quad \delta_3 = \pm 1, \quad \nu \neq 0, \quad \gamma \neq 0, \quad$ and $\nu \neq \gamma.$

$$X_7^0 = \nu x\frac{\partial}{\partial x} + \nu y\frac{\partial}{\partial y} + \gamma z\frac{\partial}{\partial z} + 2u\frac{\partial}{\partial u}$$

III. $h(u) = \delta_3, \quad \delta_3 = \pm 1,$ and $f(u)$ is an arbitrary function.

$$X_7^0 = \delta_3 z\frac{\partial}{\partial t} + t\frac{\partial}{\partial z}$$

For each of the following functions $f(u)$ there is an additional extension:

III.1. $f(u) = u^\nu, \quad \nu \neq -1, 0.$

$$X_8^0 = x\frac{\partial}{\partial x} + y\frac{\partial}{\partial y} + \frac{2}{\nu}u\frac{\partial}{\partial u}$$

III.2. $f(u) = u^{-1}.$

$$X_\infty^0 = \alpha^1(x, y)\frac{\partial}{\partial x} + \alpha^2(x, y)\frac{\partial}{\partial y} - 2\alpha_x^1(x, y)u\frac{\partial}{\partial u}$$

where $\alpha^1(x, y)$ and $\alpha^2(x, y)$ solve the system

$$\alpha_x^1 = \alpha_y^2, \quad \delta_1 \alpha_x^2 = \delta_2 \alpha_y^1.$$

III.3. $f(u) = e^u$.

$$X_8^0 = x\frac{\partial}{\partial x} + y\frac{\partial}{\partial y} + 2\frac{\partial}{\partial u},$$

$$X_\infty^0 = \alpha^1(t, z)\frac{\partial}{\partial t} + \alpha^2(t, z)\frac{\partial}{\partial z} - 2\alpha_t^1(t, z)\frac{\partial}{\partial u},$$

where $\alpha^1(t, z)$ and $\alpha^2(t, z)$ solve the system

$$\alpha_t^1 = \alpha_z^2, \quad \alpha_t^2 = \delta_3\alpha_z^1.$$

IV. $f(u) = 1$ and $h(u)$ is an arbitrary function.

$$X_7^0 = \delta_1 x\frac{\partial}{\partial t} + t\frac{\partial}{\partial x}, \quad X_8^0 = \delta_2 y\frac{\partial}{\partial t} + t\frac{\partial}{\partial y}$$

For each of the following functions $h(u)$ there is an additional extension:

IV.1. $h(u) = \delta_3 u^\gamma$, $\delta_3 = \pm1$, $\gamma \neq 4$, $-\dfrac{4}{3}$, 0.

$$X_9^0 = z\frac{\partial}{\partial z} + \frac{2}{\gamma}u\frac{\partial}{\partial u}$$

IV.2. $h(u) = \delta_3 u^4$, $\delta_3 = \pm1$.

$$X_9^0 = z\frac{\partial}{\partial z} + \frac{u}{2}\frac{\partial}{\partial u},$$

$$X_{10}^0 = (t^2 + \delta_1 x^2 + \delta_2 y^2)\frac{\partial}{\partial t} + 2tx\frac{\partial}{\partial x} + 2ty\frac{\partial}{\partial y} - tu\frac{\partial}{\partial u},$$

$$X_{11}^0 = tx\frac{\partial}{\partial t} + \frac{1}{2}(\delta_1 t^2 + x^2 - \delta_1\delta_2 y^2)\frac{\partial}{\partial x} + xy\frac{\partial}{\partial y} - \frac{1}{2}xu\frac{\partial}{\partial u},$$

$$X_{12}^0 = ty\frac{\partial}{\partial t} + xy\frac{\partial}{\partial x} + \frac{1}{2}(\delta_2 t^2 - \delta_1\delta_2 x^2 + y^2)\frac{\partial}{\partial y} - \frac{1}{2}yu\frac{\partial}{\partial u}$$

IV.3. $h(u) = \delta_3 u^{-4/3}$, $\delta_3 = \pm1$.

$$X_9^0 = z\frac{\partial}{\partial z} - \frac{3}{2}u\frac{\partial}{\partial u}, \quad X_{10}^0 = z^2\frac{\partial}{\partial z} - 3zu\frac{\partial}{\partial u}$$

IV.4. $h(u) = \delta_3 e^u$, $\delta_3 = \pm1$.

$$X_9^0 = z\frac{\partial}{\partial z} + 2\frac{\partial}{\partial u}$$

9.2.3.2.2. Perturbed equation: Exact symmetries

$$u_{tt} + \varepsilon u_t = \delta_1(f(u)u_x)_x + \delta_2(f(u)u_y)_y + (h(u)u_z)_z$$

CLASSIFICATION (LIE POINT SYMMETRIES)
(Baikov, Gazizov, and Ibragimov [1990], [1991a])

Equivalence Transformations are taken from Section 9.2.3.1.1.

Classification Result
The principal Lie algebra L_P is five-dimensional and is spanned by

$$Y_1 = \frac{\partial}{\partial t}, \quad Y_2 = \frac{\partial}{\partial x}, \quad Y_3 = \frac{\partial}{\partial y}, \quad Y_4 = \frac{\partial}{\partial z},$$

$$Y_5 = \delta_2 y \frac{\partial}{\partial x} - \delta_1 x \frac{\partial}{\partial y}.$$

The algebra L_P extends in the following cases:
I. $f(u) = e^u$ and $h(u) = \delta_3 e^{ku}$, $\delta_3 = \pm 1$, $k \neq 0, 1$.

$$Y_7 = x\frac{\partial}{\partial x} + y\frac{\partial}{\partial y} + kz\frac{\partial}{\partial z} + 2\frac{\partial}{\partial u}$$

II. $f(u) = u^\nu$ and $h(u) = \delta_3 u^\gamma$, $\delta_3 = \pm 1$, $\nu \neq 0$, $\gamma \neq 0$, and $\nu \neq \gamma$.

$$Y_6 = \nu x\frac{\partial}{\partial x} + \nu y\frac{\partial}{\partial y} + \gamma z\frac{\partial}{\partial z} + 2u\frac{\partial}{\partial u}$$

III. $h(u) = \delta_3$, $\delta_3 = \pm 1$.
For each of the following functions $f(u)$ there is an extension:
   III.1. $f(u) = u^\nu$, $\nu \neq -1, 0$.

$$Y_6 = x\frac{\partial}{\partial x} + y\frac{\partial}{\partial y} + \frac{2}{\nu}u\frac{\partial}{\partial u}$$

III.2. $f(u) = u^{-1}$.

$$Y_\infty = \alpha^1(x, y)\frac{\partial}{\partial x} + \alpha^2(x, y)\frac{\partial}{\partial y} - 2\alpha_x^1(x, y)u\frac{\partial}{\partial u},$$

where $\alpha^1(x, y)$ and $\alpha^2(x, y)$ solve the system

$$\alpha_x^1 = \alpha_y^2, \quad \delta_1\alpha_x^2 = \delta_2\alpha_y^1.$$

III.3. $f(u) = e^u$.

$$Y_6 = x\frac{\partial}{\partial x} + y\frac{\partial}{\partial y} + 2\frac{\partial}{\partial u}$$

IV. $f(u) = 1$.

For each of the following functions $h(u)$ there is an extension:

IV.1 $h(u) = \delta_3 u^\gamma$, $\delta_3 = \pm 1$, $\gamma \neq -\frac{4}{3}$, 0.

$$Y_6 = z\frac{\partial}{\partial z} + \frac{2}{\gamma}u\frac{\partial}{\partial u}$$

IV.2. $h(u) = \delta_3 u^{-4/3}$, $\delta_3 = \pm 1$.

$$Y_6 = z\frac{\partial}{\partial z} - \frac{3}{2}u\frac{\partial}{\partial u}, \quad Y_7 = z^2\frac{\partial}{\partial z} - 3zu\frac{\partial}{\partial u}$$

IV.3. $h(u) = \delta_3 e^u$, $\delta_3 = \pm 1$.

$$Y_6 = z\frac{\partial}{\partial z} + 2\frac{\partial}{\partial u}$$

9.2.3.2.3. Perturbed equation: Approximate symmetries

$$u_{tt} + \varepsilon u_t = \delta_1(f(u)u_x)_x + \delta_2(f(u)u_y)_y + (h(u)u_z)_z$$

CLASSIFICATION (APPROXIMATE POINT SYMMETRIES)
(Baikov, Gazizov, and Ibragimov [1990], [1991a])

Equivalence Transformations are taken from Section 9.2.3.1.1.

Classification Result
For arbitrary $f(u)$ and $h(u)$, the approximate symmetry algebra is given by

$$X_1 = \frac{\partial}{\partial t}, \qquad X_2 = \frac{\partial}{\partial x}, \qquad X_3 = \frac{\partial}{\partial y}, \qquad X_4 = \frac{\partial}{\partial z},$$

$$X_5 = \delta_2 y\frac{\partial}{\partial x} - \delta_1 x\frac{\partial}{\partial y}, \quad X_6 = \varepsilon\left(t\frac{\partial}{\partial t} + x\frac{\partial}{\partial x} + y\frac{\partial}{\partial y} + z\frac{\partial}{\partial z}\right).$$

This algebra extends in the following cases:

I. $f(u) = e^u$ $h(u) = \delta_3 e^{ku}$, $\delta_3 = \pm 1$, $k \neq 0$, 1.

$$X_7 = x\frac{\partial}{\partial x} + y\frac{\partial}{\partial y} + kz\frac{\partial}{\partial z} + 2\frac{\partial}{\partial u}$$

II. $f(u) = u^\nu$ $h(u) = \delta_3 u^\gamma$, $\delta_3 = \pm 1$, $\nu \neq 0$, $\gamma \neq 0$ and $\nu \neq \gamma$.

$$X_7 = \nu x \frac{\partial}{\partial x} + \nu y \frac{\partial}{\partial y} + \gamma z \frac{\partial}{\partial z} + 2u \frac{\partial}{\partial u}$$

III. $h(u) = \delta_3$, $\delta_3 = \pm 1$, and $f(u)$ is an arbitrary function.

$$X_7 = \varepsilon \left(\delta_3 z \frac{\partial}{\partial t} + t \frac{\partial}{\partial z} \right)$$

For each of the following functions $f(u)$ there is an additional extension:
III.1. $f(u) = u^\nu$, $\nu \neq -1$, 0.

$$\tilde{X}_6 = \left(t + \frac{\varepsilon \nu}{8} \left(t^2 + \delta_3 z^2 \right) \right) \frac{\partial}{\partial t} + x \frac{\partial}{\partial x} + y \frac{\partial}{\partial y} + \left(z + \frac{\varepsilon \nu}{4} tz \right) \frac{\partial}{\partial z} - \frac{\varepsilon}{2} tu \frac{\partial}{\partial u},$$

$$\tilde{X}_7 = \delta_3 z \left(1 + \frac{\varepsilon}{4} \nu t \right) \frac{\partial}{\partial t} + \left(t + \frac{\varepsilon \nu}{8} \left(t^2 + \delta_3 z^2 \right) \right) \frac{\partial}{\partial z} - \frac{\varepsilon}{2} \delta_3 z u \frac{\partial}{\partial u},$$

$$X_8 = x \frac{\partial}{\partial x} + y \frac{\partial}{\partial y} + \frac{2}{\nu} u \frac{\partial}{\partial u}$$

III.2. $f(u) = u^{-1}$.

$$\tilde{X}_6 = \left(t - \frac{\varepsilon}{8} \left(t^2 + \delta_3 z^2 \right) \right) \frac{\partial}{\partial t} + x \frac{\partial}{\partial x} + y \frac{\partial}{\partial y} + \left(z - \frac{\varepsilon}{4} tz \right) \frac{\partial}{\partial z} - \frac{\varepsilon}{2} tu \frac{\partial}{\partial u},$$

$$\tilde{X}_7 = \delta_3 z \left(1 - \frac{\varepsilon}{4} t \right) \frac{\partial}{\partial t} + \left(t - \frac{\varepsilon}{8} (t^2 + \delta_3 z^2) \right) \frac{\partial}{\partial z} - \frac{\varepsilon}{2} \delta_3 z u \frac{\partial}{\partial u},$$

$$X_\infty = \alpha^1(x, y) \frac{\partial}{\partial x} + \alpha^2(x, y) \frac{\partial}{\partial y} - 2\alpha_x^1(x, y) u \frac{\partial}{\partial u}$$

where $\alpha^1(x, y)$ and $\alpha^2(x, y)$ solve the system

$$\alpha_x^1 = \alpha_y^2, \quad \delta_1 \alpha_x^2 = \delta_2 \alpha_y^1.$$

III.3. $f(u) = e^u$.

$$X_8 = x \frac{\partial}{\partial x} + y \frac{\partial}{\partial y} + 2 \frac{\partial}{\partial u},$$

$$X_\infty = \varepsilon \left(\alpha^1(t, z) \frac{\partial}{\partial t} + \alpha^2(t, z) \frac{\partial}{\partial z} - 2\alpha_t^1(t, z) \frac{\partial}{\partial u} \right),$$

where $\alpha^1(t, z)$ and $\alpha^2(t, z)$ solve the system

$$\alpha_t^1 = \alpha_z^2, \quad \alpha_t^2 = \delta_3 \alpha_z^1.$$

IV. $f(u) = 1$ and $h(u)$ is an arbitrary function.

$$X_7 = \varepsilon\left(\delta_1 x \frac{\partial}{\partial t} + t \frac{\partial}{\partial x}\right), \quad X_8 = \varepsilon\left(\delta_2 y \frac{\partial}{\partial t} + t \frac{\partial}{\partial y}\right)$$

For each of the following functions $h(u)$ there is an additional extension:

IV.1. $h(u) = \delta_3 u^\gamma$, $\delta_3 = \pm 1$, $\gamma \neq 4, -\frac{4}{3}, 0$.

$$\tilde{X}_6 = \left(t + \frac{\varepsilon\gamma}{2(4-\gamma)}(t^2 + \delta_1 x^2 + \delta_2 y^2)\right)\frac{\partial}{\partial t} + x\left(1 + \frac{\varepsilon\gamma}{4-\gamma}t\right)\frac{\partial}{\partial x}$$

$$+ y\left(1 + \frac{\varepsilon\gamma}{4-\gamma}t\right)\frac{\partial}{\partial y} + z\frac{\partial}{\partial z} - \frac{2\varepsilon}{4-\gamma}tu\frac{\partial}{\partial u},$$

$$\tilde{X}_7 = \delta_1 x\left(1 + \frac{\varepsilon\gamma}{4-\gamma}t\right)\frac{\partial}{\partial t} + \left(t + \frac{\varepsilon\gamma}{2(4-\gamma)}(t^2 + \delta_1 x^2 - \delta_2 y^2)\right)\frac{\partial}{\partial x}$$

$$+ \frac{\varepsilon\gamma}{4-\gamma}\delta_1 xy\frac{\partial}{\partial y} - \frac{2\varepsilon\delta_1}{4-\gamma}xu\frac{\partial}{\partial u},$$

$$\tilde{X}_8 = \delta_2 y\left(1 + \frac{\varepsilon\gamma}{4-\gamma}t\right)\frac{\partial}{\partial t} + \frac{\varepsilon\gamma}{4-\gamma}\delta_2 xy\frac{\partial}{\partial x}$$

$$+ \left(t + \frac{\varepsilon\gamma}{2(4-\gamma)}(t^2 - \delta_1 x^2 + \delta_2 y^2)\right)\frac{\partial}{\partial y} - \frac{2\varepsilon\delta_2}{4-\gamma}yu\frac{\partial}{\partial u},$$

$$X_9 = z\frac{\partial}{\partial z} + \frac{2}{\gamma}u\frac{\partial}{\partial u}$$

IV.2. $h(u) = \delta_3 u^4$, $\delta_3 = \pm 1$.

$$X_9 = z\frac{\partial}{\partial z} + \frac{u}{2}\frac{\partial}{\partial u},$$

$$X_{10} = \varepsilon\left[(t^2 + \delta_1 x^2 + \delta_2 y^2)\frac{\partial}{\partial t} + 2tx\frac{\partial}{\partial x} + 2ty\frac{\partial}{\partial y} - tu\frac{\partial}{\partial u}\right],$$

$$X_{11} = \varepsilon\left[tx\frac{\partial}{\partial t} + \frac{1}{2}(\delta_1 t^2 + x^2 - \delta_1\delta_2 y^2)\frac{\partial}{\partial x} + xy\frac{\partial}{\partial y} - \frac{1}{2}xu\frac{\partial}{\partial u}\right],$$

$$X_{12} = \varepsilon\left[ty\frac{\partial}{\partial t} + xy\frac{\partial}{\partial x} + \frac{1}{2}(\delta_2 t^2 - \delta_1\delta_2 x^2 + y^2)\frac{\partial}{\partial y} - \frac{1}{2}yu\frac{\partial}{\partial u}\right]$$

IV.3. $h(u) = \delta_3 u^{-4/3}$, $\delta_3 = \pm 1$.

$$\tilde{X}_6 = \left(t - \frac{\varepsilon}{8}(t^2 + \delta_1 x^2 + \delta_2 y^2)\right)\frac{\partial}{\partial t} + \left(x - \frac{\varepsilon}{4}tx\right)\frac{\partial}{\partial x} + \left(y - \frac{\varepsilon}{4}ty\right)\frac{\partial}{\partial y}$$

$$+ z\frac{\partial}{\partial z} - \frac{3}{8}\varepsilon tu\frac{\partial}{\partial u},$$

$$\tilde{X}_7 = \delta_1 x\left(1 - \frac{\varepsilon}{4}t\right)\frac{\partial}{\partial t} + \left(t - \frac{\varepsilon}{8}(t^2 + \delta_1 x^2 - \delta_2 y^2)\right)\frac{\partial}{\partial x} - \frac{\varepsilon}{4}\delta_1 xy\frac{\partial}{\partial y} - \frac{3}{8}\varepsilon\delta_1 xu\frac{\partial}{\partial u},$$

$$\tilde{X}_8 = \delta_2 y\left(1 - \frac{\varepsilon}{4}t\right)\frac{\partial}{\partial t} - \frac{\varepsilon}{4}\delta_2 xy\frac{\partial}{\partial x} + \left(t - \frac{\varepsilon}{8}(t^2 - \delta_1 x^2 + \delta_2 y^2)\right)\frac{\partial}{\partial y} - \frac{3}{8}\varepsilon\delta_2 yu\frac{\partial}{\partial u},$$

$$X_9 = z\frac{\partial}{\partial z} - \frac{3}{2}u\frac{\partial}{\partial u}, \qquad X_{10} = z^2\frac{\partial}{\partial z} - 3zu\frac{\partial}{\partial u}$$

IV.4. $h(u) = \delta_3 e^u$, $\delta_3 = \pm 1$.

$$\tilde{X}_6 = \left(t - \frac{\varepsilon}{2}(t^2 + \delta_1 x^2 + \delta_2 y^2)\right)\frac{\partial}{\partial t} + (x - \varepsilon tx)\frac{\partial}{\partial x} + (y - \varepsilon ty)\frac{\partial}{\partial y} + 2\varepsilon t\frac{\partial}{\partial u},$$

$$\tilde{X}_7 = \delta_1 x(1 - \varepsilon t)\frac{\partial}{\partial t} + \left(t - \frac{\varepsilon}{2}(t^2 + \delta_1 x^2 - \delta_2 y^2)\right)\frac{\partial}{\partial x} - \varepsilon\delta_1 xy\frac{\partial}{\partial y} + 2\varepsilon\delta_1 x\frac{\partial}{\partial u},$$

$$\tilde{X}_8 = \delta_2 y(1 - \varepsilon t)\frac{\partial}{\partial t} - \varepsilon\delta_2 xy\frac{\partial}{\partial x} + \left(t - \frac{\varepsilon}{2}(t^2 - \delta_1 x^2 + \delta_2 y^2)\right)\frac{\partial}{\partial y} + 2\varepsilon\delta_2 y\frac{\partial}{\partial u},$$

$$X_9 = z\frac{\partial}{\partial z} + 2\frac{\partial}{\partial u}$$

AN APPROXIMATE CONSERVATION LAW
(Baikov, Gazizov, and Ibragimov [1990], [1991a])

The equation

$$u_{tt} + \varepsilon u_t = \delta_1 u_{xx} + \delta_2 u_{yy} + \delta_3(u^\gamma u_z)_z \tag{9.1}$$

has the approximate symmetry

$$X = \left(x + \frac{\varepsilon\gamma}{4-\gamma}tx\right)\frac{\partial}{\partial t} + \left[\delta_1 t + \frac{\varepsilon\gamma}{2(4-\gamma)}(\delta_1 t^2 + x^2 - \delta_1\delta_2 y^2)\right]\frac{\partial}{\partial x}$$

$$+ \frac{\varepsilon\gamma}{4-\gamma}xy\frac{\partial}{\partial y} - \frac{2\varepsilon xu}{4-\gamma}\frac{\partial}{\partial u}. \tag{9.2}$$

By setting

$$u = v_z$$

and integrating with respect to z, one transforms Equation 9.1 to the form

$$v_{tt} + \varepsilon v_t = \delta_1 v_{xx} + \delta_2 v_{yy} + \delta_3 v_z^{\gamma} v_{zz}. \tag{9.3}$$

Then Operator 9.2 becomes

$$\overline{X} = \left(x + \frac{\varepsilon\gamma}{4-\gamma}tx\right)\frac{\partial}{\partial t} + \left[\delta_1 t + \frac{\varepsilon\gamma}{2(4-\gamma)}(\delta_1 t^2 + x^2 - \delta_1\delta_2 y^2)\right]\frac{\partial}{\partial x}$$

$$+ \frac{\varepsilon\gamma}{4-\gamma}xy\frac{\partial}{\partial y} - \frac{2\varepsilon}{4-\gamma}xv\frac{\partial}{\partial v}, \tag{9.4}$$

and Equation 9.3 has the Lagrangian

$$L = \frac{1}{2}e^{\varepsilon t}\left[v_t^2 - \delta_1 v_x^2 - \delta_2 v_y^2 - \frac{2\delta_3}{(\gamma+1)(\gamma+2)}v_z^{\gamma+2}\right].$$

Operator 9.4 is an approximate Noether symmetry for Equation 9.3. Hence, Theorem 2.8 (Section 2.6.1) applies and yields the approximate conservation law for Equation 9.3:

$$D_t(C^t) + D_x(C^x) + D_y(C^y) + D_z(C^z) = o(\varepsilon)$$

where

$$C^t = -\frac{1}{2}e^{\varepsilon t}(v_t^2 + \delta_1 v_x^2 + \delta_2 v_y^2 + \frac{2\delta_3}{(\gamma+1)(\gamma+2)}v_z^{\gamma+2})(x + \frac{\varepsilon\gamma}{4-\gamma}tx)$$

$$- e^{\varepsilon t}v_t[\delta_1 tv_x + \frac{2\varepsilon xv}{4-\gamma} + \frac{\varepsilon\gamma}{2(4-\gamma)}(\delta_1 t^2 + x^2 - \delta_1\delta_2 y^2)v_x + \frac{\varepsilon\gamma}{4-\gamma}xyv_y],$$

$$C^x = \frac{1}{2}e^{\varepsilon t}[v_t^2 + \delta_1 v_x^2 - \delta_2 v_y^2 - \frac{2\delta_3}{(\gamma+1)(\gamma+2)}v_z^{\gamma+2}][\delta_1 t + \frac{\varepsilon\gamma}{2(4-\gamma)}(\delta_1 t^2 + x^2$$

$$- \delta_1\delta_2 y^2)] + \delta_1 e^{\varepsilon t} v_x(x v_t + \frac{2\varepsilon x v}{4-\gamma} + \frac{\varepsilon\gamma tx}{4-\gamma}v_t + \frac{\varepsilon\gamma xy}{4-\gamma}v_y) - \varepsilon\delta_1\frac{v^2}{4-\gamma},$$

$$C^y = \frac{\varepsilon\gamma xy}{2(4-\gamma)}xy(v_t^2 - \delta_1 v_x^2 + \delta_2 v_y^2 - \frac{2\delta_3}{(\gamma+1)(\gamma+2)}v_z^{\gamma+2}) + \delta_2 e^{\varepsilon t} v_y[x v_t$$

$$+ \frac{2\varepsilon x v}{4-\gamma} + \frac{\varepsilon\gamma tx}{4-\gamma}v_t + \delta_1 t v_x + \frac{\varepsilon\gamma}{2(4-\gamma)}(\delta_1 t^2 + x^2 - \delta_1\delta_2 y^2)v_x],$$

$$C^z = e^{\varepsilon t}\frac{\delta_3}{\gamma+1}v_z^{\gamma+1}[x v_t + \frac{2\varepsilon x v}{4-\gamma} + \frac{\varepsilon\gamma tx}{4-\gamma}v_t + \delta_1 t v_x + \frac{\varepsilon\gamma}{2(4-\gamma)}(\delta_1 t^2 + x^2$$

$$- \delta_1\delta_2 y^2)v_x + \frac{\varepsilon\gamma xy}{4-\gamma}v_y].$$

9.2.3.3.   Anysotropic case

$$u_{tt} + \varepsilon u_t = (f(u)u_x)_x + (g(u)u_y)_y + (h(u)u_z)_z$$

where f, g, h are linearly independent functions.

9.2.3.3.1. Unperturbed equation

$$u_{tt} = (f(u)u_x)_x + (g(u)u_y)_y + (h(u)u_z)_z$$

CLASSIFICATION (LIE POINT SYMMETRIES)
(Baikov, Gazizov, and Ibragimov [1990], [1991a])

Equivalence Transformations are taken from Section 9.2.3.1.1.

Classification Result
The principal Lie algebra L_P is five-dimensional and is spanned by

$$X_1^0 = \frac{\partial}{\partial t}, \quad X_2^0 = \frac{\partial}{\partial x}, \quad X_3^0 = \frac{\partial}{\partial y}, \quad X_4^0 = \frac{\partial}{\partial z},$$

$$X_5^0 = t\frac{\partial}{\partial t} + x\frac{\partial}{\partial x} + y\frac{\partial}{\partial y} + z\frac{\partial}{\partial z}.$$

The algebra $L_\mathcal{P}$ extends in the following cases:

I. $f(u) = \delta_1$, $\delta_1 = \pm 1$, $g(u)$ and $h(u)$ are arbitrary functions.

$$X_6^0 = \delta_1 x \frac{\partial}{\partial t} + t \frac{\partial}{\partial x}$$

For each of the following functions $g(u)$ and $h(u)$ there is an additional extension:

I.1. $g(u) = \delta_2 e^u$, $h(u) = \delta_3 e^{ku}$, $\delta_2, \delta_3 = \pm 1$, $k \neq 0, 1$.

$$X_7^0 = y \frac{\partial}{\partial y} + kz \frac{\partial}{\partial z} + 2 \frac{\partial}{\partial u}$$

I.2. $g(u) = \delta_2 u^\nu$, $h(u) = \delta_3 u^\gamma$, $\delta_2, \delta_3 = \pm 1$, $\nu \neq 0$, $\gamma \neq 0$, and $\nu \neq \gamma$.

$$X_7^0 = \nu y \frac{\partial}{\partial y} + \gamma z \frac{\partial}{\partial z} + 2u \frac{\partial}{\partial u}$$

II. $f(u) = \delta_1 e^u$, $g(u) = \delta_2 e^{k_1 u}$, $h(u) = \delta_3 e^{k_2 u}$, $\delta_1, \delta_2, \delta_3 = \pm 1$, $k_1 \neq 0, 1$, $k_2 \neq 0, 1$, and $k_1 \neq k_2$.

$$X_6^0 = x \frac{\partial}{\partial x} + k_1 y \frac{\partial}{\partial y} + k_2 z \frac{\partial}{\partial z} + 2 \frac{\partial}{\partial u}$$

III. $f(u) = \delta_1 u^\sigma$, $g(u) = \delta_2 u^\nu$, $h(u) = \delta_3 u^\gamma$, $\delta_1, \delta_2, \delta_3 = \pm 1$, $\sigma \neq 0$, $\nu \neq 0$, $\gamma \neq 0$, and $\sigma \neq \nu \neq \gamma \neq \sigma$.

$$X_6^0 = \sigma x \frac{\partial}{\partial x} + \nu y \frac{\partial}{\partial y} + \gamma z \frac{\partial}{\partial z} + 2u \frac{\partial}{\partial u}$$

9.2.3.3.2. Perturbed equation: Exact symmetries

$$u_{tt} + \varepsilon u_t = (f(u)u_x)_x + (g(u)u_y)_y + (h(u)u_z)_z$$

CLASSIFICATION (LIE POINT SYMMETRIES)
(Baikov, Gazizov, and Ibragimov [1990], [1991a])

Equivalence Transformations are taken from Section 9.2.3.1.1.

Classification Result
The principal Lie algebra $L_\mathcal{P}$ is four-dimensional and is spanned by

$$Y_1^0 = \frac{\partial}{\partial t}, \quad Y_2^0 = \frac{\partial}{\partial x}, \quad Y_3^0 = \frac{\partial}{\partial y}, \quad Y_4 = \frac{\partial}{\partial z}.$$

The algebra L_P extends in the following cases:

I. $f(u) = \delta_1$, $\delta_1 = \pm 1$.

For each of the following functions $g(u)$ and $h(u)$ there is an extension:

I.1. $g(u) = \delta_2 e^u$, $h(u) = \delta_3 e^{ku}$, $\delta_2, \delta_3 = \pm 1$, $k \neq 0, 1$.

$$Y_5 = y\frac{\partial}{\partial y} + kz\frac{\partial}{\partial z} + 2\frac{\partial}{\partial u}$$

I.2. $g(u) = \delta_2 u^\nu$, $h(u) = \delta_3 u^\gamma$, $\delta_2, \delta_3 = \pm 1$, $\nu \neq 0$, $\gamma \neq 0$, and $\nu \neq \gamma$.

$$Y_5 = \nu y\frac{\partial}{\partial y} + \gamma z\frac{\partial}{\partial z} + 2u\frac{\partial}{\partial u}$$

II. $f(u) = \delta_1 e^u$, $g(u) = \delta_2 e^{k_1 u}$, $h(u) = \delta_3 e^{k_2 u}$, $\delta_1, \delta_2, \delta_3 = \pm 1$, $k_1 \neq 0, 1$, $k_2 \neq 0, 1$, and $k_1 \neq k_2$.

$$Y_5 = x\frac{\partial}{\partial x} + k_1 y\frac{\partial}{\partial y} + k_2 z\frac{\partial}{\partial z} + 2\frac{\partial}{\partial u}$$

III. $f(u) = \delta_1 u^\sigma$, $g(u) = \delta_2 u^\nu$, $h(u) = \delta_3 u^\gamma$, $\delta_1, \delta_2, \delta_3 = \pm 1$, $\sigma \neq 0$, $\nu \neq 0$, $\gamma \neq 0$, and $\sigma \neq \nu \neq \gamma \neq \sigma$.

$$Y_5 = \sigma x\frac{\partial}{\partial x} + \nu y\frac{\partial}{\partial y} + \gamma z\frac{\partial}{\partial z} + 2u\frac{\partial}{\partial u}.$$

9.2.3.3.3. Perturbed equation: Approximate symmetries

$$u_{tt} + \varepsilon u_t = (f(u)u_x)_x + (g(u)u_y)_y + (h(u)u_z)_z$$

CLASSIFICATION (APPROXIMATE POINT SYMMETRIES)
(Baikov, Gazizov, and Ibragimov [1990], [1991a])

Equivalence Transformations are taken from Section 9.2.3.1.1.

Classification Result
For arbitrary functions $f(u)$, $g(u)$, and $h(u)$, the approximate symmetry algebra is given by

$$X_1 = \frac{\partial}{\partial t}, \quad X_2 = \frac{\partial}{\partial x}, \quad X_3 = \frac{\partial}{\partial y}, \quad X_4 = \frac{\partial}{\partial z},$$

$$X_5 = \varepsilon(t\frac{\partial}{\partial t} + x\frac{\partial}{\partial x} + y\frac{\partial}{\partial y} + z\frac{\partial}{\partial z}).$$

This algebra extends in the following cases:

I. $f(u) = \delta_1$, $\delta_1 = \pm 1$, $g(u)$ and $h(u)$ are arbitrary functions.

$$X_6 = \varepsilon \left(\delta_1 x \frac{\partial}{\partial t} + t \frac{\partial}{\partial x} \right)$$

For each of the following functions $g(u)$ and $h(u)$ there is an additional extension:

I.1. $g(u) = \delta_2 e^u$, $h(u) = \delta_3 e^{ku}$, $\delta_2, \delta_3 = \pm 1$, $k \neq 0, 1$.

$$X_7 = y \frac{\partial}{\partial y} + kz \frac{\partial}{\partial z} + 2 \frac{\partial}{\partial u}$$

I.2. $g(u) = \delta_2 u^v$, $h(u) = \delta_3 u^\gamma$, $\delta_2, \delta_3 = \pm 1$, $v \neq 0$, $\gamma \neq 0$, and $v \neq \gamma$.

$$X_7 = vy \frac{\partial}{\partial y} + \gamma z \frac{\partial}{\partial z} + 2u \frac{\partial}{\partial u}$$

II. $f(u) = \delta_1 e^u$, $g(u) = \delta_2 e^{k_1 u}$, $h(u) = \delta_3 e^{k_2 u}$, $\delta_1, \delta_2, \delta_3 = \pm 1$, $k_1 \neq 0, 1$, $k_2 \neq 0, 1$, and $k_1 \neq k_2$.

$$X_6 = x \frac{\partial}{\partial x} + k_1 y \frac{\partial}{\partial y} + k_2 z \frac{\partial}{\partial z} + 2 \frac{\partial}{\partial u}$$

III. $f(u) = \delta_1 u^\sigma$, $g(u) = \delta_2 u^v$, $h(u) = \delta_3 u^\gamma$, $\delta_1, \delta_2, \delta_3 = \pm 1$, $\sigma \neq 0$, $v \neq 0$, $\gamma \neq 0$, and $\sigma \neq v \neq \gamma \neq \sigma$.

$$X_6 = \sigma x \frac{\partial}{\partial x} + vy \frac{\partial}{\partial y} + \gamma z \frac{\partial}{\partial z} + 2u \frac{\partial}{\partial u}.$$

9.3. Boussinesq-type equations

Partial differential equations with a small parameter ε,

$$u_{11} - \varepsilon D_x D_y \varphi = 0, \quad \varphi \in \mathcal{A}, \tag{9.5}$$

are considered, where φ is an arbitrary differential function. Notation:

$$u_{10} = u_x, \quad u_{01} = u_y, \dots, u_{ij} = \frac{\partial^{i+j} u}{\partial x^i \partial y^j}$$

where D_x and D_y are the total derivatives with respect to x and y, respectively.

The Boussinesq equation

$$u_{tt} - u_{zz} - 2\varepsilon(\frac{2}{3}u_z u_{tz} + \frac{1}{3}u_t u_{zz} + u_{zzzz}) = 0$$

is rewritten as Equation 9.5, up to $o(\varepsilon)$, by setting

$$x = t - z, \qquad y = t + z$$

and letting

$$\varphi = \frac{1}{4}(yu_{10}^2 + xu_{01}^2 - \frac{2}{3}uu_{01} - \frac{2}{3}uu_{10} + 2yu_{30} + 2xu_{03}). \qquad (9.6)$$

For an investigation of long wave propagations described by the Boussinesq equation, one often allows the dependence of u on the new variable $\tau = \varepsilon t$ known as a "slow variable". Then, for the Boussinesq equation with three independent variables $x, y, \tau = \varepsilon(x + y)/2$, one arrives at Equation 9.5 with

$$\varphi = \frac{1}{4}(yu_{10}^2 + xu_{01}^2 - \frac{2}{3}uu_{01} - \frac{2}{3}uu_{10} + 2yu_{30} + 2xu_{03} - 2yu_\tau - 2xu_\tau). \quad (9.7)$$

In Section 9.3.1, Equation 9.5 with two independent variables x, y is considered so that φ is a differential function depending on $x, y, u, u_{10}, u_{01}, \ldots$, while in Section 9.3.2 the differential function φ depends on $x, y, \tau, u, u_{10}, u_{01}, u_\tau, \ldots$.

9.3.1. Equations with two independent variables x, y

9.3.1.1. General case

$$u_{11} - \varepsilon D_x D_y \varphi = 0$$

where $\varphi \in \mathcal{A}$ is an arbitrary function of $x, y, u, u_{10}, u_{01}, \ldots$

APPROXIMATE SYMMETRIES
(Baikov, Gazizov, and Ibragimov [1991b])

All Lie-Bäcklund (and hence Lie point) symmetries of the unperturbed equation, i.e., of the linear wave equation $u_{11} = 0$, are stable (see Section 2.5.2). The perturbed equation under consideration inherits them in the form of approximate symmetries

$$X = [(Cu + f + g) + \varepsilon(-C\varphi - f_*\varphi - g_*\varphi + X_0\varphi)]\frac{\partial}{\partial u} + \ldots$$

where

$$X_0 = [Cu + f(x, u_{10}, u_{20}, \ldots) + g(y, u_{01}, u_{02}, \ldots)]\frac{\partial}{\partial u} + \ldots$$

is a Lie-Bäcklund operator admitted by the equation $u_{11} = 0$. Here, C is an arbitrary constant, f, g are arbitrary differential functions of their arguments, and (for the notation, see Chapter 2, Section 2.8)

$$f_* = \sum_{i \geq 0} \frac{\partial f}{\partial u_{i0}} D_x^i, \qquad g_* = \sum_{i \geq 0} \frac{\partial g}{\partial u_{0i}} D_y^i.$$

APPROXIMATE BÄCKLUND TRANSFORMATION
(Baikov, Gazizov, and Ibragimov [1991b])

The perturbed equation is connected with the unperturbed equation $v_{11} = 0$ by the approximate Bäcklund transformation

$$v \approx u - \varepsilon\varphi.$$

9.3.1.2. Boussinesq equation

$$2u_{11} - \varepsilon(u_{10}u_{20} + u_{01}u_{02} - \frac{1}{3}u_{01}u_{20} - \frac{1}{3}u_{10}u_{02} + u_{40} + u_{04}) = 0 \qquad (9.8)$$

APPROXIMATE SYMMETRIES
(Baikov, Gazizov, and Ibragimov [1991b])

The approximate symmetries of Equation 9.8 are obtained from Section 9.3.1.1 by letting φ be of the form 9.6. For example, taking $f = f(x, u_{10})$, $g = g(x, u_{01})$ in the operator X of Section 9.3.1.1, one obtains the following approximate symmetry for the Boussinesq equation 9.8:

$$X = \left\{ Cu + f(x, u_{10}) + g(y, u_{01}) + \varepsilon \left[C\left(\frac{1}{4}yu_{10}^2 + \frac{1}{4}xu_{01}^2 - \frac{1}{6}uu_{10}\right.\right.\right.$$

$$\left. - \frac{1}{6}uu_{01}\right) - \frac{1}{6}u_{01}f - \frac{1}{6}u_{10}g + \frac{1}{2}yu_{10}f_x + \frac{1}{2}xu_{01}g_y - \frac{1}{6}uf_x - \frac{1}{6}ug_y$$

$$\left.\left. - f_{u_{10}}\left(\frac{1}{4}u_{01}^2 - \frac{1}{6}u_{10}u_{01} + \frac{1}{2}u_{03}\right) - g_{u_{01}}\left(\frac{1}{4}u_{10}^2 - \frac{1}{6}u_{10}u_{01} + \frac{1}{2}u_{03}\right)\right.\right.$$

$$+ \frac{1}{2}y\left(f_{xxx} + 3u_{20}f_{xxu_{10}} + 3u_{20}^2 f_{xu_{10}u_{10}} + 3u_{30}f_{xu_{10}} + u_{20}^3 f_{u_{10}u_{10}u_{10}}\right.$$

$$\left. + 3u_{20}u_{30}f_{u_{10}u_{10}}\right) + \frac{1}{2}x\left(g_{yyy} + 3u_{02}g_{yyu_{01}} + 3u_{02}^2 g_{yu_{01}u_{01}} + 3u_{03}g_{yu_{01}}\right.$$

$$\left.\left. + 3u_{02}^3 g_{u_{01}u_{01}u_{01}} + 3u_{02}u_{03}g_{u_{01}u_{01}}\right)\right] + \cdots \right\}\frac{\partial}{\partial u} + \cdots$$

APPROXIMATE BÄCKLUND TRANSFORMATION
(Baikov, Gazizov, and Ibragimov [1991b])

Equation 9.8 is connected with the linear wave equation

$$v_{11} = 0$$

by the approximate Bäcklund transformation

$$v = u - \varepsilon\left(\frac{1}{4}yu_{10}^2 + \frac{1}{4}xu_{01}^2 - \frac{1}{6}uu_{10} - \frac{1}{6}uu_{01} + \frac{1}{2}yu_{30} + \frac{1}{2}xu_{03}\right)$$

APPROXIMATE SOLUTION
By using approximate symmetries, one obtains, e.g., the following solution to the Boussinesq equation (Baikov, Gazizov, and Ibragimov [1991b]):

$$u \approx \alpha(x) + \beta(y) + \frac{\varepsilon}{2}\left[y\left(\frac{1}{2}\alpha_x^2 + \alpha_{xxx}\right)\right.$$

$$\left. + x\left(\frac{1}{2}\beta_y^2 + \beta_{yyy}\right) - \frac{1}{3}\beta\alpha_x - \frac{1}{3}\alpha\beta_y + \alpha_1(x) + \beta_1(y)\right]$$

with arbitrary functions $\alpha(x)$, $\beta(y)$, $\alpha_1(x)$, and $\beta_1(y)$.

9.3.2.   Equations with a "slow variable"

9.3.2.1.   General case

$$u_{11} - \varepsilon D_x D_y \varphi = 0$$

Here the differential function $\varphi \in \mathcal{A}$ depends on $\tau, x, y, u, u_\tau, u_{10}, u_{01}, \ldots$

APPROXIMATE SYMMETRIES
(Baikov, Gazizov, and Ibragimov [1991b])

All symmetries of the linear wave equation $u_{11} = 0$ are inherited by the equation in question in the form of approximate symmetries

$$X = [C_0(\tau)u + C_1(\tau)u_\tau + \ldots + f(x, \tau, u_{10}, u_{20}, u_{10\tau}, \ldots)$$

$$+ g(y, \tau, u_{01}, u_{02}, u_{01\tau}, \ldots)$$

$$+\varepsilon(-C_0(\tau)\varphi - C_1(\tau)D_\tau\varphi - \ldots - f_*\varphi - g_*\varphi + X_0\varphi)] \frac{\partial}{\partial u} + \ldots$$

where

$$X_0 = [C_0(\tau)u + C_1(\tau)u_\tau + \ldots + f(x, \tau, u_{10}, u_{20}, u_{10\tau}, \ldots)$$

$$+ g(y, \tau, u_{01}, u_{02}, u_{01\tau}, \ldots)] \frac{\partial}{\partial u} + \ldots$$

is a Lie-Bäcklund operator admitted by the unperturbed equation $u_{11} = 0$. Here, $C_0(\tau)$, $C_1(\tau)$, ... are arbitrary functions of the slow variable, f, g are arbitrary differential functions of their arguments, and

$$f_* = f_{u_{10}}D_x + f_{u_{20}}D_x^2 + f_{u_{10\tau}}D_x D_\tau + \ldots ,$$

$$g_* = g_{u_{01}}D_y + g_{u_{02}}D_y^2 + g_{u_{01\tau}}D_y D_\tau + \ldots .$$

APPROXIMATE BÄCKLUND TRANSFORMATION
(Baikov, Gazizov, and Ibragimov [1991b])

The perturbed equation is connected with unperturbed equation $v_{11} = 0$ by the approximate Bäcklund transformation

$$v \approx u - \varepsilon\varphi.$$

9.3.2.2. Boussinesq equation

$$2u_{11} - \varepsilon(u_{10}u_{20} + u_{01}u_{02} - \frac{1}{3}u_{10}u_{02} - \frac{1}{3}u_{01}u_{20} + u_{40} + u_{04} - u_{10\tau} - u_{01\tau}) = 0$$

$$(9.9)$$

APPROXIMATE SYMMETRIES
(Baikov, Gazizov, and Ibragimov [1991b])

See Section 9.3.2.1 with φ given by Equation 9.7.

REMARK.   Under the condition

$$u_{01} = 0, \tag{9.10}$$

Equation 9.9 yields

$$\varepsilon(u_{10\tau} - u_{10}u_{20} - u_{40}) = o(\varepsilon). \tag{9.11}$$

By setting $w = u_{10}$, one obtains from Equation 9.11 the Korteweg-de Vries equation. Hence, the exact symmetries of the Korteweg-de Vries equation can be obtained from the approximate symmetries X of the Boussinesq equation 9.9 by imposing the condition

$$X(u_{01})\Big|_{(9.9)-(9.11)} = o(\varepsilon)$$

APPROXIMATE BÄCKLUND TRANSFORMATION
(Baikov, Gazizov, and Ibragimov [1991b])

Equation 9.9 is connected with the linear wave equation $v_{11} = 0$ by the approximate Bäcklund transformation

$$v \approx u + \frac{\varepsilon}{2}[y(u_\tau - \frac{1}{2}u_{10}^2 - u_{30}) + x(u_\tau - \frac{1}{2}u_{01}^2 - u_{03}) + \frac{1}{3}uu_{10} + \frac{1}{3}uu_{01}]$$

APPROXIMATE SOLUTION
    By using approximate symmetries, one obtains, e.g., the following solution to Equation 9.9 (Baikov, Gazizov, and Ibragimov [1991b]):

$$u \approx \alpha(x, \tau) + \beta(y, \tau) + \frac{\varepsilon}{2}\left[-y\left(\alpha_\tau - \frac{1}{2}\alpha_x^2 - \alpha_{xxx}\right)\right.$$

$$\left. -x\left(\beta_\tau - \frac{1}{2}\beta_y^2 - \beta_{yyy}\right) - \frac{1}{3}\beta\alpha_x - \frac{1}{3}\alpha\beta_y + \alpha_1(x, \tau) + \beta_1(y, \tau)\right]$$

with arbitrary functions $\alpha(x)$, $\beta(y)$, $\alpha_1(x)$, and $\beta_1(y)$.

9.4. Evolution equations

Cf. Chapter 2, Section 2.8. Exact symmetries of the unperturbed and perturbed equations under consideration are presented in [H1], Chapter 11.

9.4.1. Korteweg-de Vries equation

$$u_t = uu_1 + \varepsilon u_3$$

EXAMPLES OF APPROXIMATE SYMMETRIES

$$X = \sum_{i=0}^{p} \varepsilon^i f^i \frac{\partial}{\partial u} + ..., \quad f^i \in \mathcal{A} \tag{9.12}$$

Example 1. In the second order of precision (i.e., $p = 2$), one has the following approximate symmetry:

$$f = \varphi(u)u_1 + \varepsilon \left[\varphi'(u)u_3 + 2\varphi''u_1u_2 + \frac{1}{2}\varphi'''u_1^3 \right]$$

$$+ \varepsilon^2 \left[\frac{3}{5}\varphi''u_5 + \frac{9}{5}\varphi'''u_1u_4 + \left(3\varphi'''u_2 + \frac{23}{10}\varphi^{IV}u_1^2 \right)u_3 \right.$$

$$+ \frac{31}{10}\varphi^{IV}u_1u_2^2 + \frac{8}{5}\varphi^{V}u_1^3u_2 + \frac{1}{8}\varphi^{VI}u_1^5 \right] + o(\varepsilon^2)$$

REMARK. If $\varphi(u)$ is a polynomial of degree n then the coefficients f^i of Operator 9.12 vanish when $i > n + 1$, and the approximate symmetries of order $p = n$ reduce to exact Lie-Bäcklund symmetries of order $2n + 1$. Cf. Section 2.8.2.

Example 2. In the first order of precision (i.e., $p = 1$), one has the following approximate symmetries:

$$f = \varphi(x + tu)u_1 + \varepsilon \left\{ \varphi'(u)\frac{1 + tu_1}{u_1}u_3 - \varphi'\frac{u_2^2}{u_1^2} \right.$$

$$+ \varphi'' \left[2(1 + tu_1)^2 - \frac{3}{2}(1 + tu_1) \right] \frac{u_2}{u_1} + \frac{\varphi''}{2}(1 + tu_1)^3 \right\} + o(\varepsilon),$$

$$f = \frac{u_2}{u_1^2} - \varepsilon \left(2\frac{u_2 u_4}{u_1^4} + 2\frac{u_3^2}{u_1^4} - 17\frac{u_2^2 u_3}{u_1^5} + 15\frac{u_2^4}{u_1^6} \right) + o(\varepsilon).$$

For more approximate symmetries, see Baikov, Gazizov, and Ibragimov [1987a], [1988d], [1989a].

APPROXIMATE BÄCKLUND TRANSFORMATION
(Baikov, Gazizov, and Ibragimov [1987b], [1988d], [1989a], [1989b])
The Korteweg-de Vries equation $u_t = uu_1 + \varepsilon u_3$ is connected (up to $o(\varepsilon^3)$) with the unperturbed equation $v_t = vv_1$ by the approximate Bäcklund transformation:

$$v = u + \varepsilon \left(-\frac{1}{2}\frac{u_3}{u_1} + \frac{1}{2}\frac{u_2^2}{u_1^2} \right) + \varepsilon^2 \left(\frac{1}{8}\frac{u_6}{u_1^2} - \frac{29}{40}\frac{u_2 u_5}{u_1^3} - \frac{37}{40}\frac{u_3 u_4}{u_1^3} + \frac{12}{5}\frac{u_2^2 u_4}{u_1^4} \right)$$

$$+ \frac{21}{8}\frac{u_2 u_3^2}{u_1^4} - \frac{11}{2}\frac{u_2^3 u_3}{u_1^5} + 2\frac{u_2^5}{u_1^6} \right) + \varepsilon^3 \left(-\frac{1}{48}\frac{u_9}{u_1^3} + \frac{19}{80}\frac{u_2 u_8}{u_1^4} + \frac{269}{560}\frac{u_3 u_7}{u_1^4} \right.$$

$$- \frac{863}{590}\frac{u_2^2 u_7}{u_1^5} + \frac{439}{560}\frac{u_4 u_6}{u_1^4} - \frac{3207}{560}\frac{u_2 u_3 u_6}{u_1^5} + \frac{2029}{280}\frac{u_2^3 u_6}{u_1^6} + \frac{67}{140}\frac{u_5^2}{u_1^4}$$

$$- \frac{943}{112}\frac{u_2 u_4 u_5}{u_1^5} - \frac{2679}{560}\frac{u_3^2 u_5}{u_1^5} + \frac{2949}{80}\frac{u_2^2 u_3 u_5}{u_1^6} - \frac{1079}{40}\frac{u_2^4 u_5}{u_1^7} - \frac{461}{80}\frac{u_3 u_4^2}{u_1^5}$$

$$+ \frac{799}{35}\frac{u_2^2 u_4^2}{u_1^6} + \left(53 - \frac{247}{560} \right)\frac{u_2 u_3^2 u_4}{u_1^6} - \left(158 + \frac{19}{40} \right)\frac{u_2^3 u_3 u_4}{u_1^7} + \frac{801}{10}\frac{u_2^5 u_4}{u_1^8}$$

$$+ \frac{679}{112}\frac{u_3^4}{u_1^6} - \frac{375}{4}\frac{u_2^2 u_3^3}{u_1^7} + \left(241 + \frac{3}{8} \right)\frac{u_2^4 u_3^2}{u_1^8} - \left(184 + \frac{2}{3} \right)\frac{u_2^6 u_3}{u_1^9}$$

$$+ 42\frac{u_2^8}{u_1^{10}} \right) + o(\varepsilon^3).$$

APPROXIMATE RECURSION OPERATORS
(Baikov, Gazizov, and Ibragimov [1987b], [1989a], [1989b])

$$L_1 = \beta(u) + \varepsilon \left[\frac{3}{2}\beta' D^2 - \frac{1}{2}\frac{\beta' u_2}{u_1} D + \frac{3}{2}\beta'' u_1 D - \frac{\beta' u_3}{2u_1} + \frac{\beta' u_2^2}{2u_1^2} \right.$$

$$+ \frac{1}{2}\beta''u_2 + \frac{1}{2}\beta'''u_1^2 \Big] + \varepsilon^2 \Big[\frac{9}{8}\beta''D^4 - \frac{3}{10}\frac{\beta'u_3}{u_1^2}D^3 + \frac{2}{5}\frac{\beta'u_2^2}{u_1^3}D^3$$

$$+ \frac{9}{4}\beta'''u_1D^3 - \frac{3}{4}\frac{\beta''u_2}{u_1}D^3 - \frac{3}{5}\frac{\beta'u_4}{u_1^2}D^2 + \frac{14}{5}\frac{\beta'u_2u_3}{u_1^3}D^2 - \frac{12}{5}\frac{\beta'u_2^3}{u_1^4}D^2$$

$$- \frac{39}{20}\frac{\beta''u_3}{u_1}D^2 + \frac{89}{40}\frac{\beta''u_2^2}{u_1^2}D^2 + \frac{3}{2}\beta'''u_2D^2 + \frac{15}{8}\beta^{IV}u_1^2D^2 - \frac{3}{10}\frac{\beta'u_5}{u_1^2}D$$

$$+ \frac{23}{10}\frac{\beta'u_2u_4}{u_1^3}D + \frac{17}{10}\frac{\beta'u_3^2}{u_1^3}D - \frac{48}{5}\frac{\beta'u_2^2u_3}{u_1^4}D + 6\frac{\beta'u_2^4}{u_1^5}D - \frac{69}{40}\frac{\beta''u_4}{u_1}D$$

$$+ \frac{61}{10}\frac{\beta''u_2u_3}{u_1^2}D - \frac{177}{40}\frac{\beta''u_2^3}{u_1^3}D - \frac{3}{10}\beta'''u_3D + \frac{23}{20}\frac{\beta'''u_2^2}{u_1}D + \frac{3}{2}\beta^{IV}u_1u_2D$$

$$+ \frac{3}{4}\beta^V u_1^3 D + \frac{3}{10}\frac{\beta'u_2u_5}{u_1^3} + \frac{9}{10}\frac{\beta'u_3u_4}{u_1^3} - \frac{27}{10}\frac{\beta'u_2^2u_4}{u_1^4} - \frac{9}{2}\frac{\beta'u_2u_3^2}{u_1^4}$$

$$+ 12\frac{\beta'u_2^3u_3}{u_1^5} - 6\frac{\beta'u_2^5}{u_1^6} - \frac{21}{40}\frac{\beta''u_5}{u_1} + \frac{99}{40}\frac{\beta''u_2u_4}{u_1^2} + \frac{39}{20}\frac{\beta''u_3^2}{u_1^2} - \frac{1041}{8}\frac{\beta''u_2^2u_3}{u_1^3}$$

$$+ \frac{177}{40}\frac{\beta''u_2^4}{u_1^4} - \frac{9}{20}\beta'''u_4 + \frac{9}{5}\frac{\beta'''u_2u_3}{u_1} - \frac{23}{20}\frac{\beta'''u_2^3}{u_1^2} + \frac{17}{40}\beta^{IV}u_1u_3$$

$$+ \frac{29}{40}\beta^{IV}u_2^2 + \frac{17}{20}\beta^V u_1^2u_2 + \frac{1}{8}\beta^{VI}u_1^4 \Big] + o(\varepsilon^2),$$

$$L_2 = u_1 D^{-1} + \varepsilon \Big[\frac{u_2}{u_1}D + \frac{u_3}{u_1} - \frac{u_2^2}{u_1^2} \Big]$$

$$+ \varepsilon^2 \Big[\frac{3}{5}\frac{u_3}{u_1^2}D^3 - \frac{4}{5}\frac{u_2^2}{u_1^3}D^3 + \frac{6}{5}\frac{u_4}{u_1^2}D^2 - \frac{28}{5}\frac{u_2u_3}{u_1^3}D^2 + \frac{24}{5}\frac{u_2^3}{u_1^4}D^2$$

$$+ \frac{3}{5}\frac{u_5}{u_1^2}D - \frac{23}{5}\frac{u_2u_4}{u_1^3}D - \frac{17}{5}\frac{u_3^2}{u_1^3}D + \frac{96}{5}\frac{u_2^2u_3}{u_1^4}D - 12\frac{u_2^4}{u_1^5}D - \frac{3}{5}\frac{u_2u_5}{u_1^3}$$

$$-\frac{9}{5}\frac{u_3u_4}{u_1^3}+\frac{27}{5}\frac{u_2^2u_4}{u_1^4}+9\frac{u_2u_3^2}{u_1^4}-24\frac{u_2^2u_3}{u_1^5}+12\frac{u_2^5}{u_1^6}\right]+o(\varepsilon^2).$$

REMARK. If $\beta(u)=u$, the approximate recursion operator $L=L_1+2L_2$ satisfies the break-off condition (see Section 2.8.5) and coincides with the exact recursion operator

$$L=u+2u_1D^{-1}+3\varepsilon D^2$$

for the Korteweg-de Vries equation.

APPROXIMATE SOLUTION

The following approximate solution is taken from Baikov, Gazizov, and Ibragimov [1989a]:

$$u=\mu+\frac{\varepsilon}{2}\left[\frac{\mu'''}{\mu'(1-t\mu')^3}+\frac{(3t\mu'-1)(\mu'')^2}{(\mu')^2(1-t\mu')^4}\right]+o(\varepsilon).$$

Here, μ is an arbitrary function of one variable, and $v=\mu(x+tv)$ is the general solution to the equation $v_t=vv_1$.

9.4.2. Burgers-Korteweg-de Vries equation

$$u_t=uu_1+\varepsilon(au_3+bu_2)$$

where a and b are arbitrary constants.

APPROXIMATE SYMMETRIES

$$X=\sum_{i=0}^{p}\varepsilon^i f^i\frac{\partial}{\partial u}+\dots,\quad f^i\in\mathcal{A}.$$

We give here one example of an approximate symmetry in the second order of precision ($p=2$):

$$f=\left[\varphi(u)u_1+\varepsilon\left(a\varphi'u_3+2a\varphi''u_1u_2+\frac{1}{2}a\varphi'''u_1^3+b\varphi'u_3+b\varphi''u_1^2\right)\right.$$

$$\left.+\varepsilon^2\left(\frac{3}{5}a^2\varphi''u_5+\frac{5}{4}ab\varphi''u_4+\frac{1}{10}ab\varphi''\frac{u_2u_3}{u_1}-\frac{1}{20}ab\varphi''\frac{u_2^3}{u_1^2}\right.\right.$$

$$+ \frac{2}{3}b^2\varphi'' u_3 + \frac{9}{5}a^2\varphi'' u_1 u_4 + 3a^2\varphi''' u_2 u_3 + \frac{7}{2}ab\varphi''' u_1 u_3$$

$$+ \frac{23}{10}ab\varphi''' u_2^2 + \frac{5}{3}b^2\varphi''' u_1 u_2 + \frac{23}{10}a^2\varphi^{IV} u_1^2 u_3 + \frac{31}{10}a^2\varphi^{IV} u_1 u_2^2$$

$$+ \frac{15}{4}ab\varphi^{IV} u_1^2 u_2 + \frac{1}{2}b^2\varphi^{IV} u_1^3 + \frac{8}{5}a^2\varphi^V u_1^3 u_2 + \frac{1}{2}ab\varphi^V u_1^4$$

$$+ \frac{1}{8}a^2\varphi^{VI} u_1^5 \Big) \Big] + o(\varepsilon^2).$$

For more symmetries, see Baikov, Gazizov, and Ibragimov [1987a], [1987b], [1988a], [1988d], [1989a], [1989b].

APPROXIMATE BÄCKLUND TRANSFORMATION
(Baikov, Gazizov, and Ibragimov [1987b], [1989a], [1989b])

Burgers-Korteweg-de Vries equation is approximately (up to $o(\varepsilon^2)$) connected with the equation

$$v_t = vv_1$$

by the approximate Bäcklund transformation:

$$v = u + \varepsilon \left(-\frac{a}{2}\frac{u_3}{u_1} + \frac{a}{2}\frac{u_2^2}{u_1^2} - b\frac{u_2}{u_1} \right) + \varepsilon^2 \left(\frac{a^2}{8}\frac{u_6}{u_1^2} - \frac{29}{40}a^2\frac{u_2 u_5}{u_1^3} \right.$$

$$- \frac{37}{40}a^2\frac{u_3 u_4}{u_1^3} + \frac{12}{5}a^2\frac{u_2^2 u_4}{u_1^4} + \frac{21}{8}a^2\frac{u_2 u_3^2}{u_1} - \frac{11}{2}a^2\frac{u_2^3 u_3}{u_1^5} + 2a^2\frac{u_2^5}{u_1^6}$$

$$+ \frac{1}{2}ab\frac{u_5}{u_1^2} - \frac{9}{4}ab\frac{u_2 u_4}{u_1^3} - \frac{5}{4}ab\frac{u_3^2}{u_1^3} + \frac{57}{10}ab\frac{u_2^2 u_3}{u_1^4} - \frac{13}{5}ab\frac{u_2^4}{u_1^5}$$

$$+ \frac{1}{2}b^2\frac{u_4}{u_1^2} - \frac{5}{3}b^2\frac{u_2 u_3}{u_1^3} + b^2\frac{u_2^3}{u_1^4} \right) + o(\varepsilon^2).$$

9.4.3. Generalized Korteweg-de-Vries equation

$$u_t = h(u)u_1 + \varepsilon u_3$$

where $h(u)$ is an arbitrary function.

APPROXIMATE SYMMETRIES
(Baikov, Gazizov, and Ibragimov [1987b], [1989a], [1989b])

$$X = \sum_{i=0}^{p} \varepsilon^i f^i \frac{\partial}{\partial u} + \dots, \quad f^i \in \mathcal{A}$$

An example of a second-order approximate symmetry is provided by

$$f = \varphi(u)u_1 + \varepsilon\left(\psi_1 u_3 + 2\psi_1 u_1 u_2 + \frac{1}{2}\psi_1'' u_1^3\right)$$

$$+ \varepsilon^2\left(\frac{3}{5}\psi_2 u_5 + \frac{9}{5}\psi_2' u_1 u_4 + 3\psi_2' u_2 u_3 + \left(\frac{23}{10}\psi_2'' + \frac{2}{5}\psi_3 h'' - \frac{1}{10}\psi_2 \frac{h'''}{h'}\right)u_1^2 u_3\right.$$

$$+ \left(\frac{31}{10}\psi_2'' + \frac{4}{5}\psi_3 h'' - \frac{1}{5}\psi_2 \frac{h'''}{h'}\right)u_1 u_2^2 + \left(\frac{8}{5}\psi_2''' + \frac{4}{5}\psi_3' h'' + \frac{3}{5}\psi_3 h'''\right.$$

$$- \frac{1}{5}\psi_2 \frac{h^{IV}}{h'} + \frac{1}{5}\psi_2 \frac{h'' h'''}{h'^2}\right)u_1^3 u_2 + \frac{1}{8}\left(\frac{8}{5}\psi_2^{IV} + \frac{4}{5}\psi_3'' h'' + \frac{7}{5}\psi_3 h'''\right.$$

$$+ \frac{2}{3}\psi_3 h^{IV} + \frac{1}{5}\psi_3 \frac{h'' h'''}{h'} + \frac{2}{5}\psi_2 \frac{h'' h^{IV}}{h'^2} - \frac{1}{5}\psi_2 \frac{h^V}{h'} + \frac{1}{5}\psi_2 \frac{h'''^2}{h'^2}$$

$$\left.\left. - \frac{2}{5}\psi_2 \frac{h''^2 h'''}{h'^3}\right)u_1^5\right) + o(\varepsilon^2).$$

Here, $\psi_1(u) = \dfrac{\varphi'(u)}{h'(u)}$, $\quad \psi_k(u) = \dfrac{\varphi_{k-1}'(u)}{h'(u)}$, $\quad k = 2, 3$.

9.4.4. Perturbed Korteweg-de Vries equation

$$u_t = u_3 + u u_1 + \varepsilon R, \quad R \in \mathcal{A}$$

APPROXIMATE TRANSFORMATIONS
(Baikov, Gazizov, and Ibragimov [1987b], [1989a])

1. The equation

$$u_t = u u_1 + u_3 + \varepsilon(C_1 u + C_2)$$

with arbitrary constants C_1 and C_2 is connected with Korteweg-de Vries equation

$$v_s = vv_1 + v_3, \quad v = v(s, y),$$

by the approximate point transformation

$$s = t + \frac{3}{2}\varepsilon C_1 t^2 + o(\varepsilon), \quad y = x + \varepsilon(C_1 tx + \frac{1}{2}C_2 t^2) + o(\varepsilon),$$

$$v = u - \varepsilon(2C_1 tu + C_1 x + C_2 t) + o(\varepsilon).$$

2. The equation

$$u_t = uu_1 + u_3 + \varepsilon \left\{ u_2 + \frac{1}{3}xuu_1 + \frac{1}{3}u^2 + a(t)uu_1 + b(t, x)u_1 \right.$$

$$+ [b_x(t, x) - a'(t) + c(t)]u - b_t(t, x) + b_{xxx}(t, x) + c'(t)x + d(t) \}$$

with arbitrary functions $a(t)$, $b(t, x)$, $c(t)$, and $d(t)$ is connected with the Korteweg-de Vries equation

$$v_s = vv_1 + v_3, \quad v = v(s, y)$$

by the approximate transformation

$$s = t + 3\varepsilon \int dt \int c(t)dt + o(\varepsilon), \quad y = x + \varepsilon \left[x \int c(t)dt + \int dt \int d(t)dt \right] + o(\varepsilon),$$

$$v = u + \varepsilon \left[\left(\frac{x}{3} + a(t) - 2 \int c(t)dt \right) u + b(t, x) - c(t)x - \int d(t)dt \right] + o(\varepsilon).$$

3. The equation

$$u = uu_1 + u_3 + \varepsilon \left(u_5 + \frac{4}{3}uu_3 + 3u_1u_2 + \frac{1}{3}u^2u_1 \right)$$

is connected with the Korteweg-de Vries equation

$$v_s = vv_1 + v_3, \quad v = v(s, y)$$

by the approximate transformation

$$s = t - \frac{1}{3}\varepsilon x + o(\varepsilon), \quad y = x + o(\varepsilon), \quad v = u + o(\varepsilon).$$

APPROXIMATE SOLUTIONS
(Baikov, Gazizov, and Ibragimov [1987b], [1989a])

1. The equation

$$u_t = uu_1 + u_3 + \varepsilon u$$

has the approximate solution

$$u = 3c \sec^2 \left[\frac{\sqrt{c}}{2}(x + ct) \right] + \varepsilon \left\{ x + 6ct \sec^2 \left[\frac{\sqrt{c}}{2}(x + ct) \right] \right.$$

$$\left. - 3c^{\frac{3}{2}} \operatorname{sh} \left[\frac{\sqrt{c}}{2}(x + ct) \right] \operatorname{ch}^{-3} \left[\frac{\sqrt{c}}{2}(x + ct) \right] \left(tx + \frac{3}{2}ct^2 \right) \right\} + o(\varepsilon).$$

This solution is approximately invariant under the operator

$$X = (1 - 3\varepsilon t)\frac{\partial}{\partial t} - c \left(1 - \varepsilon t + \frac{\varepsilon x}{c} \right)\frac{\partial}{\partial x} + \varepsilon(2u - c)\frac{\partial}{\partial u}.$$

2. The equation

$$u_t = uu_1 + u_3 + \varepsilon \left(u_5 + \frac{u}{3}uu_3 + 3u_1u_2 + \frac{1}{3}u^2u_1 \right)$$

has the approximate solution

$$u = 3c \sec^2 \left[\frac{\sqrt{c}}{2}\left(x - \frac{c}{3}\varepsilon x + ct\right) \right] + o(\varepsilon).$$

10

Differential Constraints

The method of differential constraints (Yanenko [1964]) is one of the possible approaches directly oriented to calculation of particular solutions of differential equations. It is based on the following idea.

Given a system of differential equations,

$$F_i(x, u, u_{(1)}) = 0, \quad i = 1, \ldots, n, \tag{10.1}$$

consider an overdetermined system obtained by attaching additional differential equations

$$\Phi_k(x, u, u_{(1)}) = 0, \quad k = 1, \ldots, s. \tag{10.2}$$

Here, $x = (x^1, \ldots, x^n)$ are independent variables, and $u = (u^1, \ldots, u^m)$ are dependent variables with the first-order derivatives $u_{(1)} = \{u_i^\alpha\}$ (for the notation, see Chapter 1). The supplemental equations 10.2 are termed *differential constraints* to Equation 10.1. It is required that the resulting overdetermined system 10.1 – 10.2 be compatible. Then the differential constraints 10.2 are *admissible* by the system 10.1. Thus, what is called in the literature "the method of differential constraints" is an integrability analysis of overdetermined systems obtained by appending arbitrary differential constraints 10.2 to a given system 10.1. One can also treat a restricted problem (Shapeev [1974a], [1974b]) by considering *involutive systems* (see, e.g., Cartan [1946], Finikov [1948], Kuranishi [1967]; for references, see also Chapter 13, Section 13.2.3, of this volume). Further restriction is considered in Kaptsov [1992], [1993] by relating involutive systems with determining equations.

Examples of differential constraints are provided by group-invariant solutions. In this sense, the method of differential constraints generalizes the Lie group analysis of particular solutions of differential equations. See also Chapter 11 of this volume.

A survey of the method is to be found in Sidorov, Shapeev, and Yanenko [1984]. See also Rozhdestvenskii and Yanenko [1978].

This chapter contains a list of several admissible differential constraints.

10.1.   Maxwell equations

Equations

$$\frac{\partial \mathbf{E}}{\partial t} = \text{curl}\mathbf{B}, \quad \frac{\partial \mathbf{B}}{\partial t} = -\text{curl}\mathbf{E}$$

admit the differential constraints given by the following two equations with arbitrary functions $f(x)$, $g(x)$:

$$\text{div}\mathbf{E} = f(\mathbf{x}), \quad \text{div}\mathbf{B} = g(\mathbf{x}).$$

Moreover, these constraints provide an involutory overdetermined system. This property has a group theoretic nature; see Chapter 1, Section 1.4.3.

10.2.   One-dimensional motion of elastic media

$$v_t = \sigma_x, \quad \varepsilon_t = v_x, \quad \sigma_t = a(\sigma, \varepsilon)\varepsilon_t + c(\sigma, \varepsilon). \qquad (10.3)$$

Here v, σ, and ε are velocity, stress, and deformation, respectively.

DIFFERENTIAL CONSTRAINTS
(Fomin, Shapeev, and Yanenko [1974])

System 10.3 admits the following differential constraints given by one additional equation in each case.
1.

$$\sigma_x = E(\sigma, \varepsilon)v_x + G(\sigma, \varepsilon)$$

where

$$a = E^2, \ E = c_3 l, \ l = c_1\sigma + c_2, \ G = \mp\sqrt{l} + \frac{c}{c_3}l,$$

$$c = \pm c_3\sqrt{l^3} + c_3(c_4 + c_1 c_3\varepsilon)l^4;$$

or

$$a = E^2, \ E = c_1\omega, \ G = c_2\omega, \ c = c_1 c_2\omega^2 + c_3\omega$$

with an arbitrary function $\omega = \omega(\sigma)$.
2.

$$\sigma_x = a\varepsilon_x + G(\sigma, \varepsilon)$$

where

$$G = c_1 \omega, \quad c = c_2 \omega, \quad a = \omega \left(f + \int \frac{\omega_\varepsilon \, d\sigma}{\omega^2} \right),$$

with arbitrary functions $\omega = \omega(\sigma, \varepsilon)$ and $f = f(\varepsilon)$.

In both cases, the solutions satisfying the differential constraints depend upon two arbitrary functions of one variable.

SOLUTIONS

In Fomin, Shapeev, and Yanenko [1974], the general solution is given for the model of standard linear body with

$$c = k_1(\sigma - k\varepsilon) + k_2, \quad a = k$$

where $k, k_1, k_2 = \text{const.}$, $k \geq 0$. The solution is as follows:

1. If $k \neq 0$, $k_1 \neq 0$, then

$$\sigma = \Phi(\eta) + \Psi(\xi) + k_1^2 \Omega(x) e^{k_1 t},$$

$$v = \frac{1}{\sqrt{k}} (\Psi(\xi) - \Phi(\eta)) + k_1 \Omega'(x) e^{k_1 t},$$

$$\varepsilon = \frac{1}{k} (\Psi(\xi) + \Phi(\eta)) + \Omega''(x) e^{k_1 t} + \frac{k_2}{k k_1}.$$

2. If $k_1 = 0$, then

$$\sigma = \Phi(\eta) + \Psi(\xi),$$

$$v = \frac{1}{\sqrt{k}} (\Psi(\xi) - \Phi(\eta)) - \frac{k_2 x}{k},$$

$$\varepsilon = \frac{1}{k} (\Psi(\xi) + \Phi(\eta) - \Omega(x) - k_2 t).$$

3. If $k = 0$, then

$$\sigma = \Omega(x) e^{k_1 t} - \frac{k_2}{k_1},$$

$$v = \frac{1}{k_1} \Omega'(x) e^{k_1 t} + \Phi(x),$$

$$\varepsilon = \frac{1}{k_1^2} \Omega''(x) e^{k_1 t} + t \Phi'(x) + \Psi(x),$$

where

$$\xi = (t + xk^{-1/2})/2, \quad \eta = (t - xk^{-1/2})/2$$

and $\Omega(x)$, $\Phi(x)$, $\Psi(x)$ are arbitrary functions.

10.3. One-dimensional gas dynamics

$$\frac{\partial u}{\partial t} + x^v \frac{\partial p}{\partial q} = 0,$$

$$\frac{\partial \tau}{\partial t} - x^v \frac{\partial u}{\partial q} - v\frac{u\tau}{x} = 0, \tag{10.4}$$

$$\frac{\partial p}{\partial t} + Ax^v \frac{\partial u}{\partial q} + v\frac{u\tau}{x}A = 0.$$

Here, u, p, and τ are velocity, pressure, and specific volume of a gas, $A = A(p, \tau)$, and $v = 0, 1, 2$, respectively, for plane, cylindrically, and spherically symmetric motions. The Euler spatial coordinate x and the Lagrange mass coordinate q are related by the equations

$$\frac{\partial x}{\partial q} = \frac{\tau}{x^v}, \quad \frac{\partial x}{\partial t} = u.$$

DIFFERENTIAL CONSTRAINTS
(Raspopov, Shapeev, and Yanenko [1974], [1979])

System 10.4 admits the following differential constraints given by one additional equation in each case.
1.

$$\frac{\partial \tau}{\partial q} + A^{-1}\frac{\partial p}{\partial q} + f = 0 \tag{10.5}$$

where $f = f(\tau, p, q)$ is defined by the equation

$$\frac{\partial f}{\partial \tau} - A\frac{\partial f}{\partial p} + fA^{-1}\frac{\partial A}{\partial \tau} = 0.$$

2.

$$\frac{\partial u}{\partial q} + C\frac{\partial p}{\partial q} + f = 0 \tag{10.6}$$

where $A = A(p)$, $C = \pm A^{-1/2}$, and $f = f(u, \tau, p, x, q, t)$ are related by the equations

$$x^{v+1}(4f_u - 4C^{-1}f_p + 2Cf_\tau - fCA_p) - v\tau(2 - uCA_p) = 0,$$

$$x^{v+1}(f_q + ff_u - Cf(A + fA_p/2)) + v\tau uC(f_\tau - Af_p + fA_p/2)$$

$$+x(Cf_t + (uC + \tau)f_x) + v\tau^2 ux^{-(v+1)} + v\tau f = 0,$$

$$x^{v+1}f_\tau - vu = 0.$$

System 10.4 admits the following differential constraints given by two additional equations in each case.

3.

$$\frac{\partial u}{\partial q} + C\frac{\partial p}{\partial q} + f_1 = 0, \qquad \frac{\partial \tau}{\partial q} + A^{-1}\frac{\partial p}{\partial q} + f_2 = 0, \tag{10.7}$$

where the following involution conditions of the overdetermined system 10.4 and 10.7 are satisfied:

$$u\frac{\partial f_2}{\partial x} + \frac{\partial f_2}{\partial t} - C(v\frac{u\tau}{x} - x^v f_1)\frac{\partial f_2}{\partial u} = 0,$$

$$C\frac{\partial f_2}{\partial \tau} + 2C_\tau f_2 + C^2\frac{\partial f_2}{\partial u} - C^3\frac{\partial f_2}{\partial p} = 0,$$

$$2\frac{x^v}{C}(C\frac{\partial f_1}{\partial \tau} + C^2\frac{\partial f_1}{\partial u} - C^3\frac{\partial f_1}{\partial p}) + C_\tau(-v\frac{u\tau}{Cx} + x^v(f_2 + \frac{f_1}{C}))$$

$$+CC_p(-x^v f_1 + v\frac{u\tau}{x}) - v\frac{u + C\tau}{x} = 0,$$

$$\frac{\partial f_1}{\partial \tau}(x^v Cf_2 + x^v f_1 - v\frac{u\tau}{x}) + x^v Cf_1\frac{\partial f_1}{\partial u} + C^2(v\frac{u\tau}{x} - x^v f_1)\frac{\partial f_1}{\partial p} \tag{10.8}$$

$$-(u + C\tau)\frac{\partial f_1}{\partial x} - x^v C\frac{\partial f_1}{\partial q} - \frac{\partial f_1}{\partial t} - C_\tau(-2x^v f_1 f_2 + 2v\frac{u\tau}{x}f_2)$$

$$-v\frac{C}{x}(2\tau f_1 + uf_2 + \frac{u\tau^2}{x^{v+1}}) = 0.$$

4.

$$\frac{\partial u}{\partial q} + f_1 = 0, \qquad \frac{\partial p}{\partial q} + f_2 = 0, \tag{10.9}$$

where $C = \alpha A^{-1/2}$, $\alpha^2 = 1$. In this case, the involution conditions have the form:

$$\frac{\partial f_2}{\partial \tau} = 0, \qquad x^v(A\frac{\partial f_1}{\partial \tau} + f_1\frac{\partial A}{\partial \tau}) - v\frac{u}{x}(\tau\frac{\partial A}{\partial \tau} + A) = 0,$$

$$\frac{\partial f_2}{\partial q} - f_1 \frac{\partial f_2}{\partial u} - f_2 \frac{\partial f_2}{\partial p} + v f_2 \frac{\tau}{x^{\nu+1}} + f_2 \frac{\partial f_1}{\partial u}$$

$$-(f_1 - v \frac{u\tau}{x^{\nu+1}}) \frac{\partial f_1}{\partial \tau} - A(v \frac{u\tau}{x^{\nu+1}} - f_1) \frac{\partial f_1}{\partial p} \qquad (10.10)$$

$$+ \frac{1}{x^\nu} (u \frac{\partial f_1}{\partial x} + \tau \frac{\partial f_2}{\partial x} + \frac{\partial f_1}{\partial t}) = 0,$$

$$Ax^\nu (\frac{\partial f_1}{\partial q} - f_1 \frac{\partial f_1}{\partial u} - f_2 \frac{\partial f_1}{\partial p}) - x^\nu f_1 f_2 \frac{\partial A}{\partial p} + v \frac{\tau}{x} A f_1$$

$$+ v \frac{\tau}{x} (u f_2 \frac{\partial A}{\partial p} + A f_1 + \frac{u\tau}{x^{\nu+1}} A) + x^\nu f_2 \frac{\partial f_2}{\partial u}$$

$$+ A(f_1 x^\nu - v \frac{u\tau}{x}) \frac{\partial f_2}{\partial p} + u \frac{\partial f_2}{\partial x} + \tau A \frac{\partial f_1}{\partial x} + \frac{\partial f_2}{\partial t} = 0.$$

In the cases 3 and 4, the solutions of the system 10.4, 10.7 and of the system 10.4, 10.9, respectively, depend upon one arbitrary function of one argument.

PARTICULAR SOLUTIONS WITH THE PLANE SYMMETRY ($\nu = 0$)
(Raspopov, Shapeev, and Yanenko [1979])

By letting $f = 0$, $A = a^2 p^n$, one obtains the following solution of the overdetermined system 10.4, 10.6:

$$u = \pm \frac{p^{1-n/2}}{a(1 - n/2)}, \quad \tau = \frac{p^{1-n}}{a^2(n - 1)} + g_1(q), \quad G(p, \mp q + a p^2 t) = 0$$

with arbitrary functions g_1, G, and arbitrary constants a, n.

The overdetermined system 10.4, 10.7 has solutions given by Riemann's waves if $f_1 = f_2 = 0$.

Consider the differential constraints 10.9. For the polytropic gas with $A = \gamma p/\tau$, Equations 10.10 are satisfied by

$$f_1 = \frac{p g_1(t)}{A}, \quad f_2 = -u(\frac{g_1(t)}{\gamma} + g_1^{-1}(t) g_1'(t)),$$

where $g_1(t)$ is determined by the equation

$$g_1'' + \frac{\gamma + 3}{\gamma} g_1 g_1' + \frac{\gamma + 1}{\gamma^2} g_1^3 = 0.$$

A particular solution of the latter equation is

$$g_1 = 2\gamma/[(\gamma + 1)t].$$

With this function g_1, one obtains the following solutions of Equations 10.4, 10.9 given in Sedov [1957]:

$$u = 2[(\gamma + 1)t]^{(1-\gamma)/(1+\gamma)}\beta'(q),$$

$$p = 2(\gamma - 1)[(\gamma + 1)t]^{-2\gamma/(1+\gamma)}\beta(q),$$

$$\tau = [(\gamma + 1)t]^{2/(1+\gamma)}\beta''(q)$$

where $\beta(q)$ is an arbitrary function.

Another solution of Equation 10.10, in the case of the polytropic gas $A = \gamma p/\tau$, is given by

$$f_1 = \frac{F_1(p, q, t)}{A}, \quad f_2 = F_2(p, q, t),$$

where the functions F_1, F_2 assume any of the following forms:

$$F_1 = \frac{\gamma p}{t}, \quad F_2 = \frac{C_1 t^{-\gamma} + p}{q};$$

$$F_1 = \frac{\gamma p}{t}, \quad F_2 = C_1 t^{-\gamma};$$

$$F_1 = C_1 q^{-1/\gamma} p^{(\gamma+1)/\gamma}, \quad F_2 = \frac{p}{q}.$$

Consequently, one obtains the following solution of Equations 10.4, 10.9:

$$u = \frac{C_2}{\gamma - 1}t^{1-\gamma} + \beta_1(q), \quad p = (C_2 q - C_1)t^{-\gamma}, \quad \tau = t\beta_1'(q).$$

INTERNAL CONTACT SYMMETRIES

It is well known that, in the case of many dependent variables, there are no Lie contact transformations different from point ones (see [H1], Section 1.7). However, certain systems of differential equations may admit non-point symmetries that are contact transformations defined only on the solutions of the differential equations under consideration. These transformations are termed *internal contact symmetries*. It was shown by Krendelev and Talyshev [1979] that the gas dynamics equation 10.4 with $\nu = 0$ do not admit nontrivial (i.e., non-point) internal contact symmetries. Differential constraints change the situation. Namely, it is shown by Talyshev [1980] that Equation 10.4 with one differential constraint admits groups of internal contact symmetries.

11

Nonclassical and Conditional Symmetries

Lie's classical theory of symmetries of differential equations is an inspiring source for various generalizations aiming to find the ways for obtaining explicit solutions, conservation laws, linearizing substitutions, etc. This chapter[1] describes one of the possible extensions of the Lie theory of invariant solutions, first considered by Bluman and Cole [1969] and named the "nonclassical method". This method, and its equivalence to direct reduction methods of Clarkson and Kruskal [1989] and Galaktionov [1990], has become the focus of much research and many applications to physically important partial differential equations. It is clear that other related topics, such as partially invariant solutions, differentially partially invariant solutions, group foliation, and so on, will give rise to efficient and elegant methods of treating differential equations.

The material of the chapter is split into two parts: theoretical background and the most important results. The first part is written mainly on the basis of the papers of Olver and Rosenau [1987], Olver [1994a], and Vorob'ev [1986], [1989], [1991], [1992].

Part I. THEORY AND EXAMPLES

In order to discuss our subject, we will employ the standard geometric approach to the theory of symmetries of differential equations. A kth order system of differential equations is naturally treated as a submanifold $E \subset J^k$ of the kth order jet space on the space of independent and dependent variables. As described in the earlier chapters of this book, the classical Lie symmetries of differential equations may be characterized by the following features. Through the process of prolongation, which requires the group transformations to preserve the intrinsic contact

---

[1]Research supported in part by NSF Grants DMS 91–16672 and DMS 92–04192 and by the Grant of Russian State Committee of Higher Education.

structure on the jet space, they define local groups of contact transformations on the kth order jet spaces J^k. Such a transformation group will be a symmetry group of the system of differential equations $E \subset J^k$ if the transformations of the symmetry group *leave E invariant*. This implies that the group transformations map solutions of E onto solutions of E. The classical Lie symmetries are sometimes called *external* symmetries.

To date, several extensions of the classical Lie approach have been proposed. Each of them relaxes one or more of the basic properties obeyed by classical symmetry groups. If we relax the restriction that the infinitesimal generators determine geometrical transformations on a finite order jet space, then we are naturally led to the class of *generalized* or *Lie-Bäcklund* symmetries, first used by Noether [1918] in her famous theorem relating symmetries and conservation laws.[2] If we only require that the symmetries preserve the restriction of the contact structure on the system of differential equations E, then we find the *internal* symmetries first discussed at length in the works of Cartan [1914], [1915]. Actually, for a wide class of differential equations all internal symmetries are generated by external symmetries (see Anderson, Kamran, and Olver [1993a,b]). Lastly, if the symmetries only leave invariant a certain submanifold of E, we find the class of *conditional* symmetries, treated by Bluman and Cole [1969], Olver and Rosenau [1987], and, subsequently, many others. The term "conditional" is explained by the fact that the submanifold

---

[2]Editor's note: Actually, Noether [1918], Section III, gave an example of such an infinitesimal symmetry for the second-order ordinary differential equation $u''(x) = 0$, namely,

$$X = -2\frac{u}{u'^2}\frac{\partial}{\partial x} + (x - 2\frac{u}{u'})\frac{\partial}{\partial u}.$$

Indeed, X is different from Lie point and contact symmetries. However, this is a seeming difference. In fact, X differs from the point symmetry generator $x\partial/\partial u$ by a trivial *differentiation operator* of the form

$$X_* = \xi(x, u, u')D_x \equiv \xi(x, u, u')\left(\frac{\partial}{\partial x} + u'\frac{\partial}{\partial u} + \cdots\right)$$

admissible by any differential equation (see Chapter 1). Namely,

$$X = x\frac{\partial}{\partial u} - 2\frac{u}{u'^2}D_x.$$

In Section III of her paper, E. Noether also remarked that the finite group transformations may depend on infinitely many derivatives if an infinitesimal transformation contains derivatives of u. But this remark is covered by Bäcklund's theorem. For a historical survey of the development of the theory of Lie-Bäcklund transformation groups starting with the works of Lie [1874] and Bäcklund [1874], [1876] and ending with the modern theory, see Chapter 6, Section 6.2.1 of this volume.

of E is determined by attaching additional differential equations (called differential constraints) to the original system E. The theory of differential constraints has its origins in the work of Yanenko [1964] on gas dynamics; see Chapter 10 in this volume and the book by Sidorov, Shapeev, and Yanenko [1984] for a survey of this method. The most popular way is to append to E a system of first-order differential equations defined by the invariant surface conditions associated with a group that is not necessarily a symmetry group of the system, and to require that the resulting overdetermined system admits the prescribed group as a symmetry group. Other types of differential constraints give rise to partially invariant solutions, or separation of variables (see Olver and Rosenau [1986]). They also can be set up so that the appended system admits *a priori* fixed group of transformations (see Fushchich, Serov, and Chopik [1988], Fushchich and Serov [1988]). Certainly, these examples do not exhaust all possible interesting classes of differential constraints, and the full applicability of the method remains unexplored.

11.1. The nonclassical method

We begin by presenting a version of the nonclassical symmetries first discussed in Bluman and Cole [1969] in their treatment of generalized self-similar solutions of the linear heat equation.

11.1.1. Theoretical background

Consider a kth order system E of differential equations

$$\Delta_\nu(x, u, u^{(k)}) = 0, \qquad \nu = 1, \ldots, l, \tag{11.1}$$

in n independent variables $x = (x_1, \ldots, x_n)$, and q dependent variables $u = (u^1, \ldots, u^q)$, with $u^{(k)}$ denoting the derivatives of the u's with respect to the x's up to order k. Suppose that $\mathbf{v}$ is a vector field on the space $R^n \times R^q$ of independent and dependent variables:

$$\mathbf{v} = \sum_{i=1}^{n} \xi^i(x, u) \frac{\partial}{\partial x_i} + \sum_{\alpha=1}^{q} \varphi^\alpha(x, u) \frac{\partial}{\partial u^\alpha}. \tag{11.2}$$

(In what follows, the derivatives $\partial/\partial x_i$, $\partial/\partial u^\alpha$ and so on will be for short denoted by ∂_{x_i}, ∂_{u^α}, and so on.) The graph of a solution

$$u^\alpha = f^\alpha(x_1, \ldots, x_n), \qquad \alpha = 1, .., q, \tag{11.3}$$

to the system defines an n-dimensional submanifold $\Gamma_f \subset R^n \times R^q$ of the space of independent and dependent variables. The solution will be invariant under the one-parameter subgroup generated by $\mathbf{v}$ if and only if Γ_f is an invariant submanifold of this group. By applying the well-known criterion of invariance of a submanifold under a vector field we get that (11.3) is invariant under $\mathbf{v}$ if and only if f satisfies the first-order system E_Q of partial differential equations,

$$Q^\alpha(x, u, u^{(1)}) = \varphi^\alpha(x, u) - \sum_{i=1}^n \xi^i(x, u) \frac{\partial u^\alpha}{\partial x_i} = 0, \qquad \alpha = 1, \ldots, q, \quad (11.4)$$

known as the *invariant surface conditions*. The q-tuple $Q = (Q^1, \ldots, Q^q)$ is known as the *characteristic* of the vector field 11.3. Since all the solutions of (11.4) are invariant under $\mathbf{v}$, the first prolongation $\mathbf{v}^{(1)}$ of $\mathbf{v}$ is tangent to E_Q. Therefore, we conclude that invariant solutions of the system 11.1 are in fact solutions of the joint overdetermined system 11.1, 11.4. In what follows, the kth prolongation of the invariant surface conditions 11.4 will be denoted by $E_Q^{(k)}$, which is a kth-order system of partial differential equations obtained by appending to (11.4) its partial derivatives with respect to the independent variables of orders $j \leq k - 1$.

For the system 11.1, 11.4 to be compatible, the kth prolongation $\mathbf{v}^{(k)}$ of the vector field $\mathbf{v}$ must be tangent to the intersection $E \cap E_Q^{(k)}$:

$$\mathbf{v}^{(k)}(\Delta_\nu)|_{E \cap E_Q^{(k)}} = 0, \qquad \nu = 1, \ldots, l. \tag{11.5}$$

If the equations 11.5 are satisfied, then the vector field 11.2 is called a *nonclassical infinitesimal symmetry* of the system 11.1. The relations 11.5 are generalizations of the relations

$$\mathbf{v}^{(k)}(\Delta_\nu)|_E = 0, \qquad \nu = 1, \ldots, l, \tag{11.6}$$

for the vector fields of the infinitesimal classical symmetries. Inserting l variables $u^{(n)}$ found from (11.1) into (11.6), taking the Taylor series of the functions $\mathbf{v}^{(k)}(\Delta_\nu)|_E$ with respect to the remaining variables, and setting the coefficients of these series equal to zero generate an overdetermined system of linear differential equations of an order not larger than k for the coefficients $\xi^i(x, u)$, $\varphi^\alpha(x, u)$ of the vector field 11.2 of the infinitesimal classical symmetries. A similar procedure is applicable to the case of the nonclassical infinitesimal symmetries with an evident difference that in general one has fewer determining equations than in the classical case. Therefore, we expect that nonclassical symmetries are much more numerous than classical ones, since any classical symmetry is clearly a nonclassical one.

The important feature of determining equations for nonclassical symmetries is that they are nonlinear. This implies that the space of nonclassical symmetries does not, in general, form a vector space. Moreover, the Lie bracket of two nonclassical symmetry vector fields is not, as a rule, a nonclassical symmetry. If $\lambda(x, u)$ is an arbitrary function, then the prolongation formulae for vector fields imply that

$$\left(\lambda\mathbf{v}\right)^{(k)}\Big|_{E_Q^{(k)}} = \lambda\mathbf{v}^{(k)}\Big|_{E_Q^{(k)}}. \tag{11.7}$$

Formula 11.7 means that if the vector field $\mathbf{v}$ is a nonclassical symmetry, then $\lambda\mathbf{v}$ is also a nonclassical symmetry yielding the same equations 11.4. This property allows us to normalize any one nonvanishing coefficient of the vector field 11.2 by setting it equal to one when finding nonclassical symmetries.

11.1.2. Nonclassical symmetries of Burgers' equation

As an example of finding the nonclassical symmetries, consider the system E of first-order equations

$$u_t + uv - v_x = 0, \qquad u_x - v = 0 \tag{11.8}$$

obtained from the well-known Burgers' equation $u_t + uu_x - u_{xx} = 0$. If we assume that the coefficient of ∂_t of the vector field 11.2 does not identically equal zero, then for the vector field

$$\mathbf{v} = \partial_t + \xi(t, x, u, v)\partial_x + \varphi(t, x, u, v)\partial_u + \psi(t, x, u, v)\partial_v, \tag{11.9}$$

the invariant surface conditions are

$$u_t + \xi u_x = \varphi, \qquad v_t + \xi v_x = \psi. \tag{11.10}$$

The equation 11.5 take the form:

$$D_t\varphi - u_x D_t\xi + v\varphi + u\psi - D_x\psi - v_x D_x\xi = 0,$$

$$D_x\varphi - u_x D_x\xi - \psi = 0, \tag{11.11}$$

where $D_t = \partial_t + u_t\partial_u + v_t\partial_v$, $D_x = \partial_x + u_x\partial_u + v_x\partial_v$. The variables u_t, u_x, v_t, v_x found from (11.8), (11.10):

$$u_t = \varphi - \xi v, \qquad u_x = v, \qquad v_x = \varphi - \xi v + uv, \qquad v_t = \psi - \xi(\varphi - \xi v + uv)$$

must be substituted into (11.11). After substituting, the latter becomes an underdetermined system of two differential equations for three unknown functions

$\xi(t, x, u, v)$, $\varphi(t, x, u, v)$, $\psi(t, x, u, v)$. It is evident that this is a generic situation for first-order systems in two independent variables. It means that there is a rich variety of nonclassical symmetries and the problem is how one can explicitly obtain them.

We will restrict ourselves to finding nonclassical symmetries 11.9 for which $\xi = \xi(t, x, u)$, $\varphi = \varphi(t, x, u)$, $\psi = \psi(t, x, u, v)$, which corresponds to the nonclassical symmetries of Burgers' equation itself. In that case the system 11.11 is written as

$$\varphi_t + \varphi_u(\varphi - \xi v) - v\big(\xi_t + \xi_u(\varphi - \xi v)\big) + \varphi v + uv$$

$$- \psi_x - \psi_u v - \psi_v(\varphi - \xi v) + (\xi_x + v\xi_u)(\varphi - \xi v + uv) = 0,$$

$$\varphi_x + v\varphi_u - v(\xi_x + v\xi_u) - \psi = 0. \tag{11.12}$$

The second equation of (11.12) implies that the coefficient ψ is at most a quadratic function of v. After inserting ψ and its derivatives, as determined by the second equation in (11.12), into the first equation we obtain the equation:

$$\varphi_t + \varphi_u(\varphi - \xi v) - v\xi_t + \varphi v + u\big(\varphi_x + v\varphi_u - v(\xi_x + v\xi_u)\big)$$

$$- \varphi_{xx} - v\varphi_{ux} + v(\xi_{xx} + v\xi_{ux}) - v\big(\varphi_{xu} + v\varphi_{uu} - v(\xi_{xu} + v\xi_{uu})\big)$$

$$- (\varphi - \xi v)(\varphi_u\xi_x - 2v\xi_u) + \xi_x(\varphi - \xi v) = 0. \tag{11.13}$$

The function in the left-hand side of (11.13) is a third-order polynomial in v; its coefficients yield the equations:

$$\xi_{uu} = 0, \qquad \varphi_{uu} = 2\xi_{xu} + 2(u - \xi)\xi_u,$$

$$2\varphi\xi_u - 2\xi\xi_x - \xi_t - 2\varphi_{xu} + \xi_{xx} + u\xi_x + \varphi = 0,$$

$$2\varphi\xi_x + \varphi_t + u\varphi_x - \varphi_{xx} = 0. \tag{11.14}$$

From the first two equations of (11.14), we find that

$$\xi = a(t, x)u + b(t, x),$$

$$\varphi = \frac{a(1 - a)}{3}u^3 + (a_x - ab)u^2 + \alpha(t, x)u + \beta(t, x), \tag{11.15}$$

where $a(t, x)$, $b(t, x)$, $\alpha(t, x)$, $\beta(t, x)$ are arbitrary functions. After inserting (11.15) into the third equation of (11.14) we get a third-order polynomial in u with the coefficient of u^3 equal to $a(1 - a)(1 + 2a)$. From this we deduce that the function $a(t, x)$ is constant and it can take three values: $a = 0, a = 1, a = -1/2$. The other three coefficients of this polynomial imply the equations:

$$ab(1 + 2a) = 0, \qquad (2a + 1)(\alpha + b_x) = 0,$$

$$b_t + 2bb_x - b_{xx} + 2\alpha_x - (1 + 2a)\beta = 0, \tag{11.16}$$

and the fourth equation of (11.14) yields the equations:

$$a(1 + 2a)b_x = 0, \qquad a(b_t + 2bb_x - b_{xx}) - \alpha_x = 0,$$

$$\alpha_t + \beta_x - \alpha_{xx} + 2\alpha b_x = 0, \qquad \beta_t - \beta_{xx} + 2\beta b_x = 0. \tag{11.17}$$

In the case $a = 0$ the functions $b(t, x), \alpha(t, x), \beta(t, x)$ satisfy the system:

$$\alpha_x = 0, \qquad \alpha + b_x = 0, \qquad b_t + 2bb_x - \beta = 0,$$

$$\alpha_t + \beta_x + 2\alpha b_x = 0, \qquad \beta_t - \beta_{xx} + 2\beta b_x = 0, \tag{11.18}$$

which can easily be solved and produces the vector field

$$\mathbf{v} = (At^2 + 2Bt + C)\partial_t + (Atx + Bx + Dt)\partial_x + (Ax - Atu - Bu + D)\partial_u$$

with A, B, C, D parameters. This vector field belongs in fact to the five-dimensional Lie algebra of the classical symmetries of Burgers' equation. In the case $a = 1$, Equations 11.16 and 11.17 yield $b = 0$, $\alpha = 0$, $\beta = 0$, so there exists only the nonclassical vector field $\mathbf{v} = \partial_t + u\partial_x$, with invariant solutions $u = (x - x_0)/(t - t_0)$.

Let $a = -1/2$, then the functions $b(t, x), \alpha(t, x), \beta(t, x)$ are found as solutions of the system:

$$b_t + 2bb_x - b_{xx} + 2\alpha_x = 0, \quad \alpha_t + \beta_x - \alpha_{xx} + 2\alpha b_x = 0, \quad \beta_t - \beta_{xx} + 2b_x\beta = 0,$$

which admits solutions $\alpha = 0$, $\beta = 0$, and $b(t, x)$ satisfying Burgers' equation $b_t + 2bb_x - b_{xx} = 0$. This allows us to successively generate solutions of Burgers' equation from the solutions of the same equation. The process reminds one of the generation of solutions to soliton equations by successive Bäcklund transformations. Setting $b(t, x) = 0$ we can get the family of vector fields of the nonclassical symmetries:

$$\mathbf{v} = 4\partial_t - 2u\partial_x + (-u^3 + (\gamma t + \delta)u + \epsilon - \gamma x)\partial_u \qquad (11.19)$$

with parameters γ, δ, ϵ.

Now assume that the coefficient of ∂_t in (11.2) equals zero and try to find the infinitesimal nonclassical symmetries of the form

$$\mathbf{v} = \partial_x + \phi(t, x, u)\partial_u + \psi(t, x, u, v)\partial_v,$$

for which the invariant surface conditions are the following ones:

$$u_x = \phi, \qquad v_x = \psi. \qquad (11.20)$$

Relation 11.5 leads to the system of equations for the functions ϕ, ψ:

$$\phi_t + u\phi\phi_u - \psi\phi_u + \phi^2 + \psi u - \psi_x - \phi\psi_u - \psi_v\psi = 0, \qquad \phi_x + \phi_u\phi - \psi = 0. \qquad (11.21)$$

These are not differential equations, since the arguments t, x, u, and v in (11.21) are tied by the relation $\phi(t, x, u) - v = 0$, which is a consequence of system 11.8, 11.20. There are no established methods to solve such systems. Severely restricting the class of solutions, one can regard (11.21) as a system of differential equations. Exact solutions of (11.21) yielding invariant solutions that are not invariant under classical symmetries have not yet been obtained.

11.1.3. Nonclassical symmetries and direct reduction methods

Clarkson and Kruskal [1989] proposed a direct method for determining ansätze which reduce the partial differential equation to a single ordinary differential equation. This method was generalized by Galaktionov [1990], who showed how to effect reductions to two (or more) coupled ordinary differential equations, and was applied to the study of the blow-up of solutions to parabolic equations. In Arrigo, Broadbridge, and Hill [1993] and Olver [1994a], it was proved that these reduction methods are equivalent to particular cases of the nonclassical symmetry method.

For simplicity, consider a partial differential equation 11.1 in two independent variables x, t. The differential equation admits a *direct reduction* if there exist functions $z = \zeta(x, t)$, $u = U(x, t, w)$, such that the Clarkson-Kruskal ansatz

$$u(x, t) = U(x, t, w(z)) = U\big(x, t, w(\zeta(x, t))\big) \qquad (11.22)$$

reduces (11.1) to a single ordinary differential equation for $w = w(z)$. Let $\mathbf{w} = \tau(x, t)\partial_t + \xi(x, t)\partial_x$ be any vector field such that $\mathbf{w}(\zeta) = 0$, i.e. $\zeta(x, t)$ is the unique (up to functions thereof) invariant of the one-parameter group generated by $\mathbf{w}$. Applying $\mathbf{w}$ to the ansatz 11.22, we find

$$\tau u_t + \xi u_x = \tau U_t + \xi U_x \equiv V(x, t, w). \tag{11.23}$$

On the other hand, assuming $U_w \neq 0$, we can solve (11.22) for $w = W(x, t, u)$ using the Implicit Function Theorem. (We avoid singular points, and note that if $U_w \equiv 0$, the ansatz would not explicitly depend on w.) Substituting this into the right-hand side of (11.23), we find that if u has the form 11.22, then it satisfies a first-order quasi-linear partial differential equation of the form

$$\mathbf{w}(u) = \tau(x, t)u_t + \xi(x, t)u_x = \varphi(x, t, u). \tag{11.24}$$

Conversely, if u satisfies an equation of the form 11.24, then it can be shown that u satisfies a direct reduction-type ansatz 11.22. Therefore, there is a one-to-one correspondence between ansätze of the direct reduction form 11.22 with $U_w \neq 0$ and quasi-linear first-order differential constraints 11.24. Solutions $u = f(x, t)$ to (11.24) are just the functions which are invariant under the one-parameter group generated by the vector field

$$\mathbf{v} = \tau(x, t)\partial_t + \xi(x, t)\partial_x + \varphi(x, t, u)\partial_u. \tag{11.25}$$

Note in particular that $\mathbf{w}$ generates a group of "fiber-preserving transformations", meaning that the transformations in x and t do not depend on the coordinate u.

In the direct method, one requires that the ansatz 11.22 reduces the partial differential equation 11.1 to an ordinary differential equation. In the nonclassical method of Bluman and Cole, one requires that the differential constraint 11.24 which requires the solution to be invariant under the group generated by $\mathbf{w}$ be compatible with the original partial differential equation 11.1, in the sense that the overdetermined system of partial differential equations defined by (11.1), (11.24) has no integrability conditions. The following result demonstrates the equivalence of the nonclassical method (with projectable symmetry generator) and the direct method.

THEOREM. *The ansatz 11.22 will reduce the partial differential equation 11.1 to a single ordinary differential equation for $w(z)$ if and only if the overdetermined system of partial differential equations defined by (11.1), (11.24) is compatible.*

The proof and generalizations to Galaktionov's "nonlinear separation" are discussed in Olver [1994]. See also Arrigo, Broadbridge, and Hill [1993] and Zidowitz [1993].

As an example, the ansatz $u = w(z) - t^2$, where $z = x - \frac{1}{2}t^2$, reduces the Boussinesq equation

$$u_{tt} + \frac{1}{2}(u^2)_{xx} + u_{xxxx} = 0$$

to a fourth-order ordinary differential equation

$$w'''' + ww'' + (w')^2 - w' = 2.$$

This reduction follows from the constraint $tu_x + u_t + 2t = 0$ arising from the nonclassical symmetry $\mathbf{v} = t\partial_x + \partial_t - 2t\partial_u$.

11.2. Multidimensional modules of nonclassical symmetries

As was mentioned above, the Lie bracket $[\mathbf{v}, \mathbf{w}]$ of two infinitesimal nonclassical symmetries $\mathbf{v}$ and $\mathbf{w}$ is not in general a nonclassical infinitesimal symmetry. The easiest way to be convinced of this fact is to consider the Lie bracket of two distinct vector fields 11.19. For nonclassical symmetries the analog of multidimensional Lie algebras of the classical symmetries is a multidimensional module or differential system. Let us consider the simplest case of the latter. Consider a two-dimensional differential system $\mathbf{g}$ (or distribution) on the space of independent and dependent variables, spanned by two independent vector fields $\mathbf{v}$ and $\mathbf{w}$. Thus $\mathbf{g}$ is defined as the set of all vector fields $\mathbf{u}$ which can be represented in the form $\mathbf{u} = \mathbf{f}(\mathbf{x}, \mathbf{u})\mathbf{v} + g(\mathbf{x}, \mathbf{u})\mathbf{w}$. Suppose that $\mathbf{g}$ is involutive, i.e., closed under the Lie bracket: $[\mathbf{v}, \mathbf{w}] \subset \mathbf{g}$. In this case, Frobenius' Theorem implies that we can find a new basis of $\mathbf{g}$ vector fields which have a vanishing Lie bracket. Denote by $E\mathbf{g}$ the union of the invariant surface conditions 11.4 for $\mathbf{v}$ and $\mathbf{w}$; the solutions of $E\mathbf{g}$ are the $\mathbf{g}$ invariant functions. As in the case of one-dimensional nonclassical modules, $E\mathbf{g}$ is invariant under $\mathbf{g}$. For $\mathbf{g}$ to be a two-dimensional module of nonclassical symmetries it is necessary that the basis vector fields $\mathbf{v}$ and $\mathbf{w}$ satisfy Equation 11.5 with $E_Q^{(k)}$ replaced by $E_{\mathbf{g}}^{(k)}$.

11.2.1. Two-dimensional modules of nonclassical symmetries of nonlinear acoustics equations

The following system of equations:

$$uu_t + u_x + v_y = 0, \qquad u_y - v_t = 0 \qquad (11.26)$$

describes a sound beam propagating in a nonlinear medium. Consider the two-dimensional involutive module **g** of vector fields with the basis

$$\mathbf{v} = \partial_t + \xi(u, v)\partial_y, \qquad \mathbf{w} = \partial_x + \zeta(u, v)\partial_y.$$

The vector fields **v** and **w** commute; therefore, the coefficients $\xi(u, v)$, $\zeta(u, v)$ are found from Equation 11.5, and the system $E\mathbf{g}$ is

$$u_t + \xi u_y = 0, \qquad u_x + \zeta u_y = 0,$$

$$v_t + \xi v_y = 0, \qquad v_x + \zeta v_y = 0. \tag{11.27}$$

The system 11.26, 11.27 can be treated as a system of six linear homogeneous algebraic equations for six first derivatives of the functions ξ, ζ. For this system to admit nontrivial solutions, the equation

$$\zeta(u, v) = -\big(1 + u\xi^2(u, v)\big)/\xi(u, v)$$

must hold. In this case all derivatives can be expressed through the derivative u_y:

$$u_t = -\xi u_y, \qquad u_x = \frac{1+u\xi^2}{\xi}u_y, \qquad v_t = u_y,$$

$$v_x = -\frac{1+u\xi^2}{\xi^2}u_y, \qquad v_y = -\frac{u_y}{\xi}. \tag{11.28}$$

Calculating $\mathbf{v}^{(1)}(uu_t + u_x + v_y)$, $\mathbf{v}^{(1)}(u_y - v_t)$, $\mathbf{w}^{(1)}(uu_t + u_x + v_y)$, $\mathbf{w}^{(1)}(u_y - v_t)$ and inserting (11.28), we find that they are identically zero. So the system 11.26 admits the two-dimensional module of nonclassical symmetries with the basis vector fields

$$\mathbf{v} = \partial_t + \xi(u, v)\partial_y, \qquad \mathbf{w} = \partial_x - \frac{1 + u\xi(u, v)^2}{\xi(u, v)}\partial_y,$$

where $\xi(u, v)$ is an arbitrary function.

11.2.2. Bäcklund transformations and two-dimensional modules of nonclassical symmetries for the sine-Gordon equation

It is well known that if $u = f(x, y)$ is any particular solution to the sine-Gordon equation,

$$2u_{xy} = \sin 2u, \tag{11.29}$$

then the system of equations

$$u_x = -f_x + \sin(u - f), \qquad u_y = f_y + \sin(u + f) \qquad (11.30)$$

determines the Bäcklund transformation for the sine-Gordon equation 11.29. Equation 11.29 can be treated as a compatibility condition for the system 11.30. The function $u(x, y)$ found from the system 11.30 is a solution of Equation 11.29. From the point of view of the nonclassical symmetries, this can be interpreted as follows.

Consider the vector fields

$$\mathbf{v} = \partial_x + (-f_x + \sin(u - f))\partial_u, \qquad \mathbf{w} = \partial_y + (f_y + \sin(u + f))\partial_u.$$

Provided the function $f(x, y)$ is a solution of (11.29), these vector fields have a vanishing Lie bracket, so they give rise to a two-dimensional involutive module $\mathbf{g}$ with $E\mathbf{g}$ given by (11.30). In the case considered, the relation $E \cap E_\mathbf{g}^{(2)} = E_\mathbf{g}^{(2)}$ holds, hence $\mathbf{g}$ is a two-dimensional module of the nonclassical symmetries for the sine-Gordon equation 11.29 with all invariant functions under $\mathbf{g}$ automatically satisfying (11.29).

11.3. Partial symmetries

A further step towards generalization of the classical symmetries is consideration of tangent transformations instead of point ones.

11.3.1. Contact transformations and modules of partial symmetries

According to Bäcklund's Theorem, in the case of one unknown function $u(x)$, transformations of J^k that preserve the contact structure (contact transformations) are prolongations of either point transformations or contact transformations on J^1. The infinitesimal contact transformations on the space J^1 are in one-to-one correspondence with their characteristic functions $Q(x, u, u^{(1)})$, which generates the contact vector field

$$\mathbf{v}_Q = -Q_{u_{x_i}}\partial_{x_i} + (Q - u_{x_i}Q_{u_{x_i}})\partial_u + (Q_{x_i} + u_{x_i}Q_u)\partial_{u_{x_i}}.$$

This vector field is the first prolongation of a point transformation if and only if Q is an affine function of the derivative coordinates $u_{x_1}, \ldots, u_{x_n}$. A function $u = f(x)$ is invariant with respect to $\mathbf{v}_Q$ if and only if it satisfies the invariant surface condition

$$Q(x, u, u^{(1)}) = 0. \tag{11.31}$$

Let E_Q denote the submanifold of J^1 determined by Equation 11.31, and $E_Q^{(k)}$ its prolongation to J^k. The contact vector field $\mathbf{v}_Q$ is called a *partial symmetry* of the differential equation

$$\Delta(x, u, u^{(k)}) = 0 \tag{11.32}$$

if $\mathbf{v}_Q^{(k)}$ is tangent to the intersection $E \cap E_Q^{(k)}$.

Partial symmetries admit the natural structure of a one-dimensional module. Clearly, if Q generates a partial symmetry, so does $\tilde{Q} = gQ$, where g is an arbitrary function on J^1, since

$$\mathbf{v}_{gQ}^{(k)}\big|_{E_Q^{(k)}} = g\mathbf{v}_Q^{(k)}\big|_{E_Q^{(k)}}. \tag{11.33}$$

Relation 11.33 implies that, as long as it depends explicitly on the derivative coordinates, the characteristic function of the infinitesimal partial symmetry can be chosen in the form $Q = -u_{x_j} + \phi(x, u, \tilde{u}^{(1)})$ for some index j with $\tilde{u}^{(1)}$ denoting the set of first derivatives of u with the derivative u_{x_j} omitted.

Now consider the r-tuple $Q = (Q_1, \ldots, Q_r)$ of functions on J^1 satisfying the relation

$$\text{rank } \|\partial Q_i / \partial u_{x_j}\| = r \le n \tag{11.34}$$

and r contact vector fields $\mathbf{v}_{Q_1}, \ldots, \mathbf{v}_{Q_r}$. Let E_Q now denote the system of differential equations

$$Q_1(x, u, u^{(1)}) = 0, \quad \ldots, \quad Q_r(x, u, u^{(1)}) = 0 \tag{11.35}$$

satisfied by functions $u = f(x)$ invariant under all of the vector fields $\mathbf{v}_{Q_1}, \ldots, \mathbf{v}_{Q_r}$. The system 11.35 is compatible iff the relations

$$(Q_i, Q_j)|_{E_Q} = 0, \quad 1 \le i \le j \le r, \tag{11.36}$$

hold true. Here, (Q_i, Q_j) is the "Lagrange" bracket of the functions Q_i and Q_j, defined as the characteristic function of the Lie bracket $[\mathbf{v}_{Q_i}, \mathbf{v}_{Q_j}]$, i.e.,

$$[\mathbf{v}_{Q_i}, \mathbf{v}_{Q_j}] = \mathbf{v}_{(Q_i, Q_j)}. \tag{11.37}$$

Since $\mathbf{v}_{Q_i}(Q_j) = (Q_i, Q_j) + Q_j \partial Q_i / \partial u$, the submanifold E_Q is invariant under $\mathbf{v}_{Q_1}, \ldots, \mathbf{v}_{Q_r}$ provided (11.36) are satisfied.

Any smooth function R defined on J^1 which vanishes on E_Q can be represented in the form $R = a_i Q_i$ for functions $a_i(x, u, u^{(1)})$. Therefore, the restriction of the vector field

$$\mathbf{v}_R = a_i \mathbf{v}_{Q_i} + Q_i \mathbf{v}_{a_i} - R \partial_u \tag{11.38}$$

to E_Q equals $a_i \mathbf{v}_{Q_i}|_{E_Q}$. Relation 11.38 makes valid the following assertion. Denote by $I(Q)$ the ideal of the functions on J^1 vanishing on E_Q and by $A(Q)$ the family of contact vector fields generated by the functions in $I(Q)$. Then the restriction of $A(Q)$ to E_Q is an r-dimensional module $\mathbf{g}$ of vector fields over the ring of smooth functions on E_Q. The basis of $\mathbf{g}$ consists of the restrictions of the vector fields $\mathbf{v}_{Q_1}, \ldots, \mathbf{v}_{Q_r}$ to E_Q. Moreover, relations 11.36 to 11.38 imply that $\mathbf{g}$ is an involutive module: $[\mathbf{g}, \mathbf{g}] \subset \mathbf{g}$.

It is easily deduced from (11.34), (11.35) that the same module $\mathbf{g}$ is generated by the functions $\tilde{Q}_1 = -u_{x_{i_1}} + \psi_1(x, u, \tilde{u}^{(1)}), \ldots, \tilde{Q}_r = -u_{x_{i_r}} + \psi_r(x, u, \tilde{u}^{(1)})$, where $i_1, \ldots, i_r$ are the indices of r linearly independent columns of the matrix 11.34. The functions $\tilde{Q}_1, \ldots, \tilde{Q}_r$ are obtained by solving (11.35) with respect to the derivative coordinates $u_{x_{i_1}}, \ldots, u_{x_{i_r}}$. They generate the contact vector fields $\mathbf{v}_{\tilde{Q}_1}, \ldots, \mathbf{v}_{\tilde{Q}_r}$, which commute when restricted to E_Q.

In the case the relations

$$\mathbf{v}_{Q_i}(\Delta)|_{E \cap E_Q} = 0, \qquad i = 1, \ldots, r, \tag{11.39}$$

are satisfied, the module $\mathbf{g}$ is called *an r-dimensional module of infinitesimal partial symmetries of* (11.32).

11.3.2. Two-dimensional modules of partial symmetries for a family of nonlinear heat equations

Consider the family of nonlinear heat equations

$$u_t = (f(u)u_x)_x + g(u) \tag{11.40}$$

in one space variable x with $f(u)$, $g(u)$ smooth functions. We try to find two-dimensional modules of partial symmetries of (11.40). The characteristic functions Q_1, Q_2 taken as

$$Q_1 = -u_t + a(t, x, u), \qquad Q_2 = -u_x + b(t, x, u) \tag{11.41}$$

imply the compatibility condition 11.36 in the form

$$a_x + a_u b - b_t - a b_u = 0. \tag{11.42}$$

The system E_Q in the case considered, precisely,

$$u_t = a(t, x, u), \qquad u_x = b(t, x, u), \qquad (11.43)$$

admits a one-parameter family of solutions. One of the six equations determining the system $E \cap E_Q^{(2)}$, which is geometrically a two-dimensional surface in the eight-dimensional space J^2, looks like

$$a = (b_x + bb_u)f + b^2 f' + g. \qquad (11.44)$$

If this equation is not a differential consequence of (11.39), we have to solve functional equations for the functions $a(t, x, u)$, $b(t, x, u)$ obtained by restricting (11.39) to the two-dimensional surface given by (11.44) in the space R^3 having coordinates t, x, u. To avoid this difficult problem, we treat (11.44) as a differential equation satisfied by $a(t, x, u)$, $b(t, x, u)$. In this approach relations 11.39 are automatically satisfied, and in order that $\mathbf{v}_{Q_1}$, $\mathbf{v}_{Q_2}$ generate the two-dimensional module $\mathbf{g}$ of partial symmetries of (11.40) the functions $a(t, x, u)$ and $b(t, x, u)$ must satisfy differential equations 11.42 and 11.44.

If we substitute the function $a(t, x, u)$ given by (11.44) into (11.42), we obtain the equation

$$b_t = (b_{xx} + 2bb_{xu} + b^2 b_{uu})f + (3bb_x + 2b^2 b_u)f' + b^3 f'' + bg' - gb_u \quad (11.45)$$

for the function $b(t, x, u)$. Let us try to find this function in the form: $b(t, x, u) = \theta(t)h(u)$, i.e., independent of x and admitting the separation of variables t, u. After substituting $b(t, u) = \theta(t)h(u)$ in (11.45) we obtain the relation

$$\dot{\theta}(t) = \theta^3(t)h(u)(f(u)h(u))'' + \theta(t)h(u)(g(u)/h(u))'.$$

This equation has nontrivial solutions provided

$$h(u)(f(u)h(u))'' = \lambda, \qquad h(u)(g(u)/h(u))' = \mu, \qquad (11.46)$$

with λ, μ constant. Equation 11.46 contains three unknown functions, and they can be treated from various points of view. For example, if we regard the function $h(u)$ as given, then (11.46) are equations for $f(u)$ and $g(u)$.

The function $b(t, u) = \theta(t)h(u)$ yields the invariant solutions

$$u(t, x) = F(\theta(t)x + \phi(t)) \qquad (11.47)$$

to (11.40) as it can be seen from the second equation in (11.43). The function $\theta(t)$ satisfies the ordinary differential equation $\dot{\theta} = \lambda\theta^3 + \mu\theta$ integrated explicitly, and the function $\phi(t)$ is a solution of the equation obtained after substituting (11.47) either into (11.40) or into the first equation in (11.41). If we take $h(u) = u^{-1}$, $f(u) = u^2 + u$, $g(u) = u/2$, we obtain the family $u(t, x) = \sqrt{2((x + c)\exp t + \exp 2t)}$ of invariant solutions to the equation $u_t = ((u^2 + u)u_x)_x + u/2$.

11.3.3. Partial symmetries and multidimensional integrable differential equations

Consider the second-order partial differential equation 11.32 in two independent variables. In this case partial symmetries allow us to associate new differential equations of the second order in four independent variables with equation 11.32. The solutions of the Cauchy problem for these associated equations can be expressed through solutions of the appropriately posed Cauchy problems for the original equation 11.32. In particular, each second-order linear equation in two independent variables gives rise to multidimensional nonlinear equations which are linearizable in the sense described below. Let us consider the case of an evolution equation, and suppose that the characteristic function of the partial symmetry is taken in the form: $Q = -u_t + \phi(t, x, u, u_x)$. Then the relation

$$\mathbf{v}_Q^{(2)}(\Delta)|_{E \cap E_Q^{(2)}} = 0 \qquad (11.48)$$

satisfied by the partial symmetry $\mathbf{v}_Q$ is actually a differential equation of the second order for the function $\phi(t, x, u, u_x)$. Indeed, by the prolongation formula, the coefficients of the contact vector field $\mathbf{v}_Q^{(2)}$ depend on the derivatives of the function ϕ at most of the second order. Moreover, the variables u_t, u_{tx}, and u_{xx} can be found as functions of the variables t, x, u, u_x from the equations determining $E_Q^{(2)}$ and from (11.32). (We leave aside the case when the equation $u_t = \phi(t, x, u, u_x)$ is an intermediate integral of (11.32).) For example, if (11.32) is the linear heat equation $u_t = u_{xx}$, then Equation 11.48 takes the form:

$$\phi_t = \phi^2\phi_{pp} + 2p\phi\phi_{up} + p^2\phi_{uu} + 2\phi\phi_{xp} + 2p\phi_{xu} + \phi_{xx}, \qquad (11.49)$$

where $p = u_x$. Suppose that the Cauchy problem :

$$\phi|_{t=0} = \phi_0(x, u, p), \qquad (11.50)$$

for (11.49) and the Cauchy problem:

$$u|_{t=0} = a(x) \qquad (11.51)$$

for (11.32) are both well posed. Let us determine which Cauchy data $a(x)$ generate solutions of (11.32) invariant under the vector field $\mathbf{v}_Q$ determined by the given solution ϕ of Equation 11.49. Since invariant solutions of Equation 11.32 are the common solutions to (11.32) and the equation $u_t - \phi(t, x, u, u_x) = 0$, we can consider this joint system at $t = 0$ and obtain the equation

$$a''(x) = \phi_0(x, a(x), a'(x))$$

for the initial function $a(x)$. The general solution of the latter equation depends on two constants c_1, c_2, so there appears the two-parameter family $u(t, x, c_1, c_2)$ of invariant solutions of the Cauchy problem 11.32, 11.51. Suppose that the system of equations

$$u(t, x, c_1, c_2) = v, \qquad u_x(t, x, c_1, c_2) = p$$

uniquely determines c_1 and c_2 as implicit functions of t, x, v, p, then

$$\phi(t, x, v, p) = u_x(t, x, c_1(t, x, v, p), c_2(t, x, v, p))$$

is a solution of the Cauchy problem 11.49, 11.50.

11.3.4. Induced classical symmetries

Suppose $\mathbf{w}$ is a vector field of infinitesimal classical symmetries of system 11.1. Denote by $\exp(\epsilon \mathbf{w})$ the one-parameter family of transformations of the space $R^n \times R^q$ of independent and dependent variables associated with $\mathbf{w}$ and by $\exp(\epsilon \mathbf{w})_*$ the differential $d(\exp(\epsilon \mathbf{w}))$ of $\exp(\epsilon \mathbf{w})$ treated as a mapping of vector fields on $R^n \times R^q$. Then if $\mathbf{g}$ is a module of the Bluman-Cole infinitesimal symmetries of (11.1), the module $\mathbf{g}_\epsilon = \exp(\epsilon \mathbf{w})_*(\mathbf{g})$ forms a one-parameter family of modules of nonclassical symmetries for the same system.

To see this, assume that $\mathbf{v}_1, \ldots, \mathbf{v}_r$ is a basis of $\mathbf{g}$. Generalizing the construction of Section 11.2.1 we can state that these vector fields must satisfy the relations

$$[\mathbf{v}_i, \mathbf{v}_j] \in \mathbf{g}, \qquad \mathbf{v}_j^{(k)}(E \cap E_{\mathbf{g}}^{(k)})|_{E \cap E_{\mathbf{g}}^{(k)}} = 0, \qquad 1 \le i < j \le r, \qquad j = 1, \ldots, r. \tag{11.52}$$

Let $\mathbf{g}_\epsilon = \exp(\epsilon \mathbf{w})_* \mathbf{g}$, which is spanned by $\mathbf{v}_{\epsilon i} = \exp(\epsilon \mathbf{w})_*(\mathbf{v}_i)$. By the general properties of the mapping $\exp(\epsilon \mathbf{w})_*$ the module $\mathbf{g}_\epsilon$ is also an r-dimensional module of vector fields closed under the Lie bracket: $[\mathbf{v}_{\epsilon i}, \mathbf{v}_{\epsilon j}] \in \mathbf{g}$. The only solutions of the system $E\mathbf{g}$ are vector-valued functions $u^{(\alpha)} = f_\alpha(x)$, $\alpha = 1, \ldots q$, invariant under $\mathbf{g}$. Indeed, if Γ_f is invariant under $\mathbf{g}$, then $\exp(\epsilon \mathbf{w})(\Gamma_f)$ is a graph of a solution invariant under $\mathbf{g}_\epsilon$, hence $E\mathbf{g}_\epsilon = \exp(\epsilon \mathbf{w})(E\mathbf{g})$. From these considerations it follows that $\mathbf{g}_\epsilon$ is tangent to the intersection $E \cap E_{\mathbf{g}_\epsilon}^{(k)}$.

Modules of partial symmetries of Equation 11.32 are treated quite similarly. The basic property of the mapping $\exp(\epsilon \mathbf{w})_*$ allows us to define the concept of vector fields of nonclassical symmetries invariant under $\mathbf{w}$. Such invariant vector field $\mathbf{v}$ satisfies the relation $[\mathbf{w}, \mathbf{v}] = \lambda \mathbf{v}$. Moreover, the flow $\exp(\epsilon \mathbf{w})$ determines the symmetry transformations of the determining equations for the coefficients of nonclassical infinitesimal symmetries.

11.3.5. Partial symmetries and differential substitutions

The differential equations for the functions ϕ in $Q = -u_t + \phi(t, x, u, u_x)$ and ψ in $Q = -u_x + \psi(t, x, u, u_t)$ —more precisely, Equation 11.48 — inherit the classical Lie symmetries of the original equation 11.32. If the Lie algebra of infinitesimal symmetries of Equation 11.32 is at least two-dimensional, then the quotient equation for solutions of Equation 11.48 invariant under two-dimensional subalgebras is a differential equation in two independent variables just as Equation 11.32. There exists a differential substitution of the group nature connecting these two equations.

We describe the origin of this differential substitution, taking as an example the linear heat equation

$$u_t = u_{xx}. \tag{11.53}$$

Equation 11.53 admits the infinite dimensional Lie algebra $\mathbf{g}$ of the classical Lie symmetries with the generators:

$$\mathbf{v}_1 = \partial_t, \quad \mathbf{v}_2 = 2t\partial_t + x\partial_x, \quad \mathbf{v}_3 = 4t^2\partial_t + 4tx\partial_x - (x^2 + 2t)u\partial_u,$$

$$\mathbf{v}_4 = \partial_x, \quad \mathbf{v}_5 = -2t\partial_x + xu\partial_u, \quad \mathbf{v}_6 = u\partial_u, \quad \mathbf{v}_\alpha = \alpha(t, x)\partial_u, \tag{11.54}$$

where $\alpha(t, x)$ is an arbitrary solution of Equation 11.53. Consider the two-dimensional subalgebra $\mathbf{g}_2 \subset \mathbf{g}$ with the generators $\mathbf{v} = \mathbf{v}_6$ and $\mathbf{w} = \partial_u$, and the solutions ϕ of Equation 11.49 invariant under $\mathbf{g}_2$. In the space R^5 of the variables t, x, u, p, ϕ, where the first four are the arguments of the function ϕ, the vector fields $\mathbf{v}$ and $\mathbf{w}$ look as follows:

$$\mathbf{v} = u\partial_u + p\partial_p + \phi\partial_\phi, \qquad \mathbf{w} = \partial_u. \tag{11.55}$$

The solutions ϕ of (11.49) invariant under $\mathbf{g}_2$ are common solutions of (11.46) and the system of invariant surface conditions for the vector fields $\mathbf{v}$, $\mathbf{w}$ given by (11.55):

$$\phi_u = 0, \qquad \phi_{tu} = \phi_{xu} = \phi_{uu} = \phi_{up} = 0,$$

$$p\phi_p = \phi, \qquad p\phi_{tp} = \phi_t, \qquad p\phi_{xp} = \phi_x, \qquad \phi_{pp} = 0. \tag{11.56}$$

We restrict the system 11.49, 11.56 to the submanifold N in the space R^5 determined by the equations $u = 0$, $p = 1$ and coordinatized by the variables t, x.

After expressing the outer derivatives ϕ_{pp}, ϕ_{up}, ϕ_{uu}, ϕ_{xp}, ϕ_{xu} through the inner derivatives ϕ_t, ϕ_x, with the aid of (11.56) we obtain the quotient equation:

$$v_t = 2vv_x + v_{xx}. \tag{11.57}$$

This is Burgers' equation for the restriction $v(t, x)$ of the function $\phi(t, x, u, u_x)$ to the submanifold N. If $v(t, x)$ is a solution of the quotient equation 11.57, then the invariant solution of Equation 11.49 has the form $\phi(t, x, u, p) = pv(t, x)$. In their turn, the solutions of the heat equation invariant under the partial symmetry v_Q of the heat equation with the characteristic function $Q = -u_t + u_x v(t, x)$ are common solutions of Equation 11.53 and the equation

$$u_t = u_x v(t, x)$$

for invariant functions. The latter relation is a variant of the Hopf-Cole substitution linearizing Burgers' equation 11.57. See also the work of Guthrie [1993] for generalizations and additional applications of this method.

11.3.6. Partial symmetries and functionally invariant solutions

Consider a linear differential equation of the second order:

$$\sum_{i,j=1}^{n} a_{ij}(x)u_{x_i x_j} + \sum_{i=1}^{n} b_i(x)u_{x_i} = 0. \tag{11.58}$$

A solution $u(t, x)$ is called *functionally invariant* if $v(t, x) = F(u(t, x))$ is also a solution of Equation 11.58 for an arbitrary function $F(u)$. Functionally invariant solutions are common solutions of Equation 11.58 and the equation for its characteristics:

$$\sum_{i,j=1}^{n} a_{ij}(x)u_{x_i}u_{x_j} = 0. \tag{11.59}$$

For the wave equation,

$$u_{tt} = u_{xx} + u_{yy}$$

Equation 11.59 takes the form $u_t^2 = u_x^2 + u_y^2$. By direct calculations it is demonstrated that the vector field v_Q with the characteristic function $Q = u_t - \sqrt{u_x^2 + u_y^2}$ is a nonclassical partial symmetry of the wave equation. We see that functionally invariant solutions are solutions invariant under the partial symmetries.

11.3.7. Nonclassical symmetries and partially invariant solutions

Let $\mathbf{g}$ be an r-dimensional involutive differential system on $R^n \times R^q$. A function $f: R^n \to R^q$ is called *partially invariant* under $\mathbf{g}$ if the orbit $G \cdot \Gamma_f$ of its graph has dimension strictly less than $n + \min(r, q)$. The quantity $\delta = \dim(G \cdot \Gamma_f) - n$ is called the *deficiency* of a partially invariant function f. Evidently, $0 \leq \delta \leq \min(q - 1, r - 1)$. The partially invariant functions satisfy the following system PE_δ of first-order differential equations:

$$\text{rank} \left\| \varphi_j^\alpha - \sum_{i=1}^n u_{x_i}^\alpha \xi_{ji} \right\| \leq \delta. \tag{11.60}$$

The system PE_δ is invariant under $\mathbf{g}$. Provided $\mathbf{g}$ is a Lie algebra of infinitesimal classical symmetries of the system E, partially invariant solutions of the system E are the common solutions of E and PE_δ. See Ondich [1994] for applications of this method.

11.3.8. Nonclassical symmetries of the second type for the equations of nonlinear acoustics

The preceding construction admits a natural generalization. A differential system $\mathbf{g}$ is called a *nonclassical symmetry of the second type* of the system E if the intersection $E \cap PE_\delta$ is invariant under $\mathbf{g}$.

The system 11.26 contains two unknown functions, so the only possible value of the deficiency index δ of partially invariant solutions of (11.26) is one. Consider the family of two-dimensional abelian Lie algebras $\mathbf{g}$ of vector fields with the generators:

$$\mathbf{v}_1 = \partial_v,$$

$$\mathbf{v}_2 = \partial_t + \xi(t, x, y, u)\partial_x + \eta(t, x, y, u)\partial_y + \varphi(t, x, y, u)\partial_u + \psi(t, x, y, u)\partial_v.$$

We prescribe the coefficients ξ, η, φ, ψ so that the vector fields $\mathbf{v}_1, \mathbf{v}_2$ are tangent to the intersection $E \cap PE_1$. The system PE_1 is determined by Equation 11.60, which now looks like

$$\det \left\| \begin{matrix} 0 & \varphi - u_t - \xi u_x - \eta u_y \\ 1 & \psi - v_t - \xi v_x - \eta v_y \end{matrix} \right\| = 0$$

or

$$\varphi - u_t - \xi u_x - \eta u_y = 0. \tag{11.61}$$

The vector field $\mathbf{v}_1$ is a classical infinitesimal symmetry of the system 11.26 and the coefficients of the vector field $\mathbf{v}_2$ are found from the relations:

$$\mathbf{v}_2^{(1)}(uu_t + u_x + v_y)|_{E\cap PE_1} = 0, \qquad \mathbf{v}_2^{(1)}(u_y - v_t)|_{E\cap PE_1} = 0. \qquad (11.62)$$

Finding the variables v_t, v_y, u_t from Equations 11.26 and 11.61 and substituting the results into (11.62) yields the determining equations for the coefficients. One of the particular solutions of these equations:

$$\xi = 0, \qquad \eta = 0, \qquad \varphi = y, \qquad \psi = t - y^3/3$$

yields the following partially invariant solutions of (11.26):

$$u(t, x, y) = ty - \frac{y^4}{12} + h(x)y + e(x), \qquad v(t, x, y) = \frac{t^2}{2} - t\left(\frac{y^3}{3} - h(x)\right)$$

$$+ \frac{y^6}{60} - h(x)\frac{y^3}{3} - [e(x) + h'(x)]\frac{y^2}{2} - e'(x)y + d(x)$$

with $h(x)$, $e(x)$, $d(x)$ arbitrary functions.

11.4. Conditional symmetries

Consider the system E of differential equations

$$\Delta_\nu(x, u, u^{(k)}) = 0, \qquad \nu = 1, \ldots, l \qquad (11.1)$$

and append it with the system of differential constraints E_Θ:

$$\Theta_\mu(x, u, u^{(k)}) = 0, \qquad \mu = 1, \ldots, m. \qquad (11.63)$$

Assume that the appended system $E \cap E_\Theta$ (11.1), (11.63) is compatible. A vector field $\mathbf{v}$ is called a *conditional* infinitesimal symmetry of the system E if the system $E \cap E_\Theta$ is invariant under $\mathbf{v}$, i.e., $\mathbf{v}$ is a Lie classical infinitesimal symmetry of $E \cap E_\Theta$. Possible approaches to constructing the appended system 11.63 are explained below, based on ideas contained in Fushchich, Serov, and Chopik [1988] and Fushchich and Serov [1988].

11.4.1. Conditional symmetries of a nonlinear heat equation

As a first example consider the nonlinear heat equation

$$\Delta \equiv u_t - uu_{xx} - u_x^2 = 0. \tag{11.64}$$

The Lie algebra of the classical symmetries of Equation 11.64 has the generators

$$\mathbf{v}_1 = \partial_t, \qquad \mathbf{v}_2 = \partial_x, \qquad \mathbf{v}_3 = t\partial_t - u\partial_u, \qquad \mathbf{v}_4 = x\partial_x + 2u\partial_u. \tag{11.65}$$

Comparison with that of the linear heat equation reveals the fact that (11.65) does not contain an infinitesimal Galilean-like transformation

$$\mathbf{v} = t\partial_x + xh(u)\partial_u,$$

which is admissible by the linear heat equation $u_t = u_{xx}$ when $h(u) = -u/2$. Indeed, if we apply the prolonged vector field $\mathbf{v}^{(2)}$ to the left-hand side of (11.64), we obtain the function

$$\Theta \equiv -\mathbf{v}^{(2)}(u_t - uu_{xx} - u_x^2)$$

$$= u_x + 2u_x h + xu_{xx}h - xu_t h_u + 2uu_x h_u + 2xu_x^2 h_u + xuu_{xx}h_u + uxu_x^2 h_{uu}$$

that does not vanish, being restricted to E:

$$\Lambda \equiv u\Theta|_E = xu_t h + uu_x + 2uu_x h - xu_x^2 h + 2u^2 u_x h_u + xuu_x^2 h_u + xu^2 u_x^2 h_{uu} \neq 0$$

So append the equation

$$\Lambda = 0 \tag{11.66}$$

to (11.64). Let us check the compatibility of (11.64) and (11.66) later on and try to find the function $h(u)$ so that the vector field $\mathbf{v}^{(2)}$ is tangent to the system 11.64, 11.66. The function $\Psi = \mathbf{v}^{(1)}(\Lambda)$ vanishes on the submanifold $\{\Delta = 0, \Lambda = 0\}$ if and only if

$$h(u) = -\frac{1}{2} + \frac{c}{u},$$

with c an arbitrary constant. For checking the compatibility we apply the contact vector field $\mathbf{v}^{(2)}_\Lambda$ to Δ and find that the function $\mathbf{v}^{(2)}_\Lambda(\Delta)$ vanishes when restricted to the submanifold $\{\Delta = 0, \Lambda = 0\}$. Therefore the vector field $\mathbf{v}_\Lambda$ is a partial symmetry of Equation 11.64 and the system

$$u_t = uu_{xx} + u_x^2, \qquad (-1 + 2c/u)u_t + u_x^2 = 0, \tag{11.67}$$

is Galilean-invariant.

It is interesting to note that if $c \neq 0$, then (11.67) admits only three-dimensional subalgebra $\mathbf{g_3} = L(\mathbf{v}_1, \mathbf{v}_2, 2\mathbf{v}_3 + \mathbf{v}_4)$ of the initially four-dimensional algebra 11.65 of the classical Lie symmetries of (11.65). Taking the latter fact into account, one can try to preserve the classical Lie symmetry group as a subgroup of the classical symmetry group of the appended system. For a generic function $f(u)$, the family of nonlinear heat equations

$$u_t = (f(u)u_x)_x \tag{11.68}$$

admits the three-dimensional Lie algebra $\mathbf{g_3}$ generated by the vector fields

$$\mathbf{v}_1 = \partial_t, \qquad \mathbf{v}_2 = \partial_x, \qquad \mathbf{v}_3 = 2t\partial_t + x\partial_x. \tag{11.69}$$

The functions u and u_t/u_x^2 are first-order differential invariants for $\mathbf{g_3}$. Thus, if we append the equation

$$u_t = g(u)u_x^2, \tag{11.70}$$

for $g(u)$ arbitrary (for the moment), the combined system 11.68, 11.70 admits $\mathbf{g_3}$.

The compatibility condition for the system 11.68, 11.70 is the following ordinary differential equation connecting the functions $f(u)$ and $g(u)$:

$$f^2 g'' + f(4g - 3f')g' + (11(f')^2 - ff'')g - 5g'f^2 + 2g^3 = 0.$$

For $f(u)$ given, the function $g(u) = f(u)/u$ is a particular solution of the latter equation, and we are led to the compatible system

$$u_t = (f(u)u_x)_x, \qquad u_t = \frac{f(u)u_x^2}{u}$$

admitting (11.69) as a subalgebra of the Lie algebra of its infinitesimal symmetries.

11.5. Weak symmetries

In Olver and Rosenau [1987], a further generalization of the nonclassical method was proposed. Since the combined system 11.1, 11.4 is an overdetermined system of partial differential equations, one should, in treating it, take into account: any integrability conditions given by equating mixed partials. (The Cartan-Kuranishi Theorem assures us that, under mild regularity conditions, the integrability conditions can all be found in a finite number of steps; differential Gröbner basis methods, as in Pankrat'ev [1989] and Topunov [1989], provide a practical means to compute them.) Therefore, one should compute the symmetry group not of just the system 11.1, 11.4 but also any associated integrability conditions. Thus, we define a *weak symmetry group* of the system 11.1 to be any symmetry group of the overdetermined system 11.1, 11.4 and all its integrability conditions.

11.5.1. An example of a weak symmetry group

For the Boussinesq equation

$$u_{tt} + uu_{xx} + u_x^2 + u_{xxxx} = 0,$$

consider the scaling group generated by the vector field $\mathbf{v} = x\partial_x + t\partial_t$. This is not a symmetry of the Boussinesq equation, nor is it a symmetry of the combined system

$$u_{tt} + uu_{xx} + u_x^2 + u_{xxxx} = 0, \qquad Q = xu_x + tu_t = 0. \tag{11.71}$$

Nevertheless, if we append the integrability conditions to (11.71), we do find that $\mathbf{v}$ satisfies the weak symmetry conditions. To compute the invariant solutions, we begin by introducing the invariants, $y = x/t$, and $w = u$. Differentiating the formula $u = w(y) = w(x/t)$ and substituting the result into the Boussinesq equation, we come to the following equation:

$$t^{-4}w'''' + t^{-2}[(y^2 + w)w'' + (w')^2 + 2yw'] = 0. \tag{11.72}$$

At this point the crucial difference between the weak symmetries and the non-classical (or classical) symmetries appears. In the latter case, any non-invariant coordinate, e.g., the t here, will factor out of the resulting equation and thereby leave a single ordinary differential equation for the invariant function $w(y)$. For weak symmetries this is no longer true, since we have yet to incorporate the integrability conditions for (11.67). However, we can separate out the coefficients of the various powers of t in Equation 11.68, leading to an overdetermined system of ordinary differential equations,

$$w'''' = 0, \qquad (y^2 + w)w'' + (w')^2 + 2yw' = 0,$$

for the unknown function w. In this particular case, the resulting overdetermined system does have solutions, namely $w(y) = -y^2$, or $w(y) = \text{constant}$. The latter are trivial, but the former yield a nontrivial similarity solution: $u(x, t) = -x^2/t^2$.

11.5.2. A particular case of the infinitesimal weak symmetries

An explication of one possible approach to treating the weak symmetries (Dzhamay and Vorob'ev [1994]) is given below. We will restrict ourselves to the case of differential equations in one unknown function such as Equation 11.32:

$$\Delta(x, u, u^{(k)}) = 0. \tag{11.32}$$

Consider the contact vector field $\mathbf{v}_Q$, the function $\Gamma(x, u, u^{(k)}) = \mathbf{v}_Q^{(k)}(\Delta)$ $(x, u, u^{(k)})$, and the system W of differential equations

$$\Delta = 0, \qquad \Gamma = 0, \qquad Q = 0. \tag{11.73}$$

DEFINITION 11.1. *A vector field $\mathbf{v}_Q$ is an infinitesimal weak symmetry of Equation 11.32 if*

(1) $\mathbf{v}_Q$ is a classical infinitesimal symmetry of system 11.73 and

(2) system 11.73 is compatible.

Property (1) can be reformulated by saying that $\mathbf{v}_Q$ is a partial infinitesimal symmetry of the system $\Delta = 0$, $\Gamma = 0$. Note that if we required that $\mathbf{v}_Q$ were a classical infinitesimal symmetry of the latter system, we would have got a conditional infinitesimal symmetry considered in Section 11.4. Property (2) means that system 11.73 implies no extra conditions that may arise by cross differentiation of the equations of system 11.73 and their differential consequences. Since $\Gamma = \mathbf{v}_Q^{(k)}(\Delta)(x, u, u^{(k)})$ and $\mathbf{v}_Q(Q) = Q_u \, Q$, the criterion of tangency of $\mathbf{v}^{(k)}$ to W takes the form:

$$\mathbf{v}_Q^{(k)}(\Gamma)|_W = 0. \tag{11.74}$$

Relation 11.74 and the compatibility conditions for W imply the determining equations for the characteristic function Q.

It is clear that relations 11.73 can be generalized by adding to W the functions $\mathbf{v}_Q^{(k)}(\Gamma)$ and so on.

11.5.3. Weak symmetries of the nonlinear heat equation

In this section, the exact solutions of the equation

$$u_t = u_{xx} + u_x^2 + u^2 \tag{11.75}$$

obtained by Galaktionov [1990] are interpreted as invariant under the weak symmetries of (11.75). Consider the vector field $\mathbf{v}_Q$ with the characteristic function
$Q = -p_x + a(t, x)$, where the function $a(t, x)$ needs to be defined. Since the
equations of the intersection $E_\Delta \cap E_Q^{(2)}$ are equivalent to the equations

$$\tilde{\Delta} \equiv -p_t + a_x + a^2 + u^2, \qquad p_x = a, \qquad p_{tx} = a_t, \qquad p_{xx} = a_x,$$

the following formula is valid: $\Gamma \equiv \mathbf{v}_Q^{(2)}(\tilde{\Delta}) = a_t + a_{xx} + 2aa_x + 2ua$. Therefore,
if the function a is fixed, a unique invariant solution $u(t, x)$ is obtained from the
equation $\Gamma = 0$:

$$u(t, x) = \frac{a_t - a_{xx} - 2aa_x}{2a}. \tag{11.76}$$

So we can conclude that the system W takes the form:

$$p_t = a_x + a^2 + u^2, \qquad p_x = a(t, x), \qquad u(t, x) = \frac{a_t - a_{xx} - 2aa_x}{2a}. \tag{11.77}$$

The compatibility conditions for system 11.77 are evident:

$$\frac{\partial}{\partial x}\left(\frac{a_t - a_{xx} - 2aa_x}{2a}\right) = a,$$

$$\frac{\partial}{\partial t}\left(\frac{a_t - a_{xx} - 2aa_x}{2a}\right) = a_x + a^2 + \left(\frac{a_t - a_{xx} - 2aa_x}{2a}\right)^2. \tag{11.78}$$

Equations 11.78 admit separation of variables; precisely, the first equation is satisfied if $a(t, x) = \phi(t)\sin x$ with $\phi(t)$ arbitrary function. Hence, the second
equation implies the relation

$$\frac{d}{dt}\left(\frac{\dot{\phi} + \phi}{2\phi}\right) = \left(\frac{\dot{\phi} + \phi}{2\phi}\right)^2 + \phi^2 \tag{11.79}$$

for the function ϕ.

After the function $\phi(t)$ is found from (11.79), we obtain the infinitesimal weak
symmetry $\mathbf{v}_Q$ of Equation 11.75 with the characteristic function $Q = -p_x + \phi(t)\sin x$ and the invariant under $\mathbf{v}_Q$ solution

$$u(t, x) = \frac{\dot{\phi} + \phi}{2\phi} - \phi\sin x.$$

Galaktionov obtained this solution by directly applying his method of generalized separation of variables in the form $u(t, x) = \theta(t) - \phi(t) \sin x$ to Equation 11.75.

11.5.4. Discussion

Weak symmetry groups, while at the outset quite promising, have some critical drawbacks. It can be shown that every group is a weak symmetry group of a given system of partial differential equations, and, moreover, every solution to the system can be derived from some weak symmetry group — see Olver and Rosenau [1987]. Therefore, the generalization is too severe. Nevertheless, it gives some hints as to how to proceed in any practical analysis of such solution methods. What is required is an appropriate theory of overdetermined systems of partial differential equations which will allow one to write down reasonable classes of groups for which the combined system 11.1, 11.4 is compatible, in the sense that is has solutions, or, more restrictively, has solutions that can be algorithmically computed. For example, restricting to scaling groups, or other elementary classes of groups, might be a useful starting point.

Part II. A SURVEY OF RESULTS

The following preliminary comments will be helpful for the reader. First, to date only particular solutions of the determining equations for the coefficients of infinitesimal nonclassical symmetries have been obtained and are therefore given below. The reason is partly explained in Section 11.1.2. Second, we do not point out which vector fields are obtained in what papers, so the results are a set theoretic union of those in separate papers. Third, not all of the known results are given, but only those important for applications or most completely investigated. And last, exact solutions invariant under the nonclassical symmetries can give rise to multiparameter families of solutions with the aid of the classical symmetry transformations.

11.6. Boussinesq equation

$$u_{tt} + uu_{xx} + (u_x)^2 + u_{xxxx} = 0$$

Lie point symmetries

(Clarkson and Kruskal [1989], Nishitani and Tajiri [1982], Rosenau and Schwarzmeier [1986])

$$X_1 = x\partial_x + 2t\partial_t - 2u\partial_u, \qquad X_2 = \partial_x, \qquad X_3 = \partial_t.$$

Nonclassical conditional symmetries

(Levi and Winternitz [1989], Fushchich and Serov [1989])

$$\mathbf{v}_1 = \partial_t + t\partial_x - 2t\partial_u,$$

$$\mathbf{v}_2 = t^3\partial_t + (\beta_1 t^7 - xt^2)\partial_x + (2t^2 u + 6x^2 - 2\beta_1 t^5 x - 4\beta_1^2 t^{10})\partial_u,$$

(by dilations and reflections, one can transform β_1 into $\beta_1 = 1$ or $\beta_1 = 0$),

$$\mathbf{v}_3 = 2t\partial_t + (x + 2t^2)\partial_x - 2(u + 2x + 4t^2)\partial_u,$$

$$\mathbf{v}_4 = 2p(t)\partial_t + (x + \beta_2 w)\dot{p}\partial_x - [2\dot{p}u + 6\dot{p}px^2 + \beta_2(1 + 12p\dot{p}w)x + \beta_2^2 w$$

$$+ 6\beta_2^2 p\dot{p}w^2]\partial_u,$$

$$\dot{p}^2 = 4p^3 - c, \qquad c = \text{const.}, \qquad w(t) = \int_0^t p(s)/\dot{p}(s)^2 \, ds,$$

$$\mathbf{v}_5 = t^2\partial_x + (t^5 - 2x)\partial_u,$$

$$\mathbf{v}_6 = \partial_x + (\lambda(t) - \frac{1}{3}\mu(t)x)\partial_u, \qquad \ddot{\mu} = \mu^2, \qquad \ddot{\lambda} = \mu\lambda,$$

$$\mathbf{v}_7 = x\partial_x + 2u\partial_u,$$

$$\mathbf{v}_8 = x^3\partial_x + 2(x^2 u + 24)\partial_u.$$

Nonclassical weak symmetries

(Olver and Rosenau [1987])

$$\mathbf{w} = x\partial_x + t\partial_t$$

Exact solutions invariant under nonclassical symmetries

$\mathbf{v}_1$: $\quad u = \phi(z) - t^2, \qquad z = x - t^2/2, \qquad \dddot{\phi} + \phi\dot{\phi} - \phi = 2z + c_1,$

$\mathbf{v}_2$: $\quad u = \phi(z)t^2 - z^2/t^2, \qquad z = xt - \beta_1 t^6/6, \qquad \ddot{\phi} + \dot{\phi}^2/2 = c_1 z + c_0$

$\qquad (\beta_1 = 0), \qquad \dddot{\phi} + \phi\dot{\phi} - 5\phi = 50z + c_0 \qquad (\beta_1 = 1),$

$\mathbf{v}_3$: $\quad u = \phi(z)/t - (x/2t + t)^2, \qquad z = x/\sqrt{t} - 2t^{3/2}/3,$

$\qquad \dddot{\phi} + \phi\ddot{\phi} + (\dot{\phi})^2 + 3z\dot{\phi}/4 + 3\phi/2 = 9z^2/8,$

$\mathbf{v}_4$: $\quad u = \phi(z)p^{-1} - (\dot{p}/2p + \beta_2 w \dot{p}/2p),$

$$z = \frac{\beta_2 c^{-1} p^{-1/2} \int_0^t p(s)\, ds}{3} + xp^{-1/2},$$

$\qquad \dddot{\phi} + \phi\ddot{\phi} + \dot{\phi}^2 - 3c\dot{\phi}/4 - 3c\phi/2 = 9c^2 z^2/8,$

$\mathbf{v}_5$: $\quad u = \phi(t) - x^2/t^2 + t^3 x, \qquad \ddot{\phi} - 2\phi/t^2 + t^6 = 0,$

$\mathbf{v}_6$: $\quad u = \phi(t) - x^2\mu/6 + \lambda x, \qquad \ddot{\phi} - 2\phi/t^2 + t^6 = 0,$

$\mathbf{v}_7$: $\quad u = x^2\phi(t), \qquad \ddot{\phi} + 6\phi^2 = 0,$

$\mathbf{v}_8$: $\quad u = x^2\phi(t) - 12x^{-2}, \qquad \ddot{\phi} + 6\phi^2 = 0,$

$\mathbf{w}$: $\quad u = -x^2/t^2.$

11.7. Burgers' equation

$$u_t + u u_x = u_{xx}$$

Lie point symmetries

$$X_1 = \partial_t, \qquad X_2 = \partial_x, \qquad X_3 = t\partial_x + \partial_u,$$

$$X_4 = 2t\partial_t + x\partial_x - u\partial_u, \qquad X_5 = t^2\partial_t + tx\partial_x + (x - tu)\partial_u.$$

Nonclassical conditional symmetries

(Arrigo, Broadbridge, and Hill [1993], Pucci [1992], Vorob'ev [1986])

The following general expressions are valid for the vector fields of the nonclassical infinitesimal symmetries:

$$\mathbf{v} = \partial_t + \xi(t, x, u)\partial_x + \phi(t, x, u)\partial_u, \qquad \text{or} \qquad \mathbf{w} = \partial_x + \psi(t, x, u)\partial_u,$$

where the functions ξ, ϕ, and ψ satisfy the equations:

$$\xi_{uu} = 0, \qquad \phi_{uu} - 2\xi_{xu} - 2(u - \xi)\xi_u = 0, \qquad \xi_{xx} - (2\xi - u)\xi_x$$

$$+ 2\phi\xi_u - \xi_t - 2\phi_{xu} + \phi = 0, \qquad \phi_{xx} + 2\phi\xi_x + u\phi_x + \phi_t = 0,$$

or

$$u\psi_x - \psi_{xx} - \psi^2\psi_{uu} - 2\psi\psi_{xu} + \psi_t + \psi^2 = 0.$$

Particular solutions of the first set of equations generate the infinitesimal nonclassical symmetries:

$$\mathbf{v}_1 = \partial_t + u\partial_x,$$

$$\mathbf{v}_2 = \partial_t + \left(-\frac{1}{2}u + a_0 t^2 + a_1 t + a_2\right)\partial_x + \left(-\frac{1}{4}u^3 + \frac{1}{2}u^2(a_0 t^2 + a_1 t + a_2)\right.$$

$$\left. - (a_0 t + \frac{1}{2}a_1)xu + (b_0 t + b_2)u + \frac{1}{2}a_0 x^2 - b_0 x + a_0 t + a_3\right)\partial_u,$$

$$\mathbf{v}_3 = \partial_t - \frac{1}{2}u\partial_x - \frac{u^3}{4}\partial_u,$$

$$\mathbf{v}_4 = \partial_t - \frac{1}{2}u\partial_x - \left(\frac{1}{4}u^3 - \frac{1}{4}a_1^2 u\right)\partial_u,$$

$$\mathbf{v}_5 = \partial_t + \left(a_2 - \frac{1}{2}u\right)\partial_x + \left(-\frac{1}{4}u^3 + \frac{1}{2}a_2 u^2\right)\partial_u,$$

$$\mathbf{v}_6 = \partial_t - \left(\frac{1}{2}u + x^{-1}\right)\partial_x - \left(\frac{1}{4}u^3 + \frac{u^2}{2x}\right)\partial_u,$$

$$\mathbf{v}_7 = \partial_t + \left(-\frac{1}{2}u + w(t,x)\right)\partial_x + \left(-\frac{1}{4}u^3 + \frac{1}{2}w(t,x)u^2\right)\partial_u,$$

with a, b parameters, and $w(t,x)$ a solution of Burgers' equation $w_t + 2ww_x - w_{xx} = 0$.

Solutions invariant under the nonclassical symmetries

$$\mathbf{v}_1: \qquad u = x/t,$$

$$\mathbf{v}_3: \qquad u = -\frac{4x}{x^2 + 2t},$$

$$\mathbf{v}_4: \qquad u = -\frac{a_1(c_1\exp(a_1 x) - 1)}{c_1\exp(a_1 x) + 1 \pm c_2\exp(a_1 x/2 - a_1^2 t/4)},$$

$$\mathbf{v}_5: \qquad u = -2\frac{1 + a_2\exp(a_2 x + a_2^2 t)}{x + \exp(a_2 x + a_2^2 t)},$$

$$\mathbf{v}_6: \qquad u = -\frac{12t + 6x^2}{6xt + x^3}.$$

11.8. Generalized Korteweg-de Vries equation

$$u_t + f(u)u_x^k + u_{xxx} = 0, \qquad k > 0$$

Lie point symmetries

k	$f(u)$	
arbitrary	arbitrary	$X_1 = \partial_t, \quad X_2 = \partial_x$
	u^n	$X_1, \quad X_2, \quad X_3 = 3t(k+n-1)\partial_t + x(k+n-1)\partial_x$ $+ (k-3)u\partial_u$
	$\exp u$	$X_1, \quad X_2, \quad X_4 = 3t\partial_t + x\partial_x + (k-3)\partial_u$
	1	$X_1, \quad X_2, \quad X_5 = 3(k-1)t\partial_t + (k-1)x\partial_x$ $+ (k-3)u\partial_u, \quad X_6 = \partial_u$
3	arbitrary	$X_1, \quad X_2, \quad X_7 = 3t\partial_t + x\partial_x$
	u^{-2}	$X_1, \quad X_2, \quad X_7, \quad X_8 = u\partial_u$
	1	$X_1, \quad X_2, \quad X_6, \quad X_7$
1	$u^n + c$	$X_1, \quad X_2, \quad X_9 = 3nt\partial_t + n(2ct+x)\partial_x - 2u\partial_u$
	u	$X_1, \quad X_2, \quad X_{10} = 3t\partial_t + x\partial_x - 2u\partial_u,$ $X_{11} = t\partial_x + \partial_u$
	$\exp u + c$	$X_1, \quad X_2, \quad X_{12} = 3t\partial_t + (2ct+x)\partial_x - 2\partial_u$
	1	$X_1, \quad X_2, \quad X_8, \quad X_{13} = 3t\partial_t + (2t+x)\partial_x,$ $X_g = g(t,x)\partial_u,$ $g(t,x)$ solves the generalized KdV equation

Nonclassical conditional symmetries

(Fushchich, Serov, and Ahmerov [1991])

$$\mathbf{v} = t^{1/k}\partial_x + F(u)\partial_u$$

$$
\begin{aligned}
f(u) &= c_1 u^{1-k/2} + c_2 u^{(1-k)/2}, & F(u) &= \left(\tfrac{kc_1}{2}\right)^{-1/k}\sqrt{u} \\
f(u) &= (c_1 \log u + c_2)(1 - u^2)^{1-k}, & F(u) &= (kc_1)^{-1/k}u \\
f(u) &= (c_1 \arcsin u + c_2)(1 - u^2)^{(1-k)/2}, & F(u) &= (kc_1)^{-1/k}\sqrt{1 - u^2} \\
f(u) &= (c_1 \sinh^{-1} u + c_2)(1 + u^2)^{(1-k)/2}, & F(u) &= (kc_1)^{-1/k}\sqrt{1 + u^2} \\
f(u) &= c_1 u, & F(u) &= (kc_1)^{-1/k}
\end{aligned}
$$

Exact solutions invariant under the nonclassical symmetries

$$u(t,x) = \left(\frac{x}{2}\left(\frac{kc_1 t}{2}\right)^{-1/k} + ct^{-1/k} - \frac{c_2}{c_1}\right)^2,$$

$$u(t,x) = \exp\left(\frac{k(kc_1)^{-3/k}}{k-2}t^{1-3/k} - \frac{c_2}{c_1} + ct^{-1/k} + (kc_1t)^{-1/k}x\right), \quad k \neq 2,$$

$$u(t,x) = \exp\left((2c_1)^{-3/2}t^{-1/2}\log t - \frac{c_2}{c_1} + ct^{-1/2} + (2c_1t)^{-1/2}x\right), \quad k = 2,$$

$$u(t,x) = \sin\left(-\frac{k(kc_1)^{-3/k}}{k-2} - \frac{c_2}{c_1} + ct^{-1/k} + (kc_1t)^{-1/k}x\right), \quad k \neq 2,$$

$$u(t,x) = \sin\left(-(2c_1)^{-3/2}\frac{\log t}{\sqrt{t}} - \frac{c_2}{c_1} + ct^{-1/2} + (2c_1t)^{-1/2}x\right), \quad k = 2,$$

$$u(t,x) = \sinh\left(\frac{k(kc_1)^{-3/k}}{k-2}t^{1-3/k} - \frac{c_2}{c_1} + ct^{-1/k} + (kc_1t)^{-1/k}x\right), \quad k \neq 2,$$

$$u(t,x) = \sinh\left((2c_1)^{-3/2}\frac{\log t}{\sqrt{t}} - \frac{c_2}{c_1} + ct^{-1/2} + (2c_1t)^{-1/2}x\right), \quad k = 2$$

$$u(t,x) = ct^{-1/k} + x(kc_1t)^{-1/k}.$$

11.9. Kadomtsev-Petviashvili equation

(Kadomtsev and Petviashvili [1970])

$$(u_t + uu_x + u_{xxx})_x + ku_{yy} = 0, \qquad k = \pm 1$$

Lie point symmetries

(Tajiri, Nishitani, and Kawamoto [1982], David, Kamran, Levi, and Winternitz [1985])

$$X_\alpha = 6\alpha(t)\partial_t + (2x\dot\alpha(t) - ky^2\ddot\alpha(t))\partial_x + 4y\dot\alpha(t)\partial_y$$

$$- (4u\dot\alpha(t) - 2x\ddot\alpha(t) + ky^2\dddot\alpha(t))\partial_u,$$

$$X_\beta = \beta(t)\partial_x + \dot{\beta}(t)\partial_u, \qquad X_\gamma = -y\dot{\gamma}(t)\partial_x + 2k\gamma(t)\partial_y - y\ddot{\gamma}(t)\partial_u,$$

where $\alpha(t)$, $\beta(t)$, and $\gamma(t)$ are arbitrary functions.

Nonclassical conditional symmetries

(Clarkson and Winternitz [1991])

$$\mathbf{v} = \partial_x + [R(y,t)x + S(y,t)]\partial_u, \qquad R \neq 0 \qquad \text{and} \qquad S_y \neq 0,$$

where the functions $R(y,t)$ and $S(y,t)$ are solutions of the system

$$kR_{yy} + 3R^2 = 0, \qquad kS_{yy} + 3RS + R_t = 0.$$

Exact solutions invariant under the nonclassical symmetries

$$u(x,y,t) = w(y,t) + x^2\gamma(y,t) + x\psi(y,t),$$

where

$$\gamma(y,t) = -kW(y + \phi_0(t); 0, g(t)),$$

$$\psi(y,t) = W(y + \phi_0(t); 0, g(t))\left(A(t) + B(t)\int^y \frac{dz}{W^2(z+\phi_0;0,g)}\right)$$

$$-\frac{\dot{g}}{12g} + y\left(\dot{\phi}_0 + \dot{g}\frac{y+2\phi_0}{2g}\right)W(y + \phi_0(t); 0, g(t))$$

with $\phi_0(t)$, $g(t)$, $A(t)$, $B(t)$ arbitrary functions, $W(z; 0, h)$ the Weierstrass elliptic function.

11.10. Kolmogorov-Petrovskii-Piskunov, or Fitzhugh-Nagumo Equation

$$u_t = u_{xx} + u(1-u)(u-a), \qquad -1 \leq a \leq 1$$

Lie point symmetries

$$X_1 = \partial_t, \qquad X_2 = \partial_x$$

Nonclassical conditional symmetries

(Nucci and Clarkson [1992], Vorob'ev [1986])

$$\mathbf{v}_1 = \partial_t \pm \frac{1}{\sqrt{2}}(3u - a - 1)\partial_x + \frac{3}{2}u(1 - u)(u - a)\partial_u,$$

$$\mathbf{v}_2 = \partial_t + \alpha(x)\partial_x - \alpha'(x)u\partial_u, \quad a = -1, \quad \alpha(x) = \frac{3}{\sqrt{2}}\frac{1 + \exp\sqrt{2}x}{1 - \exp\sqrt{2}x},$$

$$\mathbf{v}_3 = \partial_t + \alpha(x)\partial_x - \alpha'(x)(u - 1/2)\partial_u, \quad a = 1/2, \quad \alpha(x) = \frac{3}{\sqrt{2}}\frac{1 + \exp x/\sqrt{2}}{1 - \exp x/\sqrt{2}}$$

$$\mathbf{v}_4 = \partial_t + \alpha(x)\partial_x - \alpha'(x)(u - 1)\partial_u, \quad a = 2, \quad \alpha(x) = \frac{3}{\sqrt{2}}\frac{1 + \exp\sqrt{2}x}{1 - \exp\sqrt{2}x}$$

Exact solutions invariant under the nonclassical symmetries

$\mathbf{v}_1$:

$$u = \frac{ac_1 \exp((\sqrt{2}x + a^2 t)/2) + c_2 \exp((\sqrt{2}ax + t)/2)}{c_1 \exp((\sqrt{2}x + a^2 t)/2) + c_2 \exp((\sqrt{2}ax + t)/2) + c_3 \exp((\sqrt{2}(a+1)x + at)/2)},$$

$\mathbf{v}_2, \mathbf{v}_4$:

$$u = (c_1 \exp((\sqrt{2}x + 3t)/2) - c_2 \exp((-\sqrt{2}x + 3t/2)))w(z),$$

$$z = c_1 \exp((\sqrt{2}x + 3t)/2) + c_2 \exp((-\sqrt{2}x + 3t/2)) + c_3,$$

$\mathbf{v}_3$:

$$u = (c_1 \exp((2\sqrt{2}x + 3t)/8) - c_2 \exp((-2\sqrt{2}x + 3t)/8))w(z),$$

$$z = c_1 \exp((2\sqrt{2}x + 3t)/8) + c_2 \exp((-2\sqrt{2}x + 3t)/8) + c_3,$$

where $w(z)$ is the Weierstrass function satisfying the equation $w''(z) = 2w(z)^3$.

11.11.   Nonlinear wave equation

$$u_{tt} = uu_{xx}$$

Lie point symmetries

$$X_1 = \partial_t, \quad X_2 = \partial_x, \quad X_3 = t\partial_t + x\partial_x, \quad X_4 = t\partial_t - 2u\partial_u$$

Nonclassical conditional symmetries

(Fushchich and Serov [1988])

$$\mathbf{v}_1 = \partial_x + a_1\partial_u, \quad \mathbf{v}_2 = \partial_t + (a_2x + a_3)\partial_u,$$

$$\mathbf{v}_3 = \partial_t + (a_4t + a_5)\partial_x + 2a_4(a_4t + a_5)\partial_u, \quad \mathbf{v}_4 = \partial_x + [w(t) + f(t)]\partial_u,$$

$$\mathbf{v}_5 = t\partial_t + (u + a_7x + a_8)\partial_u, \quad \mathbf{v}_6 = t\partial_t + [t^3(a_9x + a_{10}) - 2u]\partial_u,$$

$$\mathbf{v}_7 = x\partial_x + (u + b_1t + b_2)\partial_u, \quad \mathbf{v}_8 = x\partial_x + [u + w(t)x^2/2 - f(t)]\partial_u,$$

$$\mathbf{v}_9 = (t^2 - 1)\partial_t + 2x\partial_x + (t + 1)u\partial_u, \quad \mathbf{v}_{10} = t^3\partial_t + (3t^2 - 15x^2 + b_3x + b_4)\partial_u,$$

$$\mathbf{v}_{11} = t^2x\partial_x + (t^2u + 3x^2 + b_2t^5 + b_6)\partial_u, \quad \mathbf{v}_{12} = w(t)\partial_t + \dot{w}(t)u\partial_u,$$

where a, b are arbitrary parameters, $w(t)$ and $f(t)$ satisfy the ODE: $\ddot{w} = w^2$, $\ddot{f} = wf$.

Exact solutions invariant under the nonclassical symmetries

$$\mathbf{v}_1: \quad u = \phi(t) + a_1x, \quad \ddot{\phi} = 0,$$

$$\mathbf{v}_2: \quad u = \phi(t) + t(a_2x + a_3), \quad \ddot{\phi} = 0,$$

$\mathbf{v}_3:$ $u = \phi(z) + 2a_4 x,$ $z = a_4 t^2/2 + a_5 t - x,$ $(\phi - 2a_4 z - a_5^2)\ddot{\phi} = a_4\dot{\phi},$

$\mathbf{v}_4:$ $u = w(t)x^2/2 + f(t)x + \phi(t),$ $\ddot{\phi} = w\phi,$

$\mathbf{v}_5:$ $u = t\phi(x) - (a_7 x + a_8),$ $\ddot{\phi} = 0,$

$\mathbf{v}_6:$ $u = t^{-2}\phi(x) + t^3(a_9 x + a_{10})/5,$ $\ddot{\phi} = 6,$

$\mathbf{v}_7:$ $u = x\phi(t) - (b_1 t + b_2),$ $\ddot{\phi} = 0,$

$\mathbf{v}_8:$ $u = w(t)x^2/2 + \phi(t)x + f(t),$ $\ddot{\phi} = w(t)\phi,$

$\mathbf{v}_9:$ $u = (t-1)\phi((t+1)x/(t-1)),$ $\ddot{\phi} = 0,$

$\mathbf{v}_{10}:$ $u = t^3\phi(x) + 3x^2 t^{-2} - (b_3 x + b_4)/5t^2,$ $\ddot{\phi} = 0,$

$\mathbf{v}_{11}:$ $u = x\phi(t) + 3x^2/t^2 - b_5 t^2 - b_6 t^{-2},$ $t^2\ddot{\phi} = 64,$

$\mathbf{v}_{12}:$ $u = w(t)\phi(x),$ $\ddot{\phi} = 1.$

11.12. Zabolotskaya-Khokhlov equation

$$u_{tx} - (uu_x)_x - u_{yy} = 0$$

The infinitesimal Lie point symmetry algebra of this equation contains the subalgebra with the generators:

$$X_1 = \partial_t, \quad X_2 = \partial_x, \quad X_3 = \partial_y, \quad X_4 = y\partial_x + 2t\partial_y,$$

$$X_5 = t\partial_t + x\partial_x + y\partial_y, \quad X_6 = 4t\partial_t + 2x\partial_x + 3y\partial_y - 2u\partial_u, \quad X_7 = t\partial_x - \partial_u$$

Nonclassical conditional symmetries

(Extracted from Fushchich, Chopik, and Mironiuk [1991] and adapted to the case of three independent variables)

The equation considered attached by the first-order equation:

$$u_t u_x - u u_x^2 - u_y^2 = 0,$$

which is its characteristic equation, forms the compatible system while such a system is incompatible in the case of four independent variables, i.e., for the equation $u_{tx} - (u_x)_x - u_{yy} - u_{zz} = 0$. The system admits the infinite-dimensional Lie algebra of classical infinitesimal symmetries whose vector fields can be presented as

$$X = \sum_{1}^{8} a_i(u) X_i$$

with $a_i(u)$ smooth functions, X_i for $1 \leq i \leq 7$ given above, and

$$X_8 = y \partial_t + 2(x + 2ut) \partial_y.$$

Acknowledgments

The authors are pleased to thank Mikhail V. Foursov for his help in checking the results of this Survey by using the "Mathematica" program SYMMAN (Dzhamay *et al.* [1994]).

12

Quasi-exactly-solvable Differential Equations

This chapter[1] is devoted to a description of a new connection between Lie algebras and linear differential equations.

The main idea is surprisingly easy. Let us consider a certain set of differential operators of the first order

$$J^\alpha(x) = a^{\alpha,\mu}(x)\partial_\mu + b^\alpha(x), \quad \partial_\mu \equiv \frac{d}{dx^\mu}$$

$\alpha = 1, 2, \ldots, k, x \in R^n, \mu = 1, 2, \ldots, n$ and $a^{\alpha,\mu}(x)$, $b^\alpha(x)$ are certain functions on R^n. Then assume that the operators form a basis of some Lie algebra g of the dimension $k = \dim g$. Now take a polynomial h in generators $J^\alpha(x)$ and ask a question:

> Does the differential operator $h(J^\alpha(x))$ have some specific properties which distinguish this operator from a general linear differential operator?

Generically, an answer is *negative*. However, if the algebra g is taken in a finite-dimensional representation, the answer becomes *positive*:

> The differential operator $h(J^\alpha(x))$ does possess a finite-dimensional invariant subspace coinciding a representation space of the finite-dimensional representation of the algebra g of differential operators of the first order. If a basis of this finite-dimensional representation space can be constructed explicitly, the operator h can be presented in the explicit block-diagonal form.

[1] Supported in Part by a CAST grant of the US National Academy of Sciences and a research grant of CONACyT, Mexico.

Such differential operators having a finite-dimensional invariant subspace with an explicit basis in functions can be named *quasi-exactly-solvable*.

Up to our knowledge and understanding, the first *explicit* examples of quasi-exactly-solvable problems were found by Razavy [1980], [1981] and by Singh et al. [1980]. In an explicit form a general idea of quasi-exactly-solvability had been formulated for the first time in Turbiner [1988a, 1988b], and it had led to a complete catalog of one-dimensional, quasi-exactly-solvable Schroedinger operators in connection with spaces of polynomials (Turbiner [1988c]). The term "quasi-exactly-solvability" has been suggested in Turbiner-Ushveridze [1987]. A connection of quasi-exactly-solvability and finite-dimensional representations of sl_2 was mentioned for the first time by Zaslavskii and Ulyanov [1984]. Later, the idea of quasi-exactly-solvability was generalized to multidimensional differential operators, matrix differential operators (Shifman and Turbiner [1989]), finite-difference operators (Ogievetski and Turbiner [1991]), and, recently, to "mixed" operators containing differential operators and permutation operators (Turbiner [1994a]).

Morozov et al. [1990] described a connection of quasi-exactly-solvability with conformal quantum field theories (see also Halpern-Kiritsis [1989]), while recent development can be found in Tseytlin [1994] and also in a brief review done by Shifman [1994]. Relationship with solid-state physics is given in Ulyanov and Zaslavskii [1992]. A general survey of a phenomenon of the quasi-exactly-solvability can be found in Turbiner [1994b] and González-Lopéz, Kamran, and Olver [1994].

This chapter will be devoted mainly to a description of quasi-exactly-solvable operators acting on functions in one real (complex) variable.

12.1. Generalities

Let us take n linearly independent functions $f_1(x)$, $f_2(x)$, ..., $f_n(x)$ and form a linear space

$$\mathcal{F}_n(x) = \langle f_1(x), f_2(x), \ldots, f_n(x) \rangle, \tag{12.1}$$

where n is a non-negative integer and $x \in \mathbf{R}$.

DEFINITION 12.1. *Two spaces $\mathcal{F}_n^{(1)}(x)$ and $\mathcal{F}_n^{(2)}(x)$ are named equivalent spaces, if one space can be obtained through another one via a change of the variable and/or the multiplication on some function (making a gauge transformation),*

$$\mathcal{F}_n^{(2)}(x) = g(x)\mathcal{F}_n^{(1)}(y(x)). \tag{12.2}$$

Choosing a certain space (12.1) and considering all possible changes of the variable, $y(x)$ and gauge functions, $g(x)$, one can describe a whole class of equivalent spaces. It allows us to introduce a certain standardization, $f_1(x) = 1$, $f_2(x) = x$, since in each class of spaces, one can find a representative satisfying the standardization. Thus, hereafter the only spaces of the following form

$$\mathcal{F}_n(x) = \langle 1, x, f_3(x), f_4(x), \ldots, f_n(x) \rangle, \tag{12.3}$$

are taken into account.

It is easy to see that once an operator $h(y, d_y)$ acts on a space $\mathcal{F}_n^{(1)}(y)$, one can construct an operator acting on an equivalent space $\mathcal{F}_n^{(2)}(y(x))$

$$\bar{h}(y, d_y) = g(x)h(x, d_x)g^{-1}(x)|_{x=x(y)}, \quad d_x \equiv \frac{d}{dx}. \tag{12.4}$$

Later it will be considered one of the most important particular cases of the space 12.3: the space of polynomials of finite order

$$\mathcal{P}_{n+1}(x) = \langle 1, x, x^2, \ldots, x^n \rangle, \tag{12.5}$$

where n is a non-negative integer and $x \in \mathbf{R}$. It is worth noting the important property of an invariance of the space 12.5:

$$x^n \mathcal{P}_{n+1}(\frac{1}{x}) \equiv \mathcal{P}_{n+1}(x). \tag{12.6}$$

It stems from an evident feature of polynomials: if $p_n(x) \in \mathcal{P}_{n+1}(x)$ is any polynomial, then $x^n p_n(1/x)$ remains a polynomial with inverse order of the coefficients then in $p_n(x)$.

12.2. Ordinary differential equations

12.2.1. General consideration

DEFINITION 12.2. *Let us name a linear differential operator of the kth order, $T_k(x, d_x)$ quasi-exactly-solvable, if it preserves the space $\mathcal{P}_{n+1}$. Correspondingly, the operator $E_k(x, d_x)$ is named exactly-solvable, if it preserves the infinite flag $\mathcal{P}_1 \subset \mathcal{P}_2 \subset \mathcal{P}_3 \subset \ldots \subset \mathcal{P}_n \subset \ldots$ of spaces of all polynomials:*
$E_k(x, d_x) : \mathcal{P}_j \mapsto \mathcal{P}_j , \ j = 0, 1, \ldots.$

LEMMA 12.1. *(Turbiner 1994b]). (i) Suppose $n > (k - 1)$. Any quasi-exactly-solvable operator T_k can be represented by a kth degree polynomial of the operators*

$$J_n^+ = x^2 d_x - nx,$$

$$J_n^0 = x d_x - \frac{n}{2}, \qquad (12.7)$$

$$J_n^- = d_x,$$

(the operators (12.7) obey the $sl_2(\mathbf{R})$ commutation relations: $[J^\pm, J^0] = \pm J^\pm$, $[J^+, J^-] = -2J^0$ [2]). If $n \leq (k - 1)$, the part of the quasi-exactly-solvable operator T_k containing derivatives up to order n can be represented by an nth degree polynomial in the generators (12.7).

(ii) Conversely, any polynomial in (12.7) is quasi-exactly-solvable.

(iii) Among quasi-exactly-solvable operators there exist exactly-solvable operators $E_k \subset T_k$.

DEFINITION 12.3. *Let us name a universal enveloping algebra U_{sl_2} of a Lie algebra sl_2 the algebra of all ordered polynomials in generators $J^{\pm,0}$. The notion ordering means that in any monomial in generators $J^{\pm,0}$ all J^+ are placed to the left and all J^- to the right.*

Comment 12.1. Notion of the universal enveloping algebra allows us to make a statement that T_k at $k < n + 1$ is simply an element of the universal enveloping algebra $U_{sl_2(\mathbf{R})}$ of the algebra $sl_2(\mathbf{R})$ taken in representation (12.7). If $k \geq n + 1$, then T_k is represented as an element of $U_{sl_2(\mathbf{R})}$ plus $B \frac{d^{n+1}}{dx^{n+1}}$, where B is any linear differential operator of an order not higher than $(k - n - 1)$. In other words, the algebra of differential operators acting on the space 12.4 coincides to the universal enveloping algebra $U_{sl_2(\mathbf{R})}$ of the algebra $sl_2(\mathbf{R})$ taken in representation (12.7) plus operators annihilating (12.4).

Comment 12.2. The algebra 12.7 has the following invariance property

$$x^{-n} J_n^{\pm,0}(x, d_x) x^n \big|_{x=\frac{1}{z}} \Rightarrow J_n^{\pm,0}(z, d_z)$$

as a consequence of the invariance (12.6) of the space $\mathcal{P}_{n+1}$. In particular, if $z = 1/x$, then

$$J_n^+(x, d_x) \Rightarrow -J_n^-(z, d_z)$$

---

[2] This realization of $sl_2(\mathbf{R})$ had been derived the first time by Sophus Lie. Generically, the representation (12.7) is one of the so-called 'projectivized' representations (see Turbiner [1992, 1994]).

$$J_n^0(x, d_x) \Rightarrow -J_n^0(z, d_z)$$

$$J_n^-(x, d_x) \Rightarrow -J_n^+(z, d_z). \tag{12.8}$$

Let us introduce the *grading* of generators (12.7) as follows. It is easy to check that any $sl_2(\mathbf{R})$-generator maps a monomial into monomial, $J_n^\alpha x^p \mapsto x^{p+d_\alpha}$, then

DEFINITION 12.4. *The number d_α is named a grading of the generator J_n^α:* $deg(J_n^\alpha) = d_\alpha.$

Following this definition

$$deg(J_n^+) = +1 \,, \ deg(J_n^0) = 0 \,, \ deg(J_n^-) = -1, \tag{12.9}$$

and

$$deg[(J_n^+)^{n_+}(J_n^0)^{n_0}(J_n^-)^{n_-}] = n_+ - n_-. \tag{12.10}$$

Notion of the grading allows us to classify the operators T_k in the Lie-algebraic sense.

LEMMA 12.2. *A quasi-exactly-solvable operator $T_k \subset U_{sl_2(\mathbf{R})}$ has no terms of positive grading, if and only if it is an exactly-solvable operator.*

Comment 12.3. Any exactly-solvable operator having a term of negative grading possesses terms of positive grading after transformation (12.8). A quasi-exactly-solvable operator always possesses terms of positive grading as in x-space representation, as in z-space representation.

THEOREM 12.1. *(Turbiner [1994b]). Let n be a non-negative integer. Take the eigenvalue problem for a linear differential operator of the kth order in one variable*

$$T_k(x, d_x)\varphi(x) = \varepsilon\varphi(x), \tag{12.11}$$

where T_k is symmetric. The problem (12.11) has $(n + 1)$ linearly independent eigenfunctions in the form of a polynomial in variable x of an order not higher than n, if and only if T_k is quasi-exactly-solvable. The problem (12.11) has an infinite sequence of polynomial eigenfunctions, if and only if the operator is exactly-solvable.

Comment 12.4. The "if" part of the first statement is obvious. The "only if" part is a direct corollary of Lemma 12.1.

This theorem gives a general classification of differential equations

$$\sum_{j=0}^{k} a_j(x)d_x^j \varphi(x) \; = \; \varepsilon\varphi(x) \tag{12.12}$$

having at least one polynomial solution in x. The coefficient functions $a_j(x)$ must have the form

$$a_j(x) \; = \; \sum_{i=0}^{k+j} a_{j,i}x^i. \tag{12.13}$$

The explicit expressions (12.13) for coefficient function in (12.12) are obtained by the substitution (12.7) into a general, kth-degree polynomial element of the universal enveloping algebra $U_{sl_2(\mathbf{R})}$. Thus the coefficients $a_{j,i}$ can be expressed through the coefficients of the kth-degree polynomial element of the universal enveloping algebra $U_{sl_2(\mathbf{R})}$. The number of free parameters of the polynomial solutions is defined by the number of parameters characterizing a general kth-degree polynomial element of the universal enveloping algebra $U_{sl_2(\mathbf{R})}$. A rather straightforward calculation leads to the following formula

$$par(T_k) = (k+1)^2 \tag{12.14}$$

where we denote the number of free parameters of operator T_k by the symbol $par(T_k)$. For the case of an infinite sequence of polynomial solutions, expression (12.13) simplifies to

$$a_j(x) \; = \; \sum_{i=0}^{j} a_{j,i}x^i \tag{12.15}$$

in agreement with the results by Krall [1938] (see also Littlejohn [1988]). In this case the number of free parameters is equal to

$$par(E_k) = \frac{(k+1)(k+2)}{2}. \tag{12.16}$$

One can show that the operators T_k with the coefficients (12.15) preserve a finite flag $\mathcal{P}_0 \subset \mathcal{P}_1 \subset \mathcal{P}_2 \subset \ldots \subset \mathcal{P}_k$ of spaces of polynomials. One can easily verify that the preservation of such a finite flag of spaces of polynomial leads to the preservation of an infinite flag of such spaces.

A class of spaces equivalent to the space of polynomials (12.5) is presented by

$$\langle \alpha(z), \alpha(z)\beta(z), \ldots, \alpha(z)\beta(z)^n \rangle, \tag{12.17}$$

where $\alpha(z)$, $\beta(z)$ are any functions. Linear differential operators acting in (12.17) are easily obtained from the quasi-exactly-solvable operators (12.11) to (12.13) (see Lemma 12.1) making the change of variable $x = \beta(z)$ and the "gauge" transformation $\tilde{T} = \alpha(z)T\alpha(z)^{-1}$ and have the form

$$\bar{T}_k = \alpha(z) \sum_{j=0}^{k} (\sum_{i=0}^{k+j} a_{j,i}\beta(z)^i)(d_x^j) \, |_{x=\beta(z)} \, \alpha(z)^{-1} \qquad (12.18)$$

where the coefficients $a_{j,i}$ are the same as in (12.13).

So the expression (12.18) gives a general form of the linear differential operator of the kth order acting on a space equivalent to (12.5). Since any one- or two-dimensional invariant sub-space can be presented in the form (12.17) and can be reduced to (12.5), the general statement takes place:

THEOREM 12.2. *There are no linear operators possessing a one- or two-dimensional invariant sub-space with an explicit basis other than given by Lemma 12.1.*

Therefore an eigenvalue problem 12.11, for which one eigenfunction can be found in an explicit form, is related to the operator

$$\mathbf{T}^{(1)} = B(x, d_x)d_x + q_0 \qquad (12.19)$$

and its all modifications 12.4, occurred after a change of the variable and a "gauge" transformation. A general differential operator possessing two eigenfunctions in an explicit form is presented as

$$\mathbf{T}^{(2)} = B(x, d_x)d_x^2 + q_2(x)d_x + q_1(x), \qquad (12.20)$$

plus its all modifications owing to a change of the variable and a "gauge" transformation. Here, $B(x, d_x)$ is any linear differential operator, $q_0 \in \mathbf{R}$ and $q_{1,2}(x)$ are the first- and second-order polynomials, respectively, with coefficients such that $q_2(x)d_x + q_1(x)$ can be expressed as a linear combination of the generators $J_1^{\pm,0}$ (see (12.7)).

12.2.2. Second-order differential equations

The second-order differential equations play an exceptionally important role in applications. Therefore, let us consider in detail the second-order differential equation (12.11) possessing polynomial solutions. From Theorem 12.1 it follows that the corresponding differential operator must be quasi-exactly-solvable and can be represented as

$$T_2 = c_{++} J_n^+ J_n^+ + c_{+0} J_n^+ J_n^0 + c_{+-} J_n^+ J_n^- + c_{0-} J_n^0 J_n^- + c_{--} J_n^- J_n^-$$

$$+ c_+ J_n^+ + c_0 J_n^0 + c_- J_n^- + c, \qquad (12.21.1)$$

where $c_{\alpha\beta}, c_\alpha, c \in \mathbf{R}$. The number of free parameters is $par(T_2) = 9$. Under the condition $c_{++} = c_{+0} = c_+ = 0$, the operator T_2 becomes exactly-solvable (see Lemma 12.2)

$$E_2 = c_{+-} J_n^+ J_n^- + c_{0-} J_n^0 J_n^- + c_{--} J_n^- J_n^- + c_0 J_n^0 + c_- J_n^- + c, \qquad (12.21.2)$$

and the number of free parameters is reduced to $par(E_2) = 6$.

LEMMA 12.3. *If the operator (12.21.1) is such that*

$$c_{++} = 0 \quad and \quad c_+ = (\frac{n}{2} - m)c_{+0}, \quad at\ some\ m = 0, 1, 2, \ldots \qquad (12.22)$$

then the operator T_2 preserves both $\mathcal{P}_{n+1}$ and $\mathcal{P}_{m+1}$. In this case the number of free parameters is $par(T_2) = 7$.

In fact, Lemma 12.3 claims that $T_2(J_n^\alpha, c_{\alpha\beta}, c_\alpha)$ can be rewritten as $T_2(J_m^\alpha, c'_{\alpha\beta}, c'_\alpha)$. As a consequence of Lemma 12.3 and Theorem 12.1, in general, among polynomial solutions of (12.12) there are polynomials of order n and order m.

REMARK.   From the Lie-algebraic point of view, Lemma 12.3 means the existence of representations of second-degree polynomials in the generators (12.7) possessing two invariant sub-spaces. In general, if n in (12.7) is a non-negative integer, then among representations of kth-degree polynomials in the generators (12.7), lying in the universal enveloping algebra, there exist representations possessing $1, 2, \ldots, k$ invariant sub-spaces. Even starting from an infinite-dimensional representation of the original algebra (n in (12.7) is *not* a non-negative integer), one can construct the elements of the universal enveloping algebra having finite-dimensional representation (e.g., the parameter n in (12.22) is non-integer; however, T_2 has the invariant sub-space of dimension $(m + 1)$). Also, this property implies the existence of representations of the polynomial elements of the universal enveloping algebra $U_{sl_2(\mathbf{R})}$, which can be obtained starting from different representations of the original algebra (12.7).
    Substituting (12.7) into (12.21.1) and then into (12.12), we obtain

$$- P_4(x)d_x^2\varphi(x) + P_3(x)d_x\varphi(x) + P_2(x)\varphi(x) = \varepsilon\varphi(x), \qquad (12.23)$$

where the $P_j(x)$ are polynomials of jth order with coefficients related to $c_{\alpha\beta}, c_\alpha$ and n. In general, problem (12.23) has $(n + 1)$ polynomial solutions of the form

of polynomials in x of nth degree. If $n = 1$, as a consequence of Theorem 12.2, a more general spectral problem than (12.23) arises (cf. (12.20))

$$- F_3(x)d_x^2\varphi(x) + q_2(x)d_x\varphi(x) + q_1(x)\varphi(x) = \varepsilon\varphi(x), \qquad (12.24)$$

possessing only two polynomial solutions of the form $(ax + b)$. Here, F_3 is an arbitrary complex function of x and $q_j(x)$, $j = 1, 2$ are polynomials of order j the same as in (12.20). For the case $n = 0$ (one polynomial solution, $\varphi = const.$) the spectral problem (12.11) becomes (cf. (12.19))

$$- F_2(x)d_x^2\varphi(x) + F_1(x)d_x\varphi(x) + q_0\varphi(x) = \varepsilon\varphi(x), \qquad (12.25)$$

where $F_{2,1}(x)$ are arbitrary complex functions and $q_0 \in R$.

Substituting (12.7) into (12.21.2) and then into (12.12), we obtain

$$- Q_2(x)d_x^2\varphi(x) + Q_1(x)d_x\varphi(x) + Q_0(x)\varphi(x) = \varepsilon\varphi(x), \qquad (12.26)$$

where the $Q_j(x)$ are polynomials of jth order with arbitrary coefficients. One can easily show that the differential operator in the r.h.s. can be always presented in the form (12.21.2). The coefficients of $Q_j(x)$ are unambiguously related to $c_{\alpha\beta}, c_\alpha$ for *any* value of the parameter n. Thus n is a fictitious parameter and, for instance, it can be put equal to zero.

12.2.3. Quasi-exactly-solvable Schroedinger equations (examples)

From the point of applications the Schroedinger equation

$$(-d_z^2 + V(z))\Psi(z) = \varepsilon\Psi(z), \qquad (12.27)$$

is one of the important among the second-order differential equations (see, e.g., Landau and Lifschitz [1974]). Here, ϵ is the spectral parameter and $\Psi(z)$ must be square-integrable function on some space. It can be the whole real line, semi-line, or a finite interval. Therefore it is quite natural to search the quasi-exactly-solvable and exactly-solvable operators of the Schroedinger type acting on a finite-dimensional linear space of square-integrable functions: $f_i(z)$, $i = 1, 2, \ldots n$.

One possible way to get the quasi-exactly-solvable and exactly-solvable Schroedinger operators is to transform the quasi-exactly-solvable T_2 and exactly-solvable E_2 operators acting on finite-dimensional spaces of polynomials (12.5) into the Schroedinger type operators. It always can be done making a change of a variable and a gauge transformation (see (12.2)) as a consequence of the one-dimensional nature of the equations studied. In practice, the realization of this transformation is nothing but a conversion of (12.23) to (12.26) into (12.27). The only open question remains: does this new basis belong to the square-integrable one or not? This question will be discussed in the end of this section. In the following consideration we restrict ourselves to the case of real functions and real variables.

Introducing a new function

$$\Psi(z) = \varphi(x(z))e^{-A(z)}, \tag{12.28}$$

and a new variable $x = x(z)$, where $A(z)$ is a certain real function, one can reduce (12.23) – (12.26) to the Sturm-Liouville-type problem (12.27) with the potential equal to

$$V(z) = (A')^2 - A'' + P_2(x(z)), \tag{12.29}$$

if

$$A = \int (\frac{P_3}{P_4})dx - logz' \, , \, z = \pm \int \frac{dx}{\sqrt{P_4}}$$

for the case of (12.23), or

$$A = \int (\frac{Q_2}{F_3})dx - logz' \, , \, z = \pm \int \frac{dx}{\sqrt{F_3}}$$

for the case of (12.24), or

$$A = \int (\frac{F_1}{F_2})dx - logz' \, , \, z = \pm \int \frac{dx}{\sqrt{F_2}}$$

for the case of (12.25) with replacement of $P_2(x(z))$ for two latter cases by $Q_1(x(z))$ or Q_0, respectively. Hereafter, the prime denotes differentiation. If the functions (12.28), obtained after transformation, belong to the $\mathcal{L}_2(\mathcal{D})$-space,[3] we arrive at the recently discovered quasi-exactly-solvable Schroedinger equations (Turbiner [1988a, 1988b, 1988c]), where a finite number of eigenstates is found algebraically.

In order to proceed with a description of concrete examples of quasi-exactly-solvable Schroedinger equations, first of all let us generalize the eigenvalue problem (12.27) inserting a weight function $\hat{\varrho}(z) \equiv \varrho(x(z))$ in the r.h.s.

$$(-d_z^2 + V(z))\Psi(z) = \varepsilon\varrho(x(z))\Psi(z), \tag{12.30}$$

---

[3]Depending on the change of variable $x = x(z)$, the space $\mathcal{D}$ can be whole real line, semi-line, and a finite interval.

where for a sake of future convenience the weight function is presented in the form of composition $\varrho(x(z))$. This equation can be obtained from (12.23) – (12.25) by taking the same gauge factor (12.28) as before, but with another change of the variable

$$z = \pm \int \frac{dx}{\sqrt{\varrho(x)P_4(x)}}$$

which leads to a slightly modified potential then in (12.29)

$$V(z) = (A')^2 - A'' + P_2(x(z))\varrho(x(z)).$$

Where prime denotes differentiation.

Below we will follow the catalog given at Turbiner [1988c]. A presentation of results is the following: firstly, we display the quadratic element T_2 of the universal enveloping algebra sl_2 in the representation (12.7) and its equivalent form of differential operator $T_2(x, d_2)$; secondly, the corresponding potential $V(z)$ and afterwards, the explicit expression for the change of the variable $x = x(z)$, the weight function $\varrho(z)$, and finally the functional form of the eigenfunctions $\Psi(z)$ of the "algebraized" part of the spectra.[4]

We begin a consideration with the quasi-exactly-solvable equations, associated to the exactly-solvable Morse oscillator; it implies that at the limit, when the number of "algebraized" eigenstates $(n + 1)$ goes to infinity, the Morse oscillator occurs.

Comment 12.5. The Morse oscillator is one of the well-known exactly-solvable quantum-mechanical problems (see, e.g., Landau and Lifschitz [1974]). It is described by the Schroedinger operator with the potential

$$V(z) = A^2 e^{-2\alpha z} - 2A e^{-\alpha z}, \quad A, \alpha > 0.$$

This potential is used to model the interaction of the atoms in diatomic molecules.

I.

$$T_2 = -\alpha^2 J_n^+ J_n^- + 2\alpha a J_n^+ - \alpha[\alpha(n + 1) + 2b]J_n^0 - 2\alpha c J_n^- \qquad (12.31)$$

$$-\frac{\alpha n}{2}[\alpha(n + 1) + 2b]$$

or as the differential operator,

$$T_2(x, d_x) = -\alpha^2 x^2 d_x^2 + \alpha[2ax^2 - (\alpha + 2b)x - 2c]d_x - 2\alpha a n x$$

---

[4]The functions $p_n(x)$ occurring in the forthcoming expressions for $\Psi(z)$ are polynomials of the nth order. They are nothing but the polynomial eigenfunctions of the operator $T_2(x, d_x)$.

leads to

$$V(z) = a^2 e^{-2\alpha z} - a[2b + \alpha(2n+1)]e^{-\alpha z} \tag{12.32}$$

$$+ c(2b - \alpha)e^{\alpha z} + c^2 e^{2\alpha z} + b^2 - 2ac$$

where

$$x = e^{-\alpha z}, \quad \varrho = 1,$$

with the eigenfunctions of the "algebraized" part of the spectra

$$\Psi(z) = p_n(e^{-\alpha z}) \exp\left(-\frac{a}{\alpha}e^{-\alpha z} - bz - \frac{c}{\alpha}e^{\alpha z}\right)$$

at $\alpha > 0$, $a, c \geq 0$, and $\forall b$.

II.

$$T_2 = -\alpha J_n^0 J_n^- + 2a J_n^+ - 2c J_n^0 - [\alpha(\frac{n}{2}+1) + 2b]J_n^- - cn \tag{12.33}$$

or as the differential operator,

$$T_2(x, d_x) = -\alpha x d_x^2 + (-2ax^2 + 2cx + \alpha + 2b)d_x + 2anx$$

leads to

$$V(z) = a^2 e^{-4\alpha z} - 2ace^{-3\alpha z} + [c^2 - 2a(b + \alpha an + \alpha)]e^{-2\alpha z} \tag{12.34}$$

$$+ c(2b + \alpha)e^{-\alpha z} + b^2,$$

where

$$x = e^{-\alpha z}, \quad \varrho = \frac{1}{\alpha}e^{-\alpha z},$$

$$\Psi(z) = p_n(e^{-\alpha z}) \exp\left(-\frac{a}{2\alpha}e^{-2\alpha z} + \frac{c}{\alpha}e^{-\alpha z} - bz\right)$$

at $\alpha > 0$, $a \geq 0$, $b > 0$, and $\forall c$.

III.

$$T_2 = -\alpha J_n^+ J_n^0 + (2b - 3\alpha n/2)J_n^+ - 2a J_n^0 - 2c J_n^- - an \tag{12.35}$$

or as the differential operator,

$$T_2(x, d_x) = -\alpha x^3 d_x^2 + [(2b - \alpha)x^2 - 2ax - 2c]d_x + (\alpha n - 2b)nx$$

leads to

$$V(z) = c^2 e^{4\alpha z} + 2ace^{3\alpha z} + [a^2 - 2c(b + \alpha)]e^{2\alpha z} \qquad (12.36)$$

$$-a(2b + \alpha)e^{\alpha z} + b^2 + \alpha n(\alpha n - 2b),$$

where

$$x = e^{-\alpha z}, \; \varrho = \frac{1}{\alpha}e^{\alpha z},$$

$$\Psi(z) = p_n(e^{-\alpha z}) \exp\left(-\frac{c}{2\alpha}e^{2\alpha z} - \frac{a}{\alpha}e^{\alpha z} + bz\right)$$

at $\alpha > 0, b > 0, c \geq 0$, and $\forall a$.

The next two potentials are associated to the one-soliton or Pöschle-Teller potential.

Comment 12.6. The Pöschle-Teller or one-soliton potential describes another well-known exactly-solvable quantum-mechanical problem (see, e.g., Landau and Lifschitz [1974]). It is given by the Schroedinger operator with the potential

$$V(z) = -A^2 \frac{1}{\cosh^2 \alpha z}.$$

This potential has a unique property of an absence of reflection of plane waves and is also one of the simplest solutions of the so-called Korteweg-de Vries equation, playing an important role in the *inverse problem method* (for detailed discussion see, e.g., the book by V.E. Zakharov et al. [1980]).

IV.

$$T_2 = -4\alpha^2 J_n^+ J_n^0 + 4\alpha^2 J_n^+ J_n^- - 2\alpha[\alpha(3n + 2k + 1) + 2c]J_n^+$$

$$+ 2\alpha[\alpha(n + 2) + 2c - 2a]J_n^0 + 4\alpha a J_n^- + \alpha n[\alpha(n + 2) + 2c - 2a] \quad (12.37)$$

or as the differential operator,

$$T_2(x, d_x) = -4\alpha^2(x^3 - x^2)d_x^2 - 2\alpha[(3\alpha + 2\alpha k + 2c)x^2 - 2(\alpha - a + c)x - 2a]d_x$$

$$+(2n + k)\alpha[(2n + k + 1)\alpha + 2c]x$$

leads to

$$V(z) = a^2 \cosh^4 \alpha z - a(a + 2\alpha - 2c) \cosh^2 \alpha z \qquad (12.38)$$

$$- [c(c + \alpha) + \alpha(2n + k)(\alpha(2n + k) + \alpha + 2c)] \cosh^{-2} \alpha z + c^2 + a\alpha - 2ac$$

where

$$x = \cosh \alpha z^{-2}, \; \varrho = 1,$$

$$\Psi(z) = (\tanh \alpha z)^k p_n(\tanh^2 \alpha z)(\cosh \alpha z)^{-c/\alpha} \exp\left(-\frac{a}{4\alpha}\cosh 2\alpha z\right)$$

at $\alpha > 0$, $a \geq 0$, $\forall c$, and $k = 0, 1$.
  V.

$$T_2 = -4\alpha^2 J_n^+ J_n^- + 4\alpha^2 J_n^0 J_n^- + 4\alpha b J_n^+ - 2\alpha[\alpha(2n + 2k + 3) + 2a + 4b]J_n^0$$

$$+ 2\alpha[\alpha(n + 2) + 2a + 2b)J_n^- - \alpha[\alpha n(2n + 2k + 3) + 2an - 2bk] \quad (12.39)$$

or as the differential operator,

$$T_2(x, d_x) = -4\alpha^2 x(x-1)d_x^2 + 2\alpha[2bx^2 - (2a+4b+2k\alpha+3\alpha)x + 2(\alpha+a+b)]d_x$$

$$-4\alpha bnx + 2\alpha b(2n + k)$$

leads to

$$V(z) = -b^2 \cosh^{-6}\alpha z + b[2a + 3b + \alpha(4n + 2k + 3)]\cosh^{-4}\alpha z \quad (12.40)$$

$$-[(a + 3b)(a + b + \alpha) + 2(2n + k)\alpha b)]\cosh^{-2}\alpha z + (a + b)^2$$

where
$$x = \cosh \alpha z^{-2}, \quad \varrho = \cosh^{-2}\alpha z,$$

$$\Psi(z) = (\tanh \alpha z)^k p_n(\tanh^2 \alpha z)(\cosh \alpha z)^{\frac{-(a+b)}{\alpha}} \exp\left(\frac{b}{2\alpha}\tanh^2 2\alpha z\right)$$

at $\alpha > 0$, $(a + b) > 0$, and $k = 0, 1$.
  The next two potentials are associated to the harmonic oscillator potential.
  VI.
  It is one of the first discovered examples of the quasi-exactly-solvable Schroedinger operator. Let us take the following non-linear combination in the generators 12.7 (Turbiner [1988c])

$$T_2 = -4J_n^0 J_n^- + 4a J_n^+ + 4b J_n^0 - 2(n + 1 + 2k)J_n^- + 2bn \quad (12.41)$$

or as the differential operator,

$$T_2(x, d_x) = -4xd_x^2 + 2(2ax^2 + 2bx - 1 - 2k)d_x - 4anx,$$

where $x \in R$ and $a > 0$, $\forall b$ or $a \geq 0$, $b > 0$. Putting $x = z^2$ and choosing the gauge phase $A = ax^2/4 + bx/2 - k/2 \ln x$, we arrive at the spectral problem 12.25 with the potential (Turbiner and Ushveridze [1987])

$$V(z) = a^2 z^6 + 2abz^4 + [b^2 - (4n + 3 + 2k)a]z^2 - b(1 + 2k), \quad (12.42)$$

for which at $k = 0$ ($k = 1$) the first $(n + 1)$ eigenfunctions, even (odd) in x, can be found algebraically. Of course, the number of those "algebraized" eigenfunctions is nothing but the dimension of the irreducible representation (12.4) of the algebra (12.7). Finally, $(n + 1)$ "algebraized" eigenfunctions of (12.25) have the form

$$\Psi(z) = z^k p_n(z^2) e^{-\frac{az^4}{4} - \frac{bz^2}{2}}, \tag{12.43}$$

where $p_n(y)$ is a polynomial of the nth degree. It is worth noting that if the parameter a goes to zero, the potential (12.41) becomes the harmonic oscillator potential and polynomials $z^k p_n(z^2)$ reconcile to the Hermite polynomials $H_{2n+k}(z)$ (see discussion below).

VII.

$$T_2 = -4J_n^0 J_n^- + 4a J_n^+ + 4b J_n^0 - 2(n + d + 2l - 2c)J_n^- + 2bn \tag{12.44}$$

or as the differential operator,

$$T_2(x, d_x) = -4xd_x^2 + 2(2ax^2 + 2bx - d - 2l + 2c)d_x - 4anx$$

leads to

$$V(z) = a^2 z^6 + 2abz^4 + [b^2 - (4n + 2l + d + 2 - 2c)a]z^2 + c(c - 2l - d + 2)z^{-2} \tag{12.45}$$
$$-b(d + 2l - 2c),$$

where

$$x = z^2, \varrho = 1,$$

$$\Psi(z) = p_n(z^2)z^{l-c} e^{-\frac{az^4}{4} - \frac{bz^2}{2}},$$

at $a > 0, \forall b$ or $a \geq 0, b > 0$, and $(d + l - c) > 1$. This case corresponds to the radial part of the d-dimensional Schroedinger equation, $z \in [0, \infty)$. At $d = 1$ the radial part coincides with the ordinary Schroedinger equation 12.25 at $z \in R$. The potential 12.45 becomes a generalized version of the potential 12.42 with an additional singular term proportional to z^{-2}.

The next two potentials are associated to the Coulomb problem.

VIII.

$$T_2 = -J_n^0 J_n^- + 2a J_n^+ + 2b J_n^0 - (n/2 + d + 2l - 2c - 1)J_n^- - an \tag{12.46}$$

or as the differential operator,

$$T_2(x, d_x) = -xd_x^2 + (2ax^2 + 2bx + 2c - d - 2l + 1)d_x - 2anx$$

leads to

$$V(z) = a^2 z^2 + 2abz - b(2l + d - 1 - 2c)z^{-1} + c(c - 2l - d + 2)z^{-2} \quad (12.47)$$

$$+b^2 + a(2c - d - 2l - 2n),$$

where

$$x = z, \; \varrho = z^{-1},$$

$$\Psi(z) = p_n(z)z^{l-c}e^{-\frac{a}{2}z^2 - bz},$$

at $a \geq 0, b > 0$, and $(d + l - c) > 1$.

IX.

$$T_2 = -J_n^+ J_n^- + 2a J_n^+ - (n+d-1+2l-2c)J_n^0 + 2b J_n^- - n(d+2l-1-2c) \quad (12.48)$$

or as the differential operator,

$$T_2(x, d_x) = -x^2 d_x^2 - [2ax^2 + (2c - d - 2l + 1)x - 2b]d_x - 4anx$$

leads to

$$V(z) = b^2 z^{-4} - b(2c - 2l - d + 3)z^{-3} + [c(c - 2l - d + 2) - 2ab]z^{-2}$$

$$- a(2n + 2l + d - 1 - 2c)z^{-1} + a^2, \quad (12.49)$$

where

$$x = z, \; \varrho = z^{-2},$$

$$\Psi(z) = p_n(z)z^{l-c}e^{-az - bz^{-1}},$$

at $a \geq 0, b > 0$ or $a > 0, b \geq 0$, and $c, l, d \in R$.

Now let us show an example of the non-singular periodic quasi-exactly-solvable potential connected with the Mathieu potential.

Comment 12.7. The Mathieu potential

$$V(z) = A \cos \alpha z$$

is one of the most important potentials appeared in many branches of physics, engineering, etc. Detailed descriptions of the properties of the corresponding Schroedinger equation can be found, for instance, in Bateman and Erdélyi [1953], Vol. 3, and also Kamke [1959], Equation 2.22.

X.

Firstly, take the following operator

$$T_2 = \alpha^2 J_{n-\mu}^+ J_{n-\mu}^- - \alpha^2 J_{n-\mu}^- J_{n-\mu}^+ + \tag{12.50}$$

$$2a\alpha J_{n-\mu}^+ + \alpha^2 (n + \mu + 1) J_{n-\mu}^0 - 2\alpha a J_{n-\mu}^- + \alpha^2 \frac{(n-\mu)(n+\mu+1)}{2}$$

or as the differential operator,

$$T_2(x, d_x) = \alpha^2 (x^2 - 1) d_x^2 + \alpha[2ax^2 + \alpha(1 + 2\mu)x - 2a] d_x - 2\alpha a(n - \mu)x.$$

The transformation (12.28) and (12.29) leads to the periodic potential

$$V(z) = a^2 \sin^2 \alpha z - \alpha a(2n + 1) \cos \alpha z + \mu \alpha^2, \tag{12.51}$$

where

$$x = \cos \alpha z, \quad \varrho = 1,$$

$$\Psi(z) = (\sin \alpha z)^\mu \, p_{n-\mu}(\cos \alpha z) \, e^{(\frac{a}{\alpha} \cos \alpha z)},$$

at $a \geq 0, \alpha \geq 0$. Here $\mu = 0, 1$. For the fixed n, $(2n + 1)$ eigenstates having a meaning of the edges of energy bands (zones) can be found algebraically.

Secondly, take the following operator

$$T_2 = \alpha^2 J_{n-1}^+ J_{n-1}^- - \alpha^2 J_{n-1}^- J_{n-1}^- \tag{12.52}$$

$$+ 2a\alpha J_{n-1}^+ + \alpha^2 (n + 1) J_{n-1}^0 - \alpha[2a + \alpha(v_1 - v_2)] J_{n-1}^- + \alpha^2 \frac{(n^2 - 1)}{2}$$

or as the differential operator,

$$T_2(x, d_x) = \alpha^2 (x^2 - 1) d_x^2 + \alpha[2ax^2 + 2\alpha x - 2a - \alpha(v_1 - v_2)] d_x - 2\alpha a(n - 1)x$$

The transformation (12.28) and (12.29) leads to the periodic potential

$$V(z) = a^2 \sin^2 \alpha z - \alpha a(2n) \cos \alpha z + \alpha a(v_1 - v_2)] - \frac{\alpha^2}{4}, \tag{12.53}$$

where

$$x = \cos \alpha z, \quad \varrho = 1,$$

$$\Psi(z) = (\cos \alpha z)^{v_1} (\sin \alpha z)^{v_2} \, p_{n-1}(\cos \alpha z) \, e^{(\frac{a}{\alpha} \cos \alpha z)},$$

at $a \geq 0, \alpha \geq 0$. Here, $\nu_{1,2} = 0, 1$, but $\nu_1 + \nu_2 = 1$. For the fixed n, $(2n + 1)$ eigenstates having a meaning of the edges of energy bands can be found algebraically.

12.2.4. Quasi-exactly-solvable Schroedinger equations (Lame equation)

In this section we consider one of the most important second-order ordinary differential equations – m-zone Lame equation

$$- d_x^2 \Psi + m(m + 1)\mathcal{P}(x)\Psi = \varepsilon \Psi \qquad (12.54)$$

where $\mathcal{P}(x)$ is the Weierstrass function in a standard notation (see, e.g. Bateman and Erd´elyi [1953]), which depends on two free parameters, and $m = 1, 2, \ldots$. In a description we mainly follow the paper of Turbiner [1989].

The Weierstrass function is a double-periodic meromorphic function for which the equation $\mathcal{P}'^2 = (\mathcal{P} - e_1)(\mathcal{P} - e_2)(\mathcal{P} - e_3)$ holds, where $\sum e_i = 0$. Introducing the new variable $\xi = \mathcal{P}(x) + \frac{1}{3} \sum a_i$ in (12.54) (see, e.g., Kamke [1959]), the new equation emerges

$$\eta'' + \frac{1}{2}\left(\frac{1}{\xi - a_1} + \frac{1}{\xi - a_2} + \frac{1}{\xi - a_3}\right)\eta' - \frac{m(m + 1)\xi + \varepsilon}{4(\xi - a_1)(\xi - a_2)(\xi - a_3)}\eta = 0$$
$$(12.55)$$

where $\eta(\xi) \equiv \psi(x)$. Here the new parameters a_i satisfy the system of linear equations $e_i = a_i - \frac{1}{3} \sum a_i$. Equation (12.55) is named by an algebraic form for the Lame equation. There exists a spectral parameter λ for which Equation (12.55) has four types of solutions:

$$\eta^{(1)} = p_k(\xi) \qquad (12.56.1)$$

$$\eta_i^{(2)} = (\xi - a_i)^{1/2} p_k(\xi), \quad i = 1, 2, 3 \qquad (12.56.2)$$

$$\eta_i^{(3)} = (\xi - a_{l_1})^{1/2}(\xi - a_{l_2})^{1/2} p_{k-1}(\xi),$$

$$l_1 \neq l_2; i \neq l_{1,2}; i, l_{1,2} = 1, 2, 3 \qquad (12.56.3)$$

$$\eta^{(4)} = (\xi - a_1)^{1/2}(\xi - a_2)^{1/2}(\xi - a_3) p_{k-1}(\xi) \qquad (12.56.4)$$

where $p_r(\xi)$ are polynomial in ξ of degree r. If the value of parameter n is fixed, there are $(2m + 1)$ linear independent solutions of the following form: if $m = 2k$ is even, then the $\eta^{(1)}(\xi)$ and $\eta^{(3)}(\xi)$ solutions arise, if $m = 2k + 1$ is odd we have solutions of the $\eta^{(2)}(\xi)$ and $\eta^{(4)}(\xi)$ types. Those eigenvalues have a meaning of the edges of the energy zones in the potential (12.54).

THEOREM 12.3. *(Turbiner [1989]). The spectral problem (12.54) at $m = 1, 2, \ldots$ with polynomial solutions (12.56.1), (12.56.2), (12.56.3), and (12.56.4)* *is equivalent to the spectral problem (12.11) for the operator T_2 (12.21.1) belonging to the universal enveloping sl_2-algebra in the representation (12.7) with the coefficients*

$$c_{+0} = 4, \quad c_{+-} = -4\sum a_i, \quad c_{0-} = 4\sum a_i a_j, \quad c_{--} = a_1 a_2 a_3 \quad (12.57)$$

before the terms quadratic in generators and the following coefficients before linear in generators $J_r^{\pm,0}$:

(1)   *For $\eta^{(1)}(\xi)$-type solutions at $m = 2k, r = k$*

$$c_+ = -6k - 2, \; c_0 = 4(k+1)\sum a_i, \; c_- = -2(k+1)\sum a_i a_j$$
$$(12.58.1)$$

(2)   *For $\eta_i^{(2)}(\xi)$-type solutions at $m = 2k + 1, r = k$*

$$c_+ = -6k - 6, \; c_0 = 4(k+2)\sum a_i - a_i,$$

$$c_- = -2(k+1)\sum a_i a_j - 4a_{l_1} a_{l_2}, \quad i \neq l_{1,2}, l_1 \neq l_2 \quad (12.58.2)$$

(3)   *For $\eta_i^{(3)}(\xi)$-type solutions at $m = 2k, r = k - 1$*

$$c_+ = -6k - 4, \; c_0 = 4(k+1)\sum a_i + 4a_i,$$

$$c_- = -2(k+2)\sum a_i a_j + 4a_{l_1} a_{l_2}, \quad i \neq l_{1,2}, l_1 \neq l_2 \quad (12.58.3)$$

(4)   *For $\eta_i^{(4)}(\xi)$-type solutions at $m = 2k + 1, r = k - 1$*

$$c_+ = -6k - 8, \; c_0 = 4(k+2)\sum a_i, \; c_- = -2(k+2)\sum a_i a_j$$
$$(12.58.4)$$

So, each type of solution (12.56.1), (12.56.2), (12.56.3), and (12.56.4) corresponds to the particular spectral problem (12.54) with a special set of parameters (12.57) plus (12.58.1), (12.58.2), (12.58.3), and (12.58.4), respectively. It can be easily shown that the calculation of eigenvalues ε of (12.54) corresponds to the solution of a characteristic equation for the four-diagonal matrix:

$$C_{l,l-1} = (l - 1 - 2j)[4(j + 1 - l) + c_+],$$

$$C_{l,l} = [l(2j + 1 - l)c_{+-} + (l - j)c_0],$$

$$C_{l,l+1} = (l + 1)(j - l)c_{0-} + (l + 1)c_-,$$

$$C_{l,l+2} = -(l+1)(l+2)c_{--}, \tag{12.59}$$

where the size of this matrix is $(k+1) \times (k+1)$ and $2j = k$ for (12.56.1), (12.56.2), and $k \times k$ and $2j = k - 1$ for (12.56.3), (12.56.4), respectively. In connection to Theorem 2.3 one can prove the following theorem (Turbiner [1989]).

THEOREM 12.4. *Let fix the parameters $e's$ ($a's$) in (12.54) (or (12.55)) except one, e.g., $e_1(a_1)$. The first $(2m+1)$ eigenvalues of (12.54) (or (12.55)) form a $(2m+1)$-sheeted Riemann surface in parameter $e_1(a_1)$. This surface contains four disconnected pieces: one of them corresponds to $\eta^{(1)}(\eta^{(4)})$ solutions and the others correspond to $\eta^{(3)}(\eta^{(2)})$. At $m = 2k$ the Riemann subsurface for $\eta^{(1)}$ has $(k+1)$ sheets and the number of sheets in each of the others is equal to k. At $m = 2k+1$ the number of sheets for $\eta^{(4)}$ is equal to k and for $\eta^{(2)}$ each subsurface contains $(k+1)$ sheets.*

It is worth emphasizing that we cannot find a relation between the spectral problem for the two-zone potential

$$V = -2\sum_{k=1}^{3} \mathcal{P}(x - x_i), \quad \sum_{i=1}^{3} x_i = 0, \tag{12.60}$$

(see Dubrovin and Novikov [1974])[5] and the spectral problem (12.11) for T_2 with the parameters (12.57) and (12.58.1) or (12.58.3) at $k = 1$. In this case eigenvalues ε and also eigenfunctions (12.55) (but not (12.54)) do not depend on parameters c_{--}.

Comment 12.8. One can generalize a meaning of isospectral deformation, saying we want to study a variety of potentials with the first several coinciding eigenvalues. It can be named *quasi-isospectral deformation.*

Now let us consider such a quasi-isospectral deformation of (12.54) at $m = 2$. It arises from the fact that the addition of the term $c_{++}J_r^+ J_r^+$ to the operator T_2 with the parameters (12.57) and (12.58.1) or (12.58.3) at $k = 1$ does not change the characteristic matrix (12.59). Making the reduction (12.28) and (12.29) from Equation (12.11) to the Schroedinger equation (12.27), we obtain

$$V(x) = c_{++}\frac{2c_{++}\xi^6 - c_{+-}\xi^4 - 2c_{0-}\xi^3 - 3c_{--}\xi^2}{P_4^2(\xi)} + P_2(\xi), \tag{12.61}$$

---

[5]The potential (12.60) and the original Lame potential (12.54) at $m = 2$ are related via the isospectral deformation.

where

$$P_4(\xi) = c_{++}\xi^4 + c_{+0}\xi^3 + c_{+-}\xi^2 + c_{0-}\xi + c_{--}, \quad P_2(\xi) = -m(m+1)\xi + \frac{c_0}{2}.$$
$$(12.62)$$

and ξ is defined via the equation

$$x = \int \frac{d\xi}{\sqrt{P_4(\xi)}}. \tag{12.63}$$

In general, the potential (12.61) contains four double poles in x and does not reduce to (12.60). It is worth noting that the first five eigenfunctions in the potential (12.61) have the form

$$\Psi(x) = \left\{ \frac{A\xi + B}{(\xi - a_i)^{1/2}(\xi - a_j)^{1/2}} \right\} \exp\left(-c_{++} \int \frac{\xi^3 d\xi}{P_4(\xi)} \right). \tag{12.64}$$

Here, ξ is given by (12.63), $i \neq j$, and $i, j = 1, 2, 3$. The first five eigenvalues of the potential (12.61) do not depend on the parameters c_{--}, c_{++}.

12.2.5. Exactly-solvable equations and classical orthogonal polynomials

One of the most important properties of the exactly-solvable operators (12.21.2) is the following: the eigenvalues of a general exactly-solvable operator E_2 are given by the quadratic polynomial in number of eigenstate

$$\epsilon_m = c_{00}m^2 + c_0m + const. \tag{12.65}$$

(for details see Turbiner [1988a, 1988b]). It can be easily verified by straightforward calculation.

Taking different exactly-solvable operators E_2 (see (12.21.2)) for the eigenvalue problem (12.11) one can reproduce the equations having the Hermite, Laguerre, Legendre, and Jacobi polynomials as eigenfunctions (Turbiner [1988a, 1988b]), which is shown below. In the definition of the above polynomials we follow the definition given in Bateman and Erd´elyi [1953].

1. Hermite polynomials.

The Hermite polynomials $H_{2m+k}(x)$, $m = 0, 1, 2, \ldots$, $k = 0, 1$ are the polynomial eigenfunctions of the operator

$$E_2(x, d_x) = d_x^2 - 2xd_x \tag{12.66}$$

which immediately can be rewritten in terms of the generators 12.7 following the Lemma 12.1:

$$E_2 = J_0^-(x)J_0^-(x) - 2J_0^0(x). \qquad (12.67)$$

However, there exists another way to represent the operators related to the Hermite polynomials. Let us notice that the parameter k has a meaning of the parity of the polynomial H_{2m+k} and

$$H_{2m+k}(x) = x^k h_m(x^2).$$

Then it is easy to find that the operator having $h_m(y)$ as the eigenfunctions

$$\bar{E}_2(y, d_y) = 4yd_y^2 - 2(2y - 1 - 2k)d_y \qquad (12.68)$$

and, correspondingly,

$$\bar{E}_2 = 4J_0^0(y)J_0^-(y) - 4J_0^0(y) + 2(1 + 2k)J_0^-(y). \qquad (12.69)$$

Of course, those two representations are equivalent; however, a quasi-exactly-solvable generalization can be implemented for the second representation only (see examples VI and VII in Section 12.2.3).

2. *Laguerre polynomials.*

The associated Laguerre polynomials $L_m^a(x)$ occur as the polynomial eigenfunctions of the generalized Laguerre operator

$$E_2(x, d_x) = xd_x^2 + (a + 1 - x)d_x \qquad (12.70)$$

where a is any real number. Of course, the operator 12.70 can be rewritten as

$$E_2 = J_0^0 J_0^- - J_0^0 + (a + 1)J_0^-. \qquad (12.71)$$

It is worth noting that the polynomials $h_m(y)$ (see item 1) are nothing but the associated Lagherre polynomials $L_m^{-1/2+k}(y)$.

3. *Legendre polynomials.*

The Legendre polynomials $P_{2m+k}(x)$ are the polynomial eigenfunctions of the operator

$$E_2(x, d_x) = (1 - x^2)d_x^2 - 2xd_x \qquad (12.72)$$

or,

$$E_2 = -J_0^0 J_0^0 + J_0^- J_0^- - J_0^0. \qquad (12.73)$$

Analogously to the Hermite polynomials there exists another way to represent the operators related to the Legendre polynomials. Let us notice that k has a meaning of the parity of the polynomial P_{2m+k} and

$$P_{2m+k}(x) = x^k p_m(x^2).$$

Then it is easy to find that the operator having $p_m(y)$ as the eigenfunctions

$$\bar{E}_2(y, d_y) = 4y(1-y)d_y^2 + 2[1+2k-(3+2k)y]d_y \qquad (12.74)$$

and, correspondingly

$$\bar{E}_2 = -4J_0^+(y)J_0^-(y) + 4J_0^0(y)J_0^-(y) - 2(3+2k)J_0^0(y)$$

$$+ 2(1+2k)J_0^-(y) \qquad (12.75)$$

4. Jacobi polynomials.

The Jacobi polynomials appear as the polynomial eigenfunctions of the Jacobi equation taking in either symmetric form with the operator

$$E_2(x, d_x) = (1-x^2)d_x^2 + [b-a-(a+b+2)x]d_x \qquad (12.76)$$

corresponding to

$$E_2 = -J_0^0 J_0^0 + J_0^- J_0^- - (1+a+b)J_0^0 + (b-a)J_0^-, \qquad (12.77)$$

or asymmetric form (see, e.g., the book by Murphy [1960] or Bateman and Erd´elyi [1953])

$$E_2(x, d_x) = x(1-x)d_x^2 + [1+a-(a+b+2)x]d_x \qquad (12.78)$$

corresponding to

$$E_2 = -J_0^0 J_0^0 + J_0^0 J_0^- - (1+a+b)J_0^0 + (a+1)J_0^-. \qquad (12.79)$$

Under special choices of the general element $E_4(E_6, E_8)$, one can reproduce all known fourth- (sixth-, eighth-) order differential equations giving rise to infinite sequences of orthogonal polynomials (see, e.g., Littlejohn [1988] and other papers in this volume).

Recently, González-Lopéz, Kamran, and Olver [1993] gave the complete description of second-order polynomial elements of $U_{sl_2(\mathbf{R})}$ in the representation (12.7) leading to the square-integrable eigenfunctions of the Sturm-Liouville problem (12.27) after Transformation (12.28) and (12.29). Consequently, for second-order ordinary differential equation (12.23) the combination of their result and Theorems 12.1 and 12.2 gives a general solution of the problem of classification of equations possessing a finite number of orthogonal polynomial solutions.

12.3. Finite-difference equations in one variable

12.3.1. General consideration

Let us define a multiplicative finite-difference operator, or a shift operator or the so-called Jackson symbol (see, e.g., Exton [1983] and Gasper and Rahman [1990])

$$Df(x) = \frac{f(x) - f(qx)}{(1-q)x} \tag{12.80}$$

where $q \in R$ and $f(x)$ is real function $x \in R$. The Leibnitz rule for the operator D is

$$Df(x)g(x) = (Df(x))g(x) + f(qx)Dg(x)$$

Now one can easily introduce the finite-difference analog of the algebra of the differential operators (12.7) based on the operator D instead of the continuous derivative (Ogievetski and Turbiner [1991])

$$\tilde{J}_n^+ = x^2 D - \{n\}x$$

$$\tilde{J}_n^0 = xD - \hat{n} \tag{12.81}$$

$$\tilde{J}_n^- = D,$$

where $\{n\} = \frac{1-q^n}{1-q}$ is a so-called q number and $\hat{n} \equiv \frac{\{n\}\{n+1\}}{\{2n+2\}}$. The operators (12.81) after multiplication by some factors

$$\tilde{j}^0 = \frac{q^{-n}}{p+1} \frac{\{2n+2\}}{\{n+1\}} \tilde{J}_n^0$$

$$\tilde{j}^\pm = q^{-n/2} \tilde{J}_n^\pm$$

(see Ogievetski and Turbiner [1991]) form a quantum algebra $sl_2(\mathbf{R})_q$ with the following commutation relations:

$$q\tilde{j}^0\tilde{j}^- - \tilde{j}^-\tilde{j}^0 = -\tilde{j}^-$$

$$q^2\tilde{j}^+\tilde{j}^- - \tilde{j}^-\tilde{j}^+ = -(q+1)\tilde{j}^0 \tag{12.82}$$

$$\tilde{j}^0\tilde{j}^+ - q\tilde{j}^+\tilde{j}^0 = \tilde{j}^+.$$

The parameter q does characterize the deformation of the commutators of the classical Lie algebra sl_2. If $q \to 1$, the commutation relations (12.82) reduce to the standard $sl_2(\mathbf{R})$ ones. A remarkable property of generators (12.81) is that, *if n is a non-negative integer, they form the finite-dimensional representation corresponding to the finite-dimensional representation space $\mathcal{P}_{n+1}$ the same as of the non-deformed sl_2* (see (12.7)). (For values of q other than root of unity this representation is irreducible.)

Comment 12.9. The algebra (12.82) is known in literature as the so-called *second Witten quantum deformation* of sl_2 in the classification of Zachos [1991].

Similarly, as for differential operators one can introduce quasi-exactly-solvable $\tilde{T}_k(x, D)$ and exactly-solvable finite-difference operators $\tilde{E}_k(x, D)$ (see Definition 12.2).

LEMMA 12.4. *(Turbiner [1994b]). (i) Suppose $n > (k - 1)$. Any quasi-exactly-solvable operator $\tilde{T}_k$ can be represented by a kth-degree polynomial of the operators 12.81. If $n \leq (k - 1)$, the part of the quasi-exactly-solvable operator $\tilde{T}_k$ containing derivatives up to order n can be represented by an nth-degree polynomial in the generators (12.81).*

(ii) Conversely, any polynomial in (12.81) is quasi-exactly-solvable.

(iii) Among quasi-exactly-solvable operators there exist exactly-solvable operators $\tilde{E}_k \subset \tilde{T}_k$.

Comment 12.10. If we define an analog of the universal enveloping algebra U_g for the quantum algebra $\tilde{g}$ as an algebra of all ordered polynomials in generators, then a quasi-exactly-solvable operator $\tilde{T}_k$ at $k < n + 1$ is simply an element of the 'universal enveloping algebra' $U_{sl_2(\mathbf{R})_q}$ of the algebra $sl_2(\mathbf{R})_q$ taken in representation (12.81). If $k \geq n + 1$, then $\tilde{T}_k$ is represented as an element of $U_{sl_2(\mathbf{R})_q}$ plus $B D^{n+1}$, where B is any linear difference operator of order not higher than $(k - n - 1)$.

Similar to $sl_2(\mathbf{R})$ (see Definition 12.4), one can introduce the grading of generators (12.81) of $sl_2(\mathbf{R})_q$ (cf. (12.9)) and, hence, the grading of monomials of the universal enveloping $U_{sl_2(\mathbf{R})_q}$ (cf. (12.10)).

LEMMA 12.5. *A quasi-exactly-solvable operator $\tilde{T}_k \subset U_{sl_2(\mathbf{R})_q}$ has no terms of positive grading, iff it is an exactly-solvable operator.*

THEOREM 12.5. *(Turbiner [1994b]). Let n be a non-negative integer. Take the eigenvalue problem for a linear difference operator of the kth order in one variable*

$$\tilde{T}_k(x, D)\varphi(x) = \varepsilon\varphi(x), \qquad (12.83)$$

where $\tilde{T}_k$ is symmetric. The problem (12.83) has $(n + 1)$ linearly independent eigenfunctions in the form of a polynomial in variable x of order not higher than n, if and only if T_k is quasi-exactly-solvable. The problem (12.83) has an infinite *sequence of polynomial eigenfunctions, if and only if the operator is exactly-solvable* $\tilde{E}_k$.

Comment 12.11. Saying the operator $\tilde{T}_k$ is symmetric, we imply that, considering the action of this operator on a space of polynomials of degree not higher than n, one can introduce a positively defined scalar product, and the operator $\tilde{T}_k$ is symmetric with respect to it.

This theorem gives a general classification of finite-difference equations

$$\sum_{j=0}^{k} \tilde{a}_j(x) D^j \varphi(x) = \varepsilon \varphi(x) \qquad (12.84)$$

having polynomial solutions in x. The coefficient functions must have the form

$$\tilde{a}_j(x) = \sum_{i=0}^{k+j} \tilde{a}_{j,i} x^i. \qquad (12.85)$$

In particular, this form occurs after substituting (12.81) into a general kth-degree polynomial element of the universal enveloping algebra $U_{sl_2(\mathbf{R})_q}$. It guarantees the existence of at least a finite number of polynomial solutions. The coefficients $\tilde{a}_{j,i}$ are related to the coefficients of the kth-degree polynomial element of the universal enveloping algebra $U_{sl_2(\mathbf{R})_q}$. The number of free parameters of the polynomial solutions is defined by the number of free parameters of a general kth-order polynomial element of the universal enveloping algebra $U_{sl_2(\mathbf{R})_q}$.[6] A rather straightforward calculation leads to the following formula:

$$par(\tilde{T}_k) = (k+1)^2 + 1$$

(for the second-order finite-difference equation $par(\tilde{T}^2) = 10$). For the case of

---

[6]For quantum $sl_2(\mathbf{R})_q$ algebra there are no polynomial Casimir operators (see, e.g., Zachos [1994b]). However, in the representation (12.81) the relationship between generators analogous to the quadratic Casimir operator

$$q\tilde{J}_n^+ \tilde{J}_n^- - \tilde{J}_n^0 \tilde{J}_n^0 + (\{n+1\} - 2\hat{n})\tilde{J}_n^0 = \hat{n}(\hat{n} - \{n+1\})$$

appears. It reduces the number of independent parameters of the second-order polynomial element of $U_{sl_2(\mathbf{R})_q}$. It becomes the standard Casimir operator at $q \to 1$.

an infinite sequence of polynomial solutions the formula 12.85 simplifies to

$$\tilde{a}_j(x) = \sum_{i=0}^{j} \tilde{a}_{j,i} x^i \tag{12.86}$$

and the number of free parameters is given by

$$par(\tilde{E}_k) = \frac{(k+1)(k+2)}{2} + 1$$

(for $k = 2$, $par(\tilde{E}_2) = 7$). The increase in the number of free parameters compared to ordinary differential equations is due to the presence of the deformation parameter q.

12.3.2. Second-order finite-difference exactly-solvable equations

In Turbiner [1992] it is implemented as a description in the present approach of the q-deformed Hermite, Laguerre, Legendre, and Jacobi polynomials (for definitions of these polynomials see Exton [1983], Gasper and Rahman [1990]). In order to reproduce the known q-deformed classical Hermite, Laguerre, Legendre, and Jacobi polynomials (for the latter, there exists the q-deformation of the asymmetric form (12.67) only; see, e.g., Exton [1983] and Gasper and Rahman [1990]), one should modify the spectral problem (12.83):

$$\tilde{T}_k(x, D)\varphi(x) = \varepsilon\varphi(qx), \tag{12.87}$$

by introducing the r.h.s. function the dependence on the argument qx (cf. (12.11) and (12.83)) as it follows from the book by Exton [1983] (see also Gasper and Rahman [1990]). Then corresponding q-difference operators having q-deformed classical Hermite, Laguerre, Legendre and Jacobi polynomials as eigenfunctions (see Equations 5.6.2, 5.5.7.1, 5.7.2.1, and 5.8.3 in Exton [1983], respectively) are given by the combinations in the generators:

$$\tilde{E}_2 = \tilde{J}_0^- \tilde{J}_0^- - \{2\}\tilde{J}_0^0, \tag{12.88.1}$$

$$\tilde{E}_2 = \tilde{J}_0^0 \tilde{J}_0^- - q^{-a-1}\tilde{J}_0^0 + (q^{-a-1}\{a+1\})\tilde{J}_0^-, \tag{12.88.2}$$

$$\tilde{E}_2 = -q\tilde{J}_0^0 \tilde{J}_0^0 + \tilde{J}^- \tilde{J}^- + (q - \{2\})\tilde{J}_0^0, \tag{12.88.3}$$

$$\tilde{E}_2 = -q^{a+b-1}\tilde{J}_0^0 \tilde{J}_0^0 + q^a \tilde{J}_0^0 \tilde{J}_0^- + [q^{a+b-1} - \{a\}q^b - \{b\}]\tilde{J}_0^0 + \{a\}\tilde{J}_0^-, \tag{12.88.4}$$

respectively.

LEMMA 12.6. *If the operator $\tilde{T}_2$ (for the definition, see (12.21.1)) is such that*

$$\tilde{c}_{++} = 0 \quad and \quad \tilde{c}_+ = (\hat{n} - \{m\})\tilde{c}_{+0} \,, \; at \; some \; m = 0, 1, 2, \ldots \quad (12.89)$$

then the operator $\tilde{T}_2$ preserves both $\mathcal{P}_{n+1}$ and $\mathcal{P}_{m+1}$, and polynomial solutions in x with eight free parameters occur.

As usual in quantum algebras, a rather outstanding situation occurs if the deformation parameter q is equal to a primitive root of unity. For instance, the following statement holds.

LEMMA 12.7. *If a quasi-exactly-solvable operator $\tilde{T}_k$ preserves the space $\mathcal{P}_{n+1}$ and the parameter q satisfies the equation*

$$q^n = 1, \quad (12.90)$$

then the operator $\tilde{T}_k$ preserves an infinite flag of polynomial spaces $\mathcal{P}_0 \subset \mathcal{P}_{n+1} \subset \mathcal{P}_{2(n+1)} \subset \ldots \subset \mathcal{P}_{k(n+1)} \subset \ldots$.

It is worth emphasizing that, in the limit as q tends to one, Lemmas 12.4 to 12.6 and Theorem 12.5 coincide with Lemmas 12.1 to 12.3 and Theorem 12.1, respectively. Thus the case of differential equations in one variable can be treated as a limiting case of finite-difference ones. Evidently, one can consider the finite-difference operators, which are a mixture of generators (12.81) with the same value of n and different q's.

12.4. 2×2 matrix differential equations on the real line

12.4.1. General consideration

This section is devoted to a description of quasi-exactly-solvable $T_k(x)$ and exactly-solvable $E_k(x)$, 2×2 matrix differential operators acting on the space of the two-component spinors with polynomial components

$$\mathcal{P}_{n+1,m+1} = \left\langle \begin{array}{c} x^0, x^1, \ldots, x^m \\ x^0, x^1, \ldots, x^n \end{array} \right\rangle. \quad (12.91)$$

This space is a natural generalization of the space $\mathcal{P}_n$ (see (12.5)). The definition of quasi- and exactly-solvable operators is also a natural generalization of the Definition 12.2.

Now let us introduce the following two sets of 2×2 matrix differential operators:

$$T^+ = x^2 \partial_x - nx + x\sigma^+\sigma^-,$$

$$T^0 = x\partial_x - \frac{n}{2} + \frac{1}{2}\sigma^+\sigma^-, \tag{12.92}$$

$$T^- = \partial_x .$$

$$J = -\frac{n}{2} - \frac{1}{2}\sigma^+\sigma^-$$

named bosonic (even) generators and

$$Q = \begin{bmatrix} \sigma^- \\ x\sigma^- \end{bmatrix}, \quad \bar{Q} = \begin{bmatrix} x\sigma^+\partial_x - n\sigma^+ \\ -\sigma^-\partial_x \end{bmatrix}. \tag{12.93}$$

named fermionic (odd) generators, where $\sigma^{\pm,0}$ are the Pauli matrices in standard notation

$$\sigma^+ = \begin{pmatrix} 0 & 1 \\ 0 & 0 \end{pmatrix}, \quad \sigma^- = \begin{pmatrix} 0 & 0 \\ 1 & 0 \end{pmatrix}, \quad \sigma^0 = \begin{pmatrix} 1 & 0 \\ 0 & -1 \end{pmatrix}.$$

It is easy to check that these generators form the algebra $osp(2, 2)$:

$$[T^0, T^\pm] = \pm T^\pm, \quad [T^+, T^-] = -2T^0, \quad [J, T^\alpha] = 0, \quad \alpha = +, -, 0$$

$$\{Q_1, \bar{Q}_2\} = -T^-, \quad \{Q_2, \bar{Q}_1\} = T^+,$$

$$\frac{1}{2}(\{\bar{Q}_1, Q_1\} + \{\bar{Q}_2, Q_2\}) = +J, \quad \frac{1}{2}(\{\bar{Q}_1, Q_1\} - \{\bar{Q}_2, Q_2\}) = T^0,$$

$$\{Q_1, Q_1\} = \{Q_2, Q_2\} = \{Q_1, Q_2\} = 0,$$

$$\{\bar{Q}_1, \bar{Q}_1\} = \{\bar{Q}_2, \bar{Q}_2\} = \{\bar{Q}_1, \bar{Q}_2\} = 0,$$

$$[Q_1, T^+] = Q_2, \quad [Q_2, T^+] = 0, \quad [Q_1, T^-] = 0, \quad [Q_2, T^-] = -Q_1,$$

$$[\bar{Q}_1, T^+] = 0, \quad [\bar{Q}_2, T^+] = -\bar{Q}_1, \quad [\bar{Q}_1, T^-] = \bar{Q}_2, \quad [\bar{Q}_2, T^-] = 0,$$

$$[Q_{1,2}, T^0] = \pm\frac{1}{2}Q_{1,2}, \quad [\bar{Q}_{1,2}, T^0] = \mp\frac{1}{2}\bar{Q}_{1,2},$$

$$[Q_{1,2}, J] = -\frac{1}{2}Q_{1,2}, \quad [\bar{Q}_{1,2}, J] = \frac{1}{2}\bar{Q}_{1,2}, \tag{12.94}$$

where $[a, b] = ab - ba$ and $\{a, b\} = ab + ba$ are commutator and anticommutator, respectively. The super-algebra (12.94) contains the algebra $sl_2(\mathbf{R}) \oplus \mathbf{R}$ as sub-algebra.

LEMMA 12.8. *(Turbiner [1994b]). Consider the space* $\mathcal{P}_{n+1,n}$.

(i) Suppose $n > (k - 1)$. *Any quasi-exactly-solvable operator* $\mathbf{T}_k(x)$ *can be represented by a kth-degree polynomial of the operators (12.92) and (12.93). If* $n \le (k - 1)$, *the part of the quasi-exactly-solvable operator* $\mathbf{T}_k(x)$ *containing derivatives in x up to order n can be represented by an nth-degree polynomial in the generators (12.92) and (12.93).*

(ii) Conversely, any polynomial in (12.92) and (12.93) is a quasi-exactly-solvable operator.

(iii) Among quasi-exactly-solvable operators there exist exactly-solvable operators $\mathbf{E}_k \subset \mathbf{T}_k(x)$.

Let us introduce the grading of the bosonic generators (12.92)

$$deg(T^+) = +1, \ deg(J, T^0) = 0, \ deg(J^-) = -1 \qquad (12.95)$$

and fermionic generators 12.93

$$deg(Q_2, \overline{Q}_1) = +\frac{1}{2}, \ deg(Q_1, \overline{Q}_2) = -\frac{1}{2}. \qquad (12.96)$$

Hence the grading of monomials of the generators 12.92 and 12.93 is equal to

$$deg[(T^+)^{n_+}(T^0)^{n_0}(J)^{\overline{n}}(T^-)^{n_-} Q_1{}^{m_1} Q_2{}^{m_2}\overline{Q}_1{}^{\overline{m}_1}\overline{Q}_2{}^{\overline{m}_2}] =$$

$$(n_+ - n_-) - (m_1 - m_2 - \overline{m}_1 + \overline{m}_2)/2. \qquad (12.97)$$

The n's can be arbitrary non-negative integers, while the m's are equal to either 0 or 1. The notion of grading allows us to classify the operators $\mathbf{T}_k(x)$ in the Lie-algebraic sense.

LEMMA 12.9. *A quasi-exactly-solvable operator* $\mathbf{T}_k(x) \subset U_{osp(2,2)}$ *has no terms of positive grading other than monomials of grading +1/2 containing the generator* Q_1 *or* Q_2, *iff it is an exactly-solvable operator.*

Take the eigenvalue problem

$$\mathbf{T}_k(x)\varphi(x) = \varepsilon\varphi(x) \qquad (12.98)$$

where

$$\varphi(x) = \begin{bmatrix} \varphi_1(x) \\ \varphi_2(x) \end{bmatrix}, \qquad (12.99)$$

is a two-component spinor.

THEOREM 12.6. *(Turbiner [1994b]). Let n be a non-negative integer. Take the eigenvalue problem (12.97), where* $\mathbf{T}_k(x)$ *is symmetric. In general, the problem 12.97 has* $(2n+1)$ *linearly independent eigenfunctions in the form of polynomials* $\mathcal{P}_{n+1,n}$, *if and only if* $\mathbf{T}_k(x)$ *is quasi-exactly-solvable. The problem (12.97) has an infinite sequence of polynomial eigenfunctions, if and only if the operator is exactly-solvable.*

As a consequence of Theorem 12.6, the 2×2 matrix quasi-exactly-solvable differential operator $\mathbf{T}_k(x)$, possessing in general $(2n+1)$ polynomial eigenfunctions of the form $\mathcal{P}_{n,n-1}$ can be written in the form

$$\mathbf{T}_k(x) = \sum_{i=0}^{i=k} \mathbf{a}_{k,i}(x)d_x^i. \tag{12.100}$$

The coefficient functions $\mathbf{a}_{k,i}(x)$ are by 2×2 matrices, and generically for the kth-order quasi-exactly-solvable operator their matrix elements are polynomials. Suppose that $k > 0$. Then the matrix elements are given by the following expressions

$$\mathbf{a}_{k,i}(x) = \begin{pmatrix} A_{k,i}^{[k+i]} & B_{k,i}^{[k+i-1]} \\ C_{k,i}^{[k+i+1]} & D_{k,i}^{[k+i]} \end{pmatrix} \tag{12.101}$$

at $k > 0$, where the superscript in square brackets displays the order of the corresponding polynomial.

It is easy to calculate the number of free parameters of a quasi-exactly-solvable operator $\mathbf{T}_k(x)$

$$par(\mathbf{T}_k(x)) = 4(k+1)^2 \tag{12.102}$$

For the case of exactly-solvable problems, the matrix elements (12.100) of the coefficient functions are modified

$$\mathbf{a}_{k,i}(x) = \begin{pmatrix} A_{k,i}^{[i]} & B_{k,i}^{[i-1]} \\ C_{k,i}^{[i+1]} & D_{k,i}^{[i]} \end{pmatrix} \tag{12.103}$$

where $k > 0$. An infinite family of orthogonal polynomials as eigenfunctions of Equations 12.98 and 12.99, if they exist, will occur, if and only if the coefficient functions have the form (12.102). The number of free parameters of an exactly-solvable operator $\mathbf{E}_k(x)$ and, correspondingly, the maximal number of free parameters of the 2×2 matrix orthogonal polynomials in one real variable, is equal to

$$par(\mathbf{E}_k(x)) = 2k(k+3) + 3. \tag{12.104}$$

Thus, the above formulas describe the coefficient functions of matrix differential equation (12.98), which can possess polynomials in x as solutions.

12.4.2. Quasi-exactly-solvable matrix Schroedinger equations (example)

Now let us take the quasi-exactly-solvable matrix operator $\mathbf{T}_2(x)$ and try to reduce Equation (12.98) to the Schroedinger equation

$$[-\frac{1}{2}\frac{d^2}{dy^2} + \mathbf{V}(y)]\Psi(y) = E\Psi(y) \tag{12.105}$$

where $\mathbf{V}(y)$ is a two-by-two *hermitian* matrix, by making a change of variable $x \mapsto y$ and "gauge" transformation

$$\Psi(y) = \mathbf{U}(y)\varphi(x(y)) \tag{12.106}$$

where $\mathbf{U}$ is an arbitrary 2×2 matrix depending on the variable y. In order to get some "reasonable" Schroedinger equation one should fulfill two requirements: (i) the potential $\mathbf{V}(y)$ must be hermitian and (ii) the eigenfunctions $\Psi(y)$ must belong to a certain Hilbert space.

Unlike the case of quasi-exactly-solvable differential operators in one real variable (see González-Lopéz, Kamran, and Olver [1993]), this problem has no complete solution so far. Therefore it seems instructive to display a particular nontrivial example of the quasi-exactly-solvable 2×2-matrix Schroedinger operator (Shifman and Turbiner 1989]).

Take the quasi-exactly-solvable operator

$$\mathbf{T}_2 = -2T^0T^- + 2T^-J - i\beta T^0Q_1 +$$

$$\alpha T^0 - (2n+1)T^- - \frac{i\beta}{2}(3n+1)Q_1 + \frac{i}{2}\alpha\beta Q_2 - i\beta\overline{Q}_1, \tag{12.107}$$

where α and β are parameters. Upon introducing a new variable $y = x^2$ and after straightforward calculations one finds the following expression for the matrix U in Equation (12.106)

$$\mathbf{U} = \exp(-\frac{\alpha y^2}{4} + \frac{i\beta y^2}{4}\sigma_1) \tag{12.108}$$

and for the potential $\mathbf{V}$ in Equation (12.105)

$$\mathbf{V}(y) = \frac{1}{8}(\alpha^2 - \beta^2)y^2 + \sigma_2[-(n+\frac{1}{4})\beta + \frac{\alpha\beta}{4}y^2 - \frac{\alpha}{4}\tan\frac{\beta y^2}{2}]\cos\frac{\beta y^2}{2}$$

$$+ \sigma_3[-(n+\frac{1}{4})\beta + \frac{\alpha\beta}{4}y^2 - \frac{\alpha}{4}\cot\frac{\beta y^2}{2}]\sin\frac{\beta y^2}{2}. \tag{12.109}$$

It is easy to see that the potential $\mathbf{V}$ is hermitian; $(2n + 1)$ eigenfunctions have the form of polynomials multiplied by the exponential factor U and they are obviously normalizable.

Recent development of matrix quasi-exactly-solvable differential equations and further explicit examples can be found in Brihaye and Kosinski [1994].

12.5.   Ordinary differential equations with the parity operator

Let us introduce the parity operator K in the following way:

$$Kf(x) = f(-x). \tag{12.110}$$

The operator K has such properties:

$$\{K, x\}_+ = 0, \; \{K, d_x\}_+ = 0, \; K^2 = 1 \tag{12.111}$$

where $\{a, b\}_+ = ab + ba$ is anticommutator. Now take the following generators:

$$J^-(n) = d_x + \frac{v}{x}(1 - K), \tag{12.112}$$

$$J_1^0(n) = xJ^-(n) = xd_x + v(1 - K), \tag{12.113}$$

$$J^0(n) = n - xd_x, \tag{12.114}$$

$$J^+(n) = xJ^0(n) = nx - x^2 d_x. \tag{12.115}$$

These operators together with the operator K form an algebra, which was named $gl(2, \mathbf{R})_K$ in Turbiner [1994a], with the commutation relations

$$[J^0(n), J_1^0(n)] = 0, \tag{12.116}$$

$$[J^\pm(n), J^0(n)] = \pm J_i^\pm(n), \tag{12.117}$$

$$[J^\pm(n), J_1^0(n)] = \mp(1 \mp 2vK)J^\pm(n), \tag{12.118}$$

$$[J^+(n), J^-(n)] = J_1^0(n) - (1 + 2vK)J^0(n) \tag{12.119}$$

and also

$$[K, J_1^0(n)] = [K, J^0(n)] = 0, \{K, J^\pm(n)\}_+ = 0. \tag{12.120}$$

If $v = 0$ the generators (12.112) – (12.115) and commutation relations (12.116) – (12.119) become those of the algebra $gl(2, \mathbf{R})$.[7]

It is easy to check that the algebra (12.116) – (12.120) possesses two Casimir operators — the operators commuting with all generators of the algebra $gl(2, \mathbf{R})_K$: the linear one

$$C_1 = J^0 + J_1^0 + vK \tag{12.121}$$

and the quadratic one

$$C_2 = \frac{1}{2}\{J^+, J^-\}_+ + \frac{1}{4}(J^0 - J_1^0)^2 - \frac{v}{2}(J^0 - J_1^0)K + \frac{v}{2}K . \tag{12.122}$$

Substituting the concrete representation (12.112) – (12.115) into (12.121) – (12.122) one can find that

$$C_1 = n + v,$$

$$C_2 = \frac{1}{4}[(n + v)(n + v + 2) - v^2]. \tag{12.123}$$

The representation (12.112) – (12.115) has an outstanding property: for generic v, when n is a non-negative integer number, there appears a finite-dimensional irreducible representation $\mathcal{P}_{n+1}$ of dimension $(n + 1)$. So the representation space remains the same for any value of parameter v.

Like it has been done before, one can introduce quasi-exactly- and exactly-solvable mixed operators, containing the differential operator and the parity operator K: $T_k(x, d_x, K)$ and $E_k(x, d_x, K)$, respectively. It is evident, that those operators are linear in K.

LEMMA 12.10. *(i) Suppose $n > (k - 1)$. Any quasi-exactly-solvable operator T_k can be represented by a kth-degree polynomial of the operators (12.112) – (12.115). If $n \leq (k - 1)$, the part of the quasi-exactly-solvable operator T_k containing derivatives up to order n can be represented by an nth degree polynomial in the generators (12.112) – (12.115).*

(ii) Conversely, any polynomial in (12.112) – (12.115) is quasi-exactly solvable.

(iii) Among quasi-exactly-solvable operators there exist exactly-solvable operators $E_k \subset T_k$.

Similarly to the case of $sl_2(\mathbf{R})$, one can introduce the grading of generators (12.112) – (12.115):

$$deg(J^+(n)) = +1 , \; deg(J^0(n), J_1^0(n), K) = 0 , \; deg(J^-(n)) = -1, \tag{12.124}$$

---

[7]Hereafter we will omit the argument n in the generators except for the cases where the representation (12.112) – (12.115) is used concretely.

and

$$deg[(J^+(n))^{k_+}(J^0(n))^{k_0}(J_1^0(n))^{k_{0,1}}(K)^k(J^-(n))^{k_-}] = k_+ - k_-. \quad (12.125)$$

(cf. (12.9) – (12.10)). The grading allows us to classify the operators T_k in the algebraic sense.

LEMMA 12.11. *A quasi-exactly-solvable operator $T_k \subset U_{gl_2(\mathbf{R})_K}$ has no terms of positive grading, if and only if it is an exactly-solvable operator.*

THEOREM 12.7. *Let n be a non-negative integer. Take the eigenvalue problem for a linear differential operator of the kth order in one variable*

$$T_k(x, d_x, K)\varphi = \varepsilon\varphi, \quad (12.126)$$

where T_k is symmetric. The problem (12.126) has an $(n + 1)$ linearly independent eigenfunctions in the form of a polynomial in variable x of order not higher than n, if and only if T_k is quasi-exactly-solvable. The problem 12.126 has an infinite sequence of polynomial eigenfunctions, if and only if the operator is exactly-solvable. If T_k has no terms of odd grading, $(k_+ - k_-)$ is an odd number (see (12.125)), T_k commutes with K, and eigenfunctions in (12.126) have definite parity with respect to $x \rightarrow -x$.

Following the Lemma 2.10 a general second-order quasi-exactly-solvable differential operator is defined by a quadratic polynomial in generators of $gl_2(\mathbf{R})_K$. Provided that the conditions (12.121) – (12.123) are taken into account[8] we arrive at

$$T_2 = c_{++}J^+(n)J^+(n) + c_{+0}J^+(n)J^0(n) + c_{+-}J^+(n)J^-(n) + c_{0-}J^0(n)J^-(n)$$

$$+ c_{--}J^-(n)J^-(n) + c_+J^+(n) + c_0J^0(n) + c_-J^-(n) + c, \quad (12.127.1)$$

(cf. (12.21.1)), where $c_{\alpha\beta} = C_{\alpha\beta}^0 + C_{\alpha\beta}^K K, c_\alpha = C_\alpha^0 + C_\alpha^K K, c = C^0 + C^K K$ and all Cs are real numbers. The number of free parameters is $par(T_2) = 18$. Nonexistence in T_2 of the terms of odd grading leads to the conditions

$$c_{+0} = c_{0-} = 0$$

---

[8] It leads to a disappearance of the terms containing, for instance, $J_1^0(n)$ and $J^0(n)J^0(n)$.

and

$$c_+ = c_- = 0$$

and, finally, a general operator having eigenfunctions of a definite parity is

$$T_2^{(e,o)} = c_{++} J^+(n) J^+(n) + c_{+-} J^+(n) J^-(n)$$

$$+ c_{--} J^-(n) J^-(n) + c_0 J^0(n) + c, \qquad (12.127.2)$$

and the number of free parameters is $par(T_2^{(e,o)}) = 10$.
   The condition of Lemma 2.11 requires that

$$c_{++} = c_{+0} = c_+ = 0,$$

and then the operator T_2 becomes exactly-solvable

$$E_2 = c_{+-} J^+(n) J^-(n) + c_{0-} J^0(n) J^-(n)$$

$$+ c_{--} J^-(n) J^-(n) + c_0 J^0(n) + c_- J^-(n) + c, \qquad (12.127.3)$$

(cf. (12.21.2)) and the number of free parameters is reduced to $par(E_2) = 12$.

Part III

Computational Methods

13

Symbolic Software for Lie Symmetry Analysis

A survey of techniques and symbolic programs for the determination of Lie symmetry groups of systems of differential equations is presented.[1] The purpose, methods, and algorithms of symmetry analysis are outlined. An exhaustive review of the literature, including old and modern books and papers presenting key concepts, is given. Special attention is paid to methods for reducing the determining equations into standard form and their subsequent integration. Several examples illustrate the use of the Lie symmetry software. Throughout the paper, new trends in the development of symbolic packages for Lie symmetry analysis are indicated.

13.1.   Introduction and literature

The Norwegian mathematician Marius Sophus Lie (1842-1899) and Felix Klein (1849-1925), a German geometer, pioneered the study of transformation groups that leave systems of equations invariant. Klein's work focused on the role of finite groups in the study of regular bodies and algebraic equations. Lie [1880a], [1884], [1888], [1891], [1922], [1971] founded the theory of continuous transformations groups and Lie groups (see also Lie [1888], [1890], [1893]). Although Lie's starting point had been geometry, the inspirational source for his group theoretic investigations was the field of differential equations. His goal was to establish a theory of integration of differential equations that would mirror Abel's theory for the solution of algebraic equations. His work brought many diverse and ad hoc integration methods for solving special classes of differential equations under a common conceptual umbrella.

Later on, the concept of *symmetry* evolved into one of the most explosive developments of mathematics and physics throughout the twentieth century. The

---

[1]Research supported in part by NSF under Grant CCR-9300978.

theory of Lie groups and Lie algebras is now applied to diverse fields of mathematics including differential geometry, algebraic topology, bifurcation theory, numerical analysis, special functions, to name a few; and to nearly any area of theoretical physics, in particular classical, continuum and quantum mechanics, fluid dynamics, relativity, and particle physics.

Lie's infinitesimal transformation method provides the most widely applicable technique to find closed form solutions of ordinary differential equations (ODEs). Standard solution methods for first-order or linear ODEs can be characterized in terms of symmetries. For nonlinear ODEs, Lie's method, when it succeeds, provides a means of reducing the solution to a series of quadratures and can be implemented in symbolic programs (Bocharov [1993]). Under certain conditions, first- and second-order ODEs can be linearized via Lie group transformations (Ibragimov [1991], [1993]). Through the group classification of ODEs, Lie also succeeded in identifying all ODEs that can either be reduced to lower-order ones, or completely integrated via group theoretic techniques.

Applied to partial differential equations (PDEs), the method leads to group-invariant solutions, conservation laws, invariant center manifolds in bifurcation theory (Cicogna and Gaeta [1992], Gaeta [1992], [1994], Sattinger [1979], Vanderbauwhede [1982]), etc. Exploiting the symmetries of PDEs, new solutions can be derived from old ones, and PDEs can be classified into equivalence classes. Special, physically significant solutions arising from symmetry methods allow one to investigate the asymptotic or physical behavior of general solutions. The group-invariant solutions obtained via Lie's approach provide insight into the physical models themselves. Explicit solutions also serve as benchmarks in the design, accuracy testing, and comparison of numerical algorithms.

Lie's original ideas greatly influenced the study of physically important systems of differential equations in classical mechanics, fluid dynamics, elasticity, and many other applied areas. Currently, Lie symmetry methods are applied to difference, differential-difference equations (see Chapter 17 in [H1], Levi and Winternitz [1993], Vinet, Levi, and Winternitz [1994]) and integro-differential equations (see Chapter 5 of this volume). A review of the present state of affairs is to be found in Clarkson [1994].

The application of Lie groups methods to concrete physical systems involves tedious, mechanical computations. Even the calculation of the continuous symmetry group of a modest system of differential equations is prone to fail, if done with pencil and paper. Programmable computer algebra systems (CAS) such as Mathematica, MACSYMA, Maple, REDUCE, AXIOM, and MuPAD, are extremely useful aids in such computations. Symbolic packages, written in the language of CAS, can find the defining (or determining) equations of the Lie symmetry group. The most sophisticated packages then reduce the determining system into an equivalent but more suitable system, subsequently solve that system in closed form, and go on to calculate the infinitesimal generators that span the Lie algebra of symmetries.

A large body of literature exists on the topic of Lie symmetries. We list some older books by Ames [1965], [1968], [1972], Belinfante and Kolman [1989], Bianchi [1918], Bluman and Cole [1974], Campbell [1903], Cartan [1946], Chevalley [1946], Cohen [1931], Eisenhart [1933], Hansen [1964], Hermann [1968], Kähler [1934], Kuranishi [1967], Miller [1968], [1972], [1977], Ovsiannikov [1962], Palais [1989], and Sedov [1957]; more modern books by Anderson and Ibragimov [1979], Axford [1983], Barenblatt [1979], Bluman and Kumei [1989], Bryant et al. [1991], Dresner [1983], Edelen [1980b], Euler and Steeb [1993], Fushchich, Barannik, and Barannik [1991], Fushchich and Nikitin [1987], [1990], Fushchich, Shtelen, and Serov [1989], [1993], Fushchich and Zhdanov [1992], Gaeta [1994], Hill [1982], [1992], Ibragimov [1983], [1989], [1991], [1994b], Kalnins [1986], Leznov and Savel'ev [1985], Levi and Winternitz [1988], Martini [1979], Olver [1986], Ovsiannikov [1978], Pirani, Robinson, and Shadwick [1979], Pommaret [1978], [1988], Rogers and Ames [1989], Rogers and Shadwick [1982], Rozhdestvenskii and Yanenko [1978], Sánchez Mondragón and Wolf [1986], Sattinger [1979], Sattinger and Weaver [1986], Sesahdri and Na [1985], Steeb [1993], Stephani [1989], Vinogradov [1989c], Warner [1983], Wolf [1990], and Yaglom [1988]; some recent conference proceedings by Allgower, Georg, and Miranda [1993], Broadbridge and Hill [1993], Fushchich [1992], Ibragimov and Mahomed [1994], Kersten and Krasil'shchik [1994], and Vinet, Levi, and Winternitz [1994]; published lecture notes by Pommaret [1992], recent Ph.D. theses by Char [1980], Gragert [1981], Herod [1994], Hood [1993], Kersten [1987], Ondich [1989], Shadwick [1979], Sherring [1993b], and Williams [1989]; and papers presenting key concepts by Akhatov, Gazizov, and Ibragimov [1989], Ames [1992], Anderson, Kamran and Olver [1993a], [1993b], Beyer [1979], Fushchich and Zhdanov [1989], Ibragimov [1992], Kostant [1977], Sastri [1986], Schwarz [1988b], Tsujishita [1982], [1983], [1991], Vinogradov [1989a], [1989b], Vinogradov and Krasil'shchik [1984], and Winternitz [1983], [1990], [1993].

Also of interest are the lecture notes and preprints published by The International Sophus Lie Center in Oslo (for example, Komrakov and Lychagin [1993]), and the notes of the courses organized by ERCIM on "Partial Differential Equations and Group Theory" (Pommaret [1992], Schwarz and Pommaret [1994]).

Extensive tables of symmetry group generators for well-known equations from mathematical physics have been gathered by Ovsiannikov [1978], Rogers and Ames [1989], and Ibragimov [1992], [1994b], [1994c].

Many conferences addressed issues related to group analysis of differential equations. We single out the latest volumes of *Group Theoretical Methods in Physics* (see, for instance, del Olmo, Santander, and Mateos Guilarte [1993], Dodonov and Man'ko [1991], wherein a list of previous colloquia may be found. As of 1994, the proceedings Ibragimov and Mahomed [1994] covers topics in Lie groups and their applications.

We highly recommend the special issues of *Acta Applicandae Mathematicae* on "Symmetries of Partial Differential Equations" (Vinogradov [1989b]), now available in book form (Vinogradov [1989c]), and the other two volumes in this series edited by Ibragimov [1994b], [1994c].

Biographies of Lie and his contemporaries are found in Baas [1994], Hawkins [1994], Ibragimov [1992], [1994a], Polishchuk [1983], and Yaglom [1988]. Delightful historical notes about symmetry research in general, and Lie's work in particular, can be found in the 2nd, updated edition of Olver [1986]. Complete references to Lie's original work are also given in this book and in Ibragimov [1994b]. Translations of some of Lie's fundamental papers are available in Ibragimov [1994c] and Lie [1880a], [1884].

Readers interested in the differential geometrical foundation of Lie symmetry analysis, including topics like Lie-Bäcklund mappings, Cartan forms, supersymmetry, graded differential geometry, and gauge theories, may want to consult Edelen and Wang [1992], Finley and McIver [1993], Gaeta [1992], Kostant [1977], Krasil'shchik, Lychagin, and Vinogradov [1986], Martini [1979], Olver [1986], [1994b], Pirani, Robinson, and Shadwick [1979], Sherring and Prince [1992], Vorob'ev [1991], [1992], Warner [1983], Zharinov [1992], amongst others, and Gragert and Kersten [1982], [1991] for related REDUCE algorithms.

In this review, we will sparsely address the computation of conservation laws, which through Noether's famous theorem (Noether [1918]) and its extensions (Ibragimov [1969b], Bluman and Kumei [1989], Sarlet and Cantrijn [1981]), is intrinsically connected with the investigation of variational symmetries. Indeed, recall that variational symmetries admitted by a Lagrangian system—symmetries that leave the action integral invariant—can be obtained by computing the Lie-Bäcklund symmetries of the corresponding Euler-Lagrange equations. Once the Lie-Bäcklund symmetries are obtained and the variational symmetries are singled out, Boyer's formulation of Noether's theorem can be used to calculate the conservation laws (see also [H1], Chapter 6).

In Section 13.2 we discuss the purpose, methods, and algorithms used in the computation of Lie symmetries. Apart from a detailed description of the methods for computing and solving the determining equations, we address the reduction of systems of PDEs into standard and passive forms. This topic, in turn, ties in with the computation of the size of the symmetry group. In this section we also address some of the newest trends in the development of symbolic software for Lie symmetries.

In Section 13.3 we look beyond Lie point symmetries, addressing contact and Lie-Bäcklund symmetries, as well as nonclassical or conditional symmetries.

Section 13.4 is devoted to a detailed review of modern symbolic packages that aid in the investigation of Lie symmetries for systems of differential equations. The software is grouped according to the underlying CAS. The review of the available code intentionally focuses on packages written after 1985. More details about pioneering efforts and software written prior to 1985 can be found in Hereman [1993]. In Section 13.5, three examples illustrate the results that can be obtained with the software packages. Finally, in Section 13.6 we draw some conclusions.

Although no survey can be considered exhaustive, our study intended to cover all the Lie symmetry software, with the exclusion of software for Lie group computations, such as LiE (van Leeuwen [1994]), which is outside the scope of this review.

13.2. Purpose, methods, and algorithms

13.2.1. Purpose

The classical "Lie symmetry group of a system of differential equations" is a local group of point transformations, meaning diffeomorphisms on the space of independent and dependent variables, which map solutions of the system to solutions. Various other types of local symmetries (Anderson, Kamran, and Olver [1993a]), and nonlocal symmetries (Akhatov, Gazizov, and Ibragimov [1989], Bluman [1993a], [1993b], Bluman and Cole [1990b], Bluman and Kumei [1989], Vinogradov [1989c]) have been studied, as well as approximate symmetries (Baikov, Gazizov, and Ibragimov [1989]). The software reviewed in this paper attempts to compute the classical Lie point symmetries. Several packages go beyond that in computing contact (or dynamical) and Lie-Bäcklund (or generalized) symmetries, nonclassical (or conditional) symmetries. Several programs could be modified to compute Cartan's dynamical symmetries (Cartan [1946], Stephani [1989]) and hidden symmetries. Two programs that utilize point symmetries to generalize special solutions or find similarity variables recently became available (Vafeades [1992a], [1992b], [1994a], Wolf [1994]). The latter programs actually try to integrate the characteristic system of first-order differential equations.

Loosely speaking, contact symmetries are Lie-Bäcklund symmetries of order one; i.e., the coefficients in the vector field include first derivatives of the dependent variables (see Chapter 1). Among the several ways of computing Lie-Bäcklund symmetries, this paper focuses on an extension of the original methods due to Lie. For a discussion of the difference with other symmetry approaches, including other computer algebra methods, we refer the reader to Fokas [1987], Fuchssteiner, Oevel, and Wiwianka [1987], Gerdt, Shvachka, and Zharkov [1985a], [1985b], Zhang, Tu, Oevel, and Fuchssteiner [1991], and Zharkov [1993].

Conditional symmetries are found by the "nonclassical method of group-invariant solutions," as introduced in 1969 by Bluman and Cole [1969]. Further generalizations of the nonclassical method lead to the less practical "weak symmetries" introduced in Olver and Rosenau [1986], [1987], and the method of differential constraints in Olver [1992] (see Chapters 11 and 10 of this volume).

Potential symmetries (see Bluman [1993a], [1993b], [1993c], Bluman and Kumei [1989], Krasil'shchik and Vinogradov [1989], Ovsiannikov [1978], Vinogradov [1989c]) for a system of PDEs are computed as follows. First, one replaces (one or more of) the PDEs in the given system by equivalent conserved forms; second, one introduces auxiliary potential variables; finally, one determines the point symmetries of the resulting auxiliary system of PDEs. The form of an infinitesimal generator then determines whether or not it defines a nonlocal symmetry. The technique of potential symmetries leads to interesting linearizations involving non-invertible mappings.

"Hidden" symmetries, as discussed in Abraham-Shrauner [1993], Abraham-Shrauner and Guo [1994], Abraham-Shrauner and Leach [1993], and Guo and Abraham-Shauner [1993], show up when, for example, the order of an ODE is increased by one and the number of symmetries (of the now higher-order ODE) is increased at least by two. Upon subsequent reduction, a symmetry may be lost if the transformation is done in non-normal subgroup variables. Hidden symmetries also occur when an ODE is reduced in order, and when this induces an additional symmetry that was not a symmetry of the original ODE.

Somewhat related to this is the following: if a higher-order ODE is rewritten as an equivalent system of first-order equations, and one analyzes the system of first-order equations, rather than the single equation of higher order, the class of symmetries (and consequently solutions) can be substantially enlarged (see Ovsiannikov [1962], [1978], Sastri [1986]). A similar situation arises with PDEs. When a PDE is reduced to a lower-dimensional PDE one can lose or gain symmetries. Clarkson [1994] provides some examples.

Discrete symmetries of differential equations and disconnected Lie point groups have been considered by Reid, Weih, and Wittkopf [1993].

Precise definitions and a discussion of internal symmetries (also called dynamical symmetries) and external symmetries are given in Anderson, Kamran, and Olver [1993a], [1993b]. It is beyond the scope of this survey.

13.2.2. Methods for computing determining equations

Although Lie's theory is geometric, one has to resort to differential-algebraic methods to compute Lie symmetries. Indeed, the criterion for a vector field to be a generator is purely geometric, namely, certain functions must vanish on a submanifold. But differential geometry gives us no tools to implement such criteria in concrete applications. The use of differential algebra imposes some restrictions on the type of problems that can be handled. For instance, the differential equations must be polynomial in their variables (so that one can work in the ring of differential polynomials); the differential equations must be solvable for some derivatives, etc.

There are three major methods to compute Lie symmetries. The first one uses prolonged vector fields, the second utilizes differential forms (wedge products) due to Cartan [1946]. The third one uses the notion of "formal symmetry" (see Bocharov [1989], [1990a], Mikhailov, Shabat, and Sokolov [1990]). Although restricted to evolution systems with two independent variables, the latter method provides a very quick way to compute canonical Lie-Bäcklund symmetries. Due to its limited scope we will not elaborate on that technique.

13.2.2.1. Prolonged vector fields The first method is used in the algorithm of our program SYMMGRP.MAX (Champagne, Hereman, and Winternitz [1991]), and in most of the other Lie symmetry packages. Instead of looking for the Lie group G, one looks for its Lie algebra $\mathcal{L}$, realized by the vector field. From the Lie algebra of symmetry generators, one can obtain the Lie group of point transformations upon integration of a system of first-order characteristic equations (see Section 13.5.1 for a simple example).

For notational simplicity, let us consider the case of Lie point symmetries. We follow the notations and terminology used in Olver [1986].

We start with a system of m differential equations,

$$\Delta^i(x, u^{(k)}) = 0, \quad i = 1, 2, ..., m, \tag{13.1}$$

of order k, with p independent and q dependent (real) variables, denoted by $x = (x_1, x_2, ..., x_p) \in \mathbf{R}^p$, $u = (u^1, u^2, ..., u^q) \in \mathbf{R}^q$. We stress that m, k, p, and q are *arbitrary* positive integers. The partial derivatives of u^l are represented using a multi-index notation, for $J = (j_1, j_2, ..., j_p) \in \mathbf{N}^p$, we denote

$$u^l_J \equiv \frac{\partial^{|J|} u^l}{\partial x_1^{j_1} \partial x_2^{j_2} ... \partial x_p^{j_p}}, \tag{13.2}$$

where $|J| = j_1 + j_2 + ... + j_p$. Finally, let $u^{(k)}$ denote a vector whose components are all the partial derivatives of order 0 up to k of all the u^l.

The group transformations have the form $\tilde{x} = \Lambda_g(x, u)$, $\tilde{u} = \Omega_g(x, u)$, where the functions Λ_g and Ω_g are to be determined. Note that the subscript g refers to the group parameters. Instead of looking for a Lie group G, we look for its Lie algebra $\mathcal{L}$, realized by vector fields of the form

$$\alpha = \sum_{i=1}^{p} \eta^i(x, u) \frac{\partial}{\partial x_i} + \sum_{l=1}^{q} \varphi_l(x, u) \frac{\partial}{\partial u^l}. \tag{13.3}$$

The problem is now reduced to finding the coefficients $\eta^i(x, u)$ and $\varphi_l(x, u)$. In essence, the computer constructs the kth prolongation $\mathrm{pr}^{(k)}\alpha$ of the vector field α, applies it to the system 13.1, and requests that the resulting expression vanishes on the solution set of (13.1).

This sounds straightforward, but the method involves tedious calculations because the length and complexity of the expressions increase rapidly as p, q, m, and especially k, increase. The number of determining equations then also rises dramatically as is shown in Ovsiannikov [1962] (see also Ovsiannikov [1978] and Rogers and Ames [1989], p. 346). Here are the details and the steps to be performed:

1. Construct the kth prolongation of the vector field α in (13.3) by means of the formula

$$\mathrm{pr}^{(k)}\alpha = \alpha + \sum_{l=1}^{q}\sum_{J}\psi_l^J(x, u^{(k)})\frac{\partial}{\partial u_J^l}, \qquad 1 \le |J| \le k, \qquad (13.4)$$

where the coefficients ψ_l^J are defined as follows. The coefficients of the first prolongation are

$$\psi_l^{J_i} = D_i\varphi_l(x, u) - \sum_{j=1}^{p}u_{J_j}^l D_i\eta^j(x, u), \qquad (13.5)$$

where J_i is a $p-$tuple with 1 on the ith position and zeros elsewhere, and D_i is the total derivative operator

$$D_i = \frac{\partial}{\partial x_i} + \sum_{l=1}^{q}\sum_{J}u_{J+J_i}^l\frac{\partial}{\partial u_J^l}, \qquad 0 \le |J| \le k. \qquad (13.6)$$

The higher-order prolongations are defined recursively as

$$\psi_l^{J+J_i} = D_i\psi_l^J - \sum_{j=1}^{p}u_{J+J_j}^l D_i\eta^j(x, u), \qquad |J| \ge 1. \qquad (13.7)$$

2. Apply the prolonged operator $\mathrm{pr}^{(k)}\alpha$ to each equation $\Delta^i(x, u^{(k)})$ and require that
$$\mathrm{pr}^{(k)}\alpha \, \Delta^i \, |_{\Delta^j=0} = 0 \quad i, j = 1, ..., m. \qquad (13.8)$$

The meaning of condition 13.8 is that $\mathrm{pr}^{(k)}\alpha$ vanishes on the solution set of the originally given system 13.1. Precisely, this condition assures that α is an infinitesimal symmetry generator of the group transformation; $\tilde{x} = \Lambda_g(x, u)$, $\tilde{u} = \Omega_g(x, u)$, i.e., that $u(x)$ is a solution of (13.1) whenever $\tilde{u}(\tilde{x})$ is one.

3. Choose, if possible, m components of the vector $u^{(k)}$, say $v^1, ..., v^m$, such that:

   (a) Each v^i is equal to a derivative of a u^l ($l = 1, ..., q$) with respect to at least one variable x_i ($i = 1, ..., p$).

   (b) None of the v^i is the derivative of another one in the set.

   (c) The system 13.1 can be solved algebraically for the v^i in terms of the remaining components of $u^{(k)}$, which we denote by w. Hence, $v^i = S^i(x, w)$, $i = 1, ..., m$.

(d)   The derivatives of v^i, $v^i_J = D_J S^i(x, w)$,   where $D_J \equiv D_1^{j_1} D_2^{j_2} \ldots$
$D_p^{j_p}$, can all be expressed in terms of the components of w and their derivatives, without ever reintroducing the v^i or their derivatives.

The requirements in step 3 put some restrictions on the system 13.1, but for many systems the choice of the appropriate v^i is quite obvious. For example, for a system of evolution equations

$$\frac{\partial u^i}{\partial t}(x_1, \ldots, x_{p-1}, t) = F^i(x_1, \ldots, x_{p-1}, t, u^{(k)}), \quad i = 1, \ldots, m, \quad (13.9)$$

where $u^{(k)}$ involves derivatives with respect to the variables x_i but not t, an appropriate choice is $v^i = \dfrac{\partial u^i}{\partial t}$.

4.   Use $v^i = S^i(x, w)$ to eliminate all v^i and their derivatives from the expression 13.8, so that all the remaining variables are now independent of each other. It is tacitly assumed that the resulting expression is now a polynomial in the u^l_J.

5.   Obtain the determining equations for $\eta^i(x, u)$ and $\varphi_l(x, u)$ by equating to zero the coefficients of all functionally independent expressions (monomials) in the remaining derivatives u^l_J.

In the above algorithm the variables x_i, u^l, and u^l_J are treated as independent; the dependent ones are η^i and φ_l.

In summary, the result of implementing (13.4) is a system of linear homogeneous PDEs for η^i and φ_l, in which x and u are independent variables. These are the so-called determining or defining equations for the symmetries of the system. Solving these by hand, interactively or automatically with a symbolic package, will give the explicit forms of the $\eta^i(x, u)$ and $\varphi_l(x, u)$.

The procedure, which is thoroughly discussed in Ovsiannikov [1962], [1978] and Olver [1986], consists of two major steps: *deriving* the determining equations, and *solving* them explicitly.

We note that in Bluman [1990], Bluman proves theorems about the forms of admitted infinitesimal transformations, which cover a large class of scalar ODEs and PDEs. In essence, the theorems in Ovsiannikov [1962] and Bluman and Kumei [1989] imply that the coefficients of the vector field are either free or depend linearly on the dependent variable. Use of these theorems can significantly simplify the tedious work involved in setting up and solving the determining equations.

13.2.2.2.   Differential forms   A differential geometric approach to invariance groups and solutions of PDEs was presented by Harrison and Estabrook [1971]. They showed how to derive infinitesimal symmetries using Cartan's exterior differential calculus. The isovector fields, which are the generators of geometric transformations with suitable algebraic invariance properties, are then used to obtain invariant solutions of several PDEs, including the heat equation, the Korteweg-de Vries equation, and the vacuum Maxwell equations.

Briefly, this method to find infinitesimal symmetries proceeds as follows. The system of differential equations is rewritten as a Pfaffian system, that is a system of one-forms. The condition for an isovector is then that the contraction of every form in the closed ideal generated by the exterior system with the isovector remains in the ideal. That is completely equivalent to the condition in the jet bundle approach. From a computational point of view there is a disadvantage: the exterior system usually consists of more forms than the PDE system of equations. To reduce the number of forms, one can opt to formulate the system of differential equations as a closed exterior differential system of two-forms. The underlying manifold is prolonged by new coordinates, the so-called prolonged variables. The exterior differential system is then prolonged by special one-forms in these prolonged variables. The condition on these added one-forms is that the prolonged exterior differential systems remain closed. This condition leads to the determining equations for the coefficients of the manifold.

For the mathematical formulation and several worked examples, the reader is referred to Carminati, Devitt, and Fee [1992], Estabrook [1990], Gragert, Kersten, and Martini [1983], Harrison and Estabrook [1971], Kersten [1989], and Sherring and Prince [1992]. Olver [1994b] addresses the application of various approaches, particularly those of Lie and Cartan.

It should also be noted that routines by Donsig [1990], Schrüfer [1987], [1990], and Wahlquist [1977], for Cartan's exterior calculus (Hartley and Tucker [1991]) are included as standard packages of CAS. For instance, CARTAN in the MAC-SYMA Share Library, EXCALC in REDUCE, and DifferentialForms (Zizza [1992], [1994]) in Mathematica. For a list of other symbolic packages that manipulate objects of differential geometry we refer the reader to Schrüfer [1990].

13.2.3. Methods for reducing determining equations

To design a reliable and powerful integration algorithm for a system of determining equations, the system needs to be brought into a canonical form. Various related concepts appear in the literature, where authors refer to the normal, orthonomic, involutive, and passive forms (see, for example, Bryant et al. [1991], Kuranishi [1967]). All of these have slightly different definitions, for which the reader should consult the later cited literature. Involution is a geometric concept, independent of coordinate systems. Passive systems are defined with respect to a given coordinate system and a given ordering of terms. The same holds for differential Gröbner bases (DGBs).

The original theory of involutive systems goes back to Cartan [1946], Janet [1920], [1929], Kähler [1934], Kolchin [1973], [1984], Riquier [1910], Ritt [1966], Thomas [1937], and Tressé [1894], [1896]. In the introduction to his lecture notes by Pommaret [1992], he discusses the basic ideas of the theory of differential elimination. His description, which contains interesting historical remarks (see also Pommaret [1983]), relates the work of the French to the formal cohomology approach advocated by D.C. Spencer and collaborators in the USA. A concise survey of the theory of involutive systems with complete proofs is also given by Finikov [1948].

Roughly speaking, the methods that reduce systems of linear homogeneous PDEs into an equivalent, but much simpler standard form, can be viewed as generalizations of triangulation by Gauss-Jordan elimination, but are applied here to systems of linear PDEs. First, the original system is appended by all its differential consequences. Second, highest derivatives are eliminated, and, if they occur, integrability conditions are added to the system. The procedure is then repeated until the new system is in involution.

Although Schwarz did valuable pioneer work on solving systems of linear homogeneous PDEs (Schwarz [1984]), he is by far not the only one working on innovative ways of classifying, subsequently reducing, and finally solving overdetermined systems of linear homogeneous PDEs. Early efforts to implement Cartan's exterior-forms approach to involutive systems, and the method of Riquier and Janet, are due to Russian teams, such as Arais, Shapeev, and Yanenko [1974], Ganzha et al. [1981], Ganzha and Yanenko [1982], and Topunov [1989].

Also, Fedorova and Kornyak [1985], [1986], [1987], Kornyak and Fushchich [1991], Bocharov [1991], Bocharov and Bronstein [1989], Ganzha and Yanenko [1982], Gerdt and Zharkov [1990], Pankrat'ev [1989], and Topunov [1989], amongst others, partially implemented algorithms to reduce systems of PDEs.

The GRBASES algorithm of Pankrat'ev and the implementation of the Riquier-Janet method by Topunov are both in REFAL. Both programs can only reduce linear systems of PDEs, this in contrast with the program DIFFGROB2 of Mansfield discussed below.

A good description of the relation between the Riquier-Janet theory and the modern implementations (Gröbner bases) is given by Topunov [1989] (see also Schwarz [1992b]). For an introduction of Gröbner bases, including Buchberger's algorithm (Buchberger [1985], [1992]), we recommend the new books by Cox, Little, and O'Shea [1992] and Becker and Weispfenning [1993]. Most of this work directly relates to Lie symmetry computations, with the exception of Pankrat'ev, who offers general algorithms for the computation of Gröbner bases in differential and difference modules.

In the West, Reid and collaborators (Reid [1990a], [1991a], Reid and Boulton [1991], Reid et al. [1992], Reid and Wittkopf [1993], Schwarz [1984], [1992a], [1992b], and Wolf [1992], [1993], and Wolf and Brand [1992], [1993], [1994], partially implemented algorithms to reduce systems of PDEs. Their work led to sophisticated symbolic code in MACSYMA, Maple, and REDUCE for that purpose.

In Schwarz [1992a], [1992b], Schwarz describes the algorithm *InvolutionSystem*, based on the theory of differential equations due to Riquier [1910] and Janet [1920], [1929], to transform a linear system of PDEs into involutive form. In modern language: the involutive form is a DGB with respect to the selected term ordering. If all consistent orderings for the terms in the system of PDEs are known, the algorithm *InvolutionSystem* may be applied repeatedly to determine a universal Gröbner base (Schwarz [1992b]). Schwarz designed his algorithm *InvolutionSystem* for a specific purpose, namely to determine the size of the Lie symmetry group of a given system of PDEs without having to integrate the determining equations. We devote Section 13.2.5 to this important topic, where details about Schwarz's program SYMSIZE can be found.

As defined by Reid [1991a], a standard form of a system is obtained by repeating the following steps: (i) write each equation in solved form with respect to its highest order derivative, (ii) replace these highest order derivatives throughout the rest of the system, (iii) add any new equations arising from integrability conditions.

As said, standard-form algorithms have their origins in the work of Riquier and Janet for passive forms. Note that the Riquier-Janet form is not a standard form, but can be fairly simply transformed into a standard form. Also, in method, design of algorithm, and in actual implementation, modern standard form algorithms are quite different from the original inefficient methods and algorithms proposed by Riquier and Janet.

A first, but brief account of Reid's algorithm *standard_form* (Reid [1991a]), which also has it roots in the classical Riquier-Janet theory, appeared in Reid [1990a], [1990b]. The algorithm was first implemented in MACSYMA, and later on a MAPLE version became available (Reid and Wittkopf [1993]).

The algorithm *standard_form* reduces systems of PDEs to a simplified standard form. Again, the procedure can be viewed as a generalization to linear differential equations of the Gaussian reduction method for matrices or linear systems. The algorithm now takes as input the system of PDEs and a matrix which specifies a complete ordering on the derivatives appearing in the system. It then reduces the system of PDEs to an equivalent simplified ordered triangular system with all integrability conditions included and all redundancies (differential and algebraic) eliminated. Reid's algorithm implements an equivalence class approach to the problem of bringing a system of PDEs into a standard form. For that purpose, Reid developed a new completion method based on a free direction index (rather than the monomials of the Riquier-Janet theory).

Within *standard_form*, Reid uses an "update strategy" based on updating lists of equations: one term equations, easy equations, hard (or yet-to-be classified) equations; with special user-defined tuning knobs (parameters), so that the user can control the flow of equations between the various lists. Thus *standard_form* works on easier parts of the system first, a strategy that becomes crucial when dealing with large systems.

Further details about Reid's algorithm and examples of its use can be found in Reid and Boulton [1991], where it is shown how directed graphs representing the dependencies amongst the system's variables can be used to simplify or numerically integrate the system. Once the system is in standard form, one can continue with the automatic determination of a Taylor series solution of the system to a specified finite degree.

Reid and Wittkopf's package (Reid and Wittkopf [1993]) facilitates automated interfacing with major symmetry packages such as DIMSYM by Sherring [1993a], LIESYMM by Carminati, Devitt, and Fee [1992], and SYMMGRP.MAX by Champagne, Hereman, and Winternitz [1991], and also with the differential Gröbner basis package DIFFGROB2 (Mansfield [1993]). A TEX interface between *standard_form* and Hickman's program (Hickman [1993], [1994]) that uses physical variable notation has been provided by Lisle. Full details and many illustrative examples of the package, which, besides the function *standard_form*, includes other powerful algorithms for symmetry analysis of PDEs, are given in Reid and Wittkopf [1993].

Reid and McKinnon developed a recursive algorithm called *Rsolve_Pdesys* (Reid and McKinnon [1993]) that builds on Reid's *standard_form* (Reid [1991a]) and on algorithms of Abramov and Kvashenko [1991] and Bronstein [1992]. Reid and McKinnon's algorithm *RSolve_Pdesys* now finds particular solutions of linear systems of PDEs using only ODE solution techniques. Applied to symmetry problems, their algorithm will find all polynomial/rational solutions of the determining equations provided the symmetry group is finite-dimensional.

Several other approaches, and consequently implementations, are possible to complete a given system of PDEs to an involutive system. Schü, Seiler, and Calmet [1993] present an algorithm in AXIOM to perform this task. Their method is highly geometrical and their implementation is based in part on the Cartan-Kuranishi theorem (see Bryant et al. [1991]), which assures that the integrability conditions for the determining equations can be found in a finite number of steps. A detailed description of their programs, called JET, is given in Seiler's thesis (Seiler [1994b]). JET can be viewed as an environment for computations within the geometric theory of PDEs based on the jet bundle formalism. Some standard tasks are put into AXIOM packages. One such package is called CartanKuranishi, which completes a given system to an involutive one. Another package, in development at the time of writing, contains a procedure to set up determining equations for classical and nonclassical symmetries.

Hartley and Tucker [1991] implemented an algorithm (in REDUCE) to analyze involutive systems of exterior forms, based on the Cartan-Kähler theory. Later on, Hartley, Tucker, and Tuckey [1991] extended the program, originally in REDUCE, now partly rewritten in Maple, to non-involutive systems. For such systems their completion procedure constructs the needed integrability conditions. Their work corresponds that of Schü, Seiler, and Calmet on involutive systems. However, Hartley and Tucker use exterior systems.

According to my sources, a student of Fackerell (Sydney) has implemented the Vessiot approach to involution. Vessiot's method can be viewed as a dual version to the exterior system approach.

In the full computer implementation of "triangulation" algorithms, one takes advantage of a "differential" generalization of Buchberger's algorithm for Gröbner bases. Buchberger's algorithm (Buchberger [1992], Helzer [1994], Melenk, Möller, and Neun [1991]), which is included as a standard package with modern CAS (also see Appendix C in Cox, Little, and O'Shea [1992]), provides a technique for canonically simplifying polynomial nonlinear systems of algebraic equations. The "differential" generalization of that algorithm allows one to reduce systems of nonlinear (and, consequently, also linear) PDEs into standard form. In the "differential" analog of Buchberger's algorithm, one has to replace cross-multiplication by cross-differentiation, algebraic reduction by differential reduction, etc.

Carrá-Ferro [1989] was the first to define DGBs for PDE systems. She gave a method based on differential reduction for the attempted construction of such DGBs, but showed that they may be infinite (unlike the case for polynomial algebraic equations and linear systems of PDEs). Subsequently, Ollivier [1991] gave a method which could, in a finite number of steps, construct a DGB up to a given order of derivation (even when the DGB was infinite). But criteria were not given for determining when the DGB had been constructed up to the given order. Recently, Mansfield [1991] has given an algorithm (and a computer-algebra implementation) that uses pseudo-reduction instead of reduction to attempt to construct DGBs. The advantage of her technique is that it always terminates; however, her algorithm may not always terminate with a DGB. In short, the differential ideal membership problem remains unsolved; and there are even disagreements on the definition of DGBs.

The Maple program DIFFGROB (Mansfield [1991], Mansfield and Fackerell [1992]) and its second version DIFFGROB2 by Mansfield [1993], are designed to compute the DGB of a finitely generated ideal of PDEs with polynomial terms. With respect to this basis, every member of the ideal pseudo-reduces to zero. In pseudo-reduction one is allowed to multiply expressions by differential coefficients of the highest derivative terms that occur in the system to be reduced. This is needed for nonlinear systems where otherwise standard reduction algorithms would not necessarily terminate. DIFFGROB2 allows one to calculate in a systematic way: the elimination ideals, integrability conditions, and compatibility conditions of a system of nonlinear PDEs of polynomial type, up to certain technical constraints fully explained in Mansfield [1993] and Mansfield and Fackerell [1992].

The fundamental tools in Mansfield's package are the Kolchin-Ritt algorithm (Ferro and Gallo [1989]), which is a differential analog of Buchberger's algorithm, with pseudo-reduction instead of reduction, and the *diffgbasis* algorithm, which takes into account algebraic as well as differential consequences of nonlinear systems. These two algorithms allow one to compute the DGB for a wide range of systems of PDEs. A detailed discussion of these algorithms is beyond the scope of this survey paper. For more information about DIFFGROB2 and illustrations of its use, the reader should consult Clarkson and Mansfield [1993c], [1994c], Mansfield and Fackerell [1992], Mansfield and Clarkson [1994a], [1994b], and in particular the manual by Mansfield [1993].

For fairly simple examples, such as the Boussinesq equation, DIFFGROB2 is able to automatically reduce the nonlinear determining equations (corresponding to nonclassical symmetries) in standard form. For more complicated cases, DIFF-

GROB2 may need to be used interactively. Nevertheless, the package has proven to be an effective tool (Clarkson [1994], Clarkson and Mansfield [1993c], [1994a], [1994c], Mansfield and Clarkson [1994a], [1994b]) in solving overdetermined systems of linear and nonlinear PDEs arising in the study of classical and nonclassical symmetries. Needless to say, DIFFGROB2 can be used in applications other than symmetry analysis. Such applications include finding the compatibility conditions for inhomogeneous systems, testing the consistency of systems of PDEs, and finding the least amount of necessary data for a formal power series solution of a linear system (the "initial data" problem). Obviously, DIFFGROB2 can also be used to bring the input equations into involutive form. For some examples this is indeed necessary to be able to compute nonclassical symmetries. In passing, within DIFFGROB2 a package called DIRMETH is available to compute the determining equations related to the symmetry reduction of PDEs via the direct method of Clarkson and Kruskal [1989].

The program CRACK by Wolf and Brand [1992], [1993], [1994] also carries out a Gröbner Basis analysis, but in slightly modified form. First, the algorithm is enriched by the integration of PDEs whenever possible, but only in such a way that the new integrated PDEs are still polynomial in the Gröbner basis. In other words, the 'critical pair completion steps' of the Gröbner basis algorithm and the integrations used within CRACK are consistent. Selective integration can reduce the complexity and aid in solving the determining equations, in particular for systems for which pure Gröbner basis methods would be unfeasible. Second, for efficiency reasons, only a restricted completion algorithm is used, although it is the authors' intention to extend it to a complete Gröbner basis algorithm in the future.

According to a recent paper by Reid, Weih, and Wittkopf [1993], Wittkopf is also developing an algorithm, called *diff_reduce*, which attempts to reduce polynomially nonlinear systems of PDEs to the form of a DGB. In essence, the algorithm is a differential analog of Buchberger's elimination algorithm for polynomial equations. Wittkopf's algorithm uses reduction rather than pseudo-reduction, and incorporates strategies for efficient memory management.

Finally, Oaku [1994] is designing and implementing software in the computer algebra system Risa/asir to automatically find the structure of the solution space of systems of linear PDEs. Oaku's method is based on the notion of Gröbner basis and the Buchberger algorithm applied to rings of differential operators with polynomial coefficients (Weyl algebra).

13.2.4.   Methods for solving determining equations

The most challenging part of Lie symmetry analysis by computer involves the design of an "integrator" for the overdetermined systems of linear homogeneous PDEs. This topic is also of importance in the study of so-called adjoint symmetries of differential equations (Sarlet and Vanden Bonne [1992]), and in many other areas where determining equations of the same type occurs. Ideally, good integration code should be applicable to generic systems of linear differential equations, which do not necessarily come from symmetry analysis.

In the context of Lie symmetry analysis, one can aim at the design of faster and more powerful algorithms that work for *large* systems of determining equations, typically a few hundred, and that automatically reduce systems to where they can be handled interactively with the computer or by hand.

Since the early developments by Head [1993], Schwarz [1982], [1983], Steinberg [1986], [1990], and Stephani [1989] of semi-heuristic methods to solve determining equations, substantial progress has been made in understanding the mathematics of this problem, and a new breed of algorithms is now available. These algorithms attempt to close the gap between solution techniques for ODEs and PDEs (consult Singer [1991] for an impressive review and large bibliography).

Two other important topics tie in with the integration of the determining equations: (i) the transformation of the determining equations into standard and passive forms and (ii) the computation of the size of the symmetry group discussed in Sections 13.2.3 and 13.2.5, respectively.

The design of algorithms and programs to bring the determining equations into standard form were a major step forward. Once systems are reduced into standard involutive form or decoupled, subsequent integration is more tractable and reliable. One could use separation of variables, standard techniques for linear differential equations, and specific heuristic rules given below. The only determining equations left for manual handling should be the "constraint" equations or any other equations whose general solutions cannot be written explicitly in closed form.

In order to be able to make the determination of certain types of Lie generators into a decision procedure, one needs an algorithm for solving linear homogeneous ODEs. Such equations are always obtained as the lowest equation of the reduced determining system, with reduction based on lexicographical term ordering. An important step towards this goal is the factorization as it is applied in SPDE (Schwarz [1988b], [1989]). An in-depth review of issues related to the implementation of this and other algorithms is given in Schwarz [1994b].

After searching the relevant literature, it is clear that many mathematical questions remain open. Despite the innovative efforts of Bocharov and Bronstein, Reid, Schwarz, Topunov, Wolf and Brand, and many others, there is no general algorithm available to integrate an arbitrary (overdetermined) system of determining equations that consists of linear homogeneous PDEs for the η's and the ϕ's.

Most integration algorithms are based on a set of heuristic rules as given in Head [1993], Kersten [1989], Schwarz [1982], [1985], and Steinberg [1979], [1986]. In the computer programs reviewed in Section 13.4, the following rules are used.

1.   Integrate single term equations of the form

$$\frac{\partial^{|I|} f(x_1, x_2, ..., x_n)}{\partial x_1^{i_1} \partial x_2^{i_2} ... \partial x_n^{i_n}} = 0, \qquad (13.10)$$

where $|I| = i_1 + i_2 + ... + i_n$, to obtain the solution

$$f(x_1, x_2, ..., x_n) = \sum_{k=1}^{n} \sum_{j=0}^{i_k-1} h_{kj}(x_1, x_2, ..., x_{k-1}, x_{k+1}, ..., x_n)(x_k)^j,$$

(13.11)

thus, introducing functions h_{kj} with fewer variables.

2. Replace equations of type

$$\sum_{j=0}^{n} f_j(x_1, x_2, ..., x_{k-1}, x_{k+1}, ..., x_n)(x_k)^j = 0, \qquad (13.12)$$

by $f_j = 0$ ($j = 0, 1, ..., n$). More generally, this method of splitting equations (via polynomial decomposition) into a set of smaller equations is also allowed when f_j are differential equations themselves, if the variable x_k is missing.

3. Integrate linear differential equations of first and second order with constant coefficients. Integrate first-order equations with variable coefficients via the integrating factor technique, provided the resulting integrals can be computed in closed form.

4. Integrate higher-order equations of type

$$\frac{\partial^n f(x_1, x_2, ..., x_n)}{\partial x_k^n} = g(x_1, x_2, ..., x_{k-1}, x_{k+1}, ..., x_n), \qquad (13.13)$$

n successive times to obtain

$$f(x_1, x_2, ..., x_n) = \frac{(x_k)^n}{n!} g(x_1, x_2, ..., x_{k-1}, x_{k+1}, ..., x_n) \quad (13.14)$$

$$+ \frac{x_k^{n-1}}{(n-1)!} h(x_1, x_2, ..., x_{k-1}, x_{k+1}, ..., x_n)$$

$$+ ... + r(x_1, x_2, ..., x_{k-1}, x_{k+1}, ..., x_n),$$

where $h, ..., r$ are arbitrary functions.

5. Solve any simple equation (without derivatives) for a function (or a derivative of a function) provided that it both (i) occurs linearly and only once and (ii) depends on all the variables that occur as arguments in the remaining terms.

6. Explicitly integrate exact equations.

7. Substitute the solutions obtained above in all the equations.

8. Add differences sums or other linear combinations of equations (with similar terms) to the system, provided these combinations are shorter than the original equations.

With these simple rules, and perhaps a few more, the determining system can often be drastically simplified. Amazingly, in many cases nothing beyond the above heuristic rules is needed to solve the determining equations completely. If that is not possible, after simplification, the programs return the remaining unsolved differential equations for further inspection. In most programs, the user can then interactively simplify and fully solve the determining equations on the computer, thereby minimizing human errors.

At least for Lie point symmetries, solving the determining equations can usually be done by hand, using elementary results from the theory of linear PDEs. Solving them on a computer may be time consuming, since the simplest approach varies greatly from case to case. Furthermore, a computer program may accidentally not catch the most general result and therefore may return an incomplete symmetry group. The author is aware of this problem, which occurred when testing some of the reviewed symmetry programs. Even worse, the computer algorithm may not be able to determine the solution of the determining equations in a finite number of steps.

13.2.5. Computation of the size of the symmetry group

Schwarz [1992a], [1992b] and Reid [1990b], [1991a], [1991b] independently developed algorithms for determining the size of the Lie symmetry groups of differential equations without integrating the determining equations explicitly.

Schwarz's algorithm SYMSIZE (Schwarz [1992a], [1992b]) is available with the computer algebra system REDUCE, as part of the package SPDE (see Section 13.4.2). Schwarz also translated SYMSIZE into the language of Scratchpad II, the predecessor of AXIOM. Use of SYMSIZE circumvents some of the shortcomings mentioned at the end of the previous section. Indeed, if a differential equation has no other than obvious symmetries or if the symmetry group is small (because all generators are algebraic and of low degree), SYMSIZE will greatly help in completely solving the symmetry problem.

In contrast to the *heuristic* algorithms for the explicit computation of the symmetry generators, the size of the symmetry group can always be determined with SYMSIZE in a finite number of steps. SYMSIZE accepts a system of PDEs as input, and allows one to compute *a priori* the number of free parameters if the group is finite and the number of unspecified functions of the group is infinite. In turn, SYMSIZE allows one to test *a posteriori* if the solution of the determining equations is complete. In cases where some, perhaps all, symmetries are known by inspection or from the physics of the problem, the knowledge of the size of the symmetry group can evade an expensive search for more, perhaps nonexistent, symmetries.

At the heart of SYMSIZE is the procedure *InvolutionSystem*, which transforms the determining system into an involutive system by means of a critical pair/completion algorithm. Similar algorithms are applied in computing Gröbner bases in polynomial ideal theory (see Section 13.2.3).

Concurrently, yet independent of Schwarz, Reid [1990b], [1991a], [1991b] realized that triangularization algorithms may allow bypassing the explicit solution of the determining equations and compute the size of the symmetry group and the commutators immediately. Reid developed the program SYMCAL (Reid [1991a]), written originally in MACSYMA, but now converted by Reid and Wittkopf into Maple (Reid and Wittkopf [1993]).

In Reid [1990a], [1991b], a non-heuristic algorithm *structure constant* is presented, based on *Taylor* and *standard-form*, which always determines (in a finite number of steps) the dimension and the structure constants of the finite part of the Lie symmetry algebra. An extension of the algorithm (Reid [1990a]) also allows one to classify differential equations (with variable coefficients) according to the structure of their symmetry groups. Furthermore, the approach advocated by Reid applies to the determination of symmetries of Lie-contact and Lie-Bäcklund types, as well as potential symmetries.

The algorithm described in Reid [1991b] provides information about the dimension and commutators of the Lie symmetry algebra. It is based on explicit Taylor expansions of the symmetry generators, and therefore is computationally expensive and restricted to finite dimensional Lie algebras. The newest Maple algorithm (Reid et al. [1992]) allows one to compute the dimension and the commutation relations without Taylor expansions; hence, it is applicable to infinite-dimensional Lie algebras.

Readers interested in the problem of determining the "size" of the solution space for arbitrary involutive systems should consult a recent paper by Seiler [1994a]. Seiler's results can be applied to linear and nonlinear determining systems.

Finally, we should mention that skillful use of the tools for reducing systems of linear homogeneous PDEs, available within the package CRACK (Wolf and Brand [1992], [1993], [1994]), can also greatly assist in the investigation of the size of the symmetry group.

13.3. Beyond Lie point symmetries

The discussion of symmetries other than point symmetries is limited here to those for which symmetry software is already available. For a general review of various types of symmetries we refer to Chapters 1 and 4, to the 2nd edition of Olver [1986], and to Clarkson [1994].

13.3.1. Contact and Lie-Bäcklund symmetries

For the computation of Lie-Bäcklund symmetries (see Chapter 1), the use of symbolic programs is even more appropriate. The procedure to determine symmetries is essentially the same as that for point symmetries, although the calculations are lengthier and more time consuming. In a generalized vector field, which still takes the form of (13.3), the functions η^i and ϕ_l may now depend on a finite number of derivatives of u, i.e.,

$$\alpha = \sum_{i=1}^{p} \eta^i(x, u^{(k)}) \frac{\partial}{\partial x_i} + \sum_{l=1}^{q} \varphi_l(x, u^{(k)}) \frac{\partial}{\partial u^l}. \tag{13.15}$$

If $k = 1$ the Lie-Bäcklund symmetry determines, in the case of one dependent variable, a classical contact symmetry (see Chapter 1, Theorem 5). The even simpler case $k = 0$, with $u^{(0)} = u$, leads to point symmetries. Chapter 1 (see also Ibragimov [1983] and Olver [1986]) discusses various possibilities to simplify the calculations, for example by putting the symmetries in canonical (evolutionary) form, or by fixing the order of derivation on which the η's and ϕ's may depend.

13.3.2. Nonclassical or conditional symmetries

Recently it was shown that the "nonclassical method of group-invariant solutions," originally introduced in Bluman and Cole [1969], can determine new solutions of various physically significant nonlinear PDEs (see Chapter 11 of this volume).

Examples include the nonlinear Schrödinger (NLS) equation (Clarkson [1992]) and its cylindrical version (Clarkson and Hood [1993a]); the Boussinesq equation (Levi and Winternitz [1989]); the Kadomtsev-Petviashvili equation (Clarkson and Winternitz [1991]) and other members of its hierarchy (Winternitz [1991]); the Burgers equation (Estévez [1994], Pucci [1992]); the telegraph equation (Nucci [1993c]); the Fitzhugh-Nagumo equation (Estévez [1992], [1994], Nucci and Clarkson [1992]), and other reaction-diffusion equations (Clarkson and Mansfield [1993a]); the Helmholtz (Nucci and Ames [1993]) and shallow water wave equations (Clarkson and Mansfield [1994a], [1994b]); and a class of nonlinear heat equations (Clarkson and Mansfield [1993b], [1993c]). Levi and Winternitz [1993] recently showed how conditional symmetries can be determined for the 2D-Toda lattice, a differential-difference equation. For a well-documented perspective on the computation of nonclassical symmetries we recommend Clarkson and Mansfield [1994c] and Clarkson [1993], [1994].

An example is the Boussinesq equation where the new reductions that follow from application of the nonclassical method were discovered earlier with a direct method (see Clarkson [1990], [1992], Clarkson and Kruskal [1989]). The direct method was also applied to the Zabolotskaya-Khokhlov equation (Clarkson and Hood [1992]), and the Davey Stewartson system (Clarkson and Hood [1993b], [1994]). Recently, Olver [1994a] proved that both methods are equivalent in the case of fiber preserving transformations, which means that the new independent variables depend only on the old independent variables, not on the original dependent variables. Arrigo, Broadbridge, and Hill [1993] explicitly derive criteria for which the direct method and the nonclassical symmetry method lead to the same results; they use the Burgers and Boussinesq equations as illustrative examples. For shallow water wave equations, Clarkson and Mansfield [1994a], [1994b] have shown that the nonclassical symmetry method can lead to particular solutions which can not be obtained via the singular manifold method. Estévez [1992], [1994] and Pucci [1992] pointed out some interesting connections between the direct method of Clarkson and Kruskal [1989]. Estévez [1992], [1994] and Estévez and Gordoa [1993] also compare the nonclassical symmetry method with the singular manifold approach from Painlevé analysis. Nucci [1993b] shows how Bäcklund transformations can be obtained via the investigation of nonclassical symmetries.

In contrast to Lie point symmetries, for example, the transformations corresponding to nonclassical (or conditional) symmetries neither leave the differential equation invariant, nor transform all the solutions into other solutions. They merely transform a subset of solutions into other solutions.

Accounting for "nonclassical symmetries," the program should automatically add the q invariant surface conditions (Bluman and Cole [1969], Olver [1992]),

$$Q^l(x, u^{(1)}) = \sum_{i=1}^{p} \eta^i(x, u) \frac{\partial u^l}{\partial x_i} - \varphi_l(x, u) = 0, \quad l = 1, ..., q, \qquad (13.16)$$

and their differential consequences, to the system 13.1. However, the inclusion of nonclassical symmetries, and perhaps other types of symmetries as discussed in Anderson, Kamran, and Olver [1993a], [1993b] and Olver [1992], requires solving systems of determining equations which are *no longer linear*. Consequently, new integration algorithms must be designed.

It should be noted that various other types of conditional symmetries could be considered. For instance, one could ask under what extra conditions a class of PDEs would admit a symmetry chosen beforehand. Extensive work on this problem, which we will not address here, was done by Fushchich [1993].

13.4. Review of symbolic software

In this section we review the most modern Lie symmetry programs, classified according to the underlying computer algebra system. Focusing on new trends, packages written before 1985 are only briefly mentioned. Researchers interested in more details about some of the pioneering work could consult Hereman [1993].

13.4.1. Computer implementation

Ideally, a fully automated software package for Lie symmetries should consist of effective, powerful algorithms and fast procedures for the following tasks:

1. derivation of the determining equations for large or complicated systems of equations;

2. reduction of determining equations into so-called standard form;

3. finding the size of the symmetry group;

4. determining any obvious symmetry generators;

5. simplifying and integrating the determining equations to compute the generators, if not all the generators have been found yet.

Then the program should be able to execute the following steps in the order relevant to the specific application:

(a) calculation of commutator tables, based on the results of 1, 2, and 3;

(b) calculation of group invariant solutions.

The program should be able to handle: calculation of nonlocal (potential) symmetries (see Bluman [1993a], [1993b], Bluman and Kumei [1989], Ovsiannikov [1978], Vinogradov [1989c]); calculation of nonclassical reductions (conditional symmetries) and resulting solutions; calculation of generalized symmetries; and calculation of equivalent conservation laws.

Furthermore, it should be able to accept systems with free unknown (classification) functions (see Lisle [1992], Reid [1991b], Sherring [1993a]) as input. If so, questions such as "for which values of these parameters or parametric functions does a given ODE or PDE have prescribed symmetries or special solutions" could be answered.

Other ideas could be incorporated in the design of faster and more efficient symbolic software for Lie symmetry analysis. Let me give a couple of examples. Lengthy calculations should be broken up into smaller pieces by consistently taking advantage of the "linear algebra" structure of the Lie symmetry problem. For instance, prolongations should be applied to vector fields of single equations or subsets of equations, and not to the whole system at once. Furthermore, full expansions of the prolonged vector fields should be halted until the explicit forms are actually needed. Avoiding lengthy, redundant expansions will make the generation of the determining equations much more efficient, particularly for nonclassical and Lie-Bäcklund symmetries.

Currently, no software handles this entire ambitious program. Many Lie symmetry programs carry out parts of the listed tasks. Making matters worse, the available Lie symmetry programs, with the exception of Lie by Head [1993], work with specific CAS, which has to be bought separately.

In Table 1 we list the most modern software packages, along with information about developers and distributors. In Table 2 we summarize the scope of these packages.

With the exception of DELiA and DIMSYM, all the programs listed in Table 1 are public domain software. Potential users can obtain the software from the developers or through the referenced sources. The main cost in using these packages is related to the cost of the underlying CAS. As a rule of thumb, individual copies of Mathematica, MACSYMA, Maple, REDUCE, and the like, cost about 10% of the price of the platform you buy them for.

13.4.2. REDUCE programs

In the early 1980s, Schwarz [1982], [1983], [1985], [1987a], [1987b], [1988a], [1988b] developed his well-documented program SPDE. The program automatically derives and often successfully solves the determining equations for Lie point symmetries with minimal intervention by the user. Since 1986 SPDE is distributed together with REDUCE for various types of computers, ranging from PCs to CRAYs. In 1994 version 1.0 of SPDE became available. Although Schwarz decided to keep the old name, the new program is drastically different. According to the documentation in Schwarz [1994a], SPDE 1.0 guarantees that all infinitesimal symmetry generators with algebraic coefficients will be obtained if the equations are nonlinear and of order higher than one. Concerning the input, the equations must be algebraic in their arguments. There is no restriction on the number of independent and dependent variables, and the equations can have any number of constant parameters (no arbitrary functions). The program computes the determining equations, then generates a Gröbner basis for the determining system in a term ordering specified by the user (total degree, lexicographic, etc.). The integration of the reduced system is carried out automatically, the symmetry generators and their commutator table are displayed in LaTeX (if so desired).

Based on Cartan's exterior calculus (Cartan [1946], Finikov [1948], Hermann [1968]), Edelen [1980a], [1980b], [1985], [1991], Gragert [1981], and Gragert, Kersten, and Martini [1983] used computer algebra systems to calculate the classical Lie symmetries of differential equations. More recently, Gragert [1989], [1992] added a package for more general Lie algebra computations, including code for higher-order and super symmetries and super prolongations. Kersten [1987], [1989] further perfected the software package for the calculation of the Lie algebra of infinitesimal symmetries (including Lie-Bäcklund symmetries) of exterior differential systems.

Eliseev, Fedorova, and Kornyak [1985], wrote code in REDUCE-2 to generate (but not solve) the system of determining equations for point and contact symmetries. Their paper discusses the algorithm and shows three worked examples. Fedorova and Kornyak [1985], [1986], [1987] generalized the algorithm to include the case of Lie-Bäcklund symmetries.

The interactive REDUCE program NUSY by Nucci [1990], [1991], [1992a], [1993a], [1995], included with this book, generates determining equations for Lie point, nonclassical, Lie-Bäcklund, and approximate symmetries and provides interactive tools to solve them. The manual by Nucci [1990], [1992a], [1995] gives a clear description of the various routines with their scope and limitations, and has several worked examples.

The package CRACK by Wolf and Brand [1992], [1993], [1994] solves overdetermined systems of differential equations with polynomial terms. To do this, it uses code for decoupling, separating, and simplifying PDEs. Integration of exact PDEs and differential factorization are also possible. CRACK has many applications that are facilitated via special tools. For instance, the function LIEORD can aid in the investigation of Lie symmetries of ODEs. With CRACK one can also construct the Lagrangian for a given second-order ODE, and find first integrals via the integrating factor method. In attempting to solve standard ODEs, the program makes use of the REDUCE package ODESOLVE written by MacCallum [1988]. The functions and tools available within CRACK allow simplification and integration of linear homogeneous PDEs, beyond those derivable via symmetry analysis.

Upon completion of CRACK, Wolf went on to develop three new REDUCE programs, called LIEPDE, QUASILINPDE, and APPLYSYM, which all make use of the tools of CRACK.

LIEPDE (Wolf [1993]) finds Lie point and contact symmetries of PDEs by deriving and solving a few simple determining equations, before continuing with the computation of the more complicated determining equations. This idea, which makes the program highly efficient, was used in Wolf's FORMAC program (Wolf [1989a], [1989b], [1992]) and is also implemented in the design of the feedback mechanism of SYMMGRP.MAX (Champagne, Hereman, and Winternitz [1991]). For solving the determining equations, LIEPDE makes use of modules of the package CRACK discussed above. The difference is that within LIEPDE the steps are carried out automatically, without intervention by the user. This approach is particularly useful when applied to large systems of PDEs, or in the computation of higher-order symmetries, where space and memory limitations come into play.

The aim of QUASILINPDE (Wolf [1994]) is to find the solutions of quasi-linear PDEs. These solutions are then used by APPLYSYM (Wolf [1994]), which applies the symmetry to lower the order of ODEs, to calculate similarity variables for PDEs, to effectively reduce the number of independent variables of a system of PDEs, and to generalize special known solutions of ODEs and PDEs. To our knowledge, APPLYSYM is one of the first symbolic programs that truly applies point symmetries that can be calculated with the program LIEPDE. The program APPLYSYM is automatic, but can also be used interactively. Thus far, APPLYSYM is only applicable to point symmetries for which the generators are at worst rational. The actual problem solving is done in all these programs through a call to the package CRACK for solving overdetermined systems of PDEs.

Gerdt [1993a], [1993b] introduced the program HSYM for the explicit computation of higher-order symmetries for PDEs. If the given system of equations has arbitrary parameters, the necessary conditions for the existence of higher order symmetries will lead to a system of algebraic equations in the parameters. Via the program ASYS, that algebraic system is reduced into standard form via a Gröbner basis algorithm. The focus in Gerdt's work is on the investigation of the integrability of polynomial type nonlinear evolution equations, by verifying the existence of higher-order symmetries and their associated conservation laws.

Sarlet and Vanden Bonne [1992] offer specific procedures to assist in the computation of adjoint symmetries of second-order ODEs. This assistance, however, is limited to the construction of determining equations for certain classes of adjoint symmetries, which are of the same nature as determining equations for (generalized) symmetries, and relies on other packages such as DIMSYM below for solving these determining equations. In addition, procedures have been written for testing whether a given adjoint symmetry can give rise to a Lagrangian or a first integral for the original second-order equations.

The program DIMSYM by Sherring [1993a], in collaboration with Prince, was inspired by Head's symmetry program LIE (Head [1993]), in turn influenced by SPDE (Schwarz [1985], [1988b]), but is much larger and grew independently of it during development. It is capable of finding various types of symmetries, currently, point symmetries, Lie-Bäcklund, and conditional symmetries. DIMSYM can isolate special cases, bring the determining equations in standard form for example, and aid in the solution of group classification problems. It attempts to determine the generators and allows one to check whether or not the generators are correct. It allows the user to specify the dependence of the symmetry vector field coefficients, which is particularly practical if one wants to compute Lie-Bäcklund symmetries. DIMSYM provides the user with a lot of flexibility: ansätze can be made, simplification routines can be called separately, manual intervention is possible, etc. Quite often such interventions indeed allow the user to complete the desired computations, whereas the DIMSYM in auto-pilot mode may not.

DIMSYM has routines that convert the input equations into standard form. Another attractive feature of the package is that the integrator for the determining equations also works for systems of linear homogeneous differential equations not necessarily obtained from symmetry analysis. The overall strategy of the solver is to put the system of determining equations into standard form based on Reid's algorithm (see Section 13.2.3), while solving explicitly any equations in the system that the algorithm is capable of solving.

Finally, we mention the programs by Ito [1986], [1994] and Ito and Kako [1985] for the determination of symmetries and conservation laws of systems of evolution equations. Ito's program does not use any of the algorithms discussed in Section 13.2.2, but uses infinitesimal symmetries to determine the form of conservation laws.

13.4.3. MACSYMA programs

One of the first programs was written by Rosenau and Schwarzmeier [1979] and Schwarzmeier and Rosenau [1988]. Their program calculates the determining equations, simplifies them a bit, but does not solve them automatically.

The MACSYMA version of the program SYMCON (Vafeades [1990b]), which was originally written in muMATH, tries to compute Lie point and Bessel-Haagen generalized symmetries (of any order) and their conservation laws. Vafeades [1992a], [1992b], [1994a], [1994b] later produced PDELIE, which is a drastically improved version of SYMCON by Vafeades [1990b]. The package PDELIE attempts to produce similarity solutions of ODEs, analyze PDEs with a multiplicative or additive scalar parameter, and compute the commutator table and the structure constants of the Lie algebra. PDELIE also allows one to compute the Noether conservation laws of variational systems.

PDELIE consists of several subroutines. Let's discuss the main ones. The function PL_SYM produces the determining equations and the generators of the Lie group. It uses a standard form algorithm by Reid and a set of heuristic rules to facilitate the integration. The function PL_SOLVE tries to find the invariants of the symmetry group. Using these invariants, it then dimensionally reduces the given differential equation. In cases where the reduced equation is an ODE, it tries to integrate explicitly, thus arriving at special similarity solutions of the original equation. The function PL_COMTAB performs computations with elements in the commutator table of structure constants of the Lie algebra. The function PL_CON computes the densities of the Noether conservation laws of systems of variational and divergence type.

Just as PDELIE, the program SYM_DE by Steinberg [1979], [1986], [1990] was recently added to the out-of-core library of MACSYMA. Steinberg's program computes infinitesimal symmetry operators and the explicit form of the infinitesimal transformations for simple systems. In cases where the program cannot automatically finish the computation, the user can intervene and, for instance, ask for infinitesimal symmetries of polynomial form. The program solves some (or all) of the determining equations automatically and, if needed, the user can (interactively) add extra information. Steinberg intends to extend his program so that it would include the calculation of generalized symmetries.

The program SYMMGRP.MAX written by Champagne, Hereman, and Winternitz [1991] and Hereman [1993] is a modification of an earlier package by

Champagne and Winternitz [1985] that has been extensively used over the last decade at the University of Montr´eal and in many places elsewhere. It has been tested on hundreds of systems of equations and has thus been solidly debugged.

The flexibility within SYMMGRP.MAX and the possibility of using it interactively allows the user to find the symmetry group of arbitrarily large and complicated systems of equations on relatively small computers. For example, whenever the prolongation can be applied successfully to the complete system, or a subset thereof, it produces a list of determining equations. This list is free of trivial factors, duplication, and differential redundancies.

To make SYMMGRP.MAX work for *large* systems of differential equations, the designers followed the path that would be taken in manual calculations. That is, obtain in as simple a manner as possible the simplest determining equations, solve them, and feed the information back to the computer. Partial information can be extracted very rapidly. For instance, one can derive a subset of the determining equations, such as those that occur as coefficients in the highest derivatives in the independent variables. These are usually single-term equations, which express that the coefficients of the vectorfield are independent of some variables or depend linearly on some of the other variables.

A feedback mechanism facilitates the solution of the determining equations step by step on the computer; hence, avoiding human error in the algebraic simplifications. Typically, users will provide information about the η's and φ's, as it becomes available from solving the determining equations step by step. The amount of interaction by the user will depend on the complexity of the system of differential equations and on the capacity of the computer used. A worked example showing the use of the feedback mechanism is given in Champagne, Hereman, and Winternitz [1991]. Needless to say, with the feedback mechanism, the program SYMMGRP.MAX can also be used to verify previously calculated solutions of the determining equations and, hence, detect errors in the literature on the subject.

Although not designed for that purpose, the program SYMMGRP.MAX can be easily adapted to compute the determining equations corresponding to nonclassical symmetries, as illustrated in Clarkson [1994], and Clarkson and Mansfield [1993b], [1993c], [1994a], [1994b], [1994c].

In Clarkson and Mansfield [1994c], a detailed explanation of such an adaptation is given. The proof of correctness of the proposed adaptation is based on the theory of Gröbner bases.

13.4.4. MAPLE programs

In Carminati, Devitt, and Fee [1992], LIESYMM is presented for creating the determining equations via the Harrison-Estabrook procedure. Within LIESYMM various interactive tools are available for integrating the determining equations and for working with Cartan's differential forms. Their program is independent of Donsig's differential forms package *difforms*, also available in Maple.

Khai T. Vu (Department of Mathematics, Monash University, Clayton, Victoria,

Australia) has translated Head's muMATH program LIE (Head [1993]), discussed in Section 13.4.7, into Maple syntax. In 1994, the Maple version of LIE, which computes Lie point symmetries, was still being tested and, therefore, was not yet released. The β-version of the program computes the determining equations, solves them, and gives the explicit forms of the (vectorfield) coefficients together with the generators.

Hickman [1993], [1994] offers a collection of Maple routines that aid in the computation of Lie point symmetries, nonlocal symmetries, and Wahlquist-Estabrook-type prolongations. The tools for symmetry analysis include user-friendly procedures to enter names of variables, to create total derivatives, to generate and prolong vector fields, and to derive and partially solve determining equations. Program and documentation are available via anonymous FTP from math.canterbury.ac.nz.

Mansfield has developed the package DIRMETH for the computation of symmetries via the direct method proposed by Clarkson and Kruskal [1989]. The program DIRMETH is part of DIFFGROB2, discussed in detail in Section 13.2.3. Other efforts in the design of packages for the direct method are given in the doctoral thesis of Williams [1989].

13.4.5.   MATHEMATICA programs

Herod [1994] developed MathSym for deriving the determining equations corresponding to Lie point symmetries, including nonclassical (or conditional) symmetries. Upon derivation of the determining equations, the program reduces these equations via an algorithm based on the method of Riquier and Janet. Herod's doctoral thesis contains the well-documented code of MathSym and applications to various equations from fluid dynamics.

Recently, the packages *Lie.m* and *Baecklund.m* have been added by Baumann [1992], [1993a], [1993b] to MathSource, the Mathematica Program Library. Baumann's program *Lie.m* (Baumann [1992]) follows the structure of our MACSYMA program SYMMGRP.MAX, as described in Champagne, Hereman, and Winternitz [1991], very closely. Users familiar with SYMMGRP.MAX will have a short learning curve with *Lie.m*. In contrast to SYMMGRP.MAX, the program *Lie.m* can handle transcendental functions in the input equations. The newest version of *Lie.m* can be used to compute point symmetries, contact symmetries, and nonclassical symmetries. *Lie.m* brings the determining equations in canonical form via the procedure of Janet and Riquier, and goes on to solve the determining equations automatically. A finite set of integration rules, similar to the ones described in Section 13.2.4, are implemented.

Once the solution of the determining equations is obtained, the program can continue with the computation of the vector basis, ideals, and commutator table of the Lie algebra, its structure constants, Casimir operators, and its metric tensor.

Baumann's package *Baecklund.m* (Baumann [1993a], [1993b]) contains functions that attempt to compute generalized symmetries for PDEs and ODEs and invariants of ODEs only. When applied to second-order ODEs, the program attempts to verify if the computed symmetries are of variational type. If so, the program calculates the corresponding invariants (integrals of motion). For the explicit calculations to be successful, quite often one has to specify that the coefficients in the vectorfield are polynomials in the coordinates and momenta. With this "ansatz" one may not be able to obtain all the generalized symmetries, but one may successfully obtain explicit forms of invariants.

Bérubé and de Montigny [1992] produced Lie symmetry code in Mathematica. Their program *symmgroup.c* computes the determining equations for Lie point symmetries. In its syntax and format *symmgroup.c* closely follows the structure of SYMMGRP.MAX. The data for the program may consist of DEs with arbitrary functions. Transcendental functions in both dependent and independent variables are also permitted. In Bérubé and de Montigny [1992], three well-chosen examples are given to illustrate the capabilities of the program.

Finally, Coult (while at Carleton College, Northfield, Minnesota) developed a Mathematica program, temporarily called *symmgroup.m*, for the computation of the determining equations corresponding to Lie point symmetries of a large class of differential equations (with polynomial terms).

13.4.6. SCRATCHPAD and AXIOM programs

Schwarz [1988a] rewrote SPDE (Schwarz [1987b], [1988b]) for use with version 1 of Scratchpad II, a symbolic manipulation program developed by IBM. Scratchpad II is now superseded by AXIOM.

Seiler and co-workers (Schü, Seiler, and Calmet [1993], Seiler [1994b]) are designing a package that will compute determining equations for classical and nonclassical symmetries. See Section 13.2.3 for a description of their program JET for geometry computations based on the jet bundle formalism.

13.4.7. MUMATH programs

The program LIE by Head [1993] is based on version 4.12 of muMATH, but is self contained and runs on IBM-compatible PCs. As a matter of fact, the program comes bundled with a limited version of muMATH. Head's program calculates and solves the determining equations (for Lie point symmetries) automatically for single equations and systems of differential equations. LIE also computes the Lie vectors and their commutators. Interventions by the user are possible, but are rarely needed. The source code of the program is available, including the heuristic routines that attempt to solve the determining equations. Due to the limitations of muMATH, the program LIE is bounded by the 256 KB of memory for program and workspace. For a program of limited size, LIE is remarkable in its achievements.

Version 4.2 of LIE is freely available by FTP from various public domain software archives such as SIMTEL and associated archives. By this printing, version 4.3 of LIE also will be available. In addition to Lie point symmetries, the new version will be able to compute contact and Lie-Bäcklund symmetries.

The SYMCON package written by Vafeades [1990b] also uses muMATH to calculate the determining equations (without solving them). The program is restricted to point symmetries. Furthermore, the program verifies whether the symmetry group is of variational or divergence type and computes the conservation laws associated with the symmetries. Unfortunately, these programs are confined to the 256-KB memory accessible by muMATH and cannot presently handle very large systems of equations. This limitation motivated Vafeades to rewrite his SYMCON program in MACSYMA syntax (Vafeades [1992a], [1992b], [1994a], [1994b]). The MACSYMA version of SYMCON can handle generalized symmetries and their conservation laws.

For completeness, Mikhailov developed software in muMATH to verify the integrability of systems of PDEs by testing for the existence of higher symmetries. The program computes special symmetries, canonical conservation laws, and carries out conformal transformations to bring PDEs into canonical form. With their PC program, Mikhailov, Shabat, and Sokolov [1990] produced an exhaustive list of integrable nonlinear Schrödinger-type equations. In this context, integrable means that the equations have infinitely many conserved quantities and infinitely many local symmetries.

13.4.8.   Programs for other systems

Fushchich [1981], Fushchich and Kornyak [1989], and Kornyak and Fushchich [1991] developed programs in Turbo C and AMP for the computation of Lie-Bäcklund symmetries. Their programs also classify equations with arbitrary parameters and functions with respect to such symmetries. It is important to note that their programs reduce the determining equations into passive form (see Section 13.2.3). All integrability conditions are then explicit and, therefore, the resulting system is in involution.

We should mention their two FORMAC programs. The first program, called *LB*, was written in the PL/1 language by Fedorova and Kornyak [1985], [1986], [1987]. The successor, called *LBF*, was developed by Fushchich and Kornyak [1989]. Both programs create the system of determining equations for Lie-Bäcklund symmetries and attempt to solve these equations. The program *LBF*, with its 1362 lines of PL/1-FORMAC code, is completely automatic and consists of 37 subroutines, one of which brings the determining equations in passive (Riquier-Janet) form. The program *LB* by Fedorova and Kornyak [1985], [1986] is available from the Computer Physics Communications Program Library in Belfast. The above programs were designed for low-memory requirements so that they could run on PCs.

The PL/1-based FORMAC package CRACKSTAR developed by Wolf [1989a], [1989b], [1992] allows one to investigate Lie symmetries of systems of PDEs, besides dealing with dynamical symmetries of ODEs (Pohle and Wolf [1989]), and the like. A good overview of the capabilities of CRACKSTAR is given in Wolf [1989b]; a description of the routines and worked examples are in Pohle and Wolf [1989]. For efficiency, CRACKSTAR generates and solves first-order determining equations early on, and then continues with the higher-order determining equations. The successor of CRACKSTAR is the REDUCE package CRACK discussed already in Section 13.4.2.

Gerdt [1993a], [1993b], Gerdt and Zharkov [1990], and Gerdt, Shvachka, and Zharkov [1985a], [1985b] used REDUCE and PL/1-FORMAC to investigate the integrability of nonlinear evolution equations. Their program FORMINT contains algorithms to calculate Lie-Bäcklund symmetries and conserved densities, but does not use the jet bundle formalism.

The calculation of the Lie group by computer was also proposed by Popov [1985], who used the program SOPHUS for the calculation of conservation laws of evolution equations.

In Bocharov and Bronstein [1989], the authors present SCoLAr, a package written in standard PASCAL that finds infinitesimal symmetries and conservation laws of arbitrary systems of differential equations. An application of SCoLAr to the Kadomtsev-Pogutse equations is given in Gusyatnikova, Samokhin, Titov, Vinogradov and Yumaguzhin [1989].

The PC package DELiA, standing for "Differential Equations with Lie Approach," is an outgrowth of the SCoLar project (see Bocharov and Bronstein [1989]). DELiA, written in Turbo PASCAL by Bocharov and his collaborators (Bocharov [1989], [1990a], [1990b], [1991]), is a stand-alone computer algebra system for investigating differential equations. It performs various tasks based on Lie's approach, such as the computation of Lie point and Lie-Bäcklund symmetries, canonical conserved densities and generalized conservation laws, simplification and partial integration of overdetermined systems of differential equations, etc. The methods used in DELiA and many examples are well described in the user guides in Bocharov [1989], [1991].

In order to be able to handle large problems, DELiA first generates and solves first-order determining equations, and then continues to generate and solve the higher-order determining equations. The analyzer/integrator, which is available as a separate tool at the user level, includes a general algorithm for passivization (see Bocharov and Bronstein [1989]), together with a set of integration rules for linear and quasi-linear systems of PDEs. Currently, a MS Windows version of DELiA, called MS Win DELiA, is under development.

Using the algorithmic language REFAL, Topunov [1989] developed a software package for symmetry analysis that contains subroutines to reduce determining systems in passive form.

13.5. Examples

In this section we give three examples that illustrate the computation of Lie point symmetries with symbolic software. The first and simplest example involves a single scalar nonlinear equation. The second example illustrates how symmetries of a nonlinear complex equation are computed by splitting the equation into a system of nonlinearly coupled equations for the real and imaginary parts of the original dependent variable. The last and most complicated example involves a system of vector equations that needs to be split into equations for its scalar components in order to compute its Lie symmetries.

13.5.1. The Harry Dym equation

Consider the Harry Dym equation (Ablowitz and Clarkson [1991]),

$$u_t - u^3 u_{xxx} = 0. \tag{13.17}$$

Clearly, this is one equation with two independent variables and one dependent variable. The assignments of the variables are as follows:

$$x \longmapsto x[1], \quad t \longmapsto x[2], \quad u \longmapsto u[1]. \tag{13.18}$$

This permits us to rewrite Equation 13.17 in a form accepted by the program SYMMGRP.MAX; i.e.,

```
e1 : u[1,[0,1]]-u[1]^3*u[1,[3,0]].
```

For PDELIE and SYM_DE the input form would be

```
'DIFF(U,T)-U^3*'DIFF(U,X,3).
```

For SPDE and LIE the program accepts

```
U(1,2)-U(1)^3*U(1,1,1,1).
```

Next, one selects the variable u_t for elimination, e.g.,

```
v1 : u[1,[0,1]].
```

Then, the programs automatically compute the determining equations for the coefficients eta[1] $= \eta^x$, eta[2] $= \eta^t$, and phi[1] $= \varphi^u$ of the vector field

$$\alpha = \eta^x \frac{\partial}{\partial x} + \eta^t \frac{\partial}{\partial t} + \varphi^u \frac{\partial}{\partial u}. \tag{13.19}$$

There are only eight determining equations,

$$\frac{\partial \text{eta2}}{\partial u[1]} = 0,$$

$$\frac{\partial \text{eta2}}{\partial x[1]} = 0,$$

$$\frac{\partial \text{eta1}}{\partial u[1]} = 0, \tag{13.20}$$

$$\frac{\partial^2 \text{phi1}}{\partial u[1]^2} = 0,$$

$$\frac{\partial^2 \text{phi1}}{\partial u[1] \partial x[1]} - \frac{\partial^2 \text{eta1}}{\partial x[1]^2} = 0,$$

$$\frac{\partial \text{phi1}}{\partial x[2]} - u[1]^3 \frac{\partial^3 \text{phi1}}{\partial x[1]^3} = 0,$$

$$3u[1]^3 \frac{\partial^3 \text{phi1}}{\partial u[1] \partial x[1]^2} + \frac{\partial \text{eta1}}{\partial x[2]} - u[1]^3 \frac{\partial^3 \text{eta1}}{\partial x[1]^3} = 0, \tag{13.21}$$

$$u[1] \frac{\partial \text{eta2}}{\partial x[2]} - 3u[1] \frac{\partial \text{eta1}}{\partial x[1]} + 3 \text{ phi1} = 0.$$

These determining equations are easily solved explicitly, either automatically with SPDE, LIE, and PDELIE, or with the feedback mechanism within SYM_DE and SYMMGRP.MAX. The general solution, rewritten in the original variables, is

$$\eta^x = k_1 + k_3\, x + k_5\, x^2, \tag{13.22}$$

$$\eta^t = k_2 - 3k_4\, t, \tag{13.23}$$

$$\varphi^u = (k_3 + k_4 + 2k_5\, x)\, u, \tag{13.24}$$

where $k_1, \ldots, k_5$ are arbitrary constants. The five infinitesimal generators then are

$$G_1 = \partial_x, \tag{13.25}$$

$$G_2 = \partial_t, \tag{13.26}$$

$$G_3 = x\partial_x + u\partial_u, \tag{13.27}$$

$$G_4 = -3t\partial_t + u\partial_u, \tag{13.28}$$

$$G_5 = x^2\partial_x + 2xu\partial_u. \tag{13.29}$$

Clearly, (13.17) is invariant under translations (G_1 and G_2) and scaling (G_3 and G_4). The flow corresponding to each of the infinitesimal generators can be obtained via simple integration. As an example, let us compute the flow corresponding to G_5. This requires integration of the first-order system

$$\frac{d\tilde{x}}{d\epsilon} = \tilde{x}^2, \qquad \tilde{x}(0) = x,$$

$$\frac{d\tilde{t}}{d\epsilon} = 0, \qquad \tilde{t}(0) = t, \tag{13.30}$$

$$\frac{d\tilde{u}}{d\epsilon} = 2\tilde{x}\tilde{u}, \qquad \tilde{u}(0) = u,$$

where ϵ is the parameter of the transformation group. One readily obtains

$$\tilde{x}(\epsilon) = \frac{x}{(1 - \epsilon x)}, \tag{13.31}$$

$$\tilde{t}(\epsilon) = t, \tag{13.32}$$

$$\tilde{u}(\epsilon) = \frac{u}{(1 - \epsilon x)^2}. \tag{13.33}$$

Therefore, we conclude that for any solution $u = f(x, t)$ of Equation 13.17, the transformed solution

$$\tilde{u}(\tilde{x}, \tilde{t}) = (1 + \epsilon\tilde{x})^2 f(\frac{\tilde{x}}{1 + \epsilon\tilde{x}}, \tilde{t}) \tag{13.34}$$

will solve $\tilde{u}_{\tilde{t}} - \tilde{u}^3 \tilde{u}_{\tilde{x}\tilde{x}\tilde{x}} = 0$.

13.5.2. The Nonlinear Schrödinger equation

In order to compute the Lie point symmetries of the celebrated nonlinear Schrödinger equation (Ablowitz and Clarkson [1991]),

$$iu_t + u_{xx} + u|u|^2 = 0, \tag{13.35}$$

one needs to replace the single complex equation by a coupled system. One way of doing that is by introducing the real and imaginary parts v, w of the complex variable u via $u(x, t) = v(x, t) + iw(x, t)$. This yields

$$v_t + w_{xx} + w(v^2 + w^2) = 0, \tag{13.36}$$

$$w_t - v_{xx} - v(v^2 + w^2) = 0.$$

One alternative is to replace (13.35) by a system consisting of the equation itself and its complex conjugate (Clarkson [1992]) and to interpret the variables u and $v = u^*$ as real,

$$k * u_t + u_{xx} + v * u^2 = 0, \tag{13.37}$$

$$-k * v_t + v_{xx} + u * v^2 = 0.$$

In order to work with real quantities throughout, the imaginary unit i was temporarily replaced by the constant k during the computations. Once the determining equations are obtained, $k = i$ should be reintroduced.

Another alternative is to write $u(x, t) = R(x, t) \exp(i\Omega(x, t))$, thus replacing (13.35) by a coupled system

$$R_t + 2R_x\Omega_x + R\Omega_{xx} = 0, \tag{13.38}$$

$$R_{xx} - R\Omega_t - R\Omega_x^2 + R^3 = 0,$$

for the real modulus $R(x, t)$ and real phase $\Omega(x, t)$.

Adhering to (13.36), SYMMGRP.MAX (or, for that matter, any other symmetry program) quickly generates the *twenty* determining equations for the coefficients of the vector field

$$\alpha = \eta^x \frac{\partial}{\partial x} + \eta^t \frac{\partial}{\partial t} + \varphi^v \frac{\partial}{\partial v} + \varphi^w \frac{\partial}{\partial w}. \qquad (13.39)$$

The first *eleven* single-term determining equations are similar to (13.20) and provide information about the dependencies of the $\eta's$ and the $\phi's$ on x, t, v, and w, and their linearity in the latter two independent variables. The remaining nine determining equations are a bit more complicated, but the entire system is readily solved.

In the original variables, the solution reads

$$\eta^x = k_1 + 2k_4\, t + k_5\, x\,, \qquad (13.40)$$

$$\eta^t = k_2 + 2k_5\, t\,, \qquad (13.41)$$

$$\varphi^v = k_3\, w - k_4\, xw - k_5\, v\,, \qquad (13.42)$$

$$\varphi^w = -k_3\, v + k_4\, xv - k_5\, w\,, \qquad (13.43)$$

where $k_1, \ldots, k_5$ are arbitrary constants. As in the previous examples, the complete symmetry algebra is spanned by five vector fields (generators):

$$G_1 = \partial_x, \qquad (13.44)$$

$$G_2 = \partial_t, \qquad (13.45)$$

$$G_3 = w\partial_v - v\partial_w, \qquad (13.46)$$

$$G_4 = 2t\partial_x - x(w\partial_v - v\partial_w), \qquad (13.47)$$

$$G_5 = x\partial_x + 2t\partial_t - v\partial_v - w\partial_w. \qquad (13.48)$$

If we had carried out the computations with (13.38), where $u(x, t) = R(x, t)$ $\exp(i\Omega(x, t))$, we would have found:

$$G_1 = \partial_x, \qquad (13.49)$$

$$G_2 = \partial_t, \tag{13.50}$$

$$G_3 = \partial_\Omega, \tag{13.51}$$

$$G_4 = 2t\partial_x - x\partial_\Omega, \tag{13.52}$$

$$G_5 = x\partial_x + 2t\partial_t - R\partial_R. \tag{13.53}$$

Either way, (13.35) is invariant under translations in space and time (G_1 and G_2). Generator G_3 corresponds to adding an arbitrary constant to the phase of u. The Galilean boost is generated by G_4. Finally, G_5 indicates invariance of the equation under scaling (or dilation). Similarity reductions can then be obtained by solving the characteristic equations,

$$\frac{dx}{\eta^x} = \frac{dt}{\eta^t} = \frac{dv}{\varphi^v} = \frac{dw}{\varphi^w}, \tag{13.54}$$

or equivalently, the invariant surface conditions

$$\eta^x v_x + \eta^t v_t - \varphi^v = 0, \tag{13.55}$$

$$\eta^x w_x + \eta^t w_t - \varphi^w = 0.$$

The actual reductions can be found in Clarkson [1992], where a quite general nonlinear Schrödinger equation is treated. It is well known (Levi and Winternitz [1989]) that all the reductions of the NLS can be obtained from G_1 through G_5; in other words, nonclassical symmetries would not lead to new symmetry reductions. To compute nonclassical symmetries of (13.36), it suffices to replace v_t and w_t from (13.55). If $\eta^t \neq 0$, we set $\eta^t = 1$ for simplicity. Thus,

$$v_t = -\eta^x v_x + \varphi^v, \tag{13.56}$$

$$w_t = -\eta^x w_x + \varphi^w. \tag{13.57}$$

The case $\eta^t = 0$ has to be considered separately. Since SYMMGRP.MAX allows the user to give information about the coefficients in the vector field, the computation can now proceed as in the classical case. For worked examples, we refer the reader to Clarkson [1994] and Clarkson and Mansfield [1994c].

13.5.3. The Magneto-Hydro-Dynamics equations

As an example of a large system of differential equations, we take the equations for Magneto-Hydro-Dynamics (MHD) (Fuchs [1991]) and carry out the search for Lie point symmetries with SYMMGRP.MAX.

The MHD equations, with or without dissipative terms, have become a bench-mark for developers of Lie symmetry packages. Nucci [1984] computed the classical symmetries in 1984 and commented that the calculations by hand took her a year. Michel Grundland undertook the same formidable task and also finished the job in about a year. He conveyed to the author that the long winter in Newfoundland (where he was at the time) helped.

If we neglect dissipative effects, and thus restrict the analysis to the *ideal* case, the MHD equations can be reduced to

$$\frac{\partial \rho}{\partial t} + (\vec{v} \cdot \nabla)\rho + \rho \nabla \cdot \vec{v} = 0, \qquad (13.58)$$

$$\rho\left(\frac{\partial \vec{v}}{\partial t} + (\vec{v} \cdot \nabla)\vec{v}\right) + \nabla(p + \frac{1}{2}\vec{H}^2) - (\vec{H} \cdot \nabla)\vec{H} = \vec{0}, \qquad (13.59)$$

$$\frac{\partial \vec{H}}{\partial t} + (\vec{v} \cdot \nabla)\vec{H} + \vec{H}\nabla \cdot \vec{v} - (\vec{H} \cdot \nabla)\vec{v} = \vec{0}, \qquad (13.60)$$

$$\nabla \cdot \vec{H} = 0, \qquad (13.61)$$

$$\frac{\partial}{\partial t}\left(\frac{p}{\rho^\kappa}\right) + (\vec{v} \cdot \nabla)\left(\frac{p}{\rho^\kappa}\right) = 0, \qquad (13.62)$$

with pressure p, mass density ρ, coefficient of viscosity κ, fluid velocity $\vec{v}$, and magnetic field $\vec{H}$. Using the first equation, we eliminate ρ from the last equation, hence replacing it by

$$\frac{\partial p}{\partial t} + \kappa p(\nabla \cdot \vec{v}) + (\vec{v} \cdot \nabla)p = 0. \qquad (13.63)$$

If we split the vector equations in scalar equations for the vector components, we have a system of nine equations, with four independent variables and eight dependent variables. For convenience, we denote the components of the vector $\vec{v}$ by v_x, v_y, and v_z, not to be confused with partial derivatives of v.

The variables to be eliminated are selected as follows: for the first seven variables and the ninth variable we pick the partial derivatives with respect to t of ρ, v_x, v_y, v_z, H_x, H_y, H_z, and p. From the eighth equation, we select $\partial H_x/\partial x$ for elimination.

We will only consider the case where $\kappa \neq 0$. We ran this case on a Digital VAX 4500 with 64 MB of RAM, and on an IBM Risc 6000 workstation with 32 MB of RAM. On the VAX it took 50 minutes of CPU time, on the IBM workstation 1 hour and 50 minutes, for SYMMGRP.MAX to create the 222 determining equations for the coefficients of the vector field

$$\alpha = \eta^x \frac{\partial}{\partial x} + \eta^y \frac{\partial}{\partial y} + \eta^z \frac{\partial}{\partial z} + \eta^t \frac{\partial}{\partial t} + \varphi^\rho \frac{\partial}{\partial \rho} + \varphi^p \frac{\partial}{\partial p}$$

$$+ \varphi^{v_x} \frac{\partial}{\partial v_x} + \varphi^{v_y} \frac{\partial}{\partial v_y} + \varphi^{v_z} \frac{\partial}{\partial v_z} + \varphi^{H_x} \frac{\partial}{\partial H_x} + \varphi^{H_y} \frac{\partial}{\partial H_y} + \varphi^{H_z} \frac{\partial}{\partial H_z}.$$

Using SYMMGRP.MAX interactively, we then integrated the determining system and obtained the solution expressed in the original variables,

$$\eta^x = k_2 + k_5\, t - k_8\, y - k_9\, z + k_{11}\, x\,, \tag{13.64}$$

$$\eta^y = k_3 + k_6\, t + k_8\, x - k_{10}\, z + k_{11}\, y\,, \tag{13.65}$$

$$\eta^z = k_4 + k_7\, t + k_9\, x + k_{10}\, y + k_{11}\, z\,, \tag{13.66}$$

$$\eta^t = k_1 + k_{12}\, t\,, \tag{13.67}$$

$$\varphi^\rho = -2\,(k_{11} - k_{12} - k_{13})\,\rho\,, \tag{13.68}$$

$$\varphi^p = 2\,k_{13}\, p\,, \tag{13.69}$$

$$\varphi^{v_x} = k_5 - k_8\, v_y - k_9\, v_z + (k_{11} - k_{12})v_x\,, \tag{13.70}$$

$$\varphi^{v_y} = k_6 + k_8\, v_x - k_{10}\, v_z + (k_{11} - k_{12})v_y\,, \tag{13.71}$$

$$\varphi^{v_z} = k_7 + k_9\, v_x + k_{10}\, v_y + (k_{11} - k_{12})v_z\,, \tag{13.72}$$

$$\varphi^{H_x} = k_{13}\, H_x - k_8 H_y - k_9 H_z\,, \tag{13.73}$$

$$\varphi^{H_y} = k_{13}\, H_y + k_8 H_x - k_{10} H_z\,, \tag{13.74}$$

$$\varphi^{H_z} = k_{13}\, H_z + k_9 H_x + k_{10} H_y\,. \tag{13.75}$$

It is clear that there is a *thirteen*-dimensional Lie algebra spanned by the generators:

$$G_1 = \partial_t , \tag{13.76}$$

$$G_2 = \partial_x , \tag{13.77}$$

$$G_3 = \partial_y , \tag{13.78}$$

$$G_4 = \partial_z , \tag{13.79}$$

$$G_5 = t\partial_x + \partial_{v_x} , \tag{13.80}$$

$$G_6 = t\partial_y + \partial_{v_y} , \tag{13.81}$$

$$G_7 = t\partial_z + \partial_{v_z} , \tag{13.82}$$

$$G_8 = x\partial_y - y\partial x + v_x\partial_{v_y} - v_y\partial_{v_x} + H_x\partial_{H_y} - H_y\partial_{H_x} , \tag{13.83}$$

$$G_9 = y\partial_z - z\partial y + v_y\partial_{v_z} - v_z\partial_{v_y} + H_y\partial_{H_z} - H_z\partial_{H_y} , \tag{13.84}$$

$$G_{10} = z\partial_x - x\partial z + v_z\partial_{v_z} - v_x\partial_{v_z} + H_z\partial_{H_x} - H_x\partial_{H_z} , \tag{13.85}$$

$$G_{11} = x\partial_x + y\partial_y + z\partial_z - 2\rho\partial_\rho + v_x\partial_{v_x} + v_y\partial_{v_y} + v_z\partial_{v_z} , \tag{13.86}$$

$$G_{12} = t\partial_t + 2\rho\partial_\rho - (v_x\partial_{v_x} + v_y\partial_{v_y} + v_z\partial_{v_z}) , \tag{13.87}$$

$$G_{13} = 2\rho\partial_\rho + 2p\partial_p + H_x\partial_{H_x} + H_y\partial_{H_y} + H_z\partial_{H_z} . \tag{13.88}$$

Thus, the MHD equations 13.58 to 13.62 are invariant under translations G_2 through G_4, Galilean boosts G_5 through G_7, rotations G_8 through G_{10}, and dilations G_{11} through G_{13}. In contrast to the results obtained for the 1+1 and the 2+1 dimensional versions of the MHD problem, the dimension of the Lie algebra for (13.58) to (13.62) in the full 3+1 dimensions (x, y, z and time t) is independent of the value of the coefficient of viscosity κ. Our results confirm those in Fuchs [1991], and of those of Grundland and Lalague [1994a], [1994b], who computed the classical and some nonclassical symmetries of the MHD, and also classified all the subalgebras in conjugacy classes. The MHD system and our results have been used by other investigators, such as Carminati, Devitt, and Fee [1992], Reid and Wittkopf [1993], and Sherring [1993a], to test their symmetry programs.

13.5.4. Other interesting examples

Champagne and Winternitz [1985] used SYMMGRP.MAX to compute the Lie point symmetries of the Korteweg-de Vries equation with variable coefficients,

$$u_t + f(x, t)uu_x + g(x, t)u_{xxx} = 0, \tag{13.89}$$

illustrating that SYMMGRP.MAX can easily handle equations involving arbitrary functions.

Also in Champagne and Winternitz [1985], the point symmetries of a modified Kadomtsev-Petviashvili equation

$$(u_{xxx} - 2u_x^3 - 4u_t)_x - 6u_{xx}u_y + 3u_{yy} = 0, \tag{13.90}$$

in 2+1 dimensions are computed with SYMMGRP.MAX. This example was chosen because it leads to an infinite-dimensional Lie algebra involving four arbitrary functions of t.

A completely worked example of the calculation of Lie point symmetries of a system of PDEs is given in Champagne, Hereman, and Winternitz [1991]. This example shows the use of the feedback mechanism within SYMMGRP.MAX to completely solve the determining equations. It involves the Karpman equations (Karpman [1989]), for which our symmetries were independently verified with REDUCE programs by Kersten and Gragert (private communication), Sherring [1993b], and Wolf [1993].

Finally, SYMMGRP.MAX was recently used to compute Lie point symmetries of two large systems of equations representing classical field theories (Hereman, Marchildon, and Grundland [1993]). Currently, Hereman is adapting SYMMGRP.MAX for the calculation of Lie point symmetries of difference-differential equations.

13.6. Conclusion

The various programs that we reviewed need very little data and are straightforward to use provided the user has access to and knows the basics of the underlying CAS, such as MACSYMA, Maple, Mathematica, and REDUCE.

Apart from the theoretical study of the underlying mathematics, there is a need for further development and implementation of effective algorithms for generating, reducing, simplifying, and fully solving the determining equations for (classical and nonclassical) Lie point symmetries and generalized or Lie-Bäcklund symmetries.

The availability of sophisticated symbolic programs certainly will accelerate the study of symmetries of physically important systems of differential equations in classical mechanics, fluid dynamics, elasticity, and other applied areas.

Acknowledgments

While writing this survey I was able to consult with many experts in this field. I am grateful for their suggestions, comments, corrections, and criticisms which greatly helped improve the manuscript.

Sincere thanks goes to M. Ablowitz, B. Abraham-Shrauner, G. Baumann, D. Bérubé, G. Bluman, A. Bocharov, F. Cantrijn, B. Champagne, P. Clarkson, R. Conte, N. Coult, B. Fuchssteiner, V. Gerdt, K. Govinder, P. Gragert, M. Grundland, F. Gungor, A. Head, B. Herbst, S. Herod, M. Hickman, D. Holm, M. Kruskal, D. Levi, M. MacCallum, F. Mahomed, E. Mansfield, L. Marchildon, A. Mikhailov, M.C. Nucci, P. Olver, J. Ondich, G. Prince, D. Rand, G. Reid, M. Roelofs, W. Sarlet, W. Seiler, F. Schwarz, J. Sherring, W.-H. Steeb, S. Steinberg, Z. Thomova, M. Torrisi, P. Vafeades, K. Vu, T. Wilcox, P. Winternitz, T. Wolf, and C. Wulfman.

Special thanks goes to S. Collart for inviting me to write a preliminary version of this survey paper for Euromath Bulletin. It is my pleasure to thank N.H. Ibragimov for giving me the opportunity to completely rework that review paper and to publish it as a chapter in this book.

I am grateful to T. Hearn and H. Melenk for providing me with a free copy of REDUCE. Research for this survey paper was supported in part by Grant # CCR-9300978 of the National Science Foundation of the United States of America.

Table 1 — List of current symmetry programs			
Name & System	Distributor	Developer's Address	Email & Anonymous FTP
CRACK LIEPDE & APPLYSYM (REDUCE)	REDUCE Network Library	T. Wolf & A. Brand T. Wolf School Math. Sci. Queen Mary & Westfield College London E1 4NS, UK	T.Wolf@maths.qmw.ac.uk galois.maths.qmw.ac.uk /ftp/pub/crack
DELiA (Pascal)	Beaver Soft 715 Ocean View Ave Brooklyn NY 11235, USA Cost: $ 300	A. Bocharov et al. A. Bocharov Wolfram Research 100 Trade Center Dr. Urbana-Champaign IL 61820-7237, USA	alexei@wri.com
DIFFGROB2 (Maple)		E. Mansfield Dept. of Maths. Univ. of Exeter Exeter EX4 4QE United Kingdom	liz@maths.exeter.ac.uk euclid.exeter.ac.uk pub/liz
DIMSYM (REDUCE)	LaTrobe University School of Maths. Cost: $ 225	J. Sherring G. Prince School of Maths. Latrobe University Bundoora, VI 3083 Australia	matjs@lure.latrobe.edu.au G.Prince@latrobe.edu.au ftp.latrobe.edu.au /ftp/pub/dimsym
LIE (REDUCE)	CPC Program Library Belfast N. Ireland Cat. No. AABS	V. Eliseev et al. V. Eliseev Lab. Comp. Tech. Aut. JINR, Dubna Moscow Region 141980 Russia	

Table 1 cont. *List of current symmetry programs*			
Name & System	Distributor	Developer's Address	Email & Anonymous FTP
LIE (muMath) (independent)	SIMTEL	A. Head CSIRO Div. Mat. Sci. & Tech. Clayton, Victoria 3168 Australia	head@rivett.mst.csiro.au wuarchive.wustl.edu /edu/math/msdos/.. ../adv.diff.equations/lie42
Lie & LieBaecklund (Mathematica)	Wolfram Research MathSource 0202-622 0204-680	G. Baumann Abt. Math. Phys. Universität Ulm D-7900 Ulm Germany	bau@theophys.physik.uni-ulm.de mathsource.wri.com /pub/PureMath/Calculus
LIEDF/INFSYM & others (REDUCE)		P. Gragert & P. Kersten P. Kersten Dept. Appl. Math. University of Twente 7500 AE Enschede The Netherlands	gragert@math.utwente.nl kersten@math.utwente.nl
Liesymm (Maple)	Waterloo Maple Software (Packages)	J. Carminati et al. G. Fee Dept. Comp. Sci. University of Waterloo Waterloo, Canada	wmsi@daisy.uwaterloo.ca wmsi@daisy.waterloo.edu
MathSym (Mathematica)		S. Herod Program Appl. Math. University of Colorado Boulder, CO 80309, USA	sherod@newton.colorado.edu newton.colorado.edu pub/mathsym
NUSY (REDUCE)		M.C. Nucci Dept. di Mathematica Università di Perugia 06100 Perugia, Italy	nucci@gauss.dipmat.unipg.it

Table 1 cont. List of current symmetry programs			
Name & System	Distributor	Developer's Address	Email & Anonymous FTP
PDELIE (MACSYMA)	MACSYMA Out-of-Core Library	P. Vafeades Dept. of Eng. Sci. Trinity University San Antonio, TX 78212 USA	peter@engr.trinity.edu gumbo.engr.trinity.edu
SPDE & SYMSIZE (REDUCE)	REDUCE Program Lib. Rand Corp.	F. Schwarz GMD, Inst. SCAI D-53731 Sankt Augustin Germany	fritz.schwarz@gmd.de reduce-netlib@rand.org
SYMCAL (Maple & MACSYMA)		G. Reid & A. Wittkopf G. Reid Math. Dept. Univ. Brit. Columbia Vancouver, BC Canada V6T IZ2	reid@math.ubc.ca math.ubc.ca pub/reid
SYM_DE (MACSYMA)	MACSYMA Out-of-Core Library	S. Steinberg Dept. Math. & Stat. Univ.New Mexico Albuquerque, NM 87131 USA	stanly@math.unm.edu
symmgroup.c (Mathematica)		D. Bérubé & M. de Montigny D. Bérubé Centre Traitement Inform. Univ. Laval, St.-Froy Canada G1K 7P4	montigny@physics.mcgill.ca berube@genesis.ulaval.ca genesis.ulaval.ca /pub/Mathematica/symgroup
SYMMGRP.MAX (MACSYMA)	CPC Program Lib. Belfast N. Ireland Cat. No. ACBI	B. Champagne et al. W. Hereman Dept. Math. Comp. Sci. Colorado Sch. of Mines Golden, CO 80401, USA	whereman@lie.mines.colorado.edu ftp.mines.colorado.edu pub/papers/math_cs_dept/symmetry or contact: cpc@v1.am.qub.ac.uk

Table 2 Scope of symmetry programs						
Name	System	Developer(s)	Point	General.	Noncl.	Solves Det. Eqs.
CRACK	REDUCE	Wolf & Brand	-	-	-	Yes
DELiA	Pascal	Bocharov et al.	Yes	Yes	No	Yes
DIFFGROB2	Maple	Mansfield	-	-	-	Reduction
DIMSYM	REDUCE	Sherring	Yes	Yes	No	Yes
LIE	REDUCE	Eliseev et al.	Yes	Yes	No	No
LIE	muMath	Head	Yes	Yes	No	Yes
Lie	Mathematica	Baumann	Yes	No	Yes	Yes
LieBaecklund	Mathematica	Baumann	No	Yes	No	Interactive
LIEDF/INFSYM	REDUCE	Gragert & Kersten	Yes	Yes	No	Interactive
LIEPDE	REDUCE	Wolf & Brand	Yes	Yes	No	Yes

Table 2 cont.		Scope of symmetry programs				
Name	System	Developer(s)	Point	General.	Noncl.	Solves Det. Eqs.
Liesymm	Maple	Carminati et al.	Yes	No	No	Interactive
MathSym	Mathematica	Herod	Yes	No	Yes	Reduction
NUSY	REDUCE	Nucci	Yes	Yes	Yes	Interactive
PDELIE	MACSYMA	Vafeades	Yes	Yes	No	Yes
SPDE	REDUCE	Schwarz	Yes	No	No	Yes
SYMCAL	Maple/ MACSYMA	Reid & Wittkopf	-	-	-	Reduction
SYM_DE	MACSYMA	Steinberg	Yes	No	No	Partially
symgroup.c	Mathematica	Bérubé & de Montigny	Yes	No	No	No
SYMMGRP.MAX	MACSYMA	Champagne et al.	Yes	No	Yes	Interactive
SYMSIZE	REDUCE	Schwarz	-	-	-	Reduction

14

Interactive REDUCE Programs for Calculating Lie Point, Non-Classical, Lie-Bäcklund, and Approximate Symmetries of Differential Equations: Manual and Floppy Disk

14.1. Introduction

It is well known that the main obstacle to the application of the Lie group theories is the extensive calculations they involve.

Many software packages which calculate the symmetries of differential equations have been developed (see Chapter 13).

However, in the experience of the author, these are often very difficult to master and require considerable "troubleshooting" skill on the part of the user to obtain satisfactory results.

The interactive programs presented herein are designed with the unsophisticated user in mind to obviate the tedious calculations which are a necessary part of these types of analysis.

They use REDUCE version 3.3 (see Hearn [1987]) or higher,[1] which is extremely user friendly.[2]

Our interactive programs do not require an in-depth knowledge of LISP (see Fitch [1978]) or REDUCE. In fact, to use them one only needs to know a single LISP command and have a very basic familiarity with REDUCE.

---

[1] At present, the latest version is 3.5 (see Hearn [1993]).

[2] An allegoric tale on computer algebra software and mathematicians can be found in Nucci [1992b].

The programs labelled GA automatically construct the determining equations for the Lie point symmetries (see [H1], [H2]); with only minor modifications, all the other programs are derived. They calculate the non-classical (SGA) (Chapter 11 of this volume; see also Bluman and Cole [1969], Levi and Winternitz [1989]), the Lie-Bäcklund (GS) (Chapter 1 of this volume), and approximate (AGA) (Chapter 2 of this volume) symmetries of any differential equation. In the next sections we will explain all the programs in detail and give some examples.

This manual is dedicated to both the beginners and experienced researchers in the application of the Lie group theories. It is the hope of the author that its use will assist them and others in their efforts to extend the knowledge of computer algebra.

14.2.   One LISP and some REDUCE commands

In using REDUCE it is important to be aware of its limitations.

You can make your own choice for the name of a variable, but you cannot use some reserved symbols, e.g., EXP, LOG, SIN, COS, etc.[3]

REDUCE recognizes an exponential, logarithm, sine, cosine, etc.

REDUCE does not make any distinction between upper and lower case.

To tell REDUCE that a command is finished type either ";" or "$". The difference is that REDUCE will usually show on the screen the result of the command ended with ";", and will not show it if you type "$".

Another feature of REDUCE is that every command is sequentially numbered starting from 1, when using interactive mode. The result of each command is stored temporarily in a (reserved) variable called **WS**. Its content will change with the next command. For example, if you would like to look back at the result of the 13th command, you type "**WS(13);**" and REDUCE will show it on the screen.

The following list contains the REDUCE commands you need to know, with a brief explanation.

DEPEND ... , ... ;

Ex: DEPEND U,X1,X2;
      U depends on X1 and X2

DF(... , ...);

Ex: DF(U,X1);
      derive U by X1 and show the result
      DF(U,X1,3,X2,2);
      derive U three times by X1 and two times by X2

---

[3]N.B. "T" is a reserved symbol in REDUCE version 3.3.

DF(... , ...)$

Ex: DF(U,X1)$
     derive U by X1 and do not show the result

... := ... ;

Ex: DF(U,X1):=UX1;
     assign the value UX1 to the derivative of
     U by X1 and show the result
N.B. REDUCE will show **UX1**.

NUM(...);

Ex: NUM(1/U);
     evaluate the numerator of the expression in between parentheses, i.e., 1,
     and show the result

DEN(...);

Ex: DEN(1/U);
     evaluate the denominator of the expression in between parentheses,
     i.e., U, and show the result

COEFFN(... , ... ,n);

Ex: COEFFN(CU,UX1,3);
      show the coefficient of the third-power of UX1 in CU
N.B. If CU was defined to be equal to
      S1*UX1**3*LOG(U)-S2*UX1*U**2*UX2,
      REDUCE will show **S1*LOG(U)**.
      Warning: CU must be a polynomial in UX1.

COEFF(... , ... , ...);

Ex: COEFF(CU,UX1,CX1K);
      evaluate all the coefficients of UX1 in CU and call each of them
      CX1Ki where i stands for the corresponding power
N.B. If CU is the above example,
      REDUCE will give the output **C1X1K3, C1X1K1**.
      Warning: CU must be a polynomial in UX1.

SUB({... = ... , ... = ...}, ...);

Ex: SUB({DF(U,X1)=0,V=U},WW);
       replace every occurrence of DF(U,X1) and V in WW
       by 0 and U, respectively, and show the result

N.B.   If WW was defined to be equal to
       V**3-DF(U,X1)/COS(V)-K*LOG(U),
       REDUCE will give the output U^3 **-K*LOG(U)**.

The only LISP command needed is part of the following problem. Suppose we have to solve:

$$\frac{DF(S1, Y1) + 2 * S5}{2} \tag{14.1}$$

with respect to S5.[4] The solution is

$$S5 := -\frac{DF(S1, Y1)}{2} \tag{14.2}$$

We would like REDUCE to find the result by the following two steps:
SOLVE(WS,S5);

and REDUCE will call in LISP and answer in the following fashion:

$$\{S5 = -\frac{DF(S1, Y1)}{2}\}$$

At this point we use the necessary LISP command:
S5:=PART(PART(WS,1),2);

and REDUCE will answer (14.2).

14.3.   Lie point symmetries: GA programs

We refer to the existing literature for a detailed explanation of the method of calculating Lie point symmetries (e.g., Ovsiannikov [1962], [1978], Ames [1972],

---

[4]See the computer session which finds the Lie point symmetries of the Burgers equation.

Ibragimov [1983], Olver [1986], Bluman and Kumei [1989], Rogers and Ames [1989], Stephani [1989]; see also [H2], Chapter 1). First, we have to find the determining equations. This is automatically done by the program. Then the user has to call them on the screen and start interacting with the computer. It will be shown how to proceed.

A warning: these programs require some expertise in solving differential equations.

14.3.1. Description of the GA programs

Each differential equation (or system) is characterized by:

1. a number of independent variables (say n);

2. a number of dependent variables (say m);

3. an order given by the highest derivative appearing in its expression (say k).

We prefer to have different GA programs for each group, e.g., GA1O4X2 is the program which calculates the determining equations for any single partial differential equation with highest derivative of order 4 and 2 independent variables. This means that all the GA programs are basically the same apart from some additions which depend on the above list.

All of them begin with a list of **DEPEND**. For example:

DEPEND U1,X1,X2;
DEPEND U2,X1,X2;
DEPEND U3,X1,X2;

tells REDUCE that the independent variables are X1 and X2 while the dependent variables are U1, U2, and U3.[5] We call Uj and Xi the dependent and independent variables, respectively. Our goal is to find the generator of the nonextended Lie group given by:

$$\sum_{i=1}^{n} Vi(X1, ..., Xn, U1, ..., Um)\, \partial_{Xi} + \sum_{j=1}^{m} Gj(X1, ..., Xn, U1, ..., Um)\, \partial_{Uj}$$

$$(14.3)$$

with Vi, Gj unknown functions of of Xi, Uj. To construct the kth-order extension of (14.3), we need to introduce the chain rule which is not in REDUCE source. The **LET** statement can be used for this purpose, but then REDUCE will not make any distinction between partial and total derivative by Xi. To overcome this limitation, we introduce some tricks. First, we define new variables Yi such that:

---

[5]See the example of the Zakharov system.

```
DEPEND Y1,X1;
DF(Y1,X1):=1$
```

⋮

```
DEPEND Yn,Xn;
DF(Yn,Xn):=1$
```

Then we define :
```
DEPEND V1,Y1,Y2,...,Yn,U1,...,Um;
```

⋮

```
DEPEND Vn,Y1,Y2,...,Yn,U1,...,Um;
DEPEND G1,Y1,Y2,...,Yn,U1,...,Um;
```

⋮

```
DEPEND Gm,Y1,Y2,...,Yn,U1,...,Um;
```

Next we change the symbolism for the derivatives of Uj with respect to Xi. There are two reasons for doing this. The first is that the output on the screen will be neater if **DF(... , ...)** is only used for the derivatives of Vi and Gj with respect to Yi and Uj. The second is that we do not want REDUCE to get confused between our new rules and what he already knows.[6] We define new variables, e.g.:
```
DEPEND U1X1,X1,...,Xn;
DF(U1,X1):=U1X1$
```

⋮

```
DEPEND U1Xn,X1,...,Xn;
DF(U1,Xn):=U1Xn$
```

⋮

```
DEPEND UmXn,X1,...,Xn;
DF(Um,Xn):=UmXn$
```

⋮

In the end the following will be true:
```
DF(Uj,X1,h1,X2,h2,...,Xn,hn):=UjX1h1X2h2...Xnhn [7]
```

Finally, we tell REDUCE that some of these derivatives are not independent. They are actually obtained from the differential equations, e.g.:
```
U1X2X2:=-I*U1X1+U3*U1$ [8]
```

We develop the programs with the assumption that the highest derivatives of Uj

---

[6] Otherwise CATASTROPHIC ERRORS result.

[7] This heavy symbolism can be avoided by using a more sophisticated command such as MATRIX.

[8] See Zakharov system.

will be determined from the differential equations. Now REDUCE is ready to learn the chain rule, and make exceptions for Uj and their renamed "independent" derivatives, and also Yj, e.g.:[9]

FORALL Z SUCH THAT NOT(Z=U1 OR Z=U2 OR Z=Y1 OR Z=U3 OR
Z=Y2 OR Z=U1X2 OR Z=U2X2 OR Z=U1X1 OR Z=U2X1
OR Z=U3X1 OR Z=U3X2)
LET DF(Z,X1)=DF(Z,Y1)*DF(Y1,X1)+DF(Z,U1)*U1X1
+DF(Z,U2)*U2X1+DF(Z,U3)*U3X1
+DF(Z,U1X1)*U1X1X1+DF(Z,U1X2)*U1X1X2
+DF(Z,U2X1)*U2X1X1+DF(Z,U2X2)*U2X1X2
+DF(Z,U3X1)*U3X1X1+DF(Z,U3X2)*U3X1X2,
DF(Z,X2)=DF(Z,Y2)*DF(Y2,X2)+DF(Z,U1)*U1X2
+DF(Z,U2)*U2X2+DF(Z,U3)*U3X2
+DF(Z,U1X1)*U1X1X2+DF(Z,U1X2)*U1X2X2
+DF(Z,U2X1)*U2X1X2+DF(Z,U2X2)*U2X2X2
+DF(Z,U3X1)*U3X1X2+DF(Z,U3X2)*U3X2X2;

REDUCE is now ready to calculate the coefficients of the kth extension of (14.3), e.g.:[10]

G3X1:=DF(G3,X1)-U3X1*DF(V1,X1)-U3X2*DF(V2,X1)$
G3X1X1:=DF(G3X1,X1)-U3X1X1*DF(V1,X1)-U3X1X2*DF(V2,X1)$

Next REDUCE has to apply this extension to the original partial differential equations and finally construct the determining equations. We define new variables Lj, LjX1,..., LjXn,..., LjX1h1X2h2,... [11] e.g.:[12]

DEPEND L1,X1,...,Xn;
DEPEND L1X1,X1,...,Xn;
$\vdots$

and name LQ1, LQ2,... the original differential equations with Uj substituted by Lj, e.g.:[13]

LQ1:=L1X2X2+I*L1X1-L3*L1$

Then we tell REDUCE to apply the kth extension of (14.3) to LQ1,..., and name each result TEQ1,..., e.g.:[14]

---

[9]See Zakharov system. Note that the symbolism for U_{X1X1} and U_{X2X2} has been changed.
[10]See Zakharov system.
[11]We introduce these new variables because certain derivatives of Uj were previously derived from the differential equations, but to construct the determining equations we now need to have all the derivatives of Uj independent from each other.
[12]See Zakharov system.
[13]See Zakharov system.
[14]See Zakharov system.

TEQ1:=DF(LQ1,L1)*G1+DF(LQ1,L3)*G3+DF(LQ1,L1X1)*G1X1
+DF(LQ1,L1X2X2)*G1X2X2$[15]

The last step will be to substitute all the Lj with the corresponding Uj, and their derivatives as well, e.g.:[16]
    L1:=U1$
    L1X1:=U1X1$

Finally, we name CU1, CU2, ... the corresponding numerators of TEQ1, TEQ2,... and tell REDUCE to construct the coefficients of the independent derivatives of Uj (i.e., the determining equations), e.g.:[17]
    COEFF(CU1,U1X2,C1U1X2K);

for which we get the following output:
    ***** ID fill no longer supported — use lists instead
    *** C1U1X2K3 C1U1X2K2 C1U1X2K1 C1U1X2K0 are non zero
    3

When any of the programs GA is loaded, REDUCE will automatically run it and construct the determining equations. At this point the user will start to interact with him. The following examples show how this is done.[18]

14.3.2.   Example: Burgers equation

Let us consider the Burgers equation:

$$U_{X1} = U_{X2X2} + UU_{X2} \qquad\qquad (14.4)$$

Our goal is to find V1(Y1,Y2,U), V2(Y1,Y2,U), and G(Y1,Y2,U). The result is known and can be found in Olver [1986]. We will also calculate the symmetry group generators (GEN1, GEN2,...). This is the corresponding computer session.
    REDUCE 3.3, 15-Jul-87 ...
    1:
in "GA1O2X2.BUR";
    DEPEND U,X1,X2;
    DEPEND V1,Y1,Y2,U;

---

[15]We did not want to impose unnecessary calculations on REDUCE. Therefore the useless terms of the kth-extension here are omitted.

[16]See Zakharov system.

[17]See Zakharov system.

[18]Note that SS1,SS2,SS3,... are defined to be arbitrary functions of Y1 and Y2; S1,S3,... arbitrary functions of Y1; S2,S4,... arbitrary functions of Y2; C1,C2,C3, ... arbitrary constants.

```
DEPEND V2,Y1,Y2,U;
DEPEND G,Y1,Y2,U;
DEPEND SS1,Y1,Y2;
DEPEND SS2,Y1,Y2;
DEPEND SS3,Y1,Y2;
DEPEND SS4,Y1,Y2;
DEPEND SS5,Y1,Y2;
DEPEND SS6,Y1,Y2;
DEPEND SS7,Y1,Y2;
DEPEND S1,Y1;
DEPEND S3,Y1;
DEPEND S5,Y1;
DEPEND S7,Y1;
DEPEND S9,Y1;
DEPEND S2,Y2;
DEPEND S4,Y2;
DEPEND S6,Y2;
DEPEND S8,Y2;
DEPEND S10,Y2;
DEPEND Y1,X1;
DEPEND Y2,X2;
DF(Y1,X1):=1$
DF(Y2,X2):=1$
DEPEND UX1,X1,X2;
DEPEND UX2,X1,X2;
DEPEND UX1X1,X1,X2;
DEPEND UX1X2,X1,X2;
DEPEND UX2X2,X1,X2;[19]
DF(U,X1):=UX1$
DF(U,X2):=UX2$
DF(UX1,X1):=UX1X1$
DF(UX1,X2):=UX1X2$
DF(UX2,X1):=UX1X2$
DF(UX2,X2):=UX2X2$
UX2X2:=UX1-U*UX2$
    FORALL Z SUCH THAT NOT(Z=U OR Z=Y1 OR Z=Y2 OR Z=UX1 OR
Z=UX2)
    LET DF(Z,X1)=DF(Z,Y1)*DF(Y1,X1)+DF(Z,U)*UX1+DF(Z,UX1)*UX1X1+
DF(Z,UX2)*UX1X2,
    DF(Z,X2)=DF(Z,Y2)*DF(Y2,X2)+DF(Z,U)*UX2+DF(Z,UX1)*UX1X2+
DF(Z,UX2)*UX2X2;
```

---

[19] Here the symbolism for U_{X2X2} has been changed.

```
GX1:=DF(G,X1)-UX1*DF(V1,X1)-UX2*DF(V2,X1)$
GX2:=DF(G,X2)-UX1*DF(V1,X2)-UX2*DF(V2,X2)$
GX1X1:=DF(GX1,X1)-UX1X1*DF(V1,X1)-UX1X2*DF(V2,X1)$
GX1X2:=DF(GX1,X2)-UX1X1*DF(V1,X2)-UX1X2*DF(V2,X2)$
GX2X2:=DF(GX2,X2)-UX1X2*DF(V1,X2)-UX2X2*DF(V2,X2)$
DEPEND L,X1,X2;
DEPEND LX1,X1,X2;
DEPEND LX2,X1,X2;
DEPEND LX1X1,X1,X2;
DEPEND LX1X2,X1,X2;
DEPEND LX2X2,X1,X2;
LQ:=-LX1+L*LX2+LX2X2$
TEQ1:=DF(LQ,Y1)*V1+DF(LQ,Y2)*V2+DF(LQ,L)*G
+DF(LQ,LX1)*GX1+DF(LQ,LX2)*GX2
+DF(LQ,LX1X1)*GX1X1+DF(LQ,LX2X2)*GX2X2
+DF(LQ,LX1X2)*GX1X2$
L:=U$
LX1:=UX1$
LX2:=UX2$
LX1X1:=UX1X1$
LX1X2:=UX1X2$
LX2X2:=UX2X2$
CU:=NUM(TEQ1)$
COEFF(CU,UX1X2,CX1X2K);
***** ID fill no longer supported — use lists instead
*** CX1X2K1 CX1X2K0 are non zero
1
COEFF(CU,UX1,CX1K);
***** ID fill no longer supported — use lists instead
*** CX1K1 CX1K0 are non zero
1
COEFF(CU,UX2,CX2K);
***** ID fill no longer supported — use lists instead
*** CX2K3 CX2K2 CX2K1 CX2K0 are non zero
3
END;
2:[20]
cx1x2k1;
 - 2*(DF(V1,U)*UX2 + DF(V1,Y2))
 3:
V1:=S1$
```

[20]The interactive session begins.

```
   4:
cx1x2k1;[21]
   0
   5:
cx1k1;
   DF(S1,Y1) - 2*DF(V2,U)*UX2 - 2*DF(V2,Y2)
   6:
V2:=SS1$
   7:
cx1k1;
   DF(S1,Y1) - 2*DF(SS1,Y2)
   8:
SS1:=DF(S1,Y1)*Y2/2+S3$
   9:
cx1k1;
   0
   10:
cx2k3;
   0
   11:
cx2k2;
   DF(G,U,2)
   12:
G:=SS2*U+SS3$
   13:
cx2k2;
   0
   14:
cx2k1;
   (2*DF(S3,Y1) + DF(S1,Y1,2)*Y2 + DF(S1,Y1)*U + 4*DF(SS2,Y2)
   + 2*U*SS2 + 2*SS3)/2
   15:
COEFFN(WS,U,1);
```

$$\frac{DF(S1, Y1) + 2 * SS2}{2}$$

```
   16:
WS;
```

$$\frac{DF(S1, Y1) + 2 * SS2}{2}$$

---

[21] It is always wise to check the result.

17:
SOLVE(WS,SS2);

$$\{SS2 = -\frac{DF(S1, Y1)}{2}\}$$

18:
SS2:=PART(PART(WS,1),2)$
19:
WS;

$$-\frac{DF(S1, Y1)}{2}$$

20:
cx2k1;

$$\frac{2 * DF(S3, Y1) + DF(S1, Y1, 2) * Y2 + 2 * SS3}{2}$$

21:
SOLVE(WS,SS3);

$$\{SS3 = -\frac{2 * DF(S3, Y1) + DF(S1, Y1, 2) * Y2}{2}\}$$

22:
SS3:=PART(PART(WS,1),2)$
23:
cx2k1;
   0
24:
CX2K0;

$$\frac{2 * DF(S3, Y1, 2) + DF(S1, Y1, 3) * Y2}{2}$$

25:
S1:=C1*Y1**2+C2*Y1+C3$
26:
CX2K0;
   DF(S3,Y1,2)
27:
S3:=C4*Y1+C5$

28:
CX2K0;
   0
29:
CU;
   0[22]
30:
V1;[23]
   $Y1^2 *C1 + Y1*C2 + C3$
31:
V2;

$$\frac{2 * Y2 * Y1 * C1 + Y2 * C2 + 2 * Y1 * C4 + 2 * C5}{2}$$

32:
G;

$$-\frac{2 * U * Y1 * C1 + U * C2 + 2 * Y2 * C1 + 2 * C4}{2}$$

33:
GEN1:=DF(V1,C1)*DY1+DF(V2,C1)*DY2+DF(G,C1)*DU;[24]
   $GEN1 := - (U*Y1*DU - Y2*Y1*DY2 + Y2*DU - Y1^2*DY1)$
34:
GEN2:=DF(V1,C2)*DY1+DF(V2,C2)*DY2+DF(G,C2)*DU;

$$GEN2 := -\frac{U * DU - Y2 * DY2 - 2 * Y1 * DY1}{2}$$

35:
GEN3:=DF(V1,C3)*DY1+DF(V2,C3)*DY2+DF(G,C3)*DU;
   $GEN3 := DY1$
36:
GEN4:=DF(V1,C4)*DY1+DF(V2,C4)*DY2+DF(G,C4)*DU;
   $GEN4 := Y1*DY2 - DU$
37:
GEN5:=DF(V1,C5)*DY1+DF(V2,C5)*DY2+DF(G,C5)*DU;
   $GEN5 := DY2$

---

[22] We have reached our goal, i.e., CU=0

[23] The following three commands ask for the final expressions of V1, V2, G, respectively.

[24] These last commands ask for the symmetry group generators GEN1, GEN2, ...

38:
BYE;[25]
  *** END OF RUN

14.3.3. Example: $u_{tt} = [f(u)u_x]_x$

We consider the following equation (see Ames, Lohner, and Adams [1981]):

$$U_{X1X1} = [F(U)U_{X2}]_{X2} \qquad\qquad (14.5)$$

In this case F(U) is an arbitrary function of U to start with. This is the corresponding computer session.
  REDUCE 3.3, 15-Jul-87 ...
  1:
IN "GA1O2X2.FU";
  DEPEND U,X1,X2;
  DEPEND V1,Y1,Y2,U;
  DEPEND V2,Y1,Y2,U;
  DEPEND G,Y1,Y2,U;
  DEPEND SS1,Y1,Y2;
  DEPEND SS2,Y1,Y2;
  DEPEND SS3,Y1,Y2;
  DEPEND SS4,Y1,Y2;
  DEPEND SS5,Y1,Y2;
  DEPEND SS6,Y1,Y2;
  DEPEND SS7,Y1,Y2;
  DEPEND S1,Y1;
  DEPEND S3,Y1;
  DEPEND S5,Y1;
  DEPEND S7,Y1;
  DEPEND S9,Y1;
  DEPEND S2,Y2;
  DEPEND S4,Y2;
  DEPEND S6,Y2;
  DEPEND S8,Y2;
  DEPEND S10,Y2;
  DEPEND Y1,X1;
  DEPEND Y2,X2;
  DF(Y1,X1):=1$
  DF(Y2,X2):=1$
  DEPEND UX1,X1,X2;

---

[25] It tells REDUCE to terminate.

```
DEPEND UX2,X1,X2;
DEPEND UX1X1,X1,X2;
DEPEND UX1X2,X1,X2;
DEPEND UX2X2,X1,X2;
DF(U,X1):=UX1$
DF(U,X2):=UX2$
DF(UX1,X1):=UX1X1$
DF(UX1,X2):=UX1X2$
DF(UX2,X1):=UX1X2$
DF(UX2,X2):=UX2X2$
DEPEND F,U;
UX1X1:=DF(F,U)*UX2**2+F*UX2X2$
FORALL Z SUCH THAT NOT(Z=U OR Z=Y1 OR
Z=Y2 OR Z=UX1 OR Z=UX2)
LET DF(Z,X1)=DF(Z,Y1)*DF(Y1,X1)+DF(Z,U)*UX1
+DF(Z,UX1)*UX1X1+DF(Z,UX2)*UX1X2,
DF(Z,X2)=DF(Z,Y2)*DF(Y2,X2)+DF(Z,U)*UX2
+DF(Z,UX1)*UX1X2+DF(Z,UX2)*UX2X2;
GX1:=DF(G,X1)-UX1*DF(V1,X1)-UX2*DF(V2,X1)$
GX2:=DF(G,X2)-UX1*DF(V1,X2)-UX2*DF(V2,X2)$
GX1X1:=DF(GX1,X1)-UX1X1*DF(V1,X1)-UX1X2*DF(V2,X1)$
GX1X2:=DF(GX1,X2)-UX1X1*DF(V1,X2)-UX1X2*DF(V2,X2)$
GX2X2:=DF(GX2,X2)-UX1X2*DF(V1,X2)-UX2X2*DF(V2,X2)$
DEPEND L,X1,X2;
DEPEND LX1,X1,X2;
DEPEND LX2,X1,X2;
DEPEND LX1X1,X1,X2;
DEPEND LX1X2,X1,X2;
DEPEND LX2X2,X1,X2;
LQ:=-LX1X1+DF(F,U)*LX2**2+F*LX2X2$
TEQ1:=DF(LQ,Y1)*V1+DF(LQ,Y2)*V2+DF(LQ,U)*G
+DF(LQ,LX1)*GX1+DF(LQ,LX2)*GX2
+DF(LQ,LX1X1)*GX1X1+DF(LQ,LX2X2)*GX2X2
+DF(LQ,LX1X2)*GX1X2$
L:=U$
LX1:=UX1$
LX2:=UX2$
LX1X1:=UX1X1$
LX1X2:=UX1X2$
LX2X2:=UX2X2$
CU:=NUM(TEQ1)$
COEFF(CU,UX1X2,CX1X2K);
***** ID fill no longer supported — use lists instead
*** CX1X2K1 CX1X2K0 are non zero
```

```
  1
COEFF(CU,UX2X2,CX2X2K);
***** ID fill no longer supported — use lists instead
*** CX2X2K1 CX2X2K0 are non zero
  1
COEFF(CU,UX1,CX1K);
***** ID fill no longer supported — use lists instead
*** CX1K3 CX1K2 CX1K1 CX1K0 are non zero
  3
COEFF(CU,UX2,CX2K);
***** ID fill no longer supported — use lists instead
*** CX2K3 CX2K2 CX2K1 CX2K0 are non zero
  3
END;
  2:
CX1X2K1;
  2*(DF(V2,U)*UX1 + DF(V2,Y1) - DF(V1,U)*F*UX2 - DF(V1,Y2)*F)
  3:
V1:=S1$
  4:
V2:=S2$
  5:
CX1X2K1;
  0
  6:
CX1K3;
  0
  7:
CX2K3;
  0
  8:
CX1K2;
  - DF(G,U,2)
  9:
G:=SS1*U+SS2$
  10:
CX1K2;
  0
  11:
CX1K1;
  DF(S1,Y1,2) - 2*DF(SS1,Y1)
  12:
SS1:=DF(S1,Y1)/2+S4$
```

```
   13:
CX1K1;
   0
   14:
CX2X2K1;
   (DF(F,U)*DF(S1,Y1)*U + 2*DF(F,U)*U*S4 + 2*DF(F,U)*SS2 - 4*DF
   (S2,Y2)*F + 4*DF(S1,Y1)*F)/2
   15:
CX2K2;
   (DF(F,U,2)*DF(S1,Y1)*U + 2*DF(F,U,2)*U*S4 + 2*DF(F,U,2)*SS2 -
   4*DF(F,U)*DF(S2,Y2) + 5*DF(F,U)*DF(S1,Y1) + 2*DF(F,U)*S4)/2
   16:
DF(CX2X2K1,U)-CX2K2;
   0
   17:
CX2K1;
   2*DF(F,U)*DF(S4,Y2)*U + 2*DF(F,U)*DF(SS2,Y2) + 2*DF(S4,Y2)*F
   -DF(S2,Y2,2)*F
   18:
CX2X2K1;
   (DF(F,U)*DF(S1,Y1)*U + 2*DF(F,U)*U*S4 + 2*DF(F,U)*SS2 - 4*DF
   (S2,Y2)*F + 4*DF(S1,Y1)*F)/2
   19:
COEFFN(CX2K0,UX2X2,0);
   (2*DF(S4,Y2,2)*F*U - DF(S1,Y1,3)*U + 2*DF(SS2,Y2,2)*F - 2*DF
   (SS2,Y1,2))/2
   20:
SOLVE(WS,F);
```

$$\{F = \frac{DF(S1,Y1,3)*U + 2*DF(SS2,Y1,2)}{2*(DF(S4,Y2,2)*U + DF(SS2,Y2,2))}\}$$

```
   21:[26]
V1;
   S1
   22:
V2;
   S2
   23:
G;
```

---

[26] At this point different choices for F(U) could be made. They are left to the diligent reader.

$$\frac{DF(S1, Y1) * U + 2 * U * S4 + 2 * SS2}{2}$$

24:
BYE;
  *** END OF RUN

14.3.4.   Example: Boussinesq equation

Let us consider the Boussinesq equation (see Levi and Winternitz [1989]):

$$U_{X1X1} + UU_{X2X2} + U_{X2}^2 + U_{X2X2X2X2} = 0 \qquad (14.6)$$

This is the corresponding computer session.
  REDUCE 3.3, 15-Jul-87 ...
  1:
IN "GA1O4X2.BOU";
  DEPEND U,X1,X2;
  DEPEND V1,Y1,Y2,U;
  DEPEND V2,Y1,Y2,U;
  DEPEND G,Y1,Y2,U;
  DEPEND SS1,Y1,Y2;
  DEPEND SS2,Y1,Y2;
  DEPEND SS3,Y1,Y2;
  DEPEND SS4,Y1,Y2;
  DEPEND SS5,Y1,Y2;
  DEPEND SS6,Y1,Y2;
  DEPEND SS7,Y1,Y2;
  DEPEND SS8,Y1,Y2;
  DEPEND SS9,Y1,Y2;
  DEPEND S1,Y1;
  DEPEND S3,Y1;
  DEPEND S5,Y1;
  DEPEND S7,Y1;
  DEPEND S9,Y1;
  DEPEND S2,Y2;
  DEPEND S4,Y2;
  DEPEND S6,Y2;
  DEPEND S8,Y2;
  DEPEND S10,Y2;
  DEPEND Y1,X1;

```
DEPEND Y2,X2;
DF(Y1,X1):=1$
DF(Y2,X2):=1$
DEPEND UX1,X1,X2;
DEPEND UX2,X1,X2;
DEPEND UX1X1,X1,X2;
DEPEND UX1X2,X1,X2;
DEPEND UX2X2,X1,X2;
DEPEND UX2X2X2,X1,X2;
DEPEND UX1X2X2,X1,X2;
DEPEND UX1X23,X1,X2;
DF(U,X1):=UX1$
DF(U,X2):=UX2$
DF(UX1,X1):=UX1X1$
DF(UX1,X2):=UX1X2$
DF(UX2,X1):=UX1X2$
DF(UX2,X2):=UX2X2$
DF(UX1X2,X2):=UX1X2X2$
DF(UX2X2,X1):=UX1X2X2$
DF(UX2X2,X2):=UX2X2X2$
DF(UX2X2X2,X2):=UX24$
UX24:=-UX1X1-UX2**2-U*UX2X2$
FORALL Z SUCH THAT NOT(Z=U OR Z=Y1 OR Z=UX1X1 OR
Z=Y2 OR Z=UX1 OR Z=UX2 OR Z=UX1X2 OR Z=UX2X2
OR Z=UX1X2X2 OR Z=UX2X2X2)
LET DF(Z,X1)=DF(Z,Y1)*DF(Y1,X1)+DF(Z,U)*UX1+DF(Z,UX1)*UX1X1
+DF(Z,UX2)*UX1X2,
DF(Z,X2)=DF(Z,Y2)*DF(Y2,X2)+DF(Z,U)*UX2+DF(Z,UX1)*UX1X2
+DF(Z,UX2)*UX2X2+DF(Z,UX1X2)*UX1X2X2+DF(Z,UX2X2)*UX2X2X2
+DF(Z,UX1X2X2)*UX1X23+DF(Z,UX2X2X2)*UX24;
GX1:=DF(G,X1)-UX1*DF(V1,X1)-UX2*DF(V2,X1)$
GX2:=DF(G,X2)-UX1*DF(V1,X2)-UX2*DF(V2,X2)$
GX1X1:=DF(GX1,X1)-UX1X1*DF(V1,X1)-UX1X2*DF(V2,X1)$
GX2X2:=DF(GX2,X2)-UX2X2*DF(V2,X2)-UX1X2*DF(V1,X2)$
GX2X2X2:=DF(GX2X2,X2)-UX2X2X2*DF(V2,X2)-UX1X2X2*DF(V1,X2)$
GX24:=DF(GX2X2X2,X2)-UX24*DF(V2,X2)-UX1X23*DF(V1,X2)$
DEPEND L,X1,X2;
DEPEND LX1,X1,X2;
DEPEND LX2,X1,X2;
DEPEND LX1X1,X1,X2;
DEPEND LX2X2,X1,X2;
DEPEND LX2X2X2,X1,X2;
DEPEND LX24,X1,X2;
LQ:=LX24+LX1X1+LX2**2+L*LX2X2$
```

```
TEQ:=DF(LQ,LX1)*GX1+DF(LQ,LX2X2X2)*GX2X2X2
+DF(LQ,L)*G+DF(LQ,LX2)*GX2+DF(LQ,LX24)*GX24
+DF(LQ,LX2X2)*GX2X2+DF(LQ,LX1X1)*GX1X1$
L:=U$
LX1:=UX1$
LX2:=UX2$
LX1X1:=UX1X1$
LX2X2:=UX2X2$
LX2X2X2:=UX2X2X2$
LX24:=UX24$
CU:=NUM(TEQ)$
COEFF(CU,UX1X23,CX1X23K);
***** ID fill no longer supported — use lists instead
*** CX1X23K1 CX1X23K0 are non zero
1
COEFF(CU,UX2X2X2,CX23K);
***** ID fill no longer supported — use lists instead
*** CX23K1 CX23K0 are non zero
1
COEFF(CU,UX1X2X2,CX1X22K);
***** ID fill no longer supported — use lists instead
*** CX1X22K1 CX1X22K0 are non zero
1
COEFF(CU,UX1X2,CX1X2K);
***** ID fill no longer supported — use lists instead
*** CX1X2K1 CX1X2K0 are non zero
1
COEFF(CU,UX2X2,CX22K);
***** ID fill no longer supported — use lists instead
*** CX22K2 CX22K1 CX22K0 are non zero
2
COEFF(CU,UX1X1,CX12K);
***** ID fill no longer supported — use lists instead
*** CX12K1 CX12K0 are non zero
1
COEFF(CU,UX1,CX1K);
***** ID fill no longer supported — use lists instead
*** CX1K3 CX1K2 CX1K1 CX1K0 are non zero
3
COEFF(CU,UX2,CX2K);
***** ID fill no longer supported — use lists instead
*** CX2K5 CX2K4 CX2K3 CX2K2 CX2K1 CX2K0 are non zero
5
END;
```

```
   2:
CX1X23K1;
   - 4*(DF(V1,U)*UX2 + DF(V1,Y2))
   3:
V1:=S1$
   4:
CX1X23K1;
   0
   5:
CX23K1;
   2*(2*DF(G,U,Y2) + 2*DF(G,U,2)*UX2 - 8*DF(V2,U,Y2)*UX2 - 5*DF
   (V2,U,2)*UX2² - 5*DF(V2,U)*UX2X2 - 3*DF(V2,Y2,2))
   6:
COEFFN(WS,UX2X2,1);
   - 10*DF(V2,U)
   7:
V2:=SS1$
   8:
CX23K1;
   2*(2*DF(G,U,Y2) + 2*DF(G,U,2)*UX2 - 3*DF(SS1,Y2,2))
   9:
G:=SS2*U+SS3$
   10:
CX23K1;
   2*(2*DF(SS2,Y2) - 3*DF(SS1,Y2,2))
   11:
CX1X22K1;
   0
   12:
CX1X2K1;
   - 2*DF(SS1,Y1)
   13:
SS1:=S2$
   14:
CX23K1;
   - 2*(3*DF(S2,Y2,2) - 2*DF(SS2,Y2))
   15:
SS2:=3*DF(S2,Y2)/2+S3$
   16:
CX23K1;
   0
   17:
CX1X2K1;
```

  0
   18:
CX22K2;
   0
   19:
CX22K1;

$$\frac{10 * DF(S2, Y2, 3) + 7 * DF(S2, Y2) * U + 2 * U * S3 + 2 * SS3}{2}$$

   20:
COEFFN(WS,U,1);

$$\frac{7 * DF(S2, Y2) + 2 * S3}{2}$$

   21:
S3:=-7*DF(S2,Y2)/2$
   22:
CX22K1;
   - (16*DF(S2,Y2,3) - SS3)
   23:
SS3:=16*DF(S2,Y2,3)$
   24:
CX22K1;
   0
   25:
CX12K1;
   2*(2*DF(S2,Y2) - DF(S1,Y1))
   26:
S1:=C1*Y1+C2$
   27:
CX12K1;
   2*(2*DF(S2,Y2) - C1)
   28:
S2:=C1*Y2/2+C3$
   29:
CX12K1;
   0
   30:
CX1K3;
   0
   31:
CX1K2;

```
   0
  32:
CX1K1;
   0
  33:
CX2K5;
   0
  34:
CX2K4;
   0
  35:
CX2K3;
   0
  36:
CX2K2;
   0
  37:
CX2K1;
   0
  38:
CX2K0;
   0
  39:
CU;
   0
  40:
V1;
   Y1*C1 + C2
  41:
V2;
```

$$\frac{2 * C3 + Y2 * C1}{2}$$

```
  42:
G;
  - U*C1
  43:
GEN1:=DF(V1,C1)*DY1+DF(V2,C1)*DY2+DF(G,C1)*DU;
```

$$GEN1 := -\frac{2 * U * DU - Y2 * DY2 - 2 * Y1 * DY1}{2}$$

```
   44:
GEN2:=DF(V1,C2)*DY1+DF(V2,C2)*DY2+DF(G,C2)*DU;
   GEN2 := DY1
   45:
GEN3:=DF(V1,C3)*DY1+DF(V2,C3)*DY2+DF(G,C3)*DU;
   GEN3 := DY2
   46:
BYE;
   *** END OF RUN
```

14.3.5. Example: Zakharov system

We consider the following system (see Goldstein and Infeld [1984]):

$$I\, U1_{X1} + U1_{X2X2} = U3\, U1$$

$$-I\, U2_{X1} + U2_{X2X2} = U3\, U2 \tag{14.7}$$

$$U3_{X1X1} - U3_{X2X2} = (U1\, U2)_{X2X2}$$

This is the corresponding computer session.
   REDUCE 3.3, 15-Jul-87 ...
   1:

```
IN "GA3O2X2.ZAK";
   DEPEND U1,X1,X2;
   DEPEND U2,X1,X2;
   DEPEND U3,X1,X2;
   DEPEND V1,Y1,Y2,U1,U2,U3;
   DEPEND V2,Y1,Y2,U1,U2,U3;
   DEPEND G1,Y1,Y2,U1,U2,U3;
   DEPEND G2,Y1,Y2,U1,U2,U3;
   DEPEND G3,Y1,Y2,U1,U2,U3;
   DEPEND SS1,Y1,Y2;
   DEPEND SS2,Y1,Y2;
   DEPEND SS3,Y1,Y2;
   DEPEND SS4,Y1,Y2;
   DEPEND SS5,Y1,Y2;
   DEPEND SS6,Y1,Y2;
   DEPEND SS7,Y1,Y2;
   DEPEND SS8,Y1,Y2;
   DEPEND SS9,Y1,Y2;
   DEPEND S1,Y1;
```

```
DEPEND S3,Y1;
DEPEND S5,Y1;
DEPEND S7,Y1;
DEPEND S9,Y1;
DEPEND S2,Y2;
DEPEND S4,Y2;
DEPEND S6,Y2;
DEPEND S8,Y2;
DEPEND Y1,X1;
DEPEND Y2,X2;
DF(Y1,X1):=1$
DF(Y2,X2):=1$
DEPEND U1X1,X1,X2;
DEPEND U2X1,X1,X2;
DEPEND U3X1,X1,X2;
DEPEND U1X2,X1,X2;
DEPEND U2X2,X1,X2;
DEPEND U3X2,X1,X2;
DEPEND U1X1X1,X1,X2;
DEPEND U1X1X2,X1,X2;
DEPEND U2X1X1,X1,X2;
DEPEND U2X1X2,X1,X2;
DEPEND U3X1X1,X1,X2;
DEPEND U3X1X2,X1,X2;
DF(U1,X1):=U1X1$
DF(U2,X1):=U2X1$
DF(U3,X1):=U3X1$
DF(U1,X2):=U1X2$
DF(U2,X2):=U2X2$
DF(U3,X2):=U3X2$
DF(U1X1,X1):=U1X1X1$
DF(U1X1,X2):=U1X1X2$
DF(U1X2,X1):=U1X1X2$
DF(U1X2,X2):=U1X2X2$
DF(U2X1,X1):=U2X1X1$
DF(U2X1,X2):=U2X1X2$
DF(U2X2,X1):=U2X1X2$
DF(U2X2,X2):=U2X2X2$
DF(U3X1,X1):=U3X1X1$
DF(U3X1,X2):=U3X1X2$
DF(U3X2,X1):=U3X1X2$
DF(U3X2,X2):=U3X2X2$
U1X2X2:=-I*U1X1+U3*U1$
U2X2X2:=I*U2X1+U3*U2$
```

```
BA:=U1*U2$
BA1:=DF(BA,X2)$
U3X2X2:=U3X1X1-DF(BA1,X2)$
FORALL Z SUCH THAT NOT(Z=U1 OR Z=U2 OR Z=Y1 OR Z=U3
OR Z=Y2 OR Z=U1X2 OR Z=U2X2 OR Z=U1X1 OR Z=U2X1
OR Z=U3X1 OR Z=U3X2)
LET DF(Z,X1)=DF(Z,Y1)*DF(Y1,X1)+DF(Z,U1)*U1X1
+DF(Z,U2)*U2X1+DF(Z,U3)*U3X1
+DF(Z,U1X1)*U1X1X1+DF(Z,U1X2)*U1X1X2
+DF(Z,U2X1)*U2X1X1+DF(Z,U2X2)*U2X1X2
+DF(Z,U3X1)*U3X1X1+DF(Z,U3X2)*U3X1X2,
DF(Z,X2)=DF(Z,Y2)*DF(Y2,X2)+DF(Z,U1)*U1X2
+DF(Z,U2)*U2X2+DF(Z,U3)*U3X2
+DF(Z,U1X1)*U1X1X2+DF(Z,U1X2)*U1X2X2
+DF(Z,U2X1)*U2X1X2+DF(Z,U2X2)*U2X2X2
+DF(Z,U3X1)*U3X1X2+DF(Z,U3X2)*U3X2X2;
G1X1:=DF(G1,X1)-U1X1*DF(V1,X1)-U1X2*DF(V2,X1)$
G2X1:=DF(G2,X1)-U2X1*DF(V1,X1)-U2X2*DF(V2,X1)$
G3X1:=DF(G3,X1)-U3X1*DF(V1,X1)-U3X2*DF(V2,X1)$
G1X2:=DF(G1,X2)-U1X1*DF(V1,X2)-U1X2*DF(V2,X2)$
G2X2:=DF(G2,X2)-U2X1*DF(V1,X2)-U2X2*DF(V2,X2)$
G3X2:=DF(G3,X2)-U3X1*DF(V1,X2)-U3X2*DF(V2,X2)$
G1X2X2:=DF(G1X2,X2)-U1X1X2*DF(V1,X2)-U1X2X2*DF(V2,X2)$
G2X2X2:=DF(G2X2,X2)-U2X1X2*DF(V1,X2)-U2X2X2*DF(V2,X2)$
G3X1X1:=DF(G3X1,X1)-U3X1X1*DF(V1,X1)-U3X1X2*DF(V2,X1)$
G3X2X2:=DF(G3X2,X2)-U3X1X2*DF(V1,X2)-U3X2X2*DF(V2,X2)$
DEPEND L1,X1,X2;
DEPEND L1X1,X1,X2;
DEPEND L1X2,X1,X2;
DEPEND L1X2X2,X1,X2;
DEPEND L2,X1,X2;
DEPEND L2X1,X1,X2;
DEPEND L2X2,X1,X2;
DEPEND L2X2X2,X1,X2;
DEPEND L3,X1,X2;
DEPEND L3X1,X1,X2;
DEPEND L3X2,X1,X2;
DEPEND L3X2X2,X1,X2;
DEPEND L3X1X1,X1,X2;
LQ1:=L1X2X2+I*L1X1-L3*L1$
LQ2:=L2X2X2-I*L2X1-L3*L2$
LQ3:=L3X2X2-L3X1X1+L1X2X2*L2+2*L1X2*L2X2+L1*L2X2X2$
TEQ1:=DF(LQ1,L1)*G1+DF(LQ1,L3)*G3+DF(LQ1,L1X1)*G1X1
+DF(LQ1,L1X2X2)*G1X2X2$
```

```
TEQ2:=DF(LQ2,L2)*G2+DF(LQ2,L3)*G3+DF(LQ2,L2X1)*G2X1
+DF(LQ2,L2X2X2)*G2X2X2$
TEQ3:=DF(LQ3,L1)*G1+DF(LQ3,L2)*G2+DF(LQ3,L1X2)*G1X2
+DF(LQ3,L2X2)*G2X2+DF(LQ3,L3X1X1)*G3X1X1
+DF(LQ3,L3X2X2)*G3X2X2+DF(LQ3,L1X2X2)*G1X2X2
+DF(LQ3,L2X2X2)*G2X2X2$
L1:=U1$
L1X1:=U1X1$
L1X2:=U1X2$
L1X2X2:=U1X2X2$
L2:=U2$
L2X1:=U2X1$
L2X2:=U2X2$
L2X2X2:=U2X2X2$
L3:=U3$
L3X1:=U3X1$
L3X2:=U3X2$
L3X1X1:=U3X1X1$
L3X2X2:=U3X2X2$
CU1:=NUM(TEQ1)$
CU2:=NUM(TEQ2)$
CU3:=NUM(TEQ3)$
COEFF(CU1,U1X1X2,C1U1X1X2K);
***** ID fill no longer supported — use lists instead
*** C1U1X1X2K1 C1U1X1X2K0 are non zero
1
COEFF(CU1,U1X1,C1U1X1K);
***** ID fill no longer supported — use lists instead
*** C1U1X1K2 C1U1X1K1 C1U1X1K0 are non zero
2
COEFF(CU1,U1X2,C1U1X2K);
***** ID fill no longer supported — use lists instead
*** C1U1X2K3 C1U1X2K2 C1U1X2K1 C1U1X2K0 are non zero
3
COEFF(CU2,U2X1X2,C2U2X1X2K);
***** ID fill no longer supported — use lists instead
*** C2U2X1X2K1 C2U2X1X2K0 are non zero
1
COEFF(CU2,U2X1,C2U2X1K);
***** ID fill no longer supported — use lists instead
*** C2U2X1K2 C2U2X1K1 C2U2X1K0 are non zero
2
COEFF(CU2,U2X2,C2U2X2K);
***** ID fill no longer supported — use lists instead
```

*** C2U2X2K3 C2U2X2K2 C2U2X2K1 C2U2X2K0 are non zero
3
COEFF(CU3,U3X1X2,C3U3X1X2K);
***** ID fill no longer supported — use lists instead
*** C3U3X1X2K1 C3U3X1X2K0 are non zero
1
COEFF(CU3,U3X1X1,C3U3X1X1K);
***** ID fill no longer supported — use lists instead
*** C3U3X1X1K1 C3U3X1X1K0 are non zero
1
COEFF(CU3,U1X1,C3U1X1K);
***** ID fill no longer supported — use lists instead
*** C3U1X1K2 C3U1X1K1 C3U1X1K0 are non zero
2
COEFF(CU3,U1X2,C3U1X2K);
***** ID fill no longer supported — use lists instead
*** C3U1X2K3 C3U1X2K2 C3U1X2K1 C3U1X2K0 are non zero
3
COEFF(CU3,U2X1,C3U2X1K);
***** ID fill no longer supported — use lists instead
*** C3U2X1K2 C3U2X1K1 C3U2X1K0 are non zero
2
COEFF(CU3,U2X2,C3U2X2K);
***** ID fill no longer supported — use lists instead
*** C3U2X2K3 C3U2X2K2 C3U2X2K1 C3U2X2K0 are non zero
3
COEFF(CU3,U3X1,C3U3X1K);
***** ID fill no longer supported — use lists instead
*** C3U3X1K3 C3U3X1K2 C3U3X1K1 C3U3X1K0 are non zero
3
COEFF(CU3,U3X2,C3U3X2K);
***** ID fill no longer supported — use lists instead
*** C3U3X2K3 C3U3X2K2 C3U3X2K1 C3U3X2K0 are non zero
3
END;
2:
C1U1X1X2K1;
   - 2*(DF(V1,Y2) + DF(V1,U3)*U3X2 + DF(V1,U2)*U2X2 + DF(V1,U1)*
   U1X2)
3:
V1:=S1$
4:
C1U1X1X2K1;
   0

```
   5:
C2U2X1X2K1;
   0
   6:
C3U3X1X2K1;
   2*(DF(V2,Y1) + DF(V2,U3)*U3X1 + DF(V2,U2)*U2X1 + DF(V2,U1)*
   U1X1)
   7:
V2:=S2$
   8:
C3U3X1X2K1;
   0
   9:
C3U3X1X1K1;
   - (2*DF(S2,Y2) - 2*DF(S1,Y1) - DF(G2,U3)*U1 - DF(G1,U3)*U2)
   10:
C1U1X1K2;
   0
   11:
C1U1X1K1;
   I*(2*DF(S2,Y2) - DF(S1,Y1) + DF(G1,U3)*U2)
   12:
C1U1X2K3;
   0
   13:
C1U1X2K2;
   DF(G1,U1,2)
   14:
C2U2X2K2;
   DF(G2,U2,2)
   15:
C3U3X2K2;
   DF(G3,U3,2) + DF(G2,U3,2)*U1 + DF(G1,U3,2)*U2
   16:
C1U1X2K1;
   - (DF(S2,Y2,2) - 2*DF(G1,Y2,U1) - 2*DF(G1,U3,U1)*U3X2 + 2*DF
   (G1,U3)*U2X2 - 2*DF(G1,U2,U1)*U2X2)
   17:
C1U1X2K2;
   DF(G1,U1,2)
   18:
C1U1X1K2;
   0
```

```
  19:
CIU1X1K1;
   I*(2*DF(S2,Y2) - DF(S1,Y1) + DF(G1,U3)*U2)
  20:
DEPEND GU2,Y1,Y2,U2;
  21:
DEPEND GU2U3,Y1,Y2,U2,U3;
  22:
G1:=GU2*U1+GU2U3$
  23:
CIU1X2K2;
  0
  24:
CIU1X2K1;
   - (DF(S2,Y2,2) + 2*DF(GU2U3,U3)*U2X2 - 2*DF(GU2,Y2) - 2*DF
   (GU2,U2)*U2X2)
  25:
DEPEND G1U2,Y1,Y2,U2;
  26:
GU2U3:=DF(GU2,U2)*U3+G1U2$
  27:
CIU1X2K1;
   - (DF(S2,Y2,2) - 2*DF(GU2,Y2))
  28:
CIU1X1K1;
   I*(2*DF(S2,Y2) - DF(S1,Y1) + DF(GU2,U2)*U2)
  29:
GU2:=(DF(S1,Y1)-2*DF(S2,Y2)*LOG(U2)+SS1$
   GU2:=(DF(S1,Y1)-2*DF(S2,Y2)*LOG(U2)+SS1$$$$
   ***** Too few right parentheses27
  30:
GU2:=(DF(S1,Y1)-2*DF(S2,Y2))*LOG(U2)+SS1$
  31:
CIU1X1K1;
  0
  32:
CIU1X2K1;
   - (4*DF(S2,Y2,2)*LOG(U2) + DF(S2,Y2,2) - 2*DF(SS1,Y2))
  33:
S2:=A1*Y2+A2$ 28
```

---

[27] This is an example of a mistyped command and REDUCE warning.

[28] A1, A2, ... are arbitrary constants.

```
   34:
C1U1X2K1;
   2*DF(SS1,Y2)
   35:
SS1:=S3$
   36:
C1U1X2K1;
   0
   37:
WS;
   0
   38:
COEFF(CU1,U2X2,C2U2X2K);
   ***** ID fill no longer supported — use lists instead
   *** C2U2X2K2 C2U2X2K1 C2U2X2K0 are non zero
   2
   39:
C2U2X2K2;
   (2*DF(S1,Y1)*U3 - DF(S1,Y1)*U2*U1 + DF(G1U2,U2,2)*U2³
   -4*U3*A1 + 2*U2*U1*A1)/U2³
   40:
DF(WS,U1);
```

$$-\frac{DF(S1,Y1) - 2*A1}{U2^2}$$

```
   41:
S1:=2*A1*Y1+A3$
   42:
WS(40);
   0
   43:
C2U2X2K2;
   DF(G1U2,U2,2)
   44:
G1U2:=SS2*U2+SS3$
   45:
G1;
   S3*U1 + SS3 + SS2*U2
   46:
C2U2X2K2;
   0
```

47:
C2U2X2K1;
  2*DF(SS2,Y2)
  48:
SS2:=S5$
  49:
C2U2X2K1;
  0
  50:
C2U2X2K0;
  DF(S5,Y1)*I*U2 + DF(S3,Y1)*I*U1 + DF(SS3,Y2,2) + DF(SS3,Y1)*I
  + 2*I*U2X1*S5 - SS3*U3 - G3*U1 - 2*U3*U1*A1
  51:
COEFFN(WS,U2X1,1);
  2*I*S5
  52:
S5:=0$
  53:
C2U2X2K0;
  DF(S3,Y1)*I*U1 + DF(SS3,Y2,2) + DF(SS3,Y1)*I - SS3*U3 - G3*U1
  - 2*U3*U1*A1
  54:
COEFF(WS,U1,C1U1K);
  ***** ID fill no longer supported — use lists instead
  *** C1U1K1 C1U1K0 are non zero
  1
  55:
C1U1K1;
  DF(S3,Y1)*I - G3 - 2*U3*A1
  56:
SOLVE(WS,G3);
  {G3=DF(S3,Y1)*I - 2*U3*A1}
  57:
CU1;
  DF(S3,Y1)*I*U1 + DF(SS3,Y2,2) + DF(SS3,Y1)*I - SS3*U3 - G3*U1
  - 2*U3*U1*A1
  58:
SOLVE(WS,G3);
  {G3=(DF(S3,Y1)*I*U1 + DF(SS3,Y2,2) + DF(SS3,Y1)*I - SS3*U3
  - 2*U3*U1*A1)/U1}
  59:
G3:=PART(PART(WS,1),2);
  G3 := (DF(S3,Y1)*I*U1 + DF(SS3,Y2,2) + DF(SS3,Y1)*I - SS3*U3
  - 2*U3*U1*A1)/U1

```
   60:
CU1;
   0
   61:
C2U2X1K2;
   0
   62:
C2U2X1K1;
   - DF(G2,U3)*I*U1
   63:
C2U2X2K3;
   0
   64:
C2U2X2K2;
   0
   65:
C2U2X2K1;
   0
   66:
C2U2X2K0;
   0
   67:
C2U2X1K1;
   - DF(G2,U3)*I*U1
   68:
C3U3X1X1K1;
   DF(G2,U3)*U1 + 2*A1
   69:
A1:=0$
   70:
C3U3X1X1K1;
   DF(G2,U3)*U1
   71:
C3U1X1K1;
```

$$(2*DF(SS3,Y2,2,Y1) + DF(SS3,Y2,2)*I + 2*DF(SS3,Y1,2)*I - 2*DF$$
$$(SS3,Y1)*U3 - DF(SS3,Y1) + DF(G2,U3)*I*U2*U1^3$$
$$-DF(G2,U1)*I*U1^3-I*S3*U2*U1^2-I*SS3*U3$$
$$-I*SS3*U2*U1-I*G2*U1^2-2*U3X1*SS3)/U1^2$$

```
   72:
SS3:=0$
   73:
C3U1X1K1;
```

$$I*(DF(G2,U3)*U2*U1 - DF(G2,U1)*U1 - S3*U2 - G2)$$

74:
C3U1X1K2;
   0
   75:
C3U1X2K3;
   0
   76:
C3U1X2K2;
   DF(G2,U1,2)*U1 + 2*DF(G2,U1)
   77:
C3U1X2K1;
   2*(DF(G2,Y2,U1)*U1 + DF(G2,Y2) + DF(G2,U3,U1)*U3X2*U1
   + DF(G2,U3)*U3X2 - DF(G2,U3)*U2X2*U1
   + DF(G2,U2,U1)*U2X2*U1+ DF(G2,U2)*U2X2 + U2X2*S3)
   78:
DEPEND GU1U2,Y1,Y2,U1,U2;
   79:
C2U2X1K1;
   - DF(G2,U3)*I*U1
   80:
G2:=GU1U2$
   81:
C2U2X1K1;
   0
   82:
CU2;
   - (DF(S3,Y1)*I*U2 - 2*DF(GU1U2,Y2,U2)*U2X2 - 2*DF(GU1U2,Y2,U1)
   *U1X2 - DF(GU1U2,Y2,2) + DF(GU1U2,Y1)*I - 2*DF(GU1U2,U2,
   U1)*U2X2*U1X2 - DF(GU1U2,U2,2)*U2X2^2-DF(GU1U2,U2)*U3*U2
   - DF(GU1U2,U1,2)*U1X2^2+2*DF(GU1U2,U1)*I*U1X1
   - DF(GU1U2,U1)*U3*U1 + U3*GU1U2)
   83:
COEFF(WS,U2X2,C2U2X2K);
   ***** ID fill no longer supported — use lists instead
   *** C2U2X2K2 C2U2X2K1 C2U2X2K0 are non zero
   2
   84:
C2U2X2K2;
   DF(GU1U2,U2,2)
   85:
C2U2X2K1;
   2*(DF(GU1U2,Y2,U2) + DF(GU1U2,U2,U1)*U1X2)

```
   86:
G1;
   S3*U1
   87:
G3;
   DF(S3,Y1)*I
   88:
DEPEND GGU1,Y1,Y2,U1;
   89:
GU1U2:=SS5*U2+GGU1$
   90:
C2U2X2K2;
   0
   91:
C2U2X2K1;
   2*DF(SS5,Y2)
   92:
SS5:=S7$
   93:
C2U2X2K1;
   0
   94:
CU2;
   - (DF(S7,Y1)*I*U2 + DF(S3,Y1)*I*U2 - 2*DF(GGU1,Y2,U1)*U1X2
   - DF(GGU1,Y2,2) + DF(GGU1,Y1)*I - DF(GGU1,U1,2)*U1X2²
   + 2*DF(GGU1,U1)*I*U1X1 - DF(GGU1,U1)*U3*U1 + U3*GGU1)
   95:
GGU1:=SS6*U1+SS7$
   96:
CU2;
   - (DF(S7,Y1)*I*U2 + DF(S3,Y1)*I*U2 - DF(SS7,Y2,2)
   + DF(SS7,Y1) *I - DF(SS6,Y2,2)*U1 - 2*DF(SS6,Y2)*U1X2
   + DF(SS6,Y1)*I* U1 + 2*I*U1X1*SS6 + SS7*U3)
   97:
SS6:=S9$
   98:
CU2;
   - (DF(S9,Y1)*I*U1 + DF(S7,Y1)*I*U2 + DF(S3,Y1)*I*U2
   - DF(SS7, Y2,2) + DF(SS7,Y1)*I + 2*I*U1X1*S9 + SS7*U3)
   99:
S9:=0$
   100:
CU2;
```

```
   - (DF(S7,Y1)*I*U2 + DF(S3,Y1)*I*U2 - DF(SS7,Y2,2)
   + DF(SS7,Y1)*I + SS7*U3)
   101:
SS7:=0$
   102:
CU2;
   - I*U2*(DF(S7,Y1) + DF(S3,Y1))
   103:
V1;
   A3
   104:
V2;
   A2
   105:
G1;
   S3*U1
   106:
G2;
   S7*U2
   107:
G3;
   DF(S3,Y1)*I
   108:
C3U3X1X2K1;
   0
   109:
C3U3X1K1;
   0
   110:
C3U3X1K2;
   0
   111:
C3U3X1K3;
   0
   112:
C3U3X2K3;
   0
   113:
C3U3X2K2;
   0
   114:
C3U3X2K1;
   0
```

```
    115:
C3U2X2K3;
   0
    116:
C3U2X2K2;
   0
    117:
C3U2X2K1;
   2*U1X2*(S7 + S3)
    118:
S7:=-S3$
    119:
C3U2X2K1;
   0
    120:
C3U2X1K2;
   0
    121:
C3U2X1K1;
   0
    122:
C3U1X1K2;
   0
    123:
C3U1X1K1;
   0
    124:
C3U1X2K3;
   0
    125:
C3U1X2K2;
   0
    126:
C3U1X2K1;
   0
    127:
CU2;
   0
    128:
CU3;
   - DF(S3,Y1,3)*I
    129:
V1;
   A3
```

```
   130:
V2;
   A2
   131:
G1;
   S3*U1
   132:
G2;
  - S3*U2
   133:
G3;
   DF(S3,Y1)*I
   134:
WS(128);
  - DF(S3,Y1,3)*I
   135:
V1;
   A3
   136:
V2;
   A2
   137:
A3:=C1$
   138:
A2:=C2$
   139:
V1;
   C1
   140:
V2;
   C2
   141:
WS(128);
  - DF(S3,Y1,3)*I
   142:
S3:=C3*Y1**2+C4*Y1+C5$
   143:
WS(128);
   0
   144:
V1;
   C1
   145:
V2;
```

```
    C2
    146:
G1;
    2 U1*(Y1 *C3 + Y1*C4 + C5)
    147:
G2;
    2 - U2*(Y1 *C3 + Y1*C4 + C5)
    148:
G3;
    I*(2*Y1*C3 + C4)
    149:
GEN1:=DF(V1,C1)*DY1+DF(V2,C1)*DY2+DF(G1,C1)*DU1+DF(G2,C1)*DU2
    +DF(G3,C1)*DU3;
    GEN1 := DY1
    150:
GEN2:=DF(V1,C2)*DY1+DF(V2,C2)*DY2+DF(G1,C2)*DU1+DF(G2,C2)*DU2
    +DF(G3,C2)*DU3;
    GEN2 := DY2
    151:
GEN3:=DF(V1,C3)*DY1+DF(V2,C3)*DY2+DF(G1,C3)*DU1
    +DF(G2,C3)*DU2 +DF(G3,C3)*DU3;
    GEN3 := Y1*(2*I*DU3 - Y1*U2*DU2 + Y1*U1*DU1)
    152:
GEN4:=DF(V1,C4)*DY1+DF(V2,C4)*DY2+DF(G1,C4)*DU1
    +DF(G2,C4)*DU2 +DF(G3,C4)*DU3;
    GEN4 := I*DU3 - Y1*U2*DU2 + Y1*U1*DU1
    153:
GEN5:=DF(V1,C5)*DY1+DF(V2,C5)*DY2+DF(G1,C5)*DU1
    +DF(G2,C5)*DU2 +DF(G3,C5)*DU3;
    GEN5 := - (U2*DU2 - U1*DU1)
    154:
BYE;
    *** END OF RUN
```

14.4. Non-classical symmetries: SGA programs

14.4.1. Description of the SGA programs

The SGA programs are derived from the GAmOkX2 programs. The only difference is that the first-order derivative of Uj by X1 is not independent (see Levi and Winternitz [1989]), i.e.:

$$UjX1 = -(V2 * UjX2 + Gj)/V1, \quad j = 1, \ldots, m \qquad (14.8)$$

The highest derivatives of Uj by X1 are also not independent. They are determined from the differential consequences of (14.8), e.g.:[29]

```
UX1:=(-V2*UX2+G)/V1$
UX1X2:=DF(UX1,X2)$
UX12:=DF(UX1,X1)$
UX1X22:=DF(UX1X2,X2)$
UX1X23:=DF(UX1X22,X2)$
```

14.4.2. Example: Boussinesq equation

The following computer session finds the non-classical symmetries of (14.6) (see Levi and Winternitz [1989]).

```
REDUCE 3.3, 15-Jul-87 ...
1:
IN "SGA1O4X2.BOU";
DEPEND U,X1,X2;
DEPEND V1,Y1,Y2,U;
DEPEND V2,Y1,Y2,U;
DEPEND G,Y1,Y2,U;
DEPEND SS1,Y1,Y2;
DEPEND SS2,Y1,Y2;
DEPEND SS3,Y1,Y2;
DEPEND SS4,Y1,Y2;
DEPEND SS5,Y1,Y2;
DEPEND SS6,Y1,Y2;
DEPEND SS7,Y1,Y2;
DEPEND SS8,Y1,Y2;
DEPEND SS9,Y1,Y2;
DEPEND S1,Y1;
DEPEND S3,Y1;
DEPEND S5,Y1;
DEPEND S7,Y1;
DEPEND S9,Y1;
DEPEND S2,Y2;
DEPEND S4,Y2;
DEPEND S6,Y2;
DEPEND S8,Y2;
```

---

[29]See the Boussinesq equation in the next subsection.

```
DEPEND S10,Y2;
DEPEND Y1,X1;
DEPEND Y2,X2;
DF(Y1,X1):=1$
DF(Y2,X2):=1$
DEPEND UX2,X1,X2;
DEPEND UX22,X1,X2;
DEPEND UX23,X1,X2;
DEPEND UX24,X1,X2;
DF(U,X1):=UX1$
DF(U,X2):=UX2$
UX1:=(-V2*UX2+G)/V1$
DF(UX2,X2):=UX22$
DF(UX22,X2):=UX23$
FORALL Z SUCH THAT NOT(Z=U OR Z=Y1 OR
Z=Y2 OR Z=UX2 OR Z=UX22 OR Z=UX23)
LET DF(Z,X1)=DF(Z,Y1)*DF(Y1,X1)+DF(Z,U)*UX1
+DF(Z,UX2)*UX1X2+DF(Z,UX22)*UX1X22
+DF(Z,UX23)*UX1X23,
DF(Z,X2)=DF(Z,Y2)*DF(Y2,X2)+DF(Z,U)*UX2
+DF(Z,UX2)*UX22 +DF(Z,UX22)*UX23
+DF(Z,UX23)*UX24;
UX1X2:=DF(UX1,X2)$
DF(UX2,X1):=UX1X2$
UX12:=DF(UX1,X1)$
UX1X22:=DF(UX1X2,X2)$
DF(UX22,X1):=UX1X22$
UX1X23:=DF(UX1X22,X2)$
DF(UX23,X1):=UX1X23$
UX24:=-(UX12+U*UX22+UX2**2)$
DF(UX23,X2):=UX24$
GX1:=DF(G,X1)-UX1*DF(V1,X1)-UX2*DF(V2,X1)$
GX2:=DF(G,X2)-UX1*DF(V1,X2)-UX2*DF(V2,X2)$
GX12:=DF(GX1,X1)-UX12*DF(V1,X1)-UX1X2*DF(V2,X1)$
GX22:=DF(GX2,X2)-UX1X2*DF(V1,X2)-UX22*DF(V2,X2)$
GX23:=DF(GX22,X2)-UX1X22*DF(V1,X2)-UX23*DF(V2,X2)$
GX24:=DF(GX23,X2)-UX1X23*DF(V1,X2)-UX24*DF(V2,X2)$
DEPEND L,X1,X2;
DEPEND LX1,X1,X2;
DEPEND LX2,X1,X2;
DEPEND LX12,X1,X2;
DEPEND LX22,X1,X2;
DEPEND LX24,X1,X2;
LQ:=LX24+LX12+L*LX22+LX2**2$
```

```
TEQ:=DF(LQ,L)*G+
+DF(LQ,LX1)*GX1+DF(LQ,LX2)*GX2
+DF(LQ,LX12)*GX12+DF(LQ,LX22)*GX22
+DF(LQ,LX24)*GX24$
L:=U$
LX1:=UX1$
LX2:=UX2$
LX12:=UX12$
LX22:=UX22$
LX24:=UX24$
CU:=NUM(TEQ)$
COEFF(CU,UX23,CX23K);
***** ID fill no longer supported — use lists instead
*** CX23K1 CX23K0 are non zero
1
COEFF(CU,UX22,CX22K);
***** ID fill no longer supported — use lists instead
*** CX22K2 CX22K1 CX22K0 are non zero
2
COEFF(CU,UX2,CX2K);
***** ID fill no longer supported — use lists instead
*** CX2K5 CX2K4 CX2K3 CX2K2 CX2K1 CX2K0 are non zero
5
END;
2:
CX23K1;
```

$2*V1^2 *(2*DF(G,U,Y2)*V1^2 + 2*DF(G,U,2)*UX2*V1^2 - 4*DF(G,U)*DF$
$(V1,U)*UX2*V1 - 2*DF(G,U)*DF(V1,Y2)*V1 - 2*DF(G,Y2)*DF$
$(V1,U)*V1 - 8*DF(V2,U,Y2)*UX2*V1^2 - 5*DF(V2,U,2)*UX2^2 *$
$V1^2 + 10*DF(V2,U)*DF(V1,U)*UX2^2 *V1 + 8*DF(V2,U)*DF(V1,$
$Y2)*UX2*V1 - 5*DF(V2,U)*UX22*V1^2 - 3*DF(V2,Y2,2)*V1^2 +$
$8*DF(V2,Y2)*DF(V1,U)*UX2*V1 + 6*DF(V2,Y2)*DF(V1,Y2)*V1$
$- 2*DF(V1,U,Y2)*G*V1 + 8*DF(V1,U,Y2)*UX2*V2*V1 - 2*DF$
$(V1,U,2)*G*UX2*V1 + 5*DF(V1,U,2)*UX2^2 *V2*V1 + 4*$
$DF(V1,U)^2 *G*UX2 - 10*DF(V1,U)^2 *UX2^2 *V2 + 4*DF(V1,U)*DF$
$(V1,Y2)*G - 16*DF(V1,U)*DF(V1,Y2)*UX2*V2 + 5*DF(V1,U)*$
$UX22*V2*V1 + 3*DF(V1,Y2,2)*V2*V1 - 6*DF(V1,Y2)^2 *V2)$

```
3:
COEFFN(WS,UX2,0);
```

$2*V1^2 *(2*DF(G,U,Y2)*V1^2 - 2*DF(G,U)*DF(V1,Y2)*V1 - 2*DF(G,Y2)*$
$DF(V1,U)*V1 - 5*DF(V2,U)*UX22*V1^2 - 3*DF(V2,Y2,2)*V1^2 +$
$6*DF(V2,Y2)*DF(V1,Y2)*V1 - 2*DF(V1,U,Y2)*G*V1 + 4*DF$
$(V1,U)*DF(V1,Y2)*G + 5*DF(V1,U)*UX22*V2*V1 + 3*DF(V1,Y2,$

```
    2)*V2*V1 - 6*DF(V1,Y2)^2 *V2)
    4:
COEFFN(WS,UX22,1);
    - 10*V1^3 *(DF(V2,U)*V1 - DF(V1,U)*V2)
    5:
V2:=SS1*V1$
    6:
WS(4);
    0
    7:
WS(3);
    2*V1^2 *(2*DF(G,U,Y2)*V1^2 - 2*DF(G,U)*DF(V1,Y2)*V1 - 2*DF(G,Y2)*
    DF(V1,U)*V1 - 3*DF(SS1,Y2,2)*V1^3 - 2*DF(V1,U,Y2)*G*V1 +
    4*DF(V1,U)*DF(V1,Y2)*G)
    8:
CX22K2;
    3*V1^2 *(DF(G,U,2)*V1^2 - 2*DF(G,U)*DF(V1,U)*V1 - DF(V1,U,2)*G*V1
    + 2*DF(V1,U)^2 *G)
    9:
G:=(SS2*U+SS3)*V1$
    10:
CX22K2;
    0
    11:
CX23K1;
    2*V1^5 *(2*DF(SS2,Y2) - 3*DF(SS1,Y2,2))
    12:
CX22K1;
    V1^5 *(6*DF(SS2,Y2,2) - 4*DF(SS1,Y2,3) + 2*DF(SS1,Y2)*U + 4*DF
    (SS1,Y2)*SS1^2 + 2*DF(SS1,Y1)*SS1 + U*SS2 + SS3)
    13:
COEFFN(WS,U,1);
    V1^5 *(2*DF(SS1,Y2) + SS2)
    14:
CX23K1;
    2*V1^5 *(2*DF(SS2,Y2) - 3*DF(SS1,Y2,2))
    15:
SS2:=-2*DF(SS1,Y2)$
    16:
WS(13);
    0
    17:
CX23K1;
```

- 14*DF(SS1,Y2,2)*V1^5
  18:
SS1:=S1*Y2+S3$
  19:
CX23K1;
  0
  20:
CX22K1;
  V1^5 *(2*DF(S3,Y1)*S3 + 2*DF(S3,Y1)*S1*Y2 + 2*DF(S1,Y1)*S3*Y2
  + 2*DF(S1,Y1)*S1*Y2^2 + 4*S3^2 *S1 + 8*S3*S1^2 *Y2 + 4*S1^3 *Y2^2
  + SS3)
  21:
SOLVE(WS,SS3);
  $SS3= - 2*(DF(S3,Y1)*S3 + DF(S3,Y1)*S1*Y2 + DF(S1,Y1)*S3*Y2 +
  DF(S1,Y1)*S1*Y2^2 + 2*S3^2 *S1 + 4*S3*S1^2 *Y2 + 2*S1^3 *
  Y2^2)
  22:
SS3:=PART(PART(WS,1),2)$
  23:
CX22K1;
  0
  24:
CX2K5;
  0
  25:
CX2K4;
  0
  26:
CX2K3;
  0
  27:
CX2K2;
  0
  28:
CX2K1;
  - V1^5 *(DF(S3,Y1,2) + 2*DF(S3,Y1)*S1 + DF(S1,Y1,2)*Y2 + 2*DF
  (S1,Y1)*S1*Y2 - 4*S3*S1^2 - 4*S1^3 *Y2)
  29:
COEFF(WS,Y2,KK);
  ***** ID fill no longer supported — use lists instead
  *** KK1 KK0 are non zero
  1

```
30:
```
KK1;

$- V1^5 * (DF(S1,Y1,2) + 2*DF(S1,Y1)*S1 - 4*S1^3)$

```
31:
```
KK0;

$- V1^5 * (DF(S3,Y1,2) + 2*DF(S3,Y1)*S1 - 4*S3*S1^2)$

```
32:
```
CX2K1;

$- V1^5 * (DF(S3,Y1,2) + 2*DF(S3,Y1)*S1 + DF(S1,Y1,2)*Y2 + 2*DF$
$(S1,Y1)*S1*Y2 - 4*S3*S1^2 - 4*S1^3 *Y2)$

```
33:
```
CX2K0;

$- 2*V1^5 * (DF(S3,Y1,3)*S3 + DF(S3,Y1,3)*S1*Y2 + 3*DF(S3,Y1,2)*$
$DF(S3,Y1) + 3*DF(S3,Y1,2)*DF(S1,Y1)*Y2 + 8*DF(S3,Y1,$
$2)*S3*S1 + 8*DF(S3,Y1,2)*S1^2 *Y2 + 6*DF(S3,Y1)^2 *S1 +$
$3*DF(S3,Y1)*DF(S1,Y1,2)*Y2 + 2*DF(S3,Y1)*DF(S1,Y1)*$
$S3 + 14*DF(S3,Y1)*DF(S1,Y1)*S1*Y2 - 4*DF(S3,Y1)*S3*$
$S1^2 - 4*DF(S3,Y1)*S1^3 *Y2 + DF(S1,Y1,3)*S3*Y2 + DF(S1,$
$Y1,3)*S1*Y2^2 + 3*DF(S1,Y1,2)*DF(S1,Y1)*Y2^2 + DF(S1,$
$Y1,2)*U + 2*DF(S1,Y1,2)*S3^2 + 12*DF(S1,Y1,2)*S3*S1*$
$Y2 + 10*DF(S1,Y1,2)*S1^2 *Y2^2 + 2*DF(S1,Y1)^2 *S3*Y2 + 8$
$*DF(S1,Y1)^2 *S1*Y2^2 + 2*DF(S1,Y1)*U*S1 - 4*DF(S1,Y1)*$
$S3^2 *S1 - 12*DF(S1,Y1)*S3*S1^2 *Y2 - 8*DF(S1,Y1)*S1^3 *$
$Y2^2 - 4*U*S1^3 - 32*S3^2 *S1^3 - 64*S3*S1^4 *Y2 - 32*S1^5 *$
$Y2^2)$

```
34:
```
COEFF(WS,U,MM);

***** ID fill no longer supported — use lists instead

*** MM1 MM0 are non zero

1
```
35:
```
MM1;

$- 2*V1^5 * (DF(S1,Y1,2) + 2*DF(S1,Y1)*S1 - 4*S1^3)$

```
36:
```
KK1;

$- V1^5 * (DF(S1,Y1,2) + 2*DF(S1,Y1)*S1 - 4*S1^3)$

```
37:
```
MM0;

$- 2*V1^5 * (DF(S3,Y1,3)*S3 + DF(S3,Y1,3)*S1*Y2 + 3*DF(S3,Y1,2)*$
$DF(S3,Y1) + 3*DF(S3,Y1,2)*DF(S1,Y1)*Y2 + 8*DF(S3,Y1,$
$2)*S3*S1 + 8*DF(S3,Y1,2)*S1^2 *Y2 + 6*DF(S3,Y1)^2 *S1 +$
$3*DF(S3,Y1)*DF(S1,Y1,2)*Y2 + 2*DF(S3,Y1)*DF(S1,Y1)*$

S3 + 14*DF(S3,Y1)*DF(S1,Y1)*S1*Y2 - 4*DF(S3,Y1)*S3*
S1^2 - 4*DF(S3,Y1)*S1^3 *Y2 + DF(S1,Y1,3)*S3*Y2 + DF(S1,
Y1,3)*S1*Y2^2 + 3*DF(S1,Y1,2)*DF(S1,Y1)*Y2^2 + 2*DF
(S1,Y1,2)*S3^2 + 12*DF(S1,Y1,2)*S3*S1*Y2 + 10*DF(S1,Y1,
2)*S1^2 *Y2^2 + 2*DF(S1,Y1)^2 *S3*Y2 + 8*DF(S1,Y1)^2 *S1*
Y2^2 - 4*DF(S1,Y1)*S3^2 *S1 - 12*DF(S1,Y1)*S3*S1^2 *Y2 -
8*DF(S1,Y1)*S1^3 *Y2^2 - 32*S3^2 *S1^3 - 64*S3*S1^4 *Y2 - 32
*S1^5 *Y2^2)
38:
COEFF(WS,Y2,M0Y);
   ***** ID fill no longer supported — use lists instead
   *** M0Y2 M0Y1 M0Y0 are non zero
   2
   39:
M0Y2;
  - 2*V1^5 *(DF(S1,Y1,3)*S1 + 3*DF(S1,Y1,2)*DF(S1,Y1) + 10*DF(S1,
Y1,2)*S1^2 + 8*DF(S1,Y1)^2 *S1 - 8*DF(S1,Y1)*S1^3 - 32*
S1^5)
   40:
M0Y1;
  - 2*V1^5 *(DF(S3,Y1,3)*S1 + 3*DF(S3,Y1,2)*DF(S1,Y1) + 8*DF(S3,
Y1,2)*S1^2 + 3*DF(S3,Y1)*DF(S1,Y1,2) + 14*DF(S3,Y1)*
DF(S1,Y1)*S1 - 4*DF(S3,Y1)*S1^3 + DF(S1,Y1,3)*S3 + 12
*DF(S1,Y1,2)*S3*S1 + 2*DF(S1,Y1)^2 *S3 - 12*DF(S1,Y1)*
S3*S1^2 - 64*S3*S1^4)
   41:
M0Y0;
  - 2*V1^5 *(DF(S3,Y1,3)*S3 + 3*DF(S3,Y1,2)*DF(S3,Y1) + 8*DF(S3,
Y1,2)*S3*S1 + 6*DF(S3,Y1)^2 *S1 + 2*DF(S3,Y1)*DF(S1,Y1)
*S3 - 4*DF(S3,Y1)*S3*S1^2 + 2*DF(S1,Y1,2)*S3^2 - 4*DF
(S1,Y1)*S3^2 *S1 - 32*S3^2 *S1^3)
   42:
KK1;
  - V1^5 *(DF(S1,Y1,2) + 2*DF(S1,Y1)*S1 - 4*S1^3)
   43:
E1:=COEFFN(WS,V1,5);
   E1 := - (DF(S1,Y1,2) + 2*DF(S1,Y1)*S1 - 4*S1^3)
   44:
KK0;
  - V1^5 *(DF(S3,Y1,2) + 2*DF(S3,Y1)*S1 - 4*S3*S1^2)
   45:
E2:=COEFFN(WS,V1,5);

E2 := - (DF(S3,Y1,2) + 2*DF(S3,Y1)*S1 - 4*S3*S1^2)
  46:
DF(E1,Y1)$
  47:
SOLVE(E1,DF(S1,Y1,2))$
  48:
D2S1:=PART(PART(WS,1),2);
  D2S1 := - 2*S1*(DF(S1,Y1) - 2*S1^2)
  49:
SOLVE(E2,DF(S3,Y1,2))$
  50:
D2S3:=PART(PART(WS,1),2);
  D2S3 := - 2*S1*(DF(S3,Y1) - 2*S3*S1)
  51:
SUB(DF(S1,Y1,2)=D2S1,DF(E1,Y1));
  - (DF(S1,Y1,3) + 2*DF(S1,Y1)^2 - 16*DF(S1,Y1)*S1^2 + 8*S1^4)
  52:
SOLVE(WS,DF(S1,Y1,3))$
  53:
D3S1:=PART(PART(WS,1),2);
  D3S1 := - 2*(DF(S1,Y1)^2 - 8*DF(S1,Y1)*S1^2 + 4*S1^4)
  54:
SUB(DF(S3,Y1,2)=D2S3,DF(E2,Y1));
  - (DF(S3,Y1,3) + 2*DF(S3,Y1)*DF(S1,Y1) - 8*DF(S3,Y1)*S1^2 - 8*
  DF(S1,Y1)*S3*S1 + 8*S3*S1^3)
  55:
SOLVE(WS,DF(S3,Y1,3))$
  56:
D3S3:=PART(PART(WS,1),2);
  D3S3 := - 2*(DF(S3,Y1)*DF(S1,Y1) - 4*DF(S3,Y1)*S1^2 - 4*DF(S1,
  Y1)*S3*S1 + 4*S3*S1^3)
  57:
M0Y2;
  - 2*V1^5 *(DF(S1,Y1,3)*S1 + 3*DF(S1,Y1,2)*DF(S1,Y1) + 10*DF(S1,
  Y1,2)*S1^2 + 8*DF(S1,Y1)^2 *S1 - 8*DF(S1,Y1)*S1^3 - 32*
  S1^5)
  58:
SUB(DF(S1,Y1,3)=D3S1,DF(S1,Y1,2)=D2S1},WS);
  SUB(DF(S1,Y1,3)=D3S1,DF(S1,Y1,2)=D2S1$$$},WS);
  ***** Too few right parentheses
  59:
SUB({DF(S1,Y1,3)=D3S1,DF(S1,Y1,2)=D2S1},WS);
  0

60:
M0Y1;
- 2*V1^5 *(DF(S3,Y1,3)*S1 + 3*DF(S3,Y1,2)*DF(S1,Y1) + 8*DF(S3,
Y1,2)*S1^2 + 3*DF(S3,Y1)*DF(S1,Y1,2) + 14*DF(S3,Y1)*
DF(S1,Y1)*S1 - 4*DF(S3,Y1)*S1^3 + DF(S1,Y1,3)*S3 + 12
*DF(S1,Y1,2)*S3*S1 + 2*DF(S1,Y1)^2 *S3 - 12*DF(S1,Y1)*
S3*S1^2 - 64*S3*S1^4)
61:
SUB({DF(S1,Y1,3)=D3S1,DF(S1,Y1,2)=D2S1,DF(S3,Y1,3)=D3S3,
DF(S3,Y1,2)=D2S3},WS);
0
62:
M0Y0;
- 2*V1^5 *(DF(S3,Y1,3)*S3 + 3*DF(S3,Y1,2)*DF(S3,Y1) + 8*DF(S3,
Y1,2)*S3*S1 + 6*DF(S3,Y1)^2 *S1 + 2*DF(S3,Y1)*DF(S1,Y1)
*S3 - 4*DF(S3,Y1)*S3*S1^2 + 2*DF(S1,Y1,2)*S3^2 - 4*DF
(S1,Y1)*S3^2 *S1 - 32*S3^2 *S1^3)
63:
SUB({DF(S3,Y1,3)=D3S3,DF(S3,Y1,2)=D2S3,DF(S1,Y1,2)=D2S1},WS);
0
64:
V1;
V1
65:
V2;
V1*(S3 + S1*Y2)
66:
G;
- 2*V1*(DF(S3,Y1)*S3 + DF(S3,Y1)*S1*Y2 + DF(S1,Y1)*S3*Y2 + DF
(S1,Y1)*S1*Y2^2 + U*S1 + 2*S3^2 *S1 + 4*S3*S1^2 *Y2 + 2*
S1^3 *Y2^2)
67:
E1;
- (DF(S1,Y1,2) + 2*DF(S1,Y1)*S1 - 4*S1^3)
68:
E2;
- (DF(S3,Y1,2) + 2*DF(S3,Y1)*S1 - 4*S3*S1^2)
69:
BYE;
** END OF RUN

14.5. Lie-Bäcklund symmetries: GS programs

14.5.1. Description of the GS programs

The GS programs are similar to the GA programs.
The only differences are:[30]

- the lowest derivative of U by X1 (and differential consequences) is not independent, e.g.:[31]
  UX1:=UX22+UX2**2$
  UX1X2:=DF(UX1,X2)$
  UX1X22:=DF(UX1X2,X2)$
  UX1X23:=DF(UX1X22,X2)$
  UX1X24:=DF(UX1X23,X2)$

- the definition of the infinitesimal generator (order l) (see Ibragimov [1983], Olver [1986], Bluman and Kumei [1989]) is:

$$V Tl(Y1, Y2, \ldots, Yn, U, U Xi_{i \neq 1}, \ldots) \, \partial_U \qquad (14.9)$$

  where VTl is a function of Yi, U, and the derivatives of U by Xi ($i \neq 1$) up to l-th order;

- our goal is to find VTl.

14.5.2. Example: Potential Burgers equation

We consider the following equation (see Olver [1986]):

$$U_{X1} - U_{X2X2} + U_{X2}^2 = 0 \qquad (14.10)$$

The following computer session computes all third-order Lie-Bäcklund symmetries of (14.10). Note that Q1, Q2, ... are the characteristics of (14.10) (see Olver [1986]), and VT3 ≡ VTTT.
    REDUCE 3.3, 15-Jul-87 ...
    1:
IN "GS1O2X2.PBU";

---

[30] We consider one differential equation.
[31] See the potential Burgers equation in the next subsection.

```
DEPEND U,X1,X2;
DEPEND VTTT,Y1,Y2,U,UX2,UX22,UX23;
DEPEND V1TT,Y1,Y2,U,UX2,UX22;
DEPEND VTT,Y1,Y2,U,UX2,UX22;
DEPEND V1T,Y1,Y2,U,UX2;
DEPEND V2T,Y1,Y2,U,UX2;
DEPEND V3T,Y1,Y2,U,UX2;
DEPEND VT,Y1,Y2,U,UX2;
DEPEND V1U,Y1,Y2,U;
DEPEND V2U,Y1,Y2,U;
DEPEND V3U,Y1,Y2,U;
DEPEND V4U,Y1,Y2,U;
DEPEND VU,Y1,Y2,U;
DEPEND SS1,Y1,Y2;
DEPEND SS2,Y1,Y2;
DEPEND SS3,Y1,Y2;
DEPEND SS4,Y1,Y2;
DEPEND SS5,Y1,Y2;
DEPEND SS6,Y1,Y2;
DEPEND SS7,Y1,Y2;
DEPEND SS8,Y1,Y2;
DEPEND SS9,Y1,Y2;
DEPEND S1,Y1;
DEPEND S3,Y1;
DEPEND S5,Y1;
DEPEND S7,Y1;
DEPEND S9,Y1;
DEPEND S2,Y2;
DEPEND S4,Y2;
DEPEND S6,Y2;
DEPEND S8,Y2;
DEPEND S10,Y2;
DEPEND Y1,X1;
DEPEND Y2,X2;
DF(Y1,X1):=1$
DF(Y2,X2):=1$
DEPEND UX2,X1,X2;
DEPEND UX22,X1,X2;
DEPEND UX23,X1,X2;
DEPEND UX24,X1,X2;
DEPEND UX25,X1,X2;
DF(U,X2):=UX2$
DF(UX2,X2):=UX22$
DF(UX22,X2):=UX23$
```

```
DF(UX23,X2):=UX24$
DF(UX24,X2):=UX25$
UX1:=UX22+UX2**2$
DF(U,X1):=UX1$
UX1X2:=DF(UX1,X2)$
DF(UX2,X1):=UX1X2$
UX1X22:=DF(UX1X2,X2)$
DF(UX22,X1):=UX1X22$
UX1X23:=DF(UX1X22,X2)$
DF(UX23,X1):=UX1X23$
UX1X24:=DF(UX1X23,X2)$
DF(UX24,X1):=UX1X24$
FORALL Z SUCH THAT NOT(Z=U OR Z=Y1 OR
Z=Y2 OR Z=UX2 OR Z=UX22 OR Z=UX23 OR Z=UX24 )
LET DF(Z,X2)=DF(Z,Y2)*DF(Y2,X2)+DF(Z,U)*UX2+DF(Z,UX2)*UX22
+DF(Z,UX22)*UX23+DF(Z,UX23)*UX24+DF(Z,UX24)*UX25,
DF(Z,X1)=DF(Z,Y1)*DF(Y1,X1)+DF(Z,U)*UX1+DF(Z,UX2)*UX1X2
+DF(Z,UX22)*UX1X22+DF(Z,UX23)*UX1X23
+DF(Z,UX24)*UX1X24;
VDX2:=DF(VTTT,X2)$
VDX1:=DF(VTTT,X1)$
VDX22:=DF(VDX2,X2)$
DEPEND L,X1,X2;
DEPEND LX1,X1,X2;
DEPEND LX2,X1,X2;
DEPEND LX22,X1,X2;
LQ:=LX22-LX1+LX2**2$
TEQ:=DF(LQ,L)*VTTT+DF(LQ,LX1)*VDX1+DF(LQ,LX2)*VDX2
+DF(LQ,LX22)*VDX22$
L:=U$
LX1:=UX1$
LX2:=UX2$
LX22:=UX22$
CU:=NUM(TEQ)$
COEFF(CU,UX25,CX25K);
***** ID fill no longer supported — use lists instead
*** CX25K0 is non zero
0
COEFF(CU,UX24,CX24K);
***** ID fill no longer supported — use lists instead
*** CX24K2 CX24K1 CX24K0 are non zero
2
END;
```

```
  2:
CX24K2;
  DF(VTTT,UX23,2)
  3:
VTTT:=V1TT*UX23+VTT$
  4:
CX24K2;
  0
  5:
CX24K1;
  2*(DF(V1TT,U)*UX2 + DF(V1TT,UX22)*UX23 + DF(V1TT,UX2)*UX22 +
  DF(V1TT,Y2))
  6:
COEFFN(WS,UX23,1);
  2*DF(V1TT,UX22)
  7:
V1TT:=V1T$
  8:
CX24K1;
  2*(DF(V1T,U)*UX2 + DF(V1T,UX2)*UX22 + DF(V1T,Y2))
  9:
COEFFN(CU,UX23,CX23K);
  ***** CX23K invalid as COEFFN index
  10:
COEFF(CU,UX23,CX23K);
  ***** ID fill no longer supported — use lists instead
  *** CX23K2 CX23K1 CX23K0 are non zero
  2
  11:
CX23K2;
  DF(VTT,UX22,2)
  12:
CX24K1;
  2*(DF(V1T,U)*UX2 + DF(V1T,UX2)*UX22 + DF(V1T,Y2))
  13:
V1T:=V1U$
  14:
CX24K1;
  2*(DF(V1U,U)*UX2 + DF(V1U,Y2))
  15:
V1U:=S1$
  16:
CX24K1;
  0
```

```
   17:
CX23K2;
   DF(VTT,UX22,2)
   18:
VTT:=V2T*UX22+VT$
   19:
CX23K2;
   0
   20:
CX23K1;
   - (DF(S1,Y1) - 2*DF(V2T,U)*UX2 - 2*DF(V2T,UX2)*UX22 - 2*DF
   (V2T,Y2) + 6*S1*UX22)
   21:
COEFFN(WS,UX22,1);
   2*(DF(V2T,UX2) - 3*S1)
   22:
V2T:=3*S1*UX2+V2U$
   23:
CX23K1;
   - (DF(S1,Y1) - 2*DF(V2U,U)*UX2 - 2*DF(V2U,Y2))
   24:
V2U:=SS1$
   25:
CX23K1;
   - (DF(S1,Y1) - 2*DF(SS1,Y2))
   26:
SS1:=DF(S1,Y1)*Y2/2+S3$
   27:
CX23K1;
   0
   28:
COEFF(CU,UX22,CX22K);
   ***** ID fill no longer supported — use lists instead
   *** CX22K2 CX22K1 CX22K0 are non zero
   2
   29:
CX22K2;
   - (DF(S1,Y1)*Y2 - DF(VT,UX2,2) + 2*S3 + 6*S1*UX2)
   30:
SOLVE(WS,DF(VT,UX2,2))$
   31:
PART(PART(WS,1),2);
   DF(S1,Y1)*Y2 + 2*S3 + 6*S1*UX2
```

32:
COEFFN(WS,UX2,ZZ);
   ***** ZZ invalid as COEFFN index
   33:
COEFF(WS,UX2,ZZ);
   ***** ID fill no longer supported — use lists instead
   *** ZZ1 ZZ0 are non zero
   1
   34:
VT:=ZZ1*UX2**3/6+ZZ0*UX2**2/2+V3U*UX2+VU$
   35:
CX22K2;
   0
   36:
CX22K1;
   - (2*DF(S3,Y1) + DF(S1,Y1,2)*Y2 - 4*DF(V3U,U)*UX2 - 4*DF(V3U,
   Y2))/2
   37:
V3U:=SS2$
   38:
CX22K1;

$$-\frac{2*DF(S3,Y1) + DF(S1,Y1,2)*Y2 - 4*DF(SS2,Y2)}{2}$$

   39:
SOLVE(WS,DF(SS2,Y2))$
   40:
PART(PART(WS,1),2);

$$\frac{2*DF(S3,Y1) + DF(S1,Y1,2)*Y2}{4}$$

   41:
COEFF(WS,Y2,ZZ);
   ***** ID fill no longer supported — use lists instead
   *** ZZ1 ZZ0 are non zero
   1
   42:
SS2:=ZZ1*Y2**2/2+ZZ0*Y2+S5$
   43:
CX22K1;
   0

44:
COEFF(CU,UX2,CX2K);
  ***** ID fill no longer supported — use lists instead
  *** CX2K2 CX2K1 CX2K0 are non zero
  2
  45:
CX2K2;
  DF(VU,U,2) + DF(VU,U)
  46:
CX2K1;
  - (8*DF(S5,Y1) + 4*DF(S3,Y1,2)*Y2 + DF(S1,Y1,3)*Y2^2 - 2*DF(S1,
  Y1,2) - 16*DF(VU,U,Y2) - 16*DF(VU,Y2))/8
  47:
SS3;
  SS3
  48:
VU:=SS3*EXP(-U)+SS4$
  49:
CX2K2;
  0
  50:
CX2K1;
  - (8*DF(S5,Y1) + 4*DF(S3,Y1,2)*Y2 + DF(S1,Y1,3)*Y2^2 - 2*DF(S1,
  Y1,2) - 16*DF(SS4,Y2))/8
  51:
SOLVE(WS,DF(SS4,Y2))$
  52:
PART(PART(WS,1),2);
  (8*DF(S5,Y1) + 4*DF(S3,Y1,2)*Y2 + DF(S1,Y1,3)*Y2^2 - 2*DF(S1,Y1,
  2))/16
  53:
COEFF(WS,Y2,ZZ);
  ***** ID fill no longer supported — use lists instead
  *** ZZ2 ZZ1 ZZ0 are non zero
  2
  54:
SS4:=ZZ2*Y2**3/3+ZZ1*Y2**2/2+ZZ0*Y2+S7$
  55:
CX2K1;
  0
  56:
CU;
  - (48*E^U *DF(S7,Y1) + 24*E^U *DF(S5,Y1,2)*Y2 + 6*E^U *DF(S3,Y1,3)*

Y2^2 - 12*E^U *DF(S3,Y1,2) + E^U *DF(S1,Y1,4)*Y2^2 - 12*E^U *DF(S1,
Y1,3)*Y2 - 48*DF(SS3,Y2,2) + 48*DF(SS3,Y1))/(48*E^U)
  57:
COEFF(NUM(WS),EXP(U),KK);
  ***** ID fill no longer supported — use lists instead
  *** KK1 KK0 are non zero
  1
  58:
KK1;
  - (48*DF(S7,Y1) + 24*DF(S5,Y1,2)*Y2 + 6*DF(S3,Y1,3)*Y2^2 - 12*
  DF(S3,Y1,2) + DF(S1,Y1,4)*Y2^3 - 12*DF(S1,Y1,3)*Y2)
  59:
S1:=C1*Y1**3+C2*Y1**2+C3*Y1+C4$
  60:
KK1;
  - 6*(8*DF(S7,Y1) + 4*DF(S5,Y1,2)*Y2 + DF(S3,Y1,3)*Y2^2 - 2*DF
  (S3,Y1,2) - 12*Y2*C1)
  61:
COEFF(WS,Y2,CC);
  ***** ID fill no longer supported — use lists instead
  *** CC2 CC1 CC0 are non zero
  2
  62:
CC2;
  - 6*DF(S3,Y1,3)
  63:
S3:=C5*Y1**2+C6*Y1+C7$
  64:
CC2;
  0
  65:
CC1;
  - 24*(DF(S5,Y1,2) - 3*C1)
  66:
S5:=3*C1*Y1**2/2+C8*Y1+C9$
  67:
CC1;
  0
  68:
KK1;
  - 24*(2*DF(S7,Y1) - C5)
  69:
S7:=C5*Y1/2+C10$

```
  70:
KK1;
  0
  71:
KK0;
  48*(DF(SS3,Y2,2) - DF(SS3,Y1))
  72:
CU;
```

$$\frac{DF(SS3, Y2, 2) - DF(SS3, Y1)}{E^U}$$

```
  73:
Q1:=DF(VTTT,C1);
```

$Q1 := (8*UX23*Y1^3 + 24*UX22*UX2*Y1^3 + 12*UX22*Y2*Y1^2 + 8*UX2^3$
$*$
$Y1^3 + 12*UX2^2 *Y2*Y1^2 + 6*UX2*Y2^2 *Y1 + 12*UX2*Y1^2 + Y2^3$
$+ 6*Y2*Y1)/8$

```
  74:
Q2:=DF(VTTT,C2);
```

$Q2 := (4*UX23*Y1^2 + 12*UX22*UX2*Y1^2$
$+ 4*UX22*Y2*Y1 + 4*UX2^3 * Y1^2 + 4*UX2^2 *Y2*Y1 + UX2*Y2^2 - Y2)/4$

```
  75:
Q3:=DF(VTTT,C3);
```

$Q3 := (2*UX23*Y1 + 6*UX22*UX2*Y1 + UX22*Y2 + 2*UX2^3 *Y1$
$+ UX2^2 * Y2)/2$

```
  76:
Q4:=DF(VTTT,C4);
```

$Q4 := UX23 + 3*UX22*UX2 + UX2^3$

```
  77:
Q5:=DF(VTTT,C5);
```

$Q5 :=$

$$\frac{4 * UX22 * Y1^2 + 4 * UX2^2 * Y1^2 + 4 * UX2 * Y2 * Y1 + Y2^2 + 2 * Y1}{4}$$

```
  78:
Q6:=DF(VTTT,C6);
```

$$Q6 := \frac{2 * UX22 * Y1 + 2 * UX2^2 * Y1 + UX2 * Y2}{2}$$

79:
Q7:=DF(VTTT,C7);

$$Q7 := UX22 + UX2^2$$

80:
Q8:=DF(VTTT,C8);

$$Q8 := \frac{2 * UX2 * Y1 + Y2}{2}$$

81:
Q9:=DF(VTTT,C9);

$$Q9 := UX2$$

82:
Q10:=DF(VTTT,C10);

$$Q10 := 1$$

83:
QVV:=DF(VTTT,SS3);

$$QVV := \frac{1}{E^U}$$

84:
BYE;
** END OF RUN

14.6.   Approximate Symmetries: AGA programs

14.6.1.   Description of the AGA programs

The AGA programs are similar to the GA programs. The differences are: [32]

- the unperturbed equation is named UQ and the perturbed part PUQ;
- the generator of the exact symmetries of UQ is given by Vi and G as in (14.3), i.e.:

$$Z^0 = \sum_{i=1}^{n} Vi(Y1, Y2, ..., U)\, \partial_{Xi} + G(Y1, Y2, ..., U)\, \partial_U \qquad (14.11)$$

---

[32] We consider one perturbed differential equation.

and the generator of the approximate symmetries for the perturbed equation:

$$UQ + \epsilon\, PUQ = 0 \tag{14.12}$$

is given by Baikov, Gazizov, and Ibragimov [1988b]:

$$Z^0 + \epsilon\, Z^1 \tag{14.13}$$

where:

$$Z^1 = \sum_{i=1}^{n} Wi(Y1, Y2, ..., U)\, \partial_{Xi} + J(Y1, Y2, ..., U)\, \partial_U \tag{14.14}$$

- Vi and G are known expressions;
- our goal is to compute Wi, J and L(Y1,Y2, ... ,U) such that:

$$Z^0[UQ] - L * UQ = 0 \tag{14.15}$$

and:

$$Z^1[UQ] + Z^0[PUQ] - L * PUQ = 0 \tag{14.16}$$

whenever $UQ = 0$.[33]

For more details on the algorithm we refer to Baikov, Gazizov, and Ibragimov [1988b].

14.6.2. Example: Perturbed Korteweg-de Vries equation

Let us consider the perturbed equation:

$$U_{X1} - U_{X2X2X2} + \epsilon\, U\, U_{X2} = 0 \tag{14.17}$$

This is the corresponding computer session.
    REDUCE 3.3, 15-Jul-87 ...
    1:
IN "AGA1O3X2.KDV";
    DEPEND U,X1,X2;
    DEPEND V1,Y1,Y2,U;
    DEPEND V2,Y1,Y2,U;
    DEPEND G,Y1,Y2,U;

---

[33] The dependent derivative is computed from the unperturbed equation and substituted into (14.16).

```
DEPEND W1,Y1,Y2,U;
DEPEND W2,Y1,Y2,U;
DEPEND J,Y1,Y2,U;
DEPEND SS1,Y1,Y2;
DEPEND SS2,Y1,Y2;
DEPEND SS3,Y1,Y2;
DEPEND SS4,Y1,Y2;
DEPEND SS5,Y1,Y2;
DEPEND SS6,Y1,Y2;
DEPEND SS7,Y1,Y2;
DEPEND SS8,Y1,Y2;
DEPEND SS9,Y1,Y2;
DEPEND S1,Y1;
DEPEND S3,Y1;
DEPEND S5,Y1;
DEPEND S7,Y1;
DEPEND S9,Y1;
DEPEND S2,Y2;
DEPEND S4,Y2;
DEPEND S6,Y2;
DEPEND S8,Y2;
DEPEND S10,Y2;
DEPEND Y1,X1;
DEPEND Y2,X2;
DF(Y1,X1):=1$
DF(Y2,X2):=1$
DEPEND UX1,X1,X2;
DEPEND UX2,X1,X2;
DEPEND UX12,X1,X2;
DEPEND UX1X2,X1,X2;
DEPEND UX22,X1,X2;
DEPEND UX23,X1,X2;
DEPEND UX13,X1,X2;
DEPEND UX12X2,X1,X2;
DEPEND UX1X22,X1,X2;
DF(U,X1):=UX1$
DF(U,X2):=UX2$
DF(UX1,X1):=UX12$
DF(UX1,X2):=UX1X2$
DF(UX2,X1):=UX1X2$
DF(UX2,X2):=UX22$
DF(UX12,X1):=UX13$
DF(UX12,X2):=UX12X2$
DF(UX1X2,X1):=UX12X2$
```

DF(UX1X2,X2):=UX1X22$
DF(UX22,X1):=UX1X22$
DF(UX22,X2):=UX23$
UQ:=-UX23+UX1$
PUQ:=U*UX2$
V1:=C1*Y1+C2$
V2:=C1*Y2/3+C3$
G:=C4*U+SS1$
DF(SS1,Y2,3):=DF(SS1,Y1)$
FORALL Z SUCH THAT NOT(Z=U OR Z=Y1 OR
Z=Y2 OR Z=UX1 OR Z=UX2 OR Z=UX12 OR Z=UX1X2 OR Z=UX22
OR Z=UX13 OR Z=UX1X22 OR Z=UX12X2)
LET DF(Z,X1)=DF(Z,Y1)*DF(Y1,X1)+DF(Z,U)*UX1+DF(Z,UX1)*UX12+
DF(Z,UX2)*UX1X2+DF(Z,UX22)*UX1X22+DF(Z,UX1X2)*UX12X2
+DF(Z,UX12)*UX13,
DF(Z,X2)=DF(Z,Y2)*DF(Y2,X2)+DF(Z,U)*UX2+DF(Z,UX1)*UX1X2
+DF(Z,UX2)*UX22+
DF(Z,UX12)*UX12X2+DF(Z,UX1X2)*UX1X22+DF(Z,UX22)*UX23;
JX1:=DF(J,X1)-UX1*DF(W1, ØØ X1)-UX2*DF(W2,X1)$
JX2:=DF(J,X2)-UX1*DF(W1,X2)-UX2*DF(W2,X2)$
JX22:=DF(JX2,X2)-UX22*DF(W2,X2)-UX1X2*DF(W1,X2)$
JX23:=DF(JX22,X2)-UX23*DF(W2,X2)-UX1X22*DF(W1,X2)$
GX1:=DF(G,X1)-UX1*DF(V1,X1)-UX2*DF(V2,X1)$
GX2:=DF(G,X2)-UX1*DF(V1,X2)-UX2*DF(V2,X2)$
GX22:=DF(GX2,X2)-UX22*DF(V2,X2)-UX1X2*DF(V1,X2)$
GX23:=DF(GX22,X2)-UX23*DF(V2,X2)-UX1X22*DF(V1,X2)$
TEQ1:=DF(UQ,UX1)*GX1+DF(UQ,UX23)*GX23+DF(UQ,U)*G
+DF(UQ,UX2)*GX2+DF(UQ,UX22)*GX22-L*UQ$
TEQ2:=DF(UQ,UX1)*JX1+DF(UQ,UX23)*JX23+DF(UQ,U)*J
+DF(UQ,UX2)*JX2+DF(UQ,UX22)*JX22-L*PUQ+
+DF(PUQ,UX1)*GX1+DF(PUQ,UX23)*GX23+DF(PUQ,U)*G
+DF(PUQ,UX2)*GX2+DF(PUQ,UX22)*GX22$
SOLVE(TEQ1,L)$
L:=PART(PART(WS,1),2);
L := C4 - C1
CCU:=SUB(UX23=UX1,TEQ2)$
CU:=NUM(CCU)$
COEFF(CU,UX1X22,CX1X22K);
***** ID fill no longer supported — use lists instead
*** CX1X22K1 CX1X22K0 are non zero
1
COEFF(CU,UX1X2,CX1X2K);
***** ID fill no longer supported — use lists instead
*** CX1X2K1 CX1X2K0 are non zero

```
   1
COEFF(CU,UX22,CX22K);
***** ID fill no longer supported — use lists instead
*** CX22K2 CX22K1 CX22K0 are non zero
   2
COEFF(CU,UX1,CX1K);
***** ID fill no longer supported — use lists instead
*** CX1K1 CX1K0 are non zero
   1
COEFF(CU,UX2,CX2K);
***** ID fill no longer supported — use lists instead
*** CX2K4 CX2K3 CX2K2 CX2K1 CX2K0 are non zero
   4
END;
   2:
CX1X22K1;
  9*(DF(W1,U)*UX2 + DF(W1,Y2))
   3:
W1:=S1$
   4:
CX1X22K1;
   0
   5:
CX1X2K1;
   0
   6:
CX22K2;
  9*DF(W2,U)
   7:
W2:=SS2$
   8:
CX22K2;
   0
   9:
CX22K1;
  - 9*(DF(J,U,Y2) + DF(J,U,2)*UX2 - DF(SS2,Y2,2))
   10:
J:=SS3*U+SS4$
   11:
CX22K1;
  - 9*(DF(SS3,Y2) - DF(SS2,Y2,2))
   12:
CX1K1;
  - 3*(DF(S1,Y1) - 3*DF(SS2,Y2))
```

```
 13:
SS2:=DF(S1,Y1)*Y2/3+S3$
 14:
CX1K1;
  0
 15:
CX22K1;
  - 9*DF(SS3,Y2)
 16:
SS3:=S5$
 17:
CX22K1;
  0
 18:
CX2K4;
  0
 19:
CX2K3;
  0
 20:
CX2K2;
  0
 21:
CX2K1;
  - (3*DF(S3,Y1) + DF(S1,Y1,2)*Y2 - 3*U*C4 - 2*U*C1 - 3*SS1)
 22:
COEFFN(WS,U,1);
  3*C4 + 2*C1
 23:
C4:=-2*C1/3$
 24:
WS(22);
  0
 25:
CX2K1;                    .
  - (3*DF(S3,Y1) + DF(S1,Y1,2)*Y2 - 3*SS1)
 26:
SOLVE(WS,SS1)$
 27:
SS1:=PART(PART(WS,1),2);
```

$$SS1 := \frac{3 * DF(S3, Y1) + DF(S1, Y1, 2) * Y2}{3}$$

```
    28:
CX2K1;
    0
    29:
CX2K0;
    3*DF(S5,Y1)*U + DF(S1,Y1,2)*U - 3*DF(SS4,Y2,3) + 3*DF(SS4,Y1)
    30:
COEFF(WS,U,MM);
    ***** ID fill no longer supported — use lists instead
    *** MM1 MM0 are non zero
    1
    31:
MM1;
    3*DF(S5,Y1) + DF(S1,Y1,2)
    32:
S5:=-DF(S1,Y1)/3+A1$
    33:
MM1;
    0
    34:
MM0;
    - 3*(DF(SS4,Y2,3) - DF(SS4,Y1))
    35:
CU;
    - 3*(DF(SS4,Y2,3) - DF(SS4,Y1))
    36:34
DF(SS1,Y1)-DF(SS1,Y2,3);
```

$$\frac{DF(S1, Y1, 3) * Y2 + 3 * DF(S3, Y1, 2)}{3}$$

```
    37:
S1:=A2*Y1**2+A3*Y1+A4$
    38:
S3:=A5*Y1+A6$
    39:
WS(36);
    0
    40:
V1;
    Y1*C1 + C2
```

---

[34]Do not forget that SS1 must satisfy this equation.

41:
V2;

$$\frac{Y2 * C1 + 3 * C3}{3}$$

42:
G;

$$\frac{3 * A5 + 2 * Y2 * A2 - 2 * U * C1}{3}$$

43:
W1;

$$Y1^2 * A2 + Y1 * A3 + A4$$

44:
W2;

$$\frac{2 * Y1 * Y2 * A2 + 3 * Y1 * A5 + Y2 * A3 + 3 * A6}{3}$$

45:
J;

$$\frac{-2 * U * Y1 * A2 - U * A3 + 3 * U * A1 + 3 * SS4}{3}$$

46:
BYE;
** END OF RUN

14.7. About the floppy disk

We provide the user with all the programs shown in the examples, and several others which cover a wide range of possible equations. The instructions given in this manual should help the user to construct his own program, if his equation cannot be run by using any of the files in the floppy disk.

The floppy disk contains the following files divided into four directories (GA, SGA, GS, AGA):

\GA	\SGA	\GS	\AGA
GA1O1X1	READ.ME	GS1O2X1	AGA1O3X2.KDV
GA1O1X2	SGA1O4X2.BOU	GS1O2X2	
GA1O1X3	SGA1O2X2	GS1O2X3	
GA1O2X1	SGA1O2X3	GS1O3X1	
GA1O2X2	SGA1O3X2	GS1O4X1	
GA1O2X3	SGA1O3X3	GS2O2X2	
GA1O2X4	SGA1O4X2	GS1O2X2.PBU	
GA1O2X5	SGA1O4X3		
GA1O2X6	SGA2O1X2		
GA1O3X1	SGA2O2X2		
GA1O3X2	SGA2O2X3		
GA1O3X3	SGA3O1X2		
GA1O3X4	SGA3O1X3		
GA1O4X1	SGA3O2X2		
GA1O4X2			
GA1O4X3			
GA1O5X1			
GA2O1X1			
GA2O1X2			
GA2O1X3			
GA1O2X2.BUR			
GA2O2X1			
GA2O2X2			
GA2O2X3			
GA2O2X4			
GA2O3X1			
GA2O3X4			
GA1O4X2.BOU			
GA3O1X1			
GA3O1X2			
GA3O1X3			
GA3O2X1			
GA3O2X2			
GA3O2X2.ZAK			
GA3O2X3			
GA1O2X2.FU			
GA7O2X4.MHD			
GA4O1X1			
GA4O2X4.NS			
READ.ME			

Apart from the three-letter-extension files which correspond to a particular equation (e.g., GA1O2X2.BUR), the user has to provide the equation by editing the corresponding file and following the instructions given therein.

The files \GA\READ.ME and \SGA\READ.ME should be read before using the GA and SGA programs, respectively.

References

Ablowitz, M.J. and Clarkson, P.A.

[1991] *Solitons, Nonlinear Evolution Equations and Inverse Scattering*, London Mathematical Society Lecture Notes on Mathematics, Vol. 149, Cambridge University Press, Cambridge, U.K.

Ablowitz, M.J., Kaup, D.J., Newell, A.C., and Segur, H.

[1974] The inverse scattering transform — Fourier analysis for nonlinear problems, *Stud. Appl. Math.*, 53, 249.

Ablowitz, M.J. and Segur, H.

[1981] *Solitons and the Inverse Scattering Transform,* SIAM, Philadelphia.

Abraham-Shrauner, B.

[1993] Hidden symmetries and linearization of the modified Painlevé-Ince equation, *J. Math. Phys.,* 34, 4809.

Abraham-Shrauner, B. and Guo, A.

[1994] Hidden symmetries of differential equations, *Contemp. Math.*, 160, 1.

Abraham-Shrauner, B. and Leach, P.G.L.

[1993] Hidden symmetries of nonlinear ordinary differential equations, in *Exploiting Symmetry in Applied and Numerical Analysis, Proc. AMS-SIAM Summer Seminar, Fort Collins, Colorado*, Allgower, E., Georg, K., and Miranda, R., Eds., Lectures in Applied Mathematics, Vol. 29, American Mathematical Society, Providence, RI, 1.

Abramov, S.A. and Kvashenko, K.Yu.

[1991] Fast algorithms to search for the rational solutions of linear differential equations with polynomial coefficients, in *Proc. ISSAC '91, Bonn, Germany*, Watt, S.M., Ed., ACM Press, New York, 267.

Akhatov, I.Sh., Gazizov, R.K., and Ibragimov, N.H.

[1987a] Bäcklund transformations and nonlocal symmetries, *Dokl. Akad. Nauk S.S.S.R.,* 297(1), 11. (English translation in *Sov. Math. Dokl.,* 35(2), 384, 1987.)

[1987b] Quasi–local symmetries of mathematical physics equations, in *Mathematical Modelling. Nonlinear Differential Equations of Mathematical Physics*, Samarskii, A.A., Kurdyumov, S.P., and Mazhukin, V.I., Eds., Nauka, Moscow, 22.

[1989] Nonlocal symmetries. Heuristic approach, in *Itogi Nauki i Tekhniki, Ser. Sovremennye Problemy Matematiki, Noveishie Dostizheniya*, 34, VINITI, Moscow, 3. (English translation in *J. Sov. Math.*, 55, 1401, 1991.)

Allgower, E., Georg, K., and Miranda, R.

[1993] *Exploiting Symmetry in Applied and Numerical Analysis, Proc. AMS-SIAM Summer Seminar, Fort Collins, Colorado*, 1992, Lectures in Applied Mathematics, Vol. 29, American Mathematical Society, Providence, RI.

Ames, W.F.

[1968] *Nonlinear Ordinary Differential Equations in Transport Processes*, Academic Press, New York.

[1965] *Nonlinear Partial Differential Equations in Engineering, Vol. I*, Academic Press, New York.

[1972] *Nonlinear Partial Differential Equations in Engineering, Vol. II*, Academic Press, New York.

[1992] Symmetry in nonlinear mechanics, in *Nonlinear Equations in the Applied Sciences*, Ames, W.F. and Rogers, C., Eds., Academic Press, Boston, 33.

Ames, W.F., Lohner, R.J., and Adams, E.

[1981] Group properties of $u_{tt} = [f(u)u_x]_x$, *Int. J. Non-Linear Mech.*, 16(5-6), 439.

Ames, W.F. and Rogers, C.

[1992] *Nonlinear Equations in the Applied Sciences*, Academic Press, Boston. MA.

Anderson, R.L. and Fokas, A.S.

[1982] Comments on the symmetry structure of bi-Hamiltonian systems, *Czech. J. Phys.*, B32, 365.

Anderson, R.L. and Ibragimov, N.H.

[1978] Bianchi-Lie, Bäcklund and Lie-Bäcklund transformations, in *Proc. Joint IUTAM/IMU Symp. Group Theoretical Methods in Mechanics*, Ibragimov, N.H. and Ovsiannikov, L.V., Eds., Institute of Hydrodynamics and Computing Center, Academy of Sciences, U.S.S.R., Siberian Branch, Novosibirsk, 34.

[1979] *Lie-Bäcklund Transformations in Applications*, Studies in Applied Mathematics, Vol. 1, Society for Industrial and Applied Mathematics, Philadelphia.

[1994] Invariants of Lie-Bäcklund transformation groups generated by formal power series, in *Proc. Int. Conf. Modern Group Analysis V: Theory and Applications in Mathematical Modelling, Johannesburg, January 16-22, 1994*, Ibragimov, N.H. and Mahomed, F.M., Eds. Published in: *Lie Groups and Their Applications*, 1(1), 11.

Anderson, I.M., Kamran, N., and Olver, P.J.

[1993a] Internal, external and generalized symmetries, *Adv. Math.*, 100, 53.

[1993b] Internal symmetries of differential equations, in *Modern Group Analysis: Advanced*

Analytical and Computational Methods in Mathematical Physics. Proc. Int. Workshop Acireale, Catania, Italy, 1992, Ibragimov, N.H., Torrisi, M., and Valenti, A., Eds., Kluwer Academic Publishers, Dordrecht, 7.

Arais, E.A., Shapeev, V.P., and Yanenko, N.N.

[1974] Computer implementation of Cartan's exterior forms, *Dokl. Akad. Nauk S.S.S.R.*, 214(4), 737.

Arik, M., Neyzi, F., Nutku, Y., Olver, P. J., and Verosky, J. M.

[1989] Multi-Hamiltonian structure of the Born-Infeld equation, *J. Math. Phys.*, 30, 1338.

Arrigo, D.J., Broadbridge, P., and Hill, J.M.

[1993] Nonclassical symmetry solutions and the methods of Bluman-Cole and Clarkson-Kruskal, *J. Math. Phys.*, 34, 4692.

Axford, R.

[1983] *Lectures on Lie Groups and Systems of Ordinary and Partial Differential Equations*, Los Alamos, NM.

Baas, N.A.

[1994] Sophus Lie, *The Mathematical Intelligencer*, 16 (1), 16.

Bäcklund, A.V.

[1874] Einiges über Kurven-und Flächen-Transformationen, *Lunds Univ. Arsskr.*, 10, 1.

[1876] Ueber Flächentransformationen, *Math. Ann.*, 9, 297.

[1877a] Ueber partielle Differentialgleichungen höherer Ordnung, die intermediäre erste Integrale besitzen, *Math. Ann.*, 11, 199.

[1877b] Ueber Systeme partieller Differentialgleichungen erster Ordnung, *Math. Ann.*, 11, 412.

[1878a] Ueber partielle Differentialgleichungen höherer Ordnung, die intermediäre erste Integrale besitzen. Zweite Abteilung, *Math. Ann.*, 13, 69.

[1878b] Zur Theorie der Charakteristiken der partiellen Differentialgleichungen zweiter Ordnung, *Math. Ann.*, 13, 411.

[1879] Zur Theorie der partiellen Differentialgleichungen zweiter Ordnung, *Math. Ann.*, 15, 39.

[1880] Zur Theorie der partiellen Differentialgleichungen erster Ordnung, *Math. Ann.*, 17, 285.

[1882] Zur Theorie der Flächentransformationen, *Math. Ann.*, 19, 387.

[1883] Om ytor med konstant negative kröking, *Lunds Univ. Arsskr.*, 19. (In Swedish, Summary in French.)

Baikov, V.A.

[1990] Approximate Group Analysis of Nonlinear Models of Continuum Mechanics, Dr. Sci. Thesis, Institute of Applied Mathematics, Academy of Sciences, U.S.S.R., Moscow.

[1994] Approximate symmetries of Van der Pol type equations, *Differentsial'nye uravneniya*, 30(10), 51.

Baikov, V.A. and Gazizov, R.K.

[1994] Quasi-local symmetries of nonlinear wave equations, in *CRC Handbook of Lie Group Analysis of Differential Equations. Volume 1. Symmetries, Exact Solutions, and Conservation Laws*, Ibragimov, N.H., Ed., CRC Press, Boca Raton, FL, 1994, Section 12.4.4.

Baikov, V.A., Gazizov, R.K., and Ibragimov, N.H.

[1987a] Approximate symmetries of equations with a small parameter, Preprint 150, Institute of Applied Mathematics, Academy of Sciences, U.S.S.R., Moscow.

[1987b] Formal symmetries and Bäcklund transformations, Preprint 226, Institute of Applied Mathematics, Academy of Sciences, U.S.S.R., Moscow.

[1988a] Approximate symmetries, *Mat. Sb.*, 136(3), 435. (English translation in *Math. U.S.S.R. Sb.*, 64(2), 427, 1989.)

[1988b] Approximate group analysis of nonlinear equation $u_{tt} - (f(u)u_x)_x + \varepsilon\varphi(u)u_t = 0$, *Differentsial'nye Uravneniya*, 24(7), 1127. (English translation in *Differential Equations*, 24, 1988.)

[1988c] Approximate group analysis of equation $u_{tt} - (f(u)u_x)_x + \varepsilon\varphi(u)u_t = 0$, Preprint 68, Institute of Applied Mathematics, Academy of Sciences, U.S.S.R., Moscow.

[1988d] Linearization and formal symmetries of Korteweg-de Vries equation, *Dokl. Akad. Nauk S.S.S.R.*, 303(4), 781. (English translation in *Sov. Math. Dokl.*, 38(3), 588, 1989.)

[1989a] Perturbation methods in group analysis, in *Itogi Nauki i Tekhniki, Ser. Sovremennye Problemy Matematiki, Noveishie Dostizheniya*, 34, VINITI, Moscow, 85. (English translation in *J. Sov. Math.*, 55(1), 1450, 1991.)

[1989b] Approximate symmetries and formal linearization, *Zh. Prikl. Mekh. Tekh. Fiz.*, 2, 40. (English translation in *J. Appl. Mech. Tech. Phys.*, 30, 1989.)

[1990] Classification of multi-dimensional wave equations with respect to exact and approximate symmetries, Preprint 51, Institute of Applied Mathematics, Academy of Sciences, U.S.S.R., Moscow.

[1991a] Approximate symmetries and conservation laws, in *Number Theory, Algebra, Analysis and Their Applications*, Proc. V.A. Steclov Inst. of Mathematics, vol. 200, Nauka, Moscow, 35. (English translation by AMS, Providence, RI, 35, 1993.)

[1991b] Method of multiple scales in approximate group analysis: Boussinesq and Korteweg-de Vries equations, Preprint 31, Institute of Applied Mathematics, Academy of Sciences, U.S.S.R., Moscow.

[1993] Approximate groups of transformations , *Differentsial'nye Uravneniya*, 29(10), 1712. (English translation in *Differential Equations*, 29(10), 1487.)

Baikov, V.A., Gazizov, R.K., Ibragimov, N.H., and Mahomed, F.M.

[1994] Closed orbits and their stable symmetries, 35(12),6525, *J. Math. Phys.*

Barenblatt, G.I.

[1979] *Similarity, Self-similarity and Intermediate Asymptotics*, Consultants Bureau, New York.

Bateman, H. and Erdélyi, A.

[1953] *Higher Transcendental Functions*, vol. 1–3. The Bateman Project, McGraw-Hill, New York, 1953.

Baumann, G.

[1992] Lie Symmetries of Differential Equations: a Mathematica program to determine Lie symmetries, Wolfram Research Inc., Champaign, IL, MathSource 0202-622.

[1993a] Generalized Symmetries: a Mathematica Program to Determine Lie-Bäcklund Symmetries, Wolfram Research Inc., Champaign, IL., MathSource 0204-680.

[1993b] Applications of the generalized symmetry method, in *Modern Group Analysis: Advanced Analytical and Computational Methods in Mathematical Physics. Proc. Int. Workshop Acireale, Catania, Italy*, 1992, Ibragimov, N.H., Torrisi, M., and Valenti, A., Eds., Kluwer Academic Publishers, Dordrecht, 43.

Becker, T. and Weispfenning, V.

[1993] *Gröbner Bases*, Springer-Verlag, Berlin.

Belinfante, J.G. and Kolman, B.

[1989] *A Survey of Lie Groups and Lie Algebras with Applications and Computational Methods*, Classics in Applied Mathematics, Vol. 2, Society for Industrial and Applied Mathematics, Philadelphia.

Berest, Yu.Yu.

[1991] Construction of fundamental solutions for Huygens equations as invariant solutions, *Sov. Math. Dokl.*, 43(2), 496.

[1993] Group analysis of linear differential equations in distributions and the construction of fundamental solutions, *Differentsial'nye Uravneniya*, 29(11), 1958. (English translation in *Differential Equations*, 29(11), 1561.)

Berest, Yu.Yu. and Ibragimov, N.H.

[1994] Group theoretic determination of fundamental solutions, *Lie Groups and Their Applications*, 1(2), 65.

Bérubé, D. and de Montigny, M.

[1992] A Mathematica Program for the Calculation of Lie Point Symmetries of Systems of Differential Equations, Preprint CRM-1822, Centre de Recherches Mathématiques, Montréal.

Beyer, W.A.

[1979] Lie-group theory for symbolic integration of first-order differential equations, in *Proc. 1979 MACSYMA User's Conference, Washington, D.C.*, Lewis, V.E., Ed., MIT Press, Boston, 362.

Bianchi, L.

[1918] *Lezioni sulla Teoria dei Gruppi Continui Finiti di Transformatizioni*, Enrici Spoerri, Pisa.

Bilyalov, R.F.

[1963] Conformal transformation groups in gravitational field, *Dokl. Akad. Nauk S.S.S.R.*, 152(3), 570. (English translation in *Sov. Phys. Dokl.*, 8(9), 878, 1963.)

Bluman, G.W.

[1990] Simplifying the form of Lie groups admitted by a given differential equation, *J. Math. Anal. Appl.*, 145, 52.

[1993a] Potential symmetries and linearization, in *Applications of Analytic and Geometric Methods to Nonlinear Differential Equations. NATO ASI Series C: Math. and Phys. Scs.*, Clarkson, P.A., Ed., Kluwer Academic Publishers, Dordrecht, 363.

[1993b] Potential symmetries and equivalent conservation laws, in *Modern Group Analysis: Advanced Analytical and Computational Methods in Mathematical Physics. Proc. Int. Workshop Acireale, Catania, Italy*, 1992, Ibragimov, N.H., Torrisi, M., and Valenti, A., Eds., Kluwer Academic Publishers, Dordrecht, 71.

[1993c] An overview of potential symmetries, in *Exploiting Symmetry in Applied and Numerical Analysis, Proc. AMS-SIAM Summer Seminar, Fort Collins, Colorado*, 1992, Allgower, E., Georg, K., and Miranda, R., Eds., Lectures in Applied Mathematics, Vol. 29, American Mathematical Society, Providence, RI, 97.

Bluman, G.W. and Cole, J.D.

[1969] The general similarity solution of the heat equation, *J. Math. Mech.*, 18, 1025.

[1974] *Similarity Methods for Differential Equations*, Applied Mathematical Sciences, Vol. 13, Springer-Verlag, New York.

Bluman, G.W. and Kumei, S.

[1989] *Symmetries and Differential Equations*, Applied Mathematical Science, Vol. 81, Springer-Verlag, New York.

Bobylev, A.V.

[1975a] On exact solutions of the Boltzmann equations, *Dokl. Akad. Nauk S.S.S.R.*, 225(6), 1296–1299.

[1975b] Properties of the Boltzmann equation for Maxwell molecules, *Preprint 51*, Institute of Applied Mathematics, Academy of Sciences, U.S.S.R., Moscow.

[1976] A class of invariant solutions for the Boltzmann equation, *Dokl. Acad. Nauk S.S.S.R.*, 231(3), 571–574.

[1984] Exact solutions of the nonlinear Boltzmann equation and a theory of relaxation of the Maxwell gas, *Teor. Math. Phys.*, 60, 280.

Bobylev, A.V. and Ibragimov, N.H.

[1989] Relationships between the symmetry properties of the equations of gas kinetics and hydrodynamics, *Matematicheskoe Modelirovanie*, 1(3), 100. (English translation in *Mathematical Modeling and Computational Experiment*, 1(3), 291, 1993.)

Bocharov, A.V.

[1989] DEliA: a System of Exact Analysis of Differential Equations using S. Lie Approach, Report by Joint Venture OWIMEX, Program Systems Institute, Academy of Sciences, U.S.S.R., Pereslavl-Zalessky.

[1990a] DEliA: project presentation, *SIGSAM Bulletin*, 24, 37.

[1990b] Will DEliA grow into an expert system?, in *Design and Implementation of Symbolic*

Computation Systems, Proc. Int. Symposium, Capri, Italy, Miola, A., Ed., Lecture Notes in Computer Science, Vol. 429, Springer-Verlag, Berlin, 266.

[1991] DEliA: a System of Exact Analysis of Differential Equations using S. Lie Approach. DELiA 1.5.1 User Guide, Beaver Soft Programming Team, Brooklyn, NY.

[1993] Symbolic Solvers for Nonlinear Differential Equations, *The Mathematica Journal*, 3, 63.

Bocharov, A.V. and Bronstein, M.L.

[1989] Efficiently implementing two methods of the geometrical theory of differential equations: an experience in algorithm and software design, *Acta Appl. Math.*, 16, 143.

Bock, T.L. and Kruskal, M.K.

[1979] A two-parameter Miura transformation of the Benjamin-Ono equation, *Phys. Lett.*, 74A, 173.

Boisvert, R.E., Ames, W.F., and Srivastava, U.N.

[1981] Group properties and new solutions of Navier-Stokes equations, *J. Eng. Math.*, 17, 203.

Boiti, M. and Konopelchenko, B.G.

[1985] Nonformal recursion operators and nonlinear integrable evolution equations. Preprint of Laboratoire de Physique Mathématique, Université des Sciences et Techniques du Languedoc, Montpellier.

Boiti, M., Laddomana, C., and Pempinelli, F.

[1981] An equivalent real form of the nonlinear Schrödinger equation and the permutability for Bäcklund transformations, *Nuovo Cimento B, (11)*, 62, 315.

Brihaye, Y. and Kosinski, P.

[1994] Quasi-exactly-solvable 2x2 matrix equations, *J. Math. Phys.*, 35, 3089.

Broadbridge, P. and Hill, J.M.

[1993] Similarity, symmetry and solutions of nonlinear boundary value problems, *Special Issue of Mathl. Comput. Modelling*, 18, 1.

Bronstein, M.

[1992] On solutions of linear ordinary differential equations in their coefficient field, *J. Symb. Comp.*, 13, 413.

Bryant, R.L., Chern, S.S., Gardner, R.B., Goldschmidt, H.L., and Griffiths, P.A.

[1991] *Exterior Differential Systems.* Mathematical Sciences Research Institute Publications, Vol. 18, Springer-Verlag, New York.

Buchberger, B.

[1985] Gröbner bases: an algorithmic method in polynomial ideal theory, Chapter 6, in *Multidimensional Systems Theory*, Bose, N.K., Ed., D. Reidel, Dordrecht, 184.

[1992] An Implementation of Gröbner Bases in Mathematica, Wolfram Research Inc., Champaign, IL, MathSource 0205-300.

Bunimovich, A.I. and Krasnoslobodtsev, A.V.

[1982] Group-invariant solutions of kinetic equations, *Mekh. Zhidkosti Gaza*, 4, 135.

[1983]   On invariance transformations of kinetic equations, *Vestn. Mosk. Univ., Ser. 1., Mat. Mekh.*, 4, 69.

Burnat, M.

[1962]   Causchy's problem for compressible flows of the simple wave type, *Arch. Mech. Stosow.*, 14, 3/4, 313.

Calogero, F. and Degasperis, A.

[1978]   Solution by the spectral transform method of a nonlinear evolution equation including as a special case the cylindrical KdV equation, *Lett. Nuovo Cimento*, 23, 150.

Campbell, J.E.

[1903]   *Introductory Treatise on Lie's Theory of Finite Continuous Transformation Groups*, Clarendon Press, Oxford. (Reprinted by Chelsea Publishing Company, New York, 1966.)

Carminati, J., Devitt, J.S., and Fee, G.J.

[1992]   Isogroups of differential equations using algebraic computing, *J. Sym. Comp.*, 14, 103.

Carrà-Ferro, G.

[1989]   Gröbner bases and differential ideals, in *Applied Algebra, Algebraic Algorithms, and Error-Correcting Codes. Proc. 5th Int. Conf. AAECC-5*, Huguet, L. and Poli, A., Eds., Lecture Notes in Computer Science, Vol. 356, Springer-Verlag, Berlin, 129.

Cartan, E.

[1914]   Sur l'équvalence absolue de certaines systèmes d'équationes différentielles et sur certaines familles de courbes, *Bull. Soc. Math. France*, 42, 12. Reprinted in *Oeuvres Completes*, part II, vol. 2, Gauthier-Villars, Paris, 133, 1953.

[1915]   Sur l'intégration de certaines systèmes indéterminés d'équations différentielles, *J. Reine Angew. Math.*, 145, 86. Reprinted in *Oeuvres Completes*, part II, vol. 2, Gauthier-Villars, Paris, 1169, 1953.

[1946]   *Les Systèmes Différentiels Extérieur et leurs Applications Géometrique*, Hermann, Paris.

Champagne, B., Hereman, W., and Winternitz, P.

[1991]   The computer calculation of Lie point symmetries of large systems of differential equations, *Comput. Phys. Commun.*, 66, 319.

Champagne, B. and Winternitz, P.

[1985]   A MACSYMA Program for Calculating the Symmetry Group of a System of Differential Equations, Preprint CRM-1278, Centre de Recherches Mathématiques, Montréal.

Char, B.W.

[1980]   Algorithms Using Lie Transformation Groups to Solve First Order Ordinary Differential Equations Algebraically, Ph.D thesis, Department of Computer Science, University of California, Berkeley.

Chevalley, C.C.

[1946]   *Theory of Lie Groups*, Vol. 1, Princeton University Press, Princeton, NJ.

Chupakhin, A.P.

[1979]   Nontrivial conformal groups in Riemannian spaces, *Dokl. Akad. Nauk S.S.S.R.*, 246(5), 1056. (English translation in *Sov. Math. Dokl.*, 20(3), 596, 1979.)

Cicogna, G. and Gaeta, G.

[1992] Lie-point symmetries in bifurcation problems, *Ann. Inst. Henri Poincaré, Physique Théorique*, 56, 375.

Clarkson, P.A.

[1990] New exact solutions of the Boussinesq Equation, *Eur. J. Appl. Math.*, 1, 279.

[1992] Dimensional reductions and exact solutions of a generalized nonlinear Schrödinger equation, *Nonlinearity*, 5, 453.

[1993] Nonclassical symmetry reductions of nonlinear partial differential equations, *Math. Comput. Modelling*, 18, 45.

[1994] Nonclassical symmetry reductions of the Boussinesq equation, *Chaos, Solitons and Fractals*, 4 [in press].

Clarkson, P.A. and Hood, S.

[1992] Nonclassical symmetry reductions and exact solutions of the Zabolotskaya-Khokhlov equation, *Eur. J. Appl. Math.*, 3, 381.

[1993a] Symmetry reductions of a generalized, cylindrical nonlinear Schrödinger equation, *J. Phys. A: Math. Gen.*, 26, 133.

[1993b] New symmetry reductions and exact solutions of the Davey Stewartson system. II. Reductions to partial differential equations, Preprint No. 170, Program in Applied Mathematics, University of Colorado at Boulder, Boulder.

[1994] New symmetry reductions and exact solutions of the Davey Stewartson system. I. Reductions to ordinary differential equations, *J. Math. Phys.*, 35, 255.

Clarkson, P.A. and Kruskal, M.D.

[1989] New similarity solutions of the Boussinesq equation, *J. Math. Phys.*, 30, 2201.

Clarkson, P.A. and Mansfield, E.L.

[1993a] Nonclassical symmetry reductions and exact solutions of nonlinear reaction-diffusion equations, in *Applications of Analytic and Geometric Methods to Nonlinear Differential Equations. NATO ASI Series C: Math. and Phys. Scs.*, Clarkson, P.A., Ed., Kluwer Academic Publishers, Dordrecht, 375.

[1993b] Symmetries of the nonlinear heat equation, in *Modern Group Analysis: Advanced Analytical and Computational Methods in Mathematical Physics. Proc. Int. Workshop Acireale, Catania, Italy*, 1992, Ibragimov, N.H., Torrisi, M., and Valenti, A., Eds., Kluwer Academic Publishers, Dordrecht, 155.

[1993c] Symmetry reductions and exact solutions of a class of nonlinear heat equations, *Physica D*, 70, 250.

[1994a] On a shallow water wave equation, *Nonlinearity*, 7, 975.

[1994b] *Reductions and exact solutions of a shallow water wave equation*, in *Proc. 14th IMACS World Congress on Computational and Applied Mathematics, Atlanta, Georgia*, Ames, W.F., Ed., International Association for Mathematics and Computers in Simulation (IMACS), New Brunswick, NJ, Vol. 1, 97.

[1994c] Algorithms for the nonclassical method of symmetry reductions, *SIAM J. Appl. Math.* [in press].

Clarkson, P.A. and Winternitz, P.

[1991] Nonclassical symmetry reductions for the Kadomtsev-Petviashvili equation, *Physica D*, 49, 257.

Cohen, A.

[1931] *An Introduction to the Lie Theory of One-Parameter Groups*, Stechert and Co., New York.

Cox, D., Little, J., and O'Shea, D.

[1992] *Ideals, Varieties and Algorithms. An Introduction to Computational Algebraic Geometry and Commutative Algebra.* Undergraduate Texts in Mathematics, Springer-Verlag, Berlin.

David, D., Kamran, N., Levi, D., and Winternitz, P.

[1985] Subalgebras of loop algebras and symmetries of the Kadomtsev-Petviashvili equation, *Phys. Rev. Lett.*, 55, 2111.

del Olmo, M.A., Santander, M., and Mateos Guilarte, J.

[1993] Group Theoretical Methods in Physics. *Proc. XIX Int. Colloquium, Salamanca, Spain,* 1992. Anales de Física. Monografías, Vol. 1, EDITORIAL CIEMAT, Real Sociedad Expañola de Física.

Dodonov, V.V. and Man'ko, V.I.

[1991] *Group Theoretical Methods in Physics. Proc. XVIII Int. Colloquium Group Theoretical Methods in Physics, Moscow, U.S.S.R.,* 1990, Lecture Notes in Physics, Vol. 382, Springer-Verlag, Berlin.

Donsig, A.

[1990] The Differential Forms Package, available in Maple 4.2., The Symbolic Computation Group, University of Waterloo, Ontario.

Dresner, L.

[1983] *Similarity Solutions of Nonlinear Partial Differential Equations.* Research Notes in Mathematics, Vol. 88, Pitman, Boston.

Dubrovin, B. A. and Novikov, S. P.

[1974] Periodic and conditionally periodic analogs of the many-soliton solutions of the Korteweg-de Vries equation, *Zh. Eksp. Teor. Fiz.*, 67, 2131. (English translation in *Sov. Phys.–JETP*, 40, 1058, 1974.)

[1983] Hamiltonian formalism of one-dimensional systems of hydrodynamic type and the Bogolyubov-Whitham averaging method, *Sov. Math. Dokl.*, 27, 665.

[1989] Hydrodynamics of weakly deformed soliton lattices. Differential geometry and Hamiltonian theory, *Russ. Math. Surveys*, 44, 35.

Dzhamay, A.V. and Vorob'ev, E.M.

[1994] Infinitesimal weak symmetries of differential equations in two independent variables, *J. Phys. A*, 27, 5541.

Dzhamay, A.V., Foursov, M.V., Grishin, O.V., Vorob'ev, E.M., and Zhikharev, V.N.

[1994] A "Mathematica" program SYMMAN for symmetry analysis of systems of partial differential equations, in *New Computer Technologies in Control Systems*, Proceedings of the International Workshop, Pereslavl-Zalessky (Russia), 25.

Edelen, D.G.B.

[1980a] Programs for computer implementation of the exterior calculus with applications to isovector calculations, *Comp. & Maths. with Appls.*, 6, 415.

[1980b] *Isovector Methods for Equations of Balance*, Sijthoff & Noordhoff, Alphen aan de Rijn (The Netherlands).

[1985] *Applied Exterior Calculus*, Wiley-Interscience, New York.

[1991] Exterior calculus for vectors and forms, *The Mathematica Journal*, 1 (3), 92.

Edelen, D.G.B. and Wang, J.

[1992] *Transformation Methods for Nonlinear Partial Differential Equations*, World Scientific, Singapore.

Eisenhart, L.P.

[1933] *Continuous Groups of Transformations*, Princeton University Press, Princeton, NJ.

Eliseev, V.P., Fedorova, R.N., and Kornyak, V.V.

[1985] A REDUCE program for determining point and contact Lie symmetries of differential equations, *Comput. Phys. Commun.*, 36, 383.

Ermakov, V.

[1880] Second-order differential equations. Conditions of complete integrability, *Universitetskie Izvestia,* Kiev, Ser. III, 9, 1.

Ernst, M.H. and Hendriks, E.M.

[1979] An exactly solvable non-linear Boltzmann equation, *Phys. Lett. A*, 70, 183.

Estabrook, F.

[1990] Differential geometry techniques for sets of nonlinear partial differential equations, in *Partially Integrable Nonlinear Evolution Equations and their Physical Applications. NATO ASI Series C: Math. and Phys. Scs.*, Vol. 310, Conte, R. and Boccara, N., Eds., Kluwer Academic Publishers, Dordrecht, 413.

Estévez, P.G.

[1992] The direct method and the singular manifold method for the Fitzhugh-Naguma equation, *Phys. Lett. A*, 171, 259.

[1994] Non-classical symmetries and the singular manifold method: the Burgers equation and the Burgers-Huxley equation, *J. Phys. A.: Math. Gen.*, 27, 2113.

Estévez, P.G. and Gordoa, P.R.

[1993] Non Classical Symmetries and the Singular Manifold Method: Theory and Six Examples, Preprint, Departmento di Fisica Teórica, Universidad de Salamanca, Salamanca.

Euler, N. and Steeb, W.-H.

[1993] *Continuous Symmetries, Lie Algebras and Differential Equations*, Bibliographisches Institut Wissenschaftsverlag, Mannheim.

Exton, H.

[1983] *q-Hypergeometrical functions and applications*, Horwood Publishers, Chichester, U.K.

Fedorova, R.N. and Kornyak, V.V.

[1985] Applying computer algebra systems for determination of the Lie and Lie-Bäcklund symmetries of differential equations, in *Proc. Int. Conf. on Computer Algebra and its Applications in Theoretical Physics*, Joint Institute for Nuclear Research, Dubna (U.S.S.R.), 248.

[1986] Determination of Lie-Bäcklund symmetries of differential equations using FORMAC, *Comput. Phys. Commun.*, 39, 93.

[1987] A REDUCE Program for Computing Determining Equations of Lie-Bäcklund Symmetries of Differential Equations, Report R11-87-19, Joint Institute for Nuclear Research, Dubna (U.S.S.R.).

Ferapontov, E.V. and Tsarev, S.P.

[1991] Equations of hydrodynamic type arising in gas chromatography. Riemann invariants and exact solutions, *Mat. Modelirovanie*, 3(2), 82.

Ferro, A. and Gallo, G.

[1989] Gröbner Bases, Ritt's algorithms and decision procedures for algebraic theories, in *Applied Algebra, Algebraic Algorithms, and Error-Correcting Codes. Proc. 5th Int. Conf. AAECC-5*, Huguet, L. and Poli, A., Eds., Lecture Notes in Computer Science, Vol. 356, Springer-Verlag, Berlin, 230.

Finikov, S.P.

[1948] *Cartan's Exterior Forms Method*, Gostechizdat, Moscow-Leningrad.

Finley III, J.D., and McIver, J.K.

[1993] Prolongations to higher jets of Estabrook-Wahlquist coverings for PDEs, *Acta Appl. Math.*, 32, 197.

Fitch, J.

[1978] *Manual for Standard LISP on IBM System 360 and 370*, University of Utah, Salt Lake City.

Flato, M., Pinczon, G., and Simon, J.

[1977] Nonlinear representations of Lie groups, *Ann. Sci. Ec. Norm. Sup.*, 10, 405.

Fokas, A.S.

[1987] Symmetries and integrability, *Stud. Appl. Math.*, 77, 253.

Fokas, A.S. and Anderson, R.L.

[1979] Group theoretical nature of Bäcklund transformation, *Lett. Math. Phys.*, 3, 117.

[1982] On the use of isospectral eigenvalue problems for obtaining hereditary symmetries for Hamiltonian systems, *J. Math. Phys.*, 23, 1066.

Fokas, A.S. and Fuchssteiner, B.

[1980] On the structure of symplectic operators and hereditary symmetries, *Lett. Nuovo Cimento*, 28, 299.

[1981a] Bäcklund transformations for hereditary symmetries, *Nonlinear Analysis TMA*, 5, 423.

[1981b] The hierarchy of the Benjamin-Ono equation, *Phys. Lett.*, 86A, 341.

Fokas, A.S. and Santini, P.M.

[1986] The recursion operator of the Kadomtsev-Petviashvili equation and the squared eigenfunctions of the Schrödinger operator, *Stud. Appl. Math.*, 75, 179.

[1988] Bi-Hamiltonian formulation of the Kadomtsev-Petviashvili and Benjamin-Ono equations, *J. Math. Phys.*, 29, 604.

Fomin, V.M., Shapeev, V.P., and Yanenko, N.N.

[1974] D-Properties of equations in a one-dimensional nonelastic continuous media dynamics, *Dokl. Akad. Nauk S.S.S.R.*, 215(5), 1067.

Forsyth, A. R.

[1899] Invariants, covariants and quotient-derivatives associated with linear differential equations, *Philos. Trans. R. Soc. London, Ser. A*, 179, 377.

Fuchs, J.C.

[1991] Symmetry groups and similarity solutions of MHD equations, *J. Math. Phys.*, 32, 1703.

Fuchssteiner, B.

[1979] Application of hereditary symmetries to nonlinear evolution equations, *Nonlinear Analysis TMA*, 3, 849.

[1980] Application of spectral-gradient methods to nonlinear soliton equations. Preprint of Universität Paderborn, Paderborn.

[1983] Mastersymmetries, higher order time-dependent symmetries and conserved densities of nonlinear evolution equations, *Prog. Theor. Phys.*, 70, 1508.

Fuchssteiner, B. and Fokas, A.S.

[1981] Symplectic structures, their Bäcklund transformations and hereditary symmetries, *Physica*, 4D, 47.

Fuchssteiner, B., Oevel, W., and Wiwianka, W.

[1987] Computer-algebra methods for investigation of hereditary operators of higher order soliton equations, *Comput. Phys. Commun.*, 44, 47.

Fushchich, W.I.

[1993] Conditional symmetries of the equations of mathematical physics, in *Modern Group Analysis: Advanced Analytical and Computational Methods in Mathematical Physics. Proc. Int. Workshop Acireale, Catania, Italy*, 1992, Ibragimov, N.H., Torrisi, M., and Valenti, A., Eds., Kluwer Academic Publishers, Dordrecht, 231.

[1981] Symmetry in the problems of mathematical physics, in *Algebraic-Theoretical Studies in Mathematical Physics*, Institute of Mathematics, Akad. Nauk Ukr. S.S.R., Kiev.

[1992] *Symmetry Analysis of Equations of Mathematical Physics*, Institute of Mathematics, Akad. Nauk Ukr. S.S.R., Kiev.

Fushchich, W.I., Barannik, L.F., and Barannik, A.F.

[1991]   *Subgroup Analysis of the Galilei and Poincaré Groups and Reduction of Nonlinear Equations*, Naukova Dumka, Kiev.

Fushchich, W.I., Chopik, V.I., and Mironiuk, P.I.

[1991]   Conditional symmetry and explicit solutions of the three-dimensional acoustic equation, *Rep. Ukr. Acad. Sci.*, A9, 25.

Fushchich, W.I. and Kornyak, V.V.

[1989]   Computer algebra application for determining Lie and Lie-Bäcklund symmetries of differential equations, *J. Symb. Comp.*, 7, 611.

Fushchich, W.I. and Nikitin, A.G.

[1987]   *Symmetries of Maxwell's Equations*, D. Reidel, Dordrecht.

[1990]   *Symmetry in Equations of Quantum Mechanics*, Nauka, Moscow.

Fushchich, W.I. and Serov, N.I.

[1988]   Conditional symmetries and explicit solutions of the nonlinear acoustic equation, *Rep. Ukr. Acad. Sci.*, A10, 27.

[1989]   Conditional symmetry and explicit solutions of the Boussinesq equation, in *Symmetries and solutions of differential equations of mathematical physics*, Institute of Mathematics, Akad. Nauk Ukr. S.S.R., Kiev.

Fushchich, W.I., Serov, N.I., and Ahmerov, T.K.

[1991]   On the conditional symmetry of the generalized KdV equation, *Rep. Ukr. Acad. Sci.*, A12.

Fushchich, W.I., Serov, N.I., and Chopik, V.I.

[1988]   Conditional symmetry and nonlinear heat equations, *Rep. Ukr. Acad. Sci.*, A9, 17.

Fushchich, W.I., Shtelen, V.M., and Serov, N.I.

[1989]   *Symmetry Analysis and Exact Solutions of Nonlinear Equations of Mathematical Physics*, Naukova Dumka, Kiev.

[1993]   *Symmetry Analysis and Exact Solutions of Equations of Nonlinear Mathematical Physics*, Mathematics and Its Applications, Vol. 246, Kluwer, Dordrecht.

Fushchich, W.I. and Zhdanov, R.Z.

[1989]   Symmetry and exact solutions of nonlinear spinor equations, *Phys. Rep.*, 172, 123.

[1992]   *Nonlinear Spinor Equations: Symmetry and Exact Solutions*, Naukova Dumka, Kiev.

Gaeta, G.

[1992]   Reduction and equivariant branching lemma: dynamical systems, evolution PDEs, and Gauge theories, *Acta Appl. Math.*, 28, 43.

[1994]   *Geometrical Symmetries of Nonlinear Equations and Physics*, in preparation.

Galaktionov, V.A.

[1990]   On new exact blow-up solutions for nonlinear heat conduction equations with source and applications, *Diff. Int. Eqs.*, 3, 863.

Ganzha, V.G., Meleshko, S.V., Murzin, F.A., Shapeev, V.P., and Yanenko, N.N.

[1981] Computer realization of an algorithm for investigating the compatibility conditions for systems of partial differential equations, *Dokl. Akad. Nauk S.S.S.R.*, 261(5), 1044.

Ganzha, V.G. and Yanenko, N.N.

[1982] Compatibility Analysis for Systems of Differential Equations on a Computer, Preprint # 20-82, Institute of Theoretical and Appl. Mech., Novosibirsk.

Gardner, C.S., Greene, J.M., Kruskal, M.D., and Miura, R.M.

[1974] The Korteweg-de Vries equation and generalizations. VI. Methods for exact solution, *Comm. Pure Appl. Math.*, 27, 97.

Gasper, G. and Rahman, M.

[1990] *Basic Hypergeometric Series*, Cambridge University Press, Cambridge.

Gat, O.

[1992] Symmetry algebras of third-order ordinary differential equations, *J. Math. Phys.*, 33(9), 2966.

Gel'fand, I.M. and Dikii, L.A.

[1976] Fractional degrees of operators and Hamiltonian systems, *Func. Anal. Appl.*, 10, 13.

Gel'fand, I.M. and Dorfman, I.Y.

[1979] Hamiltonian operators and algebraic structures related to them, *Func. Anal. Appl.*, 13, 13.

[1980] The Shouten bracket and Hamiltonian operators, *Func. Anal. Appl.*, 14, 71.

Gel'fand, I.M. and Shilov, G.E.

[1959] *Generalized Functions*, vol. 1, Fizmatgiz, Moscow. (English translation published by Academic Press, New York, 1964.)

Gerdt, V.P.

[1993a] Computer algebra tools for higher symmetry analysis of nonlinear evolution equations, in *Programming Environments for High-Level Problem Solving*, Gaffney, P.W. and Houstis, E.N., Eds., Elsevier Science Publishers, Amsterdam, 107.

[1993b] Computer algebra, symmetry analysis and integrability of nonlinear evolution equations, *Int. J. Mod. Phys. C*, 4, 279.

Gerdt, V.P., Shvachka, A.B., and Zharkov, A.Yu.

[1985a] FORMINT—a program for classification of integrable nonlinear evolution equations, *Comput. Phys. Commun.*, 34, 303.

[1985b] Computer algebra application for classification of integrable non-linear evolution equations, *J. Symb. Comp.*, 1, 101.

Gerdt, V.P. and Zharkov, A.Yu.

[1990] Computer generation of necessary integrability conditions for polynomial-nonlinear evolution systems, in *Proc. ISSAC '90, Tokyo, Japan*, Watanabe, S. and Nagata, M., Eds., ACM Press, New York, 250.

Goldstein, P. and Infeld, E.

[1984]   The Zakharov system: a non-Painlevé system with exact N soliton solutions, *Phys. Lett. A*, 103, 8.

González-Lopéz, A., Kamran, N., and Olver, P.J.

[1992]   Lie algebras of vector fields in the real plane, *Proc. London Math. Soc.*, 64(3), 339.

[1993]   Normalizability of one-dimensional quasi-exactly-solvable Schrödinger operators, *Comm. Math. Phys.*, 153, 117.

[1994]   Quasi-exact solvability, in: *Lie Algebras, Cohomologies and New Findings in Quantum Mechanics*, N. Kamran and P. Olver, Eds., *Contemporary Mathematics*, AMS, v.160, 113.

Gragert, P.K.H.

[1981]   Symbolic Computations in Prolongation Theory, Ph.D. thesis, Department of Mathematics, Twente University of Technology, Enschede (The Netherlands).

[1989]   Lie algebra computations, *Acta Math. Appl.*, 16, 231.

[1992]   Prolongation algebras of nonlinear PDEs, in *Proc. ERCIM Advanced Course on Partial Differential Equations and Group Theory, Part II, Bonn*, Pommaret, J.F., Ed., Gesellschaft für Mathematik und Datenverarbeitung, Sankt Augustin (Germany), 1.

Gragert, P.K.H. and Kersten, P.H.M.

[1982]   Implementation of differential geometry objects and functions with an application to extended Maxwell equations, in *Proc. EUROCAM '82, Marseille, France*, Calmet, J., Ed., Lecture Notes in Computer Science, Vol. 144, Springer-Verlag, New York, 181.

[1991]   Graded differential geometry in REDUCE: Supersymmetric structures of the modified KdV equation, *Acta Appl. Math.*, 24, 211.

Gragert, P.K.H., Kersten, P.H.M., and Martini, A.

[1983]   Symbolic computations in applied differential geometry, *Acta Appl. Math.*, 1, 43.

Grigoriev, Yu.N. and Meleshko, S.V.

[1986]   Investigation of invariant solutions for the Boltzmann kinetic equation and for its models, *Preprint of Institute of Theor. and Appl. Mechanics*, Novosibirsk.

[1987]   Group analysis of an integro-differential Boltzman equation, *Dokl. Akad. Nauk S.S.S.R.*, 297(2), 323.

[1990]   Group theoretical analysis of the kinetic Boltzman equation and of its models, *Arch. Mech.*, 42(6), 693.

Grissom, C., Thompson, G., and Wilkens, G.

[1989]   Linearization of second-order ordinary differential equations via Cartan's equivalence method, *J. Differential Equations*, 77, 1.

Grundland, A.M. and Lalague, L.

[1994a] Lie subgroups of the symmetry group of the equations describing a nonstationary and isotropic flow I, *Can. J. Phys.* [in press].

[1994b] Lie subgroups of symmetry groups of the fluid dynamics and the magnetohydrodynamics equations II, *Can. J. Phys.* [in press].

Gümral, H. and Nutku, Y.

[1990] Multi-Hamiltonian structure of equations of hydrodynamic type, *J. Math. Phys.*, 31, 2606.

Guo, A. and Abraham-Shrauner, B.

[1993] Hidden symmetries of energy-conserving differential equations, *IMA J. Appl. Math.*, 51, 147.

Gusyatnikova, V.N., Samokhin, A.V., Titov, V.S., Vinogradov, A.M., and Yumaguzhin, V.A.

[1989] Symmetries and Conservation Laws of Kadomtsev-Pogutse Equations, *Acta Appl. Math.*, 15, 23.

Guthrie, G.

[1993] Miura *Transformations and Symmetry Groups of Differential Equations*, Ph.D. thesis, University of Canterbury, Christchurch, New Zealand.

Hadamard, J.

[1923] *Lectures on Cauchy's Problem in Linear Partial Differential Equations*, Yale University Press, New Haven. (Reprinted in Dover Publications, New York, 1952.)

Halpern, M.B. and Kiritsis, E.

[1989] General Virasoro construction on affine g, *Mod. Phys. Lett.*, A4, 1373; *ibid*, Erratum, 1797.

Hansen, A.G.

[1964] *Similarity Analyses of Boundary Value Problems in Engineering*, Prentice Hall, Englewood Cliffs, NJ.

Hardy, G.H.

[1949] *Divergent Series*, Oxford University Press, London.

Harrison, B.K. and Estabrook, F.B.

[1971] Geometric approach to invariance groups and solution of partial differential equations, *J. Math. Phys.*, 12, 653.

Hartley, D. and Tucker, R.W.

[1991] A constructive implementation of the Cartan-Kähler theory of exterior differential systems, *J. Symb. Comp.*, 12, 655.

Hartley, D., Tucker, R.W., and Tuckey, P.A.

[1991] Constrained dynamics and exterior differential systems, *J. Phys. A*, 24, 5253.

Hawkins, T.

[1994] The birth of Lie's theory of groups, *The Mathematical Intelligencer*, 16 (2), 6.

Head, A.K.

[1993] LIE: a PC Program for Lie Analysis of Differential Equations, *Comput. Phys. Commun.*, 77, 241.

Hearn, A.C.

[1987] *REDUCE 3.3 User's Manual*, RAND Corp., Santa Monica, CA.

[1993] *REDUCE User's Manual Version 3.5*, RAND Publication CP78 (Rev. 10/93), Santa Monica, CA.

Helzer, G.

[1994] Gröbner Basis Package, Wolfram Research Inc., Champaign, IL, MathSource 0206-794.

Hereman, W.

[1993] SYMMGRP.MAX: a symbolic program for symmetry analysis of partial differential equations, in *Exploiting Symmetry in Applied and Numerical Analysis, Proc. AMS-SIAM Summer Seminar, Fort Collins, Colorado*, 1992, Allgower, E., Georg, K., and Miranda, R., Eds., Lectures in Applied Mathematics, Vol. 29, American Mathematical Society, Providence, RI, 241.

[1994] Review of symbolic software for the computation of Lie symmetries of differential equations, *Euromath Bull.*, 2 (1), 45.

Hereman, W., Marchildon, L., and Grundland, M.

[1993] Lie point symmetries of classical field theories, in *Group Theoretical Methods in Physics, Proc. XIX International Colloquium, Salamanca, Spain*, 1992, del Olmo, M.A., Santander, M., and Mateos Guilarte, J., Eds., Anales de Física. Monografías, Vol. 1, CIEMAT, Real Sociedad Española de Física, 402.

Hermann, R.

[1968] *Differential Geometry and the Calculus of Variations*, Academic Press, New York.

Herod, S.

[1994] Computer Assisted Determination of Lie Point Symmetries with Application to Fluid Dynamics, Ph.D Thesis, Program in Applied Mathematics, The University of Colorado, Boulder.

Hickman, M.

[1993] The Use of Maple in the Search for Symmetries, Research Report no. 77, Department of Mathematics, University of Canterbury, Christchurch, New Zealand.

[1994] The use of Maple in the search for symmetries, in *Proc. 14th IMACS World Congress on Computational and Applied Mathematics, Atlanta, Georgia*, Ames, W.F., Ed., International Association for Mathematics and Computers in Simulation (IMACS), New Brunswick, NJ, Vol. 1, 226.

Hill, J.M.

[1982] *Solution of Differential Equations by Means of One-Parameter Groups*, Research Notes in Mathematics, Vol. 63, Pitman, Boston.

[1992] *Differential Equations and Group Methods for Scientists and Engineers*, CRC Press, Boca Raton, FL.

Hood, S.

[1993] Nonclassical Symmetry Reductions and Exact Solutions of Nonlinear Partial Differential Equations, Ph.D thesis, Department of Mathematics, University of Exeter, Exeter, U.K.

Ibragimov, N.H.

[1968] On the group classification of differential equations of second order, *Dokl. Akad. Nauk S.S.S.R.*, 183(2), 174. (English translation in *Sov. Math. Dokl.* 9(6), 1365, 1968.)

[1969a] The wave equation in a Riemannian space, *Dinamika Sploshnoi Sredi*, Inst. of Hydrodynamics, Novosibirsk, 1, 36.

[1969b] Invariant variational problems and conservation laws (Remarks to the Noether theorem), *Teor. Mat. Fiz.*, 1(3), 350. (English translation in *Theor. Math. Phys.*, 1(3), 267, 1969.)

[1970] Conformal invariance and Huygens' principle, *Dokl. Akad. Nauk S.S.S.R.*, 194(1), 24. (English translation in *Sov. Math. Dokl.*, 11(5), 1153, 1970.)

[1972] *Lie Groups in Several Problems of Mathematical Physics*, Novosibirsk University, Novosibirsk.

[1977] Group theoretical nature of conservation theorems, *Lett. Math. Phys.*, 1, 423.

[1979] On the theory of Lie-Bäcklund transformation groups, *Mat. Sb.*, 109(2), 229. (English translation in *Math. U.S.S.R. Sb.*, 37(2), 205, 1980.)

[1981] Sur l' équivalence des équations d'évolution, qui admettent une algèbre de Lie-Bäcklund infinie, *C.R. Acad. Sci. Paris, Sér. I*, 293, 657.

[1983] *Transformation Groups Applied to Mathematical Physics*, Nauka, Moscow. (English translation published by D. Reidel, Dordrecht, 1985.)

[1989] *Primer on Group Analysis*, Znanie, Moscow.

[1991] *Essay on Group Analysis of Ordinary Differential Equations*, Znanie, Moscow.

[1992a] Group theoretical treatment of fundamental solutions, in *Physics on Manifolds*: International colloquim in honour of Yvonne Choquet-Bruhat, Paris, June 3–5, 1992. The proceedings published by Kluwer Academic Publishers, Dordrecht, 1994.

[1992b] Group analysis of ordinary differential equations and the invariance principle in mathematical physics (to the 150th anniversary of Sophus Lie), *Uspekhi Mat. Nauk*, 47(4), 83. (English translation in *Russ. Math. Surv.*, 47(4), 89, 1992.)

[1992c] Lie group analysis of differential equations: a mosaic of results and open problems, in *The Sophus Lie Memorial Conference*, Oslo, 17–21 August 1992. The proceedings published by Scandinavian University Press, Oslo, 1994, O.A. Laudal and B. Jahren, Eds., 53.

[1993] Seven miniatures on group analysis, *Differentsial'nye Uravneniya*, 29(10), 1739. (English translation in *Differential Equations*, 29(10), 1511, 1993.)

[1994a] Sophus Lie and harmony in mathematical physics, on the 150th anniversary of his birth, *The Mathematical Intelligencer*, 16 (1), 20.

[1994b] *CRC Handbook of Lie Group Analysis of Differential Equations. Vol. 1: Symmetries, Exact Solutions, and Conservation Laws*, Ibragimov, N.H., Ed., CRC Press, Boca Raton, FL.

[1994c] *CRC Handbook of Lie Group Analysis of Differential Equations. Vol. 2: Applications in Engineering and Physical Sciences*, Ibragimov, N.H., Ed., CRC Press, Boca Raton, FL.

[1994d] Small effects in physics hinted by the Lie group philosophy: are they observable? I.

From Galilean principle to heat diffusion, in: Proceedings of Int. Conf. on Modern Group Analysis V, Johannesburg, South Africa, January 16–22, 1994. Published in *Lie Groups and Their Applications*, 1(1), 70.

Ibragimov, N.H. and Anderson, R.L.

[1977] Lie-Bäcklund tangent transformations, *J. Math. Anal. Appl.*, 59(1), 145.

Ibragimov, N.H. and Mahomed, F.M., Eds.

[1994] *Modern Group Analysis: Theory and Application in Mathematical Modelling*, Proceedings of Int. Conf. on Modern Group Analysis V, Johannesburg, South Africa, January 16–22, 1994. Published in *Lie Groups and Their Applications*, 1(1).

Ibragimov, N.H. and Mamontov, E.V.

[1970] Sur le problème de H. Hadamard relatif à la diffusion des ondes, C.R. Acad. Sci. Paris, Sér. A, 270, 456.

[1977] On the Cauchy problem for the equation $u_{tt} - u_{xx} - \sum_{i,j=1}^{n-1} a_{ij}(x-t)u_{y_i y_j} = 0$, *Mat. Sb.*, 102(3), 391. (English translation in *Math. USSR Sb.*, 31(3), 347, 1977.)

Ibragimov, N.H. and Nucci, M.C.

[1994] Integration of third-order ordinary differential equations by Lie's method: equations admitting three-dimensional Lie algebras, *Lie Groups and Their Applications*, 1(2), 49.

Ibragimov, N.H. and Shabat, A.B.

[1979] Korteweg-de Vries equation from the group-theoretical point of view, *Dokl. Akad. Nauk S.S.S.R.*, 244(1), 57. (English translation in *Sov. Phys. Dokl.*, 24(1), 15.)

[1980a] Evolution equations with non-trivial Lie-Bäcklund group, *Func. Anal. Appl.*, 14(1), 19.

[1980b] Lie-Bäcklund Infinite algebras, *Func. Anal. Appl.*, 14(4), 313.

[1980c] L–A pairs and infinity of L–B groups and integrals for nonlinear evolution equations, *Lecture at the Workshop on Nonlinear Evolution Equations and Dynamical Systems*, Chania (Crete), 9–23 July, 1980.

Ibragimov N.H., Torrisi, M., and Valenti, A., Eds.

[1992] *Modern Group Analysis: Advanced Analytical and Computational Methods in Mathematical Physics*, Proc. of Int. Workshop, Acireale (Catania), Italy, October 27–31, 1992. Published by Kluwer Academic Publishers, Dordrecht, 1993.

Ito, M.

[1986] A REDUCE program for finding symmetries of nonlinear evolution equations with uniform rank, *Comput. Phys. Commun.*, 42, 351.

[1994] SYMCD—a REDUCE program for finding symmetries and conserved densities of systems of nonlinear evolution equations, *Comput. Phys. Commun.*, 79, 547.

Ito, M. and Kako, F.

[1985] A REDUCE program for finding conserved densities of partial differential equations with uniform rank, *Comput. Phys. Commun.*, 38, 415.

Janet, M.

[1920] Les systèmes d'équations aus dérivées partielles, *J. Math. Pure Appl.*, 3, 65.

[1929] *Le̦çons sur les Systèmes d'Equations aux Dérivées Partielles*, Gauthier-Villars, Paris.

Kadomtsev, V.V. and Petviashvili, V.I.

[1970] On the stability of solitary waves in weakly dispersing media, *Sov. Phys. Dokl.*, 15, 539.

Kähler, E.

[1934] *Einführung in die Theorie der Systeme von Differentialgleichungen*, B.G. Teubner, Leipzig.

Kalnins, E.G.

[1986] *Separation of Variables for Riemannian Spaces of Constant Curvature*, Longman, Essex, U.K.

Kamke, E.

[1959] *Differentialgleichungen (Lösungsmethoden und Lösungen), I. Gewöhnliche Differentialgleichungen*, 6, Verbesserte Auflage, Leipzig.

Kaptsov, O.V.

[1992] Invariant sets of evolution equations, *Nonlinear Anal. Theory, Methods Appl.*, 19(8), 753.

[1993] Construction of Exact Solutions for Equations of Hydrodynamics, *Dr. Sci. thesis*, Novosibirsk.

Kara, A.H. and Mahomed, F.M.

[1992] Equivalent Lagrangians and the solution of some classes of non-linear equations $\ddot{q} + p(t)\dot{q} + r(t)q = \mu\dot{q}^2 q^{-1} + f(t)q^n$, *Int. J. Non-Linear Mech.*, 27(6), 919.

Karpman, V.I.

[1989] On the equations describing modulational instabilities due to nonlinear interaction between high and low frequency waves, *Phys. Lett. A*, 136, 216.

Kersten, P.H.M.

[1987] Infinitesimal Symmetries: a Computational Approach, Ph.D. thesis, Department of Mathematics, Twente University of Technology, Enschede, The Netherlands, CWI Tract 34, Center for Mathematics and Computer Science, Amsterdam.

[1989] Software to compute infinitesimal symmetries of exterior differential systems, with applications, *Acta Appl. Math.*, 16, 207.

Kersten, P.H.M. and Krasil'shchik, I.S.

[1994] Algebraic and Geometric Structures in Differential Equations, *Proc. Meeting, Twente University of Technology, Enschede, The Netherlands*. Special Issue of *Acta Appl. Math.* [in press].

Klein, F.

[1910] Über die geometrischen Grundlagen der Lorentzgruppe, *Jahresber. Dtsch. Math. Vereinigung*, 19(9/10), 281.

[1926] *Vorlesungen über höhere Geometrie*, 3rd ed., Springer-Verlag, Berlin. Academic Press, New York.

Kolchin, E.R.

[1973] *Differential Algebra and Algebraic Groups*, Academic Press, New York.

[1984] *Differential Algebraic Groups*, Academic Press, New York.

Komrakov, B. and Lychagin, V.

[1993] Symmetries and Integrals, Preprint 15, International Sophus Lie Center & Matematisk Institutt, Universitetet i Oslo, Oslo.

Kornyak, V.V. and Fushchich, W.I.

[1991] A program for symmetry analysis of differential equations, in *Proc. ISSAC '91, Bonn, Germany*, Watt, S.M., Ed., ACM Press, New York, 315.

Kosmann-Schwarzbach, Y.

[1986] Géométrie des systémes bihamiltoniens, in *Systémes Dynamiques Non Linéaires: Intégrabilité et Comportement Qualitatif*, Winternitz, P., Ed., Les Presses de l'Université de Montreal, 186.

Kosmann-Schwarzbach, Y. and Magri, F.

[1988] Poisson-Lie groups and complete integrability, Part I. Drinfeld algebras, dual extensions and their canonical representations, *Ann. Inst. Henri Poincaré, sér. A*, 49(4), 433.

[1989] Poisson-Nijenhuis Structures, *Ann. Inst. Henri Poincaré*, July 1989, 1.

Kostant, B.

[1977] Graded manifolds, graded Lie theory and prequantization, in *Proc. Symposium on Differential Geometrical Methods in Mathematical Physics, Bonn, Germany*, 1975, Bleuler, K. and A. Reetz, R., Eds., Lecture Notes in Mathematics, Vol. 570, Springer-Verlag, New York, 177.

Krall, H.L.

[1938] Certain differential equations for Chebyshev polynomials, *Duke Math. J.*, 4, 705.

Krasil'shchik, I.S., Lychagin, V.V., and Vinogradov, A.M.

[1986] Geometry of Jet Spaces and Nonlinear Partial Differential Equations. Adv. Stud. Contemp. Math., Vol. 1, Gordon & Breach, New York.

Krasil'shchik, I.S. and Vinogradov, A.M.

[1989] Nonlocal trends in the geometry of DEs: symmetries, conservation laws and Bäcklund transformations, *Acta Appl. Math.*, 15, 161.

Krause, J. and Michel, L.

[1991] Classification of the symmetries of ordinary differential equations, *Lect. Notes Phys.*, 382, 251.

Krendelev, S.F. and Talyshev, A.A.

[1979] Systems in partial derivatives admitting nontrivial tangent transformations, *Dinamika Splosh. Sredi*, Inst. of Hydrodynamics, Novosibirsk, 40, 134.

Krook, M. and Wu, T.T.

[1977] Exact solutions of the Boltzmann equation, *Phys. Fluids*, 20(10), 1589.

Kucharczyk, P., Peradzynski, Z., and Zavistowska, E.

[1973] Unsteady multidimensional isentropic flows described by linear Riemann invariants, *Arch. Mech*, 25(2), 319.

Kupershmidt, B.A. and Wilson, G.

[1981]   Modifying Lax equations and the second Hamiltonian Structure, *Invent. Math.*, 62, 403.

Kuranishi, M.

[1967]   *Lectures on Involutive Systems of Partial Differential Equations*, Publ. Soc. Math., Sao Paulo, Brasil.

Laguerre, E.

[1879]   Sur les équations différentielles linéares du troisièm ordre, *C. R. Acad. Sci. Paris*, 88, 116.

Landau, L.D. and Lifschitz, E.M.

[1967]   *The Theory of Fields. Course of Theoretical Physics, vol. 2*, 5th revised ed., Nauka, Moscow. (English translation *The Classical Theory of Fields* published by Pergamon Press, New York, 1971.)

[1974]   *Quantum Mechanics. Course of Theoretical Physics, vol. 3*, Nauka, Moscow. (English translation *Quantum Mechanics – Non Relativistic Theory* is published by Pergamon Press, New York.)

Laplace, P.S.

[1773]   Recherches sur le calcul intégral aux différences partielles, *Mém. de l'Acad. de Paris*, pp. 341–402. (Reprinted in Oevres complètes, t. IX, Gauthier-Villars, Paris, 1893, p. 5–68; English translation, New York, 1966.)

Lax, P.D.

[1968]   Integrals of nonlinear equations of evolution and solitary waves, *Comm. Pure Appl. Math.*, 21, 467.

Leray, J.

[1953]   *Hyperbolic Differential Equations*, Mimeographed, The Institute for Advanced Study, Princeton. (Published in Russian translation by N.H. Ibragimov, Nauka, Moscow, 1984.)

Levi, D. and Winternitz, P.

[1988]   *Symmetries and Nonlinear Phenomena*, World Scientific, Singapore.

[1989]   Non-classical symmetry reductions: example of the Boussinesq equation, *J. Phys. A: Math. Gen.*, 22, 2915.

[1993]   Symmetries and conditional symmetries of differential-difference equations, *J. Math. Phys.*, 34, 3713.

Lewis, H.R., Jr.

[1967]   Classical and quantum systems with time-dependent harmonic-oscillator-type Hamiltonians, *Phys. Rev. Lett.*, 18, 510.

Leznov, A.N. and Savel'ev, M.V.

[1985]   *Group Methods in Integrating Nonlinear Dynamical Systems*, Nauka, Moscow.

Lichnerowicz, A.

[1977]   Les variétés de Poisson et leurs algébres de Lie associées, *J. Diff. Geom.*, 12, 253.

Lie, S.

[1874] Begründung einer Invariantentheorie der Berührungstransformationen, *Math. Ann.*, 8, 235. (Reprinted in Lie [1922], Vol. 4, pp. 1–96.)

[1880a] Theorie der Transformationsgruppen, *Math. Ann.*, 16, 441. (Reprinted in Lie [1922], Vol. 6, pp. 1–94. English translation *Sophus Lie's 1880 Transformation Group Paper* by M. Ackerman with comments by R. Hermann published in *Lie Groups: History, Frontiers on Applications*, Vol. 1, Mathematical Sciences Press, Brookline, MA, 1975.)

[1880b] Zur Theorie der Flächen konstanter Krümmung, III, IV, *Arch. Math. og Naturvidenskab*, 5, Heft 3, 282, 328. (Reprinted in Lie [1922], Vol. 3, pp. 398–446.)

[1881] On integration of a class of linear partial differential equations by means of definite integrals, *Arch. Math.*, VI(3), 328. (English translation by N.H. Ibragimov published in [H2], Part C.)

[1883] Klassifikation und Integration von gewönlichen Differentialgleichungen zwischen x, y, die eine Gruppe von Transformationen gestatten, *Arch. Math.*, VIII, IX, 187. (Reprinted in Lie [1922], Vol. 5, Papers X,XI, XIV, XVI.)

[1884] Sophus Lie's 1884 Differential Invariants Paper. Translated by M. Ackerman, comments by R. Hermann, in: *Lie Groups: History, Frontiers on Applications*, Vol. 3, Mathematical Sciences Press, Brookline, MA, 1976.

[1888] *Theorie der Transformationsgruppen* (Bearbeitet unter Mitwirkung von F. Engel), Vol. I, Teubner, Leipzig.

[1889] Die Infinitesimalen Berührungstransformationen der Mechanik, *Leipz. Berichte*. (Reprinted in Lie [1922], Vol. 6, Paper XXIV.)

[1890] *Theorie der Transformationsgruppen* (Bearbeitet unter Mitwirkung von F. Engel), Vol. II, Teubner, Leipzig.

[1891] *Vorlesungen über Differentialgleichungen mit Bekannten Infinitesimalen Transformationen* (Bearbeitet und herausgegeben von Dr.G. Scheffers), B.G. Teubner, Leipzig. (Reprinted by Chelsea Publishing Company, New York, 1967.)

[1893] *Theorie der Transformationsgruppen* (Bearbeitet unter Mitwirkung von F. Engel), Vol. III, Teubner, Leipzig.

[1896] *Geometrie der Berührungstransformationen* (Dargestellt von Sophus Lie und Georg Scheffers), B.G. Teubner, Leipzig.

[1922] *Gesammelte Abhandlungen*, Vol. 1–7, B.G. Teubner, Leipzig, 1922–1929.

[1971] *Vorlesungen über Continuierliche Gruppen*. Abteilung III, Chelsea Publishing Company, New York, 1971.

Lisle, I.G.

[1992] Equivalence Transformations for Classes of Differential Equations, Ph.D. thesis, Department of Mathematics, University of British Columbia, Vancouver.

Littlejohn, L.L.

[1988] Orthogonal polynomial solutions to ordinary and partial differential equations, Proceedings of Int. Symp. on Orthogonal Polynomials and their Applications, Segovia, Spain,

September 22–27, 1986. *Lecture Notes in Mathematics*, No. 1329, M. Alfaro et al., Eds., Springer-Verlag, 98.

MacCallum, M.A.H.

[1988] An ordinary differential equation solver for REDUCE, in *Proc. ISSAC '88, Rome, Italy*, Gianni, G., Ed., Lecture Notes in Computer Science, Vol. 358, Springer-Verlag, Berlin, 196.

Magri, F.

[1978] A simple model of the integrable Hamiltonian equation, *J. Math. Phys.*, 19, 1156.

[1980] A geometrical approach to the nonlinear solvable equations, in *Nonlinear Evolution Equations and Dynamical Systems*, Boiti, M, Pempinelli, F., and Soliani, G., Eds., *Lecture Notes in Physics*, Vol. 120, Springer-Verlag, Berlin, 233.

Mahomed, F.M.

[1989] *Symmetry Lie Algebras of nth Order Ordinary Differential Equations*, Ph.D thesis, Faculty of Science, University of the Witwatersrand, Johannesburg.

Mahomed, F.M. and Leach, P.G.L.

[1988] Normal forms for third-order equations, *Proc. of the Workshop on Finite Dimensional Integrable Nonlinear Dynamical Systems*, Leach, P.G.L. and Steeb, W.H., Eds., World Scientific Publ., Singapore, 178.

[1989a] Lie Algebras associated with scalar second-order ordinary differential equations, *J. Math. Phys.*, 30, 2770.

[1989b] Lie Algebra $sl(3, R)$ and linearization, *Quaest. Math.*, 12, 121.

[1990] Symmetry Lie algebras of nth order ordinary differential equations, *J. Math. Anal. Appl.*, 151(1), 80.

[1991] Contact symmetry algebras of scalar second-order ordinary differential equations, *J. Math. Phys.*, 32(8), 2051.

Mamontov, E.V.

[1984] Fundamental solution of the Cauchy problem for a certain class of Huygens' equations, *Dinamika Sploshnoi Sredi*, Inst. of Hydrodynamics, Novosibirsk, 67, 86.

Manin, Yu.I.

[1979] Algebraic aspects of nonlinear differential equations, *J. Sov. Math.*, 11, 1.

Mansfield, E.L.

[1991] Differential Gröbner Bases, Ph.D thesis, Department of Computer Science, University of Sydney, Sydney.

[1993] DIFFGROB2: a Symbolic Algebra Package for Analyzing Systems of PDE using Maple, User's Manual for Release 2, Preprint, Department of Mathematics, University of Exeter, Exeter, U.K.

Mansfield, E.L. and Clarkson, P.A.

[1994a] Applications of the differential algebra package Diffgrob2 to reductions of PDE, in *Proc. 14th IMACS World Congress on Computational and Applied Mathematics, Atlanta, Geor-*

gia, Ames, W.F., Ed., International Association for Mathematics and Computers in Simulation (IMACS), New Brunswick, NJ, Vol. 1, 336.

[1994b] Applications of the Differential Algebra Package Diffgrob2 to Classical Symmetries of Differential Equations, Preprint M94/18, Department of Mathematics, University of Exeter, Exeter, U.K.

Mansfield, E.L. and Fackerell, E.D.

[1992] Differential Gröbner Bases, Preprint 92-10, School of Mathematics, Physics, Computer Science, and Electronics, Macquarie University, Sydney.

Martini, R.

[1979] *Geometric Approaches to Differential Equations*, Lecture Notes in Mathematics, Vol. 810, Springer-Verlag, New York.

Melenk, H., Möller, H.M., and Neun, W.

[1991] GROEBNER: a Gröbner basis package, in *REDUCE User's Manual, Version 3.4*, Hearn, A.C., Ed., Rand Corporation, Santa Monica, CA.

Mikhailov, A.V. and Shabat, A.B.

[1985] Integrability conditions for systems of two equations of the form $\vec{u}_t = A(\vec{u})\vec{u}_{xx} + \vec{F}(\vec{u}, \vec{u}_x)$. I, *Theor. Math. Phys.*, 62(2), 163.

[1986] Integrability conditions for systems of two equations of the form $\vec{u}_t = A(\vec{u})\vec{u}_{xx} + \vec{F}(\vec{u}, \vec{u}_x)$. II, *Theor. Math. Phys.*, 66(1), 47.

Mikhailov, A.V., Shabat, A.B., and Sokolov, V.V.

[1990] The symmetry approach to classification of integrable equations, in *What is Integrability?*, Springer Lecture Notes in Nonlinear Dynamics, Zakharov, V.I., Ed., Springer-Verlag, New York, 115.

Mikhailov, A.V., Shabat, A.B., and Yamilov, R.I.

[1987] Symmetry approach to classification of nonlinear equations. Complete lists of integrable systems, *Usp. Mat. Nauk*, 42(4), 3.

Miller, W., Jr.

[1968] *Lie Theory and Special Functions*, Academic Press, New York.

[1972] *Symmetry Groups and Their Applications*, Pure and Applied Mathematics, Vol. 50, Academic Press, New York.

[1977] *Symmetry and Separation of Variables*, Addison-Wesley, Reading, MA.

Minkowski, H.

[1908] Address delivered at the 80th Assembly of German Natural Scientists and Physicians, Cologne. English translation 'Space and time' is published in *The Principle of Relativity*, Dover Publications, New York.

Miura, R.M.

[1968] Korteweg-de Vries equation and generalizations. I. A remarkable explicit nonlinear transformation, *J. Math. Phys.*, 9, 1202.

[1976] *Bäcklund Transformations, the Inverse Scattering Method, Solitons, and Their Applications*, Lecture Notes in Mathematics, Vol. 515, Springer-Verlag, New York.

Morozov, A.Yu., Perelomov, A.M., Rosly, A.A., Shifman, M.A., and Turbiner, A.V.

[1990] Quasi-exactly-solvable problems: one-dimensional analogue of rational conformal field theories, *Int. J. Mod. Phys.*, A5, 803.

Muncaster, R.G.

[1979] On generating exact solutions of the Maxwell-Boltzmann equation, *Arch. for Rational Mech. Anal.*, 70, 79.

Murphy, G.M.

[1960] *Ordinary Differential Equations and Their Solutions*, van Nostrand, New York.

Neuman, F.

[1991] *Global Properties of Linear Ordinary Differential Equations*, Kluwer Academic Publishers, Dordrecht.

Nikol'skii, A.A.

[1963a] The simplest exact solutions of the Boltzmann equation for the rarefied gas motion, *Dokl. Akad. Nauk S.S.S.R.*, 151(2), 299.

[1963b] A three-dimensional homogeneous expansion-contraction of the rarefied gas with power interaction functions, *Dokl. Akad. Nauk S.S.S.R.*, 151(3), 522.

[1965] Homogeneous displacement motion of the monatomic rarefied gas, *Inzhenernyi J.*, 5(4), 752.

Nishitani, T. and Tajiri, M.

[1982] On similarity solutions of the Boussinesq equation, *Phys. Lett. A*, 89, 379.

Noether, E.

[1918] Invariante Variationsprobleme, *Nachr. König. Gesell. Wissen. Göttingen, Math.-Phys. Kl.*, 235. (English translation in *Transport Theory and Stat. Phys.*, 1, 186, 1971.)

Nonenmacher, T.F.

[1984] Application of the similarity method to the nonlinear Boltzmann equation, *J. Appl. Math. Phys. (ZAMP)*, 35(9), 680.

Nucci, M.C.

[1984] Group analysis of M.H.D. equations, *Atti. Sem. Mat. Fis. Univ. Modena*, 33, 21.

[1990] Interactive REDUCE programs for calculating, classical, non-classical, and Lie-Bäcklund symmetries of differential equations, Manual of the Program, Preprint GT Math:062090-051, School of Mathematics, Georgia Institute of Technology, Atlanta.

[1991] Interactive REDUCE programs for calculating, classical, non-classical, and approximate symmetries of differential equations, in *Proc. 13th IMACS World Congress, Dublin, Ireland*, Vichnevetsky, R. and Miller, J.J.H., Eds., Criterion Press, Dublin, 349.

[1992a] Interactive REDUCE programs for calculating, classical, non-classical, and Lie-Bäcklund symmetries of differential equations, in *Computational and Applied Mathematics, II: Differential Equations. Selected and revised papers from the IMACS 13th World Congress, Dublin, Ireland*, 1991, Ames, W.F. and Van der Houwen, P.J., Eds., Elsevier Science Publishers, Amsterdam, 345.

[1992b] Tales of Gods and Heroes: "The nectar of the Gods", *Notices A.M.S.*, 39, 427.

[1993a] Symmetries and symbolic computation, in *Differential Equations with Applications to Mathematical Physics*. Mathematics in Science and Engineering, 192, Ames, W.F., Harrell II, E.M., and Herod, J.V., Eds., Academic Press, New York, 249.

[1993b] Nonclassical symmetries and Bäcklund transformations, *J. Math. Anal. Appl.*, 178, 294.

[1993c] Iterating the Nonclassical Symmetry Method, Tech. Rep. 1993-14, Departimento de Matematica, Università di Perugia, Perugia.

[1995] Interactive REDUCE programs for calculating Lie-point, non-classical, Lie-Bäcklund, and approximate symmetries of differential equations: manual and floppy disk, in *CRC Handbook of Lie Group Analysis of Differential Equations. Vol. 3: New Trends*, CRC Press, Boca Raton, FL, chap. 14.

Nucci, M.C. and Ames, W.F.

[1993] Classical and nonclassical symmetries for the Helmholtz equation, *J. Math. Anal. Appl.*, 178, 584.

Nucci, M.C. and Clarkson, P.A.

[1992] The nonclassical method is more general than the direct method for symmetry reductions. An example of the Fitzhugh-Nagumo equation, *Phys. Lett. A*, 164, 49.

Oaku, T.

[1994] Algorithms for finding the structure of solutions of a system of linear partial differential equations, in *Proc. ISSAC '94, Oxford, United Kingdom*, von zur Gathen, J. and Giesbrecht, M., Eds., ACM Press, New York, 216.

Oevel, W. and Fokas, A.S.

[1982] Infinitely many commuting symmetries and constants of motion in involution for explicitly time-dependent evolution equations. Preprint, Univ. of Paderborn.

Ogievetsky, O. and Turbiner, A.

[1991] $sl(2, \mathbf{R})_q$ and quasi-exactly-solvable problems, Preprint CERN-TH 6212/91.

Okubo, S.

[1989] Nijenhuis tensor, BRST cohomology, and related algebras. Preprint, University of Rochester.

Ollivier, F.

[1991] Standard bases of differential ideals, in *Applied Algebra, Algebraic Algorithms, and Error-Correcting Codes. Proc. 8th Int. Conf. AAECC-8*, Huguet, L. and Poli, A., Eds., Lecture Notes in Computer Science, Vol. 508, Springer-Verlag, Berlin, 304.

Olver, P.J.

[1977] Evolution equations possessing infinitely many symmetries, *J. Math. Phys.*, 18, 1212.

[1980] On the Hamiltonian structure of evolution equations, *Math. Proc. Camb. Phil. Soc.*, 88, 71.

[1984] Hamiltonian perturbation theory and water waves, *Contemp. Math.*, 28, 231.

[1986] *Applications of Lie Groups to Differential Equations*, Springer-Verlag, New York; 2nd ed., Springer-Verlag, New York, 1993.

[1987] Bi-Hamiltonian systems, in *Ordinary and Partial Differential Equations*, Sleeman, B.D.

and Jarvis, R.J., Eds., *Pitman Research Notes in Mathematics Series*, No. 157, Longman Scientific and Technical, New York, 176.

[1992] Symmetry and explicit solutions of partial differential equations, *Appl. Num. Math.*, 10, 307.

[1994a] Direct reduction and differential constraints, *Proc. R. Soc. London A.*, 444, 509.

[1994b] *Equivalence, Invariants and Symmetry*, Cambridge University Press, Cambridge.

Olver, P.J. and Nutku, Y.

[1988] Hamiltonian structures for systems of hyperbolic conservation laws, *J. Math. Phys.*, 29, 1610.

Olver, P.J. and Rosenau, P.

[1986] The construction of special solutions to partial differential equations, *Phys. Lett. A*, 114, 107.

[1987] Group-invariant solutions of differential equations, *SIAM J. Appl. Math.*, 47, 263.

Ondich, J.R.

[1989] Partially Invariant Solutions of Differential Equations, Ph.D thesis, Department of Mathematics, University of Minnesota, Minneapolis, Minnesota.

[1994] The reducibility of partially invariant solutions of partial differential equations, *Eur. J. Appl. Math.*, to appear.

Oron, A. and Rosenau, P.

[1986] Some symmetries of nonlinear heat and wave equations, *Phys. Lett. A*, 118(4), 172.

Oseen, C.W.

[1929] Albert Victor Bäcklund, *Jahresber. Dtsch. Math. Vereinigung*, 38, 113.

Ovsiannikov, L.V.

[1959] Group properties of nonlinear heat equation, *Dokl. Akad. Nauk S.S.S.R.*, 125(3), 492.

[1962] *Group Properties of Differential Equations*, Academy of Sciences, Siberian Branch, U.S.S.R., Novosibirsk. (English translation by Bluman, G.W., 1967, unpublished.)

[1965] Group-invariant solutions of hydrodynamics equations, *Tr. II Vsesoiuzn. S'ezda Teor. i Prikl. Mekhan.*, No. 2, Nauka, Moscow, 302.

[1971] Nonlinear Cauchy problem in the scale of Banach spaces, *Dokl. Akad. Nauk S.S.S.R.*, 200, 789. (English translation in *Sov. Math. Dokl.*, 12.)

[1978] *Group Analysis of Differential Equations*, Nauka, Moscow. (English translation published by Academic Press, New York, 1982).

[1981] *Lectures on Foundations of Gas Dynamics*, Nauka, Moscow.

Palais, R.S.

[1989] *A Global Formulation of the Lie Theory of Transformation Groups*, Memoirs of the American Mathematical Society, Vol. 22, American Mathematical Society, Providence, RI.

Pankrat'ev, E.V.

[1989] Computations in differential and difference modules, *Acta Appl. Math.*, 16, 167.

Pavlov, M. V.

[1987] Hamiltonian formalism of the electrophoresis equation. Integrable equations of hydrodynamics. Preprint 17 of Landau Inst. Theoret. Phys., Moscow.

Penrose, R.

[1964] Conformal treatment of infinity, in: *Relativité, Groups et Topologie, Les Houches Lectures, 1963 Summer School of Theor. Phys.*, Univ. Grenoble, Gordon and Breach, New York, 565.

Petrov, A.Z.

[1966] *New Methods in General Relativity*, Nauka, Moscow. (English translation: *Einstein Spaces*, Pergamon Press, Oxford, 1969.)

Pinney, E.

[1950] The nonlinear differential equation $y'' + p(x)y + Cy^{-3} = 0$, *Proc. Am. Math. Soc.*, 1, 681.

Pirani, F.A.E., Robinson, D.C., and Shadwick, W.A.

[1979] *Local Jet Bundle Formulation of Bäcklund Transformations*. Mathematical Physics Studies, Vol. 1, D. Reidel, Dordrecht.

Pohle, H.J. and Wolf, T.

[1989] Automatic determination of dynamical symmetries of ordinary differential equations, *Computing*, 41, 297.

Polishchuk, E.M.

[1983] *Sophus Lie*, Nauka, Leningrad.

Pommaret, J.F.

[1978] *Systems of Partial Differential Equations and Lie Pseudogroups*, Gordon & Breach, New York.

[1983] *Differential Galois Theory*, Gordon & Breach, New York.

[1988] *Lie Pseudogroups and Mechanics*, Gordon & Breach, New York.

[1992] *Lecture Notes of 'Advanced Course on Partial Differential Equations and Group Theory'. New Perspectives in Computer Algebra and Applications to Engineering Sciences. Proc. ERCIM Advanced Course, Gesellschaft für Mathematik und Datenverarbeitung, Bonn.*

Popov, M.D.

[1985] Automized computing of defining equations for a Lie group, *Izv. Akad Nauk B. S.S.R., Ser. Phys.-Math.*, 2, 33.

Pucci, E.

[1992] Similarity reductions of partial differential equations, *J. Phys. A: Math. Gen.*, 25, 2631.

Raspopov, V.E., Shapeev, V.P., and Yanenko, N.N.

[1974] Application of the method of differential constraints to one-dimensional gasdynamics equations, *Izv. Vuzov. Matemat.*, 11(150), 69.

[1979] D-properties of the system of symmetrical gas flow equations, *Dokl. Akad. Nauk S.S.S.R.*, 224(2), 308.

Ray, J.R. and Reid, J.L.

[1979]   More exact invariants for the time-dependent harmonic oscillator, *Phys. Lett. A*, 71, 317.

Razavy, M.

[1980]   An exactly soluble Schrödinger equation with a distable potential, *Am. J. Phys.*, 48, 285.

[1981]   A potential model for torsional vibrations of molecules, *Phys. Lett.*, A82, 7.

Reid, G.J.

[1990a] A triangularization algorithm which determines the Lie symmetry algebra of any system of PDEs, *J. Phys. A: Math. Gen.*, 23, L853.

[1990b] Algorithmic determination of Lie symmetry algebras of differential equations, in *Lie Theory, Differential Equations and Representation Theory, Proc. Annual Seminar of the Canadian Math. Soc., Montréal*, Hussin, V., Ed., Les Publications de Centre de Recherches Mathématiques, Montréal, Canada, 363.

[1991a] Algorithms for reducing a system of PDEs to standard form, determining the dimension of its solution space and calculating its Taylor series solution, *Eur. J. Appl. Math.*, 2, 293.

[1991b] Finding abstract Lie symmetry algebras of differential equations without integrating determining equations, *Eur. J. Appl. Math.*, 2, 319.

Reid, G.J. and Boulton, A.

[1991]   Reduction of systems of differential equations to standard form and their integration using directed graphs, in *Proc. ISSAC '91, Bonn, Germany*, Watt, S.M., Ed., ACM Press, New York, 308.

Reid, G.J., Lisle, I.G., Boulton, A., and Wittkopf, A.D.

[1992]   Algorithmic determination of commutation relations for Lie symmetry algebras of PDEs, in *Proc. ISSAC '92, Berkeley, California*, Trager, B. and Lazard, D., Eds., ACM Press, New York, 63.

Reid, G.J. and McKinnon, D.K.

[1993]   Solving systems of linear PDEs in their coefficient field by recursively decoupling and solving ODEs, Preprint, Department of Mathematics, University of British Columbia, Vancouver.

Reid, G.J., Weih, D.T., and Wittkopf, A.D.

[1993]   A point symmetry group of a differential equation which cannot be found using infinitesimal methods, in *Modern Group Analysis: Advanced Analytical and Computational Methods in Mathematical Physics. Proc. Int. Workshop Acireale, Catania, Italy*, 1992, Ibragimov, N.H., Torrisi, M., and Valenti, A., Eds., Kluwer Academic Publishers, Dordrecht, 311.

Reid, G.J. and Wittkopf, A.D.

[1993]   Long Guide to the Standard Form Package, Department of Mathematics, University of British Columbia, Vancouver.

Riquier, C.

[1910]   *Les Systèmes d'Equations aux Dérivées Partielles*, Gauthier-Villars, Paris.

Ritt, J.F.

[1966] *Differential Algebra*, Dover, New York.

Rogers, C. and Ames, W.F.

[1989] *Nonlinear Boundary Value Problems in Science and Engineering*. Mathematics in Science and Engineering, Vol. 183, Academic Press, New York.

Rogers, C. and Shadwick, W.A.

[1982] *Bäcklund Transformations and Their Applications*, Academic Press, New York.

Rosenau, P. and Schwarzmeier, J.L.

[1979] Similarity Solutions of Systems of Partial Differential Equations using MACSYMA, Report COO-3077-160 MF-94, Magneto-Fluid Dynamics Division. Courant Institute of Mathematical Sciences, New York University, New York.

[1986] On similarity solutions of Boussinesq-type equations, *Phys. Lett. A*, 115, 75.

Rozhdestvenskii, B.L. and Yanenko, N.N.

[1978] *Systems of Quasilinear Equations and Their Applications to Gas Dynamics*, 2nd Ed., Nauka, Moscow. (Translations of Mathematical Monographs, Vol. 55, American Mathematical Society, Providence, RI, 1983.)

Sánchez Mondragón, J. and Wolf, K.B.

[1986] *Lie Methods in Optics, Proc. CIFMO-CIO Workshop, Léon, Guanajuato, México*, 1985. Lecture Notes in Physics, Vol. 250, Springer-Verlag, New York.

Sarlet, W. and Cantrijn, F.

[1981] Generalizations of Noether's theorem in classical mechanics, *SIAM Rev.*, 23, 467.

Sarlet, W., Mahomed, F.M., and Leach, P.G.L.

[1987] Symmetries of non-linear differential equations and linearization, *J. Phys. A*, 20, 277.

Sarlet, W. and Vanden Bonne, J.

[1992] REDUCE-procedures for the study of adjoint symmetries of second-order differential equations, *J. Sym. Comp.*, 13, 683.

Sastri, C.C.A.

[1986] Group analysis of some partial differential equations arising in applications, *Contemp. Math.*, 54, 35.

Sattinger, D.H.

[1979] *Group Theoretic Methods in Bifurcation Theory*, Lecture Notes in Mathematics, Vol. 762, Springer-Verlag, New York.

Sattinger, D.H. and Weaver, O.L.

[1986] *Lie Groups and Algebras with Applications to Physics, Geometry, and Mechanics*, Applied Mathematical Sciences, Vol. 61, Springer-Verlag, New York.

Schrüfer, E.

[1987] EXCALC, a system for doing calculations in the calculus of modern differential geometry—User's Manual, distributed with REDUCE 3.3.

[1990] EXCALC—a package for calculations in modern differential geometry, in *Computer*

Algebra in Physical Research. Proc. IV Int. Conf. Dubna, U.S.S.R., Shirkov, D.V., Ros-tovtsev, V.A., and Gerdt, V.P., Eds., World Scientific, Singapore, 71.

Schü, J., Seiler, W.M., and Calmet, J.

[1993] Algorithmic methods for Lie pseudogroups, in *Modern Group Analysis: Advanced Analytical and Computational Methods in Mathematical Physics. Proc. Int. Workshop Acireale, Catania, Italy*, 1992, Ibragimov, N.H., Torrisi, M., and Valenti, A., Eds., Kluwer Academic Publishers, Dordrecht, 337.

Schwarz, F.

[1982] A REDUCE package for determining Lie symmetries of ordinary and partial differential equations, *Comput. Phys. Commun.*, 27, 179.

[1983] Automatically determining symmetries of ordinary differential equations, in *Proc. EU-ROCAL '83, London, United Kingdom*, van Hulzen, J.A., Ed., Lecture Notes in Computer Science, Vol. 162, Springer-Verlag, New York, 45.

[1984] The Riquier-Janet theory and its application to non-linear evolution equations, *Phys. D*, 11, 243.

[1985] Automatically determining symmetries of partial differential equations, *Computing*, 34, 91; Addendum: *Computing*, 36 (1986), 279.

[1987a] Symmetries and involution systems: some experiments in computer algebra, in *Topics in Soliton Theory and Exactly Solvable Nonlinear Equations. Proc. Meeting on Nonlinear Evolution Equations, Oberwolfach, Germany*, 1986, Ablowitz, M., Fuchssteiner, B., and Kruskal, M., Eds., World Scientific, Singapore, 290.

[1987b] The Package SPDE for Determining Symmetries of Partial Differential Equations. User's Manual. Distributed with REDUCE 3.3, Rand Corporation, Santa Monica, CA.

[1988a] Programming with abstract data types: the symmetry package SPDE in SCRATCHPAD, in *Trends in Computer Algebra, Proc. Int. Symposium, Bad Neuenahr, Germany*, 1987, Janßen, R., Ed., Lecture Notes in Computer Science, Vol. 296, Springer-Verlag, New York, 167.

[1988b] Symmetries of differential equations from Sophus Lie to computer algebra, *SIAM Rev.*, 30, 450.

[1989] A factorization algorithm for linear ordinary differential equations, in *Proc. ACM-SIGSAM 1989 Int. Symp. on Symbolic and Algebraic Computation, ISSAC '89, Portland, Oregon*, Gonnet, G., Ed., ACM Press, New York, 17.

[1992a] An Algorithm for Determining the Size of Symmetry Groups, *Computing*, 49, 95.

[1992b] Reduction and completion algorithms for partial differential equations, in *Proc. ACM-SIGSAM 1992 Int. Symp. on Symbolic and Algebraic Computation, ISSAC '92, Berkeley, California*, Trager, B. and Lazard, D., Eds., ACM Press, New York, 49.

[1994a] SPDE: Symmetries of Partial Differential Equations, Guidelines for the use of SPDE 1.0, Institut SCAI, GMD, Postfach 1316, D-53731 Sankt Augustin, Germany, 4 pages.

[1994b] Computer Algebra Software for Scientific Applications, in *Proc. 1993 CISM Advanced School on Computerized Symbolic Manipulation in Mechanics, Udine, Italy*, 1993, Kreuzer, E., Ed., Springer-Verlag, Berlin [to appear].

Schwarz, F. and Pommaret, J.F.

[1994]   Partial Differential Equations and Group Theory, 'Advanced Course on Partial Differential Equations and Group Theory'. New Perspectives in Computer Algebra and Applications to Engineering Sciences. ERCIM Course, Gesellschaft für Mathematik und Datenverarbeitung, Sankt-Augustin, Germany.

Schwarzmeier, J.L. and Rosenau, P.

[1988]   Using MACSYMA to Calculate Similarity Transformations of Partial Differential Equations, Report LA-UR 88-4157, Los Alamos National Laboratory, Los Alamos, NM.

Sedov, L.I.

[1957]   *Similarity and Dimensional Methods in Mechanics*, Nauka, Moscow. (English translation by Academic Press, New York, 1959.)

Seiler, W.M.

[1994a]   On the arbitrariness of the general solution of an involutive partial differential equation, *J. Math. Phys.*, 35, 486.

[1994b]   Analysis and Application of the Formal Theory of Partial Differential Equations, Ph.D thesis, Department of Physics, University of Lancaster, Lancaster, U.K.

Sesahdri, R. and Na, T.Y.

[1985]   *Group Invariance in Engineering Boundary Value Problems*, Springer-Verlag, New York.

Shabat, A.B.

[1989]   A canonical series of conservation laws, in *Russian translation of Olver [1986], Supplement I*, Mir, Moscow, 582.

Shabat, A.B. and Sokolov, V.V.

[1982]   Necessary conditions of nontriviality of Lie-Bäcklund algebra and existence of conservation laws. Preprint of the report to Presidium of Bashkir branch of the Academy of Sciences of U.S.S.R., Ufa.

Shadwick, W.A.

[1979]   The Bäcklund Problem, Symmetries and Conservation Laws for Some Nonlinear Differential Equations, Ph.D thesis, Department of Mathematics, King's College, London.

Shapeev, V.P.

[1974a]   Application of the Method of Differential Constraints to the One-Dimensional Dynamics Equations in Continuous Media, Ph.D thesis, Novosibirsk.

[1974b]   The problem of continuous joining of DP–solutions for one-dimensional equations of nonelastic continuous media, *Chisl. Metody Mekh. Splosh. Sredy*, Novosibirsk, 4(3), 39.

Sheftel', M. B.

[1983]   Infinite-dimensional non-commutative Lie-Bäcklund algebra associated with equations of one-dimensional gas dynamics, *Theor. Math. Phys.*, 56, 878.

[1993]   Higher integrals and symmetries of semi-Hamiltonian systems, *Differentsial'nye Uravneniya*, 29(10), 1782. (English translation in *Differential Equations*, 29(10), 1548.)

[1994a]   Group analysis of determining equations: a method for finding recursion operators, *Differentsial'nye Uravneniya*, 30(3), 444.

[1994b] Symmetries, Recursions and Integrals for Hydrodynamic-type Systems, Dr. Sci. thesis, Tomsk University, Tomsk.

Sherring, J.

[1993a] DIMSYM—Symmetry Determination and Linear Differential Equations Package, Preprint, Department of Mathematics, LaTrobe University, Bundoora, Australia.

[1993b] Symmetry and Computer Algebra Techniques for Differential Equations, Ph.D thesis, Department of Mathematics, LaTrobe University, Bundoora, Australia.

Sherring, J. and Prince, G.

[1992] Geometric Aspects of reduction of order, *Trans. Am. Math. Soc.*, 314, 433.

Shifman, M.A.

[1994] Quasi-exactly-solvable spectral problems and conformal field theory, in: *Lie Algebras, Cohomologies and New Findings in Quantum Mechanics*, Kamran, N. and Olver, P., Eds., *Contemporary Mathematics*, AMS, v. 160, 237

Shifman, M.A. and Turbiner, A.V.

[1989] Quantal problems with partial algebraization of the spectrum, *Comm. Math. Phys.* 126, 347.

Sidorov, A.F., Shapeev, V.P., and Yanenko, N.N.

[1984] *Method of Differential Constraints and Its Applications in Gasdynamics*, Nauka, Novosibirsk.

Singer, M.F.

[1991] Formal solutions of differential equations, *J. Symb. Comp.*, 10, 59.

Singh, V., Rampal, A., Biswas, S.N., and Datta, K.

[1980] A class of exact solutions for doubly anharmonic oscillators, *Lett. Math. Phys.*, 4, 131.

Sokolov, V.V.

[1988] Symmetries of evolution equations, *Usp. Mat. Nauk*, 43(5), 133.

Steeb, W.-H.

[1993] *Invertible Point Transformations and Nonlinear Differential Equations*, World Scientific, Singapore.

Steinberg, S.

[1979] Symmetry operators, in *Proc. 1979 MACSYMA User's Conference, Washington, D.C.*, Lewis, V.E., Ed., MIT Press, Boston, 408.

[1986] Lie series, Lie transformations, and their applications, in *Lie Methods in Optics. Proc. CIFMO-CIO Workshop, Leon, Mexico, 1985*, Sánchez Mondragón, J. and Wolf, K.B., Eds., Lecture Notes in Physics, Vol. 250, Springer-Verlag, New York, 45.

[1990] Symmetries of differential equations, *MACSYMA Newsl.*, 7, 3.

Stephani, H.

[1989] *Differential Equations: Their Solution using Symmetries*, Cambridge University Press, Cambridge.

Tajiri, M., Nishitani, T., and Kawamoto, S.

[1982]   On solutions of the Kadomtsev-Petviashvili equation, *J. Phys. Soc. Jpn.*, 51, 2350.

Takhtadjan, L.A. and Faddeev, L.D.

[1986]   *Hamiltonian Approach to Soliton Theory*, Nauka, Moscow.

Talyshev, A.A.

[1980]   Tangent transformations of one-dimensional gasdynamics equations with one differential constraint, *Dynamika Splosh. Sredy*, Novosibirsk, 44, 121.

Taranov, V.B.

[1976]   Symmetry of the one-dimentional high frequency motion of collisionless plasma, *Zh. Tekh. Fiz.*, 46(6), 1271.

Teshukov, V.M.

[1989]   Hyperbolic systems, which admit nontrivial Lie-Bäcklund groups, in *Lie-Bäcklund Groups and Quasilinear Systems*, LIIAN preprint No. 106, Leningrad, 25.

Thomas, J.

[1937]   *Differential Systems*. Colloquium Publications, Vol. 21, American Mathematical Society, New York.

Titov, S.S.

[1990]   Analycity of linear one-parameter Lie-Bäcklund groups, *Differents. Uravn.*, 26(4), 699.

Topunov, V.L.

[1989]   Reducing systems of linear differential equations to a passive form, *Acta Appl. Math.*, 16, 191.

Tresse, A.M.

[1894]   Sur les invariants différentiels des groupes de transformations, *Acta Math.*, 18, 1.

[1896]   *Détermination des Invariants Ponctuel de l'Equation Différentielle Ordinaire du Second Ordre* $y'' = \omega(x, y, y')$, S. Hirzel, Leipzig.

Tsarev, S.P.

[1985]   On Poisson brackets and one-dimensional systems of hydrodynamic type, *Sov. Math. Dokl.*, 31, 488.

[1991]   Geometry of Hamiltonian systems of hydrodynamic type. Generalized hodograph method, *Math. in the U.S.S.R. Izvestiya*, 37, 397.

Tschebotaröw, N.G.

[1940]   *Theory of Lie Groups*, GITTL, Moscow-Leningrad.

Tseytlin, A.A.

[1994]   Conformal sigma models corresponding to gauged Wess-Zumino-Novikov-Witten theories, *Nucl. Phys.*, B411, 509.

Tsujishita, T.

[1982]   On variation bicomplexes of differential equations, *Osaka J. Math.*, 19, 311.

[1983]   Formal geometry of systems of differential equations, in *Sügaku Expositions*, Vol. 3, American Mathematical Society, Providence, RI, 1990, 25.

[1991]   *Diff. Geom. Appl.*, 1, 3.

Turbiner, A.V.

[1988a] Spectral Riemannian surfaces of the Sturm-Liouville operators and quasi-exactly-solvable problems, *Sov. Math.-Funk. Analysis i ego Prilogenia*, 22, 92.

[1988b] Quantum mechanics: problems intermediate between exactly-solvable and non-solvable, *Zh. Eksp. Teor. Fiz.*, 94, 33. (English translation in *Sov. Phys.-JETP* 67, 230.)

[1988c] Quasi-exactly-solvable problems and $sl(2, \mathbf{R})$ algebra, *Comm. Math. Phys.* 118, 467 (Preprint ITEP-197, 1987).

[1989]   Lamé equation, sl_2 and isospectral deformation, *J. Phys.*, A22, L1-L3.

[1992]   On polynomial solutions of differential equations, *J. Math. Phys.*, 33, 3989.

[1994a] Hidden algebra of Calogero model, *Phys. Lett. B*, B320, 281.

[1994b] Lie algebras and linear operators with invariant subspace, in: *Lie Algebras, Cohomologies and New Findings in Quantum Mechanics*, Kamran, N. and Olver, P., Eds., *Contemporary Mathematics*, AMS, v. 160, 263.

Turbiner, A.V. and Ushveridze, A.G.

[1987]   Spectral singularities and the quasi-exactly-solvable problem, *Phys. Lett.*, A126, 181.

Ulyanov, V.V. and Zaslavskii, O.B.

[1992]   New methods in the theory of quantum spin systems, *Phys. Reps.*, 216, 179.

Vafeades, P.

[1990a] Computer algebraic determination of symmetries and conservation laws of PDEs, in *Proc. ISMM Int. Symposium on Computer Applications in Design, Simulation and Analysis, New Orleans, Louisiana*, Park, E.K., Ed., Acta Press, Anaheim, CA, 310.

[1990b] SYMCON: A MACSYMA Package for the Determination of Symmetries and Conservation Laws of PDEs, Preprint, Department of Engineering Sciences, Trinity University, San Antonio, TX.

[1992a] PDELIE: A partial differential equation solver, *MACSYMA Newsl.*, 9 (1), 1.

[1992b] PDELIE: A partial differential equation solver II, *MACSYMA Newsl.*, 9 (2–4), 5.

[1994a] PDELIE: A partial differential equation solver III, *MACSYMA Newsl.*, 11 [to appear].

[1994b] PDELIE: Symbolic Software for the Analysis of Partial Differential Equations by Lie Group Methods, User's Manual, Preprint, Department of Engineering Sciences, Trinity University, San Antonio, TX.

Vanderbauwhede, A.

[1982]   *Local Bifurcation and Symmetry*, Pitman, London.

van Leeuwen, M.A.A.

[1994]   *LiE, A software package for Lie group computations*, Euromath Bull., 1 (2), 83.

Vawda, F.F. and Mahomed, F.M.

[1994a] Closed orbits and their exact symmetries, *Lie Groups and Their Applications*, 1(2), 81.

[1994b] Bertrand's theorem and exact symmetries, in *Proc. of the 14th IMACS World Congress on Computational and Applied Mathematics*, Vol. 1, Georgia Institute of Technology, Atlanta, 474.

Vinet, L., Levi, D., and Winternitz, P.

[1994] Symmetries and Integrability of Difference Equations, *Proc. Workshop, Estérel, Québec, Canada*, CRM Proceedings and Lecture Notes, American Mathematical Society, Providence, RI.

Vinogradov, A.M.

[1984] Local symmetries and conservation laws, *Acta Appl. Math.*, 2, 21.

[1989a] Symmetries and conservation laws of partial differential equations: basic notations and results, *Acta Appl. Math.*, 16, 3.

[1989b] Symmetries of partial differential equations, Part I, *Acta Appl. Math.*, 15 (1–2); Part II, *ibid.*, 16 (1); Part III, *ibid.*, 16 (2).

[1989c] *Symmetries of Partial Differential Equations. Conservation laws - Applications - Algorithms*, Kluwer Academic Publications, Dordrecht.

Vinogradov, A.M. and Krasil'shchik, I.S.

[1984] Nonlocal symmetries and the theory of coverings, *Acta Appl. Math.*, 2, 79.

Vorob'ev, E.M.

[1986] Partial symmetries of systems of differential equations, *Dokl. Akad. Nauk S.S.S.R.*, 287, 408. (English transl. in *Sov. Math. Dokl.*, 33, 408, 1986.)

[1989] Partial symmetries and multidimensional integrable differential equations, *Differentsial'nye Uravneniya*, 25, 322.

[1991] Reduction and quotient equations for differential equations with symmetries, *Acta Appl. Math.*, 23, 1.

[1992] Symmetries of compatibility conditions for systems of differential equations, *Acta Appl. Math.*, 26, 61.

Wahlquist, H.D.

[1977] Differential form analysis using MACSYMA, in *Proc. 1977 MACSYMA User's Conference, Berkeley, California*, Andersen, E.M., Ed., NASA CP-2012, Washington, D.C., 71.

Warner, F.W.

[1983] *Foundations of Differential Manifolds and Lie Groups*, Springer-Verlag, New York.

Williams, S.M.

[1989] Similarity Methods with Reference to a High-Order Nonlinear Diffusion Equation, Ph.D thesis, Mathematical Institute, University of Oxford, Oxford.

Winternitz, P.

[1983] Lie groups and solutions of nonlinear differential equations, in *Nonlinear Phenomena*, Wolf, K.B., Ed., Lecture Notes in Physics, Vol. 189, Springer-Verlag, New York, 263.

[1990] Group theory and exact solutions of partially integrable differential systems, in *Partially Integrable Nonlinear Evolution Equations and Their Physical Applications. NATO ASI*

Series C: Math. and Phys. Scs., Vol. 310, Conte, R. and Boccara, N., Eds., Kluwer Academic Publishers, Dordrecht, 515.

[1991] Conditional symmetries and conditional integrability for nonlinear systems, in *Group Theoretical Methods in Physics. Proc. XVIII International Colloquium, Moscow, U.S.S.R.*, 1990, Dodonov, V.V. and Man'ko, V.I., Eds., Lecture Notes in Physics, Vol. 382, Springer-Verlag, Berlin, 263.

[1993] Lie group and solutions of nonlinear partial differential equations, in *Integrable Systems, Quantum Groups, and Quantum Field Theories. NATO ASI Series C: Math. and Phys. Scs.*, Vol. 409, Ibort, L.A. and Rodríguez, M.A., Eds., Kluwer Academic Publishers, Dordrecht, 429.

Wolf, K.B.

[1990] *Lie Methods in Optics II, Proc. CIFMA-CIO Workshop, Cocoyoc, México*, 1988. Lecture Notes in Physics, Vol. 352, Springer-Verlag, New York.

Wolf, T.

[1985] An analytic algorithm for decoupling and integrating systems of nonlinear partial differential equations, *J. Comp. Phys.*, 60, 437.

[1989a] Zur analytischen Untersuching und exakten Lösung von Differentialgleichungen mit Computeralgebrasystemen. Dissertation B, Friedrich Schiller Universität, Jena.

[1989b] A package for the analytic investigation and exact solutions of differential equations, in *Proc. EUROCAL '87, Leipzig, GDR*, 1987, Davenport, J.H., Ed., Lecture Notes in Computer Science, Vol. 378, Springer-Verlag, Berlin, 479.

[1992] The symbolic integration of exact PDEs, Preprint, School of Mathematical Sciences, Queen Mary and Westfield College, University of London, London.

[1993] An efficiency improved program LIEPDE for determining Lie-symmetries of PDEs, in *Modern Group Analysis: Advanced Analytical and Computational Methods in Mathematical Physics. Proc. Int. Workshop Acireale, Catania, Italy*, 1992, Ibragimov, N.H., Torrisi, M., and Valenti, A., Eds., Kluwer Academic Publishers, Dordrecht, 377.

[1994] A Program for Applying Symmetries of PDEs, Presentation at CATHODE Workshop on Computer Algebra Tools for Handling ODEs, London.

Wolf, T. and Brand, A.

[1992] The computer algebra package CRACK for investigating PDEs, in *Proc. ERCIM Advanced Course on Partial Differential Equations and Group Theory, Bonn*, Pommaret, J.F., Ed., Gesellschaft für Mathematik und Datenverarbeitung, Sankt Augustin, Germany; also: Manual for CRACK added to the REDUCE Network Library, School of Mathematical Sciences, Queen Mary and Westfield College, University of London, London, 1.

[1993] Heuristics for overdetermined systems of PDEs, in *Proc. Meeting on Artificial Intelligence in Mathematics, Glasgow, United Kingdom*, 1991, Johnson, J.H., McKee, S., and Vella, A., Eds., Clarendon Press, Oxford [in press].

[1994] Investigating symmetries and other analytical properties of ODEs with the computer algebra package CRACK, Preprint, School of Mathematical Sciences, Queen Mary and Westfield College, University of London, London.

Yaglom, I.M.

[1988]   *Felix Klein and Sophus Lie*, Birkhäuser, Basel.

Yanenko, N.N.

[1964]   Theory of consistency and methods of integrating systems of nonlinear partial differential equations, in *Proceedings of the Fourth All–Union Mathematics Congress*, Nauka, Leningrad, 613.

Zachos, C.

[1991]   Elementary paradigms of quantum algebras, Proceedings of Conf. on Deformation Theory of Algebras and Quantization with Applications to Physics, *Contemporary Mathematics*, Stasheff, J. and Gerstenhaber, M., Eds., AMS.

Zakharov, V.E. and Konopelchenko, B.G.

[1984]   On the theory of recursion operator, *Comm. Math. Phys.*, 94, 483.

Zakharov, V.E., Manakov, S.B., Novikov, S.P., and Pitaevsky, L.P.

[1980]   *Theory of Solitons: Inverse Problem Method*, Nauka, Moscow, 1980.

Zaslavskii, O.B. and Ulyanov, V.V.

[1984]   New classes of exact solutions for the Schrödinger equation and a description of spin systems by means of potential fields, *Zh. Eksp. Teor. Fiz.*, 87, 1724.

Zhang, H., Tu, G., Oevel, W., and Fuchssteiner, B.

[1991]   Symmetries, conserved quantities, and hierarchies for some lattice systems with soliton structure, *J. Math. Phys.*, 32, 1908.

Zharinov, V.V.

[1992]   *Lecture Notes on Geometrical Aspects of Partial Differential Equations*. Series on Soviet and East European Mathematics, Vol. 9, World Scientific, Singapore.

Zharkov, A. Yu.

[1993]   Computer classification of the integrable coupled KdV-like systems with unit main matrix, *J. Symb. Comp.*, 15, 85.

Zidowitz, S.

[1993]   Conditional symmetries and the direct reduction of partial differential equations, in *Modern Group Analysis: Advanced Analytical and Computational Methods in Mathematical Physics*, Ibragimov, N.H., Torrisi, M., and Valenti, A., Eds., Kluwer Academic Publishers, Dordrecht, 387.

Zizza, F.

[1992]   Algebraic programming and differential forms, *Math. J.*, 2 (1), 88.

[1994]   Differential Forms Package, Wolfram Research Inc., Champaign, IL, MathSource 0205-221.

Author Index

Subject Index

A

Acoustic equations, nonlinear, 302
Adiabatic gas flow, one-dimensional
    equations, 134
  Euler coordinates, 134
  intermediate system, 135
  Lagrangian mass variables, 134
  polytropic, 135
Approximate
  Bäcklund transformation, 66
  Cauchy problem, 32
    solution of, 33
  commutator, 39
  conservation law, 59
  differential equations, 32
    completely integrable, 33
  equalities, 32
  equivalence transformations, 55
    infinitesimal generators of, 57
  Euler–Lagrange equation, 58
  generators
    one-parameter groups of, 35
    r-parameter groups of, 43
  group
    one-parameter, 35
    r-parameter, 43
  initial conditions, 32
  invariants, 60
  Lie algebra, 40
    basis of, 41
    essential operators of, 41

    structure constants of, 42
  Lie equations
    given approximate Lie alge-
      bra for, 46
    one-parameter groups for, 36
    multi-parameter groups for, 44
  operators, 39
    linearly independent, 41
  Noether symmetry, 58
  symmetries, 49
    AGA programs for calcula-
      tion of, 474
    calculation algorithm of, 50
  transformations, 39
Approximately invariant solutions, 60
Auto-Bäcklund transformation, 27
  Bonnet equation of (*alias* Bianchi–
    Lie transformation), 28
  Korteweg-de Vries equation of,
    28

B

Bäcklund transformation, 130, 162–
    164
  approximate, 66
  formal, 66
  Burgers and heat equations of
    (*alias* Hopf–Cole transfor-
    mation), 27

H

I